Teacher Edition

Eureka Math® Grade 4 Module 5

Special thanks go to the Gordon A. Cain Center and to the Department of Mathematics at Louisiana State University for their support in the development of *Eureka Math*.

For a free *Eureka Math* Teacher Resource Pack, Parent Tip Sheets, and more please visit
https://eurekamath.greatminds.org/teacher-resource-pack

Published by Great Minds.

Copyright © 2015 Great Minds®. No part of this work may be reproduced, sold, or commercialized, in whole or in part, without written permission from Great Minds. Non-commercial use is licensed pursuant to a Creative Commons Attribution-NonCommercial-ShareAlike 4.0 license; for more information, go to http://greatminds.net/maps/math/copyright. "Great Minds" and "Eureka Math" are registered trademarks of Great Minds.

Printed in the U.S.A.
This book may be purchased from the publisher at eureka-math.org
BAB 10 9 8 7 6 5 4 3 2
ISBN 978-1-63255-374-4

Eureka Math: A Story of Units® Contributors

Katrina Abdussalaam, Curriculum Writer
Tiah Alphonso, Program Manager—Curriculum Production
Kelly Alsup, Lead Writer / Editor, Grade 4
Catriona Anderson, Program Manager—Implementation Support
Debbie Andorka-Aceves, Curriculum Writer
Eric Angel, Curriculum Writer
Leslie Arceneaux, Lead Writer / Editor, Grade 5
Kate McGill Austin, Lead Writer / Editor, Grades PreK–K
Adam Baker, Lead Writer / Editor, Grade 5
Scott Baldridge, Lead Mathematician and Lead Curriculum Writer
Beth Barnes, Curriculum Writer
Bonnie Bergstresser, Math Auditor
Bill Davidson, Fluency Specialist
Jill Diniz, Program Director
Nancy Diorio, Curriculum Writer
Nancy Doorey, Assessment Advisor
Lacy Endo-Peery, Lead Writer / Editor, Grades PreK–K
Ana Estela, Curriculum Writer
Lessa Faltermann, Math Auditor
Janice Fan, Curriculum Writer
Ellen Fort, Math Auditor
Peggy Golden, Curriculum Writer
Maria Gomes, Pre-Kindergarten Practitioner
Pam Goodner, Curriculum Writer
Greg Gorman, Curriculum Writer
Melanie Gutierrez, Curriculum Writer
Bob Hollister, Math Auditor
Kelley Isinger, Curriculum Writer
Nuhad Jamal, Curriculum Writer
Mary Jones, Lead Writer / Editor, Grade 4
Halle Kananak, Curriculum Writer
Susan Lee, Lead Writer / Editor, Grade 3
Jennifer Loftin, Program Manager—Professional Development
Soo Jin Lu, Curriculum Writer
Nell McAnelly, Project Director

Ben McCarty, Lead Mathematician / Editor, PreK–5
Stacie McClintock, Document Production Manager
Cristina Metcalf, Lead Writer / Editor, Grade 3
Susan Midlarsky, Curriculum Writer
Pat Mohr, Curriculum Writer
Sarah Oyler, Document Coordinator
Victoria Peacock, Curriculum Writer
Jenny Petrosino, Curriculum Writer
Terrie Poehl, Math Auditor
Robin Ramos, Lead Curriculum Writer / Editor, PreK–5
Kristen Riedel, Math Audit Team Lead
Cecilia Rudzitis, Curriculum Writer
Tricia Salerno, Curriculum Writer
Chris Sarlo, Curriculum Writer
Ann Rose Sentoro, Curriculum Writer
Colleen Sheeron, Lead Writer / Editor, Grade 2
Gail Smith, Curriculum Writer
Shelley Snow, Curriculum Writer
Robyn Sorenson, Math Auditor
Kelly Spinks, Curriculum Writer
Marianne Strayton, Lead Writer / Editor, Grade 1
Theresa Streeter, Math Auditor
Lily Talcott, Curriculum Writer
Kevin Tougher, Curriculum Writer
Saffron VanGalder, Lead Writer / Editor, Grade 3
Lisa Watts-Lawton, Lead Writer / Editor, Grade 2
Erin Wheeler, Curriculum Writer
MaryJo Wieland, Curriculum Writer
Allison Witcraft, Math Auditor
Jessa Woods, Curriculum Writer
Hae Jung Yang, Lead Writer / Editor, Grade 1

Board of Trustees

Lynne Munson, President and Executive Director of Great Minds
Nell McAnelly, Chairman, Co-Director Emeritus of the Gordon A. Cain Center for STEM Literacy at Louisiana State University
William Kelly, Treasurer, Co-Founder and CEO at ReelDx
Jason Griffiths, Secretary, Director of Programs at the National Academy of Advanced Teacher Education
Pascal Forgione, Former Executive Director of the Center on K-12 Assessment and Performance Management at ETS
Lorraine Griffith, Title I Reading Specialist at West Buncombe Elementary School in Asheville, North Carolina
Bill Honig, President of the Consortium on Reading Excellence (CORE)
Richard Kessler, Executive Dean of Mannes College the New School for Music
Chi Kim, Former Superintendent, Ross School District
Karen LeFever, Executive Vice President and Chief Development Officer at ChanceLight Behavioral Health and Education
Maria Neira, Former Vice President, New York State United Teachers

A STORY OF UNITS

GRADE 4

Mathematics Curriculum

GRADE 4 • MODULE 5

Table of Contents
GRADE 4 • MODULE 5
Fraction Equivalence, Ordering, and Operations

Module Overview	2
Topic A: Decomposition and Fraction Equivalence	15
Topic B: Fraction Equivalence Using Multiplication and Division	91
Topic C: Fraction Comparison	156
Topic D: Fraction Addition and Subtraction	212
Mid-Module Assessment and Rubric	288
Topic E: Extending Fraction Equivalence to Fractions Greater Than 1	302
Topic F: Addition and Subtraction of Fractions by Decomposition	385
Topic G: Repeated Addition of Fractions as Multiplication	462
Topic H: Exploring a Fraction Pattern	531
End-of-Module Assessment and Rubric	543
Answer Key	561

Grade 4 • Module 5

Fraction Equivalence, Ordering, and Operations

OVERVIEW

In this 45-day module, students build on their Grade 3 work with unit fractions as they explore fraction equivalence and extend this understanding to mixed numbers. This leads to the comparison of fractions and mixed numbers and the representation of both in a variety of models. Benchmark fractions play an important part in students' ability to generalize and reason about relative fraction and mixed number sizes. Students then have the opportunity to apply what they know to be true for whole number operations to the new concepts of fraction and mixed number operations.

Students begin Topic A by decomposing fractions and creating tape diagrams to represent them as sums of fractions with the same denominator in different ways (e.g., $\frac{3}{5} = \frac{1}{5} + \frac{1}{5} + \frac{1}{5} = \frac{1}{5} + \frac{2}{5}$) (**4.NF.3b**). They proceed to see that representing a fraction as the repeated addition of a unit fraction is the same as multiplying that unit fraction by a whole number. This is already a familiar fact in other contexts.

For example, just as 3 twos = 2 + 2 + 2 = 3 × 2, so does 3 fourths = $\frac{1}{4} + \frac{1}{4} + \frac{1}{4} = 3 \times \frac{1}{4}$.

The introduction of multiplication as a record of the decomposition of a fraction (**4.NF.4a**) early in the module allows students to become familiar with the notation before they work with more complex problems. As students continue working with decomposition, they represent familiar unit fractions as the sum of smaller unit fractions. A folded paper activity allows them to see that, when the number of fractional parts in a whole increases, the size of the parts decreases.
They proceed to investigate this concept with the use of tape diagrams and area models. Reasoning enables them to explain why two different fractions can represent the same portion of a whole (**4.NF.1**).

In Topic B, students use tape diagrams and area models to analyze their work from earlier in the module and begin using multiplication to create an equivalent fraction that comprises smaller units, e.g., $\frac{2}{3} = \frac{2 \times 4}{3 \times 4} = \frac{8}{12}$ (**4.NF.1**). Based on the use of multiplication, they reason that division can be used to create a fraction that comprises larger units (or a single unit) equivalent to a given fraction (e.g., $\frac{8}{12} = \frac{8 \div 4}{12 \div 4} = \frac{2}{3}$). Their work is justified using area models and tape diagrams and, conversely, multiplication is used to test for and/or verify equivalence. Students use the tape diagram to transition to modeling equivalence on the number line.

They see that, by multiplying, any unit fraction length can be partitioned into *n* equal lengths and that doing so multiplies both the total number of fractional units (the denominator) and number of selected units (the numerator) by *n*. They also see that there are times when fractional units can be grouped together, or divided, into larger fractional units. When that occurs, both the total number of fractional units and number of selected units are divided by the same number.

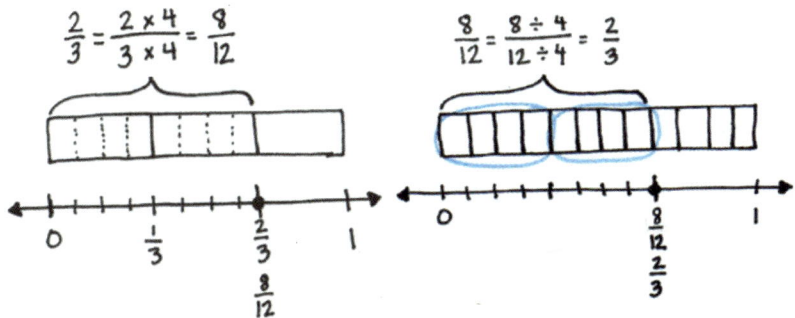

In Grade 3, students compared fractions using fraction strips and number lines with the same denominators. In Topic C, they expand on comparing fractions by reasoning about fractions with unlike denominators. Students use the relationship between the numerator and denominator of a fraction to compare to a known benchmark (e.g., 0, $\frac{1}{2}$, or 1) on the number line. Alternatively, students compare using the same numerators. They find that the fraction with the greater denominator is the lesser fraction since the size of the fractional unit is smaller as the whole is decomposed into more equal parts (e.g., $\frac{1}{5} > \frac{1}{10}$; therefore $\frac{3}{5} > \frac{3}{10}$). Throughout the process, their reasoning is supported using tape diagrams and number lines in cases where one numerator or denominator is a factor of the other, such as $\frac{1}{5}$ and $\frac{1}{10}$ or $\frac{2}{3}$ and $\frac{5}{6}$. When the units are unrelated, students use area models and multiplication, the general method pictured below to the left, whereby two fractions are expressed in terms of the same denominators. Students also reason that comparing fractions can only be done when referring to the same whole, and they record their comparisons using the comparison symbols <, >, and = (**4.NF.2**).

Comparison Using Like Denominators

$$\frac{2}{3} < \frac{3}{4}$$

Comparison Using Like Numerators

$$\frac{2}{5} < \frac{4}{9}$$

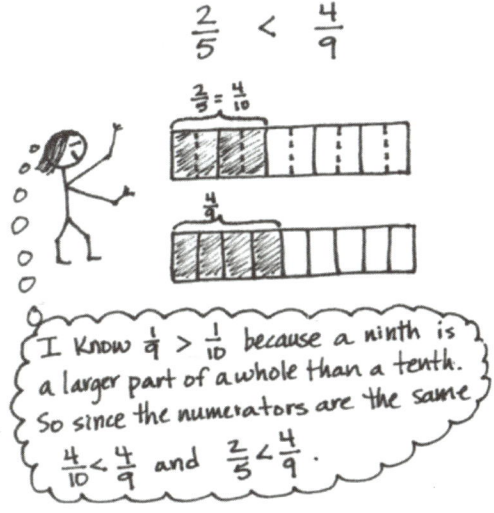

A STORY OF UNITS
Module Overview 4•5

In Topic D, students apply their understanding of whole number addition (the combining of like units) and subtraction (finding an unknown part) to work with fractions (**4.NF.3a**). They see through visual models that, if the units are the same, computation can be performed immediately, e.g., 2 bananas + 3 bananas = 5 bananas and 2 eighths + 3 eighths = 5 eighths. They see that, when subtracting fractions from one whole, the whole is decomposed into the same units as the part being subtracted, e.g., $1 - \frac{3}{5} = \frac{5}{5} - \frac{3}{5} = \frac{2}{5}$. Students practice adding more than two fractions and model fractions in word problems using tape diagrams (**4.NF.3d**). As an extension of the Grade 4 standards, students apply their knowledge of decomposition from earlier topics to add fractions with related units using tape diagrams and area models to support their numerical work. To find the sum of $\frac{1}{2}$ and $\frac{1}{4}$, for example, one simply decomposes 1 half into 2 smaller equal units, fourths, just as in Topics A and B. Now the addition can be completed: $\frac{2}{4} + \frac{1}{4} = \frac{3}{4}$. Though not assessed, this work is warranted because, in Module 6, students are asked to add tenths and hundredths when working with decimal fractions and decimal notation.

At the beginning of Topic E, students use decomposition and visual models to add and subtract fractions less than 1 to or from whole numbers, e.g., $4 + \frac{3}{4} = 4\frac{3}{4}$ and $4 - \frac{3}{4} = (3 + 1) - \frac{3}{4}$. They use addition and multiplication to build fractions greater than 1 and represent them on the number line.

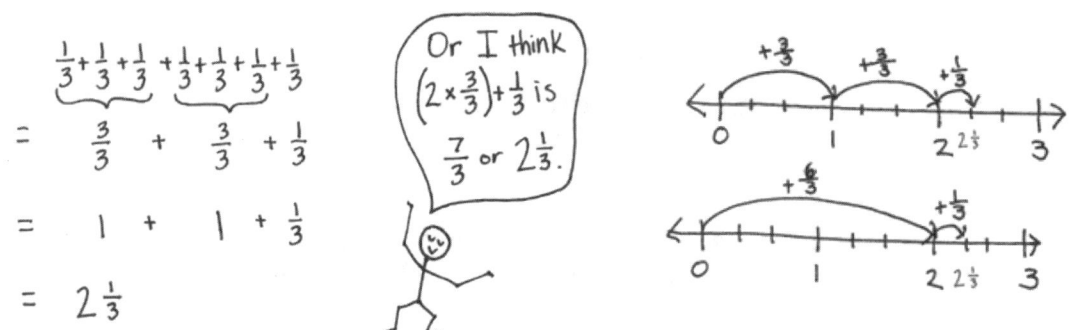

Students then use these visual models and decompositions to reason about the various forms in which a fraction greater than or equal to 1 may be presented, both as fractions and mixed numbers. They practice converting between these forms and begin understanding the usefulness of each form in different situations. Through this understanding, the common misconception that every improper fraction must be converted to a mixed number is avoided. Next, students compare fractions greater than 1, building on their rounding skills and using understanding of benchmarks to reason about which of two fractions is greater (**4.NF.2**). This activity continues to build understanding of the relationship between the numerator and denominator of a fraction. Students progress to finding and using like denominators or numerators to compare and order mixed numbers. They apply their skills of comparing numbers greater than 1 by solving word problems (**4.NF.3d**) requiring the interpretation of data presented in line plots (**4.MD.4**). Students use addition and subtraction strategies to solve the problems, as well as decomposition and modeling to compare numbers in the data sets.

Module 5: Fraction Equivalence, Ordering, and Operations

In Topic F, students estimate sums and differences of mixed numbers, rounding before performing the actual operation to determine what a reasonable outcome is. They proceed to use decomposition to add and subtract mixed numbers (**4.NF.3c**). This work builds on their understanding of a mixed number being the sum of a whole number and fraction.

$$3\frac{2}{5} + 2\frac{4}{5} = 3 + \frac{2}{5} + 2\frac{4}{5} = 3 + 2 + \frac{2}{5} + \frac{4}{5}$$

I can add the parts in any order without changing the sum.

Using unit form, students add and subtract like units first (e.g., ones and ones, fourths and fourths). Students use decomposition, shown with number bonds, in mixed number addition to make one from fractional units before finding the sum. When subtracting, students learn to decompose the minuend or subtrahend when there are not enough fractional units from which to subtract. Alternatively, students can rename the subtrahend, giving more units to the fractional units, which connects to whole number subtraction when renaming 9 tens 2 ones as 8 tens 12 ones.

$$3\frac{1}{5} - \frac{3}{5} = 2\frac{1}{5} + \frac{2}{5} = 2\frac{3}{5}$$
Take one out to subtract from one!

$$3\frac{1}{5} - \frac{3}{5} = 3 - \frac{2}{5} = 2\frac{3}{5}$$
Just like subtracting from one!

$$3\frac{1}{5} - \frac{3}{5} = 2\frac{6}{5} - \frac{3}{5} = 2\frac{3}{5}$$
Rename to make more fifths!

In Topic G, students build on the concept of representing repeated addition as multiplication, applying this familiar concept to work with fractions (**4.NF.4a**, **4.NF.4b**). They use the associative property and their understanding of decomposition. Just as with whole numbers, the unit remains unchanged.

$$4 \times \frac{3}{5} = 4 \times \left(3 \times \frac{1}{5}\right) = (4 \times 3) \times \frac{1}{5} = \frac{4 \times 3}{5} = \frac{12}{5}$$

This understanding connects to students' work with place value and whole numbers. Students proceed to explore the use of the distributive property to multiply a whole number by a mixed number. They recognize that they are multiplying each part of a mixed number by the whole number and use efficient strategies to do so. The topic closes with solving multiplicative comparison word problems involving fractions (**4.NF.4c**) as well as problems involving the interpretation of data presented on a line plot.

$$5 \times 3\frac{3}{4} = 5 \times \left(3 + \frac{3}{4}\right)$$
$$= (5 \times 3) + \left(5 \times \frac{3}{4}\right)$$
$$= 15 + \frac{15}{4}$$
$$= 15 + 3\frac{3}{4}$$
$$= 18\frac{3}{4}$$

Topic H comprises an exploration lesson where students find the sum of all like denominators from $\frac{0}{n}$ to $\frac{n}{n}$. Students first work in teams with fourths, sixths, eighths, and tenths. For example, they might find the sum of all sixths from $\frac{0}{6}$ to $\frac{6}{6}$. Students discover that they can make pairs with a sum of 1 to add more efficiently, e.g., $\frac{0}{6} + \frac{6}{6}, \frac{1}{6} + \frac{5}{6}, \frac{2}{6} + \frac{4}{6}$, and there is one fraction, $\frac{3}{6}$, without a pair. They then extend this to similarly find sums of thirds, fifths, sevenths, and ninths, observing patterns when finding the sum of odd and even denominators (**4.OA.5**).

The Mid-Module Assessment follows Topic D, and the End-of-Module Assessment follows Topic H.

A STORY OF UNITS

Module Overview 4•5

Notes on Pacing for Differentiation

For Module 5, consider the following modifications and omissions. Study the objectives and the sequence of problems within Lessons 1, 2, and 3, and then consolidate the three lessons. Omit Lesson 4. Instead, in Lesson 5, embed the contrast of the decomposition of a fraction using the tape diagram versus using the area model. Note that the area model's cross hatches are used to transition to multiplying to generate equivalent fractions, add related fractions in Lessons 20 and 21, add decimals in Module 6, add/subtract all fractions in Grade 5's Module 3, and multiply a fraction by a fraction in Grade 5's Module 4. Omit Lesson 29, and embed estimation within many problems throughout the module. Omit Lesson 40. Line plots are part of Lesson 28 and can also be reinforced within social studies or science. Be aware that there is a line plot question on the End-of-Module Assessment.

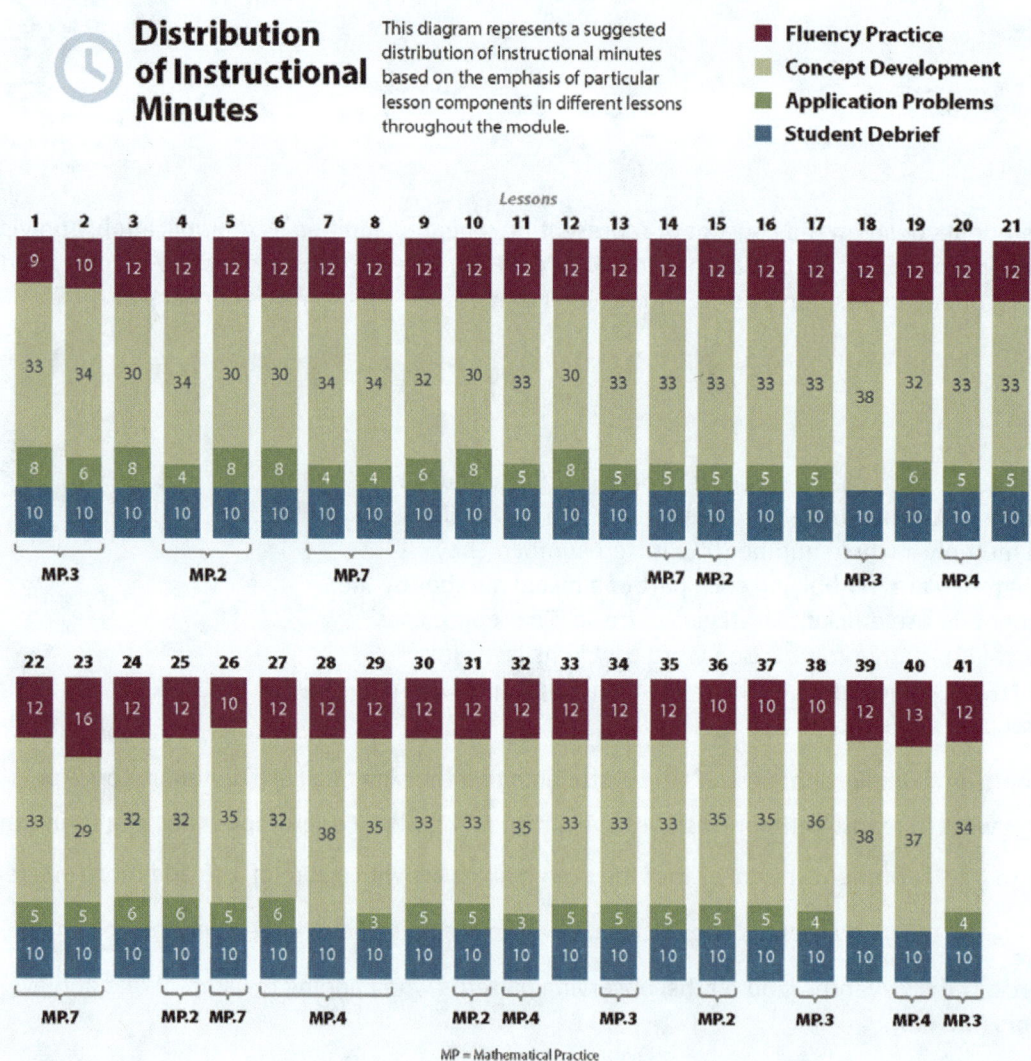

Module 5: Fraction Equivalence, Ordering, and Operations

Focus Grade Level Standards

Generate and analyze patterns.

4.OA.5 Generate a number or shape pattern that follows a given rule. Identify apparent features of the pattern that were not explicit in the rule itself. *For example, given the rule "Add 3" and the starting number 1, generate terms in the resulting sequence and observe that the terms appear to alternate between odd and even numbers. Explain informally why the numbers will continue to alternate in this way.*

Extend understanding of fraction equivalence and ordering.

4.NF.1 Explain why a fraction a/b is equivalent to a fraction $(n \times a)/(n \times b)$ by using visual fraction models, with attention to how the number and size of the parts differ even though the two fractions themselves are the same size. Use this principle to recognize and generate equivalent fractions.

4.NF.2 Compare two fractions with different numerators and different denominators, e.g., by creating common denominators or numerators, or by comparing to a benchmark fraction such as 1/2. Recognize that comparisons are valid only when the two fractions refer to the same whole. Record the results of comparisons with symbols >, =, or <, and justify the conclusions, e.g., by using a visual fraction model.

Build fractions from unit fractions by applying and extending previous understandings of operations on whole numbers.

4.NF.3 Understand a fraction a/b with $a > 1$ as a sum of fractions $1/b$.

  a. Understand addition and subtraction of fractions as joining and separating parts referring to the same whole.

  b. Decompose a fraction into a sum of fractions with the same denominator in more than one way, recording each decomposition by an equation. Justify decompositions, e.g., by using a visual fraction model. *Examples: 3/8 = 1/8 + 1/8 + 1/8; 3/8 = 1/8 + 2/8; 2 1/8 = 1 + 1 + 1/8 = 8/8 + 8/8 + 1/8.*

  c. Add and subtract mixed numbers with like denominators, e.g., by replacing each mixed number with an equivalent fraction, and/or by using properties of operations and the relationship between addition and subtraction.

  d. Solve word problems involving addition and subtraction of fractions referring to the same whole and having like denominators, e.g., by using visual fraction models and equations to represent the problem.

4.NF.4 Apply and extend previous understandings of multiplication to multiply a fraction by a whole number.

  a. Understand a fraction a/b as a multiple of $1/b$. *For example, use a visual fraction model to represent 5/4 as the product 5 × (1/4), recording the conclusion by the equation 5/4 = 5 × (1/4).*

b. Understand a multiple of *a/b* as a multiple of 1/*b*, and use this understanding to multiply a fraction by a whole number. *For example, use a visual fraction model to express 3 × (2/5) as 6 × (1/5), recognizing this product as 6/5. (In general, n × (a/b) = (n × a)/b.)*

c. Solve word problems involving multiplication of a fraction by a whole number, e.g., by using visual fraction models and equations to represent the problem. *For example, if each person at a party will eat 3/8 of a pound of roast beef, and there will be 5 people at the party, how many pounds of roast beef will be needed? Between what two whole numbers does your answer lie?*

Represent and interpret data.

4.MD.4 Make a line plot to display a data set of measurements in fractions of a unit (1/2, 1/4, 1/8). Solve problems involving addition and subtraction of fractions by using information presented in line plots. *For example, from a line plot find and interpret the difference in length between the longest and shortest specimens in an insect collection.*

Foundational Standards

3.NF.1 Understand a fraction 1/*b* as the quantity formed by 1 part when a whole is partitioned into *b* equal parts; understand a fraction *a/b* as the quantity formed by *a* parts of size 1/*b*.

3.NF.2 Understand a fraction as a number on the number line; represent fractions on a number line diagram.

    a. Represent a fraction 1/*b* on a number line diagram by defining the interval from 0 to 1 as the whole and partitioning it into *b* equal parts. Recognize that each part has size 1/*b* and that the endpoint of the part based at 0 locates the number 1/*b* on the number line.

    b. Represent a fraction *a/b* on a number line diagram by marking off *a* lengths 1/*b* from 0. Recognize that the resulting interval has size *a/b* and that its endpoint locates the number *a/b* on the number line.

3.NF.3 Explain equivalence of fractions in special cases, and compare fractions by reasoning about their size.

    a. Understand two fractions as equivalent (equal) if they are the same size, or the same point on a number line.

    b. Recognize and generate simple equivalent fractions, e.g., 1/2 = 2/4, 4/6 = 2/3. Explain why the fractions are equivalent, e.g., by using a visual fraction model.

    c. Express whole numbers as fractions, and recognize fractions that are equivalent to whole numbers. *Examples: Express 3 in the form 3 = 3/1; recognize that 6/1 = 6; locate 4/4 and 1 at the same point of a number line diagram.*

    d. Compare two fractions with the same numerator or the same denominator by reasoning about their size. Recognize that comparisons are valid only when the two fractions refer to the same whole. Record the results of comparisons with the symbols >, =, or <, and justify the conclusions, e.g., by using a visual fraction model.

3.MD.4	Generate measurement data by measuring lengths using rulers marked with halves and fourths of an inch. Show the data by making a line plot, where the horizontal scale is marked off in appropriate units—whole numbers, halves, or quarters.
3.G.2	Partition shapes into parts with equal areas. Express the area of each part as a unit fraction of the whole. *For example, partition a shape into 4 parts with equal area, and describe the area of each part as 1/4 of the area of the shape.*

Focus Standards for Mathematical Practice

MP.2	**Reason abstractly and quantitatively.** Students reason both abstractly and quantitatively throughout this module. They draw area models, number lines, and tape diagrams to represent fractional quantities, as well as word problems.
MP.3	**Construct viable arguments and critique the reasoning of others.** Much of the work in this module is centered on multiple ways to solve fraction and mixed number problems. Students explore various strategies and participate in many *turn and talk* and *explain to your partner* activities. By doing so, they construct arguments to defend their choice of strategy, as well as think about and critique the reasoning of others.
MP.4	**Model with mathematics.** Throughout this module, students represent fractions with various models. Area models are used to investigate and prove equivalence. The number line is used to compare and order fractions, as well as model addition and subtraction of fractions. Students also use models in problem solving as they create line plots to display given sets of fractional data and solve problems requiring the interpretation of data presented in line plots.
MP.7	**Look for and make use of structure.** As students progress through this fraction module, they search for and use patterns and connections that help them build understanding of new concepts. They relate and apply what they know about operations with whole numbers to operations with fractions.

Overview of Module Topics and Lesson Objectives

Standards		Topics and Objectives		Days
4.NF.3b 4.NF.4a 4.NF.3a	A	**Decomposition and Fraction Equivalence**		6
		Lessons 1–2:	Decompose fractions as a sum of unit fractions using tape diagrams.	
		Lesson 3:	Decompose non-unit fractions and represent them as a whole number times a unit fraction using tape diagrams.	
		Lesson 4:	Decompose fractions into sums of smaller unit fractions using tape diagrams.	
		Lesson 5:	Decompose unit fractions using area models to show equivalence.	
		Lesson 6:	Decompose fractions using area models to show equivalence.	
4.NF.1 4.NF.3b	B	**Fraction Equivalence Using Multiplication and Division**		5
		Lessons 7–8:	Use the area model and multiplication to show the equivalence of two fractions.	
		Lessons 9–10:	Use the area model and division to show the equivalence of two fractions.	
		Lesson 11:	Explain fraction equivalence using a tape diagram and the number line, and relate that to the use of multiplication and division.	
4.NF.2	C	**Fraction Comparison**		4
		Lessons 12–13:	Reason using benchmarks to compare two fractions on the number line.	
		Lessons 14–15:	Find common units or number of units to compare two fractions.	

Standards		Topics and Objectives	Days
4.NF.3ad 4.NF.1 4.MD.2	D	**Fraction Addition and Subtraction** Lesson 16: Use visual models to add and subtract two fractions with the same units. Lesson 17: Use visual models to add and subtract two fractions with the same units, including subtracting from one whole. Lesson 18: Add and subtract more than two fractions. Lesson 19: Solve word problems involving addition and subtraction of fractions. Lessons 20–21: Use visual models to add two fractions with related units using the denominators 2, 3, 4, 5, 6, 8, 10, and 12.	6
		Mid-Module Assessment: Topics A–D (assessment ½ day, return ½ day, remediation or further applications 1 day)	2
4.NF.2 **4.NF.3** **4.MD.4** 4.NBT.6 4.NF.1 4.NF.4a	E	**Extending Fraction Equivalence to Fractions Greater Than 1** Lesson 22: Add a fraction less than 1 to, or subtract a fraction less than 1 from, a whole number using decomposition and visual models. Lesson 23: Add and multiply unit fractions to build fractions greater than 1 using visual models. Lessons 24–25: Decompose and compose fractions greater than 1 to express them in various forms. Lesson 26: Compare fractions greater than 1 by reasoning using benchmark fractions. Lesson 27: Compare fractions greater than 1 by creating common numerators or denominators. Lesson 28: Solve word problems with line plots.	7
4.NF.3c 4.MD.2	F	**Addition and Subtraction of Fractions by Decomposition** Lesson 29: Estimate sums and differences using benchmark numbers. Lesson 30: Add a mixed number and a fraction. Lesson 31: Add mixed numbers. Lesson 32: Subtract a fraction from a mixed number. Lesson 33: Subtract a mixed number from a mixed number. Lesson 34: Subtract mixed numbers.	6

Module 5: Fraction Equivalence, Ordering, and Operations

Standards		Topics and Objectives	Days
4.NF.4 4.OA.2 4.MD.2 4.MD.4	G	**Repeated Addition of Fractions as Multiplication** Lessons 35–36: Represent the multiplication of n times a/b as $(n \times a)/b$ using the associative property and visual models. Lessons 37–38: Find the product of a whole number and a mixed number using the distributive property. Lesson 39: Solve multiplicative comparison word problems involving fractions. Lesson 40: Solve word problems involving the multiplication of a whole number and a fraction including those involving line plots.	6
4.OA.5	H	**Exploring a Fraction Pattern** Lesson 41: Find and use a pattern to calculate the sum of all fractional parts between 0 and 1. Share and critique peer strategies.	1
		End-of-Module Assessment: Topics A–H (assessment ½ day, return ½ day, remediation or further applications 1 day)	2
Total Number of Instructional Days			**45**

Terminology

New or Recently Introduced Terms

- Benchmark (standard or reference point by which something is measured)
- Common denominator (when two or more fractions have the same denominator)
- Denominator (e.g., the 5 in $\frac{3}{5}$ names the fractional unit as fifths)
- Fraction greater than 1 (a fraction with a numerator that is greater than the denominator)
- Line plot (display of data on a number line, using an x or another mark to show frequency)
- Mixed number (number made up of a whole number and a fraction)
- Numerator (e.g., the 3 in $\frac{3}{5}$ indicates 3 fractional units are selected)

Familiar Terms and Symbols[1]

- =, <, > (equal to, less than, greater than)
- Compose (change a smaller unit for an equivalent of a larger unit, e.g., 2 fourths = 1 half, 10 ones = 1 ten; combining 2 or more numbers, e.g., 1 fourth + 1 fourth = 2 fourths, 2 + 2 + 1 = 5)

---

[1] These are terms and symbols students have seen previously.

- Decompose (change a larger unit for an equivalent of a smaller unit, e.g., 1 half = 2 fourths, 1 ten = 10 ones; partition a number into 2 or more parts, e.g., 2 fourths = 1 fourth + 1 fourth, 5 = 2 + 2 + 1)
- Equivalent fractions (fractions that name the same size or amount)
- Fraction (e.g., $\frac{1}{3}, \frac{2}{3}, \frac{3}{3}, \frac{4}{3}$)
- Fractional unit (e.g., half, third, fourth)
- Multiple (product of a given number and any other whole number)
- Non-unit fraction (fractions with numerators other than 1)
- Unit fraction (fractions with numerator 1)
- Unit interval (e.g., the interval from 0 to 1, measured by length)
- Whole (e.g., 2 halves, 3 thirds, 4 fourths)

Suggested Tools and Representations

- Area model
- Fraction strips (made from paper, folded, and used to model equivalent fractions)
- Line plot
- Number line
- Rulers
- Tape diagram

Scaffolds[2]

The scaffolds integrated into *A Story of Units*® give alternatives for how students access information as well as express and demonstrate their learning. Strategically placed margin notes are provided within each lesson elaborating on the use of specific scaffolds at applicable times. They address many needs presented by English language learners, students with disabilities, students performing above grade level, and students performing below grade level. Many of the suggestions are organized by Universal Design for Learning (UDL) principles and are applicable to more than one population. To read more about the approach to differentiated instruction in *A Story of Units*, please refer to "How to Implement *A Story of Units*."

---

[2]Students with disabilities may require Braille, large print, audio, or special digital files. Please visit the website www.p12.nysed.gov/specialed/aim for specific information on how to obtain student materials that satisfy the National Instructional Materials Accessibility Standard (NIMAS) format.

Assessment Summary

Type	Administered	Format	Standards Addressed
Mid-Module Assessment Task	After Topic D	Constructed response with rubric	4.NF.1 4.NF.2 4.NF.3abd 4.NF.4a
End-of-Module Assessment Task	After Topic H	Constructed response with rubric	4.OA.5 4.NF.1 4.NF.2 4.NF.3 4.NF.4 4.MD.4

A STORY OF UNITS

GRADE 4

Mathematics Curriculum

GRADE 4 • MODULE 5

Topic A
Decomposition and Fraction Equivalence

4.NF.3b, 4.NF.4a, 4.NF.3a

Focus Standards:	4.NF.3b	Understand a fraction a/b with $a > 1$ as a sum of fractions $1/b$.
		b. Decompose a fraction into a sum of fractions with the same denominator in more than one way, recording each decomposition by an equation. Justify decompositions, e.g., by using a visual fraction model. *Examples: 3/8 = 1/8 + 1/8 + 1/8; 3/8 = 1/8 + 2/8; 2 1/8 = 1 + 1 + 1/8 = 8/8 + 8/8 + 1/8.*
	4.NF.4a	Apply and extend previous understandings of multiplication to multiply a fraction by a whole number.
		a. Understand a fraction a/b as a multiple of $1/b$. *For example, use a visual fraction model to represent 5/4 as the product 5 × (1/4), recording the conclusion by the equation 5/4 = 5 × (1/4).*
Instructional Days:	6	
Coherence -Links from:	G3–M5	Fractions as Numbers on the Number Line
-Links to:	G5–M3	Addition and Subtraction of Fractions

Topic A builds on Grade 3 work with unit fractions. Students explore fraction equivalence through the decomposition of non-unit fractions into unit fractions, as well as the decomposition of unit fractions into smaller unit fractions. They represent these decompositions, and prove equivalence, using visual models.

In Lesson 1, students use paper strips to represent the decomposition of a whole into parts. In Lessons 1 and 2, students decompose fractions as unit fractions, drawing tape diagrams to represent them as sums of fractions with the same denominator in different ways, e.g., $\frac{3}{5} = \frac{1}{5} + \frac{1}{5} + \frac{1}{5} = \frac{1}{5} + \frac{2}{5}$.

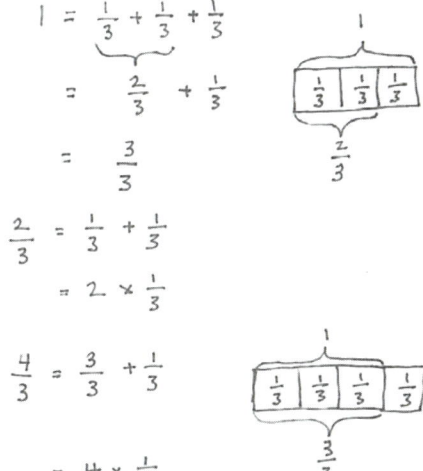

Topic A: Decomposition and Fraction Equivalence

In Lesson 3, students see that representing a fraction as the repeated addition of a unit fraction is the same as multiplying that unit fraction by a whole number. This is already a familiar fact in other contexts. An example is as follows:

$$3 \text{ bananas} = 1 \text{ banana} + 1 \text{ banana} + 1 \text{ banana} = 3 \times 1 \text{ banana}$$

$$3 \text{ twos} = 2 + 2 + 2 = 3 \times 2$$

$$3 \text{ fourths} = 1 \text{ fourth} + 1 \text{ fourth} + 1 \text{ fourth} = 3 \times 1 \text{ fourth}$$

$$\frac{3}{4} = \frac{1}{4} + \frac{1}{4} + \frac{1}{4} = 3 \times \frac{1}{4}$$

By introducing multiplication as a record of the decomposition of a fraction early in the module, students are accustomed to the notation by the time they work with more complex problems in Topic G.

Students continue with decomposition in Lesson 4, where they use tape diagrams to represent fractions, e.g., $\frac{1}{2}, \frac{1}{3}$, and $\frac{2}{3}$, as the sum of smaller unit fractions. Students record the results as a number sentence, e.g., $\frac{1}{2} = \frac{1}{4} + \frac{1}{4} = \left(\frac{1}{8} + \frac{1}{8}\right) + \left(\frac{1}{8} + \frac{1}{8}\right) = \frac{4}{8}$.

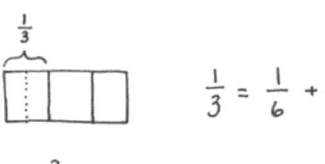

In Lesson 5, this idea is further investigated as students represent the decomposition of unit fractions in area models. In Lesson 6, students use the area model for a second day, this time to represent fractions with different numerators. They explain why two different fractions represent the same portion of a whole.

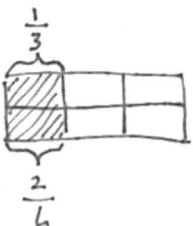 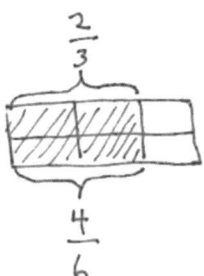

Topic A

A Teaching Sequence Toward Mastery of Decomposition and Fraction Equivalence

Objective 1: Decompose fractions as a sum of unit fractions using tape diagrams.
(Lessons 1–2)

Objective 2: Decompose non-unit fractions and represent them as a whole number times a unit fraction using tape diagrams.
(Lesson 3)

Objective 3: Decompose fractions into sums of smaller unit fractions using tape diagrams.
(Lesson 4)

Objective 4: Decompose unit fractions using area models to show equivalence.
(Lesson 5)

Objective 5: Decompose fractions using area models to show equivalence.
(Lesson 6)

Topic A: Decomposition and Fraction Equivalence

A STORY OF UNITS   Lesson 1 4•5

Lesson 1

Objective: Decompose fractions as a sum of unit fractions using tape diagrams.

Suggested Lesson Structure

- Fluency Practice   (9 minutes)
- Application Problem   (8 minutes)
- Concept Development   (33 minutes)
- Student Debrief   (10 minutes)
- **Total Time**   **(60 minutes)**

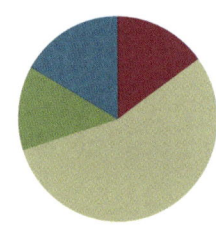

Fluency Practice (9 minutes)

- Read Tape Diagrams **3.OA.3**   (5 minutes)
- Addition of Fractions in Unit Form **3.NF.1** (4 minutes)

Read Tape Diagrams (5 minutes)

Materials: (S) Personal white board

Note: This fluency activity prepares students for Lesson 1.

T: (Project a tape diagram partitioned into 2 equal parts. Write 10 at the top.) Say the value of the whole.
S: 10.
T: Write the value of one unit as a division problem.
S: (Write $10 \div 2 = 5$.)
T: (Write 5 in both units.) Write the whole as a repeated addition sentence.
S: (Write $5 + 5 = 10$.)

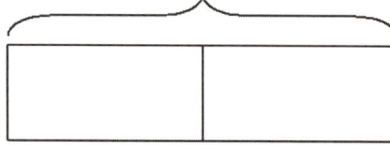

Continue with the following possible sequence: $6 \div 2$, $15 \div 3$, $6 \div 3$, $12 \div 4$, and $24 \div 4$.

Addition of Fractions in Unit Form (4 minutes)

Materials: (S) Personal white board

Note: This fluency activity prepares students for Lesson 1.

T: (Project a circle partitioned into 2 equal parts with 1 part shaded.) How many circles do you see?
S: 1.
T: How many equal parts are in the circle?
S: 2.
T: What fraction of the circle is shaded?
S: $\frac{1}{2}$.
T: (Write 1 half + 1 half = 2 halves = 1.) True or false?
S: True.
T: Explain why it is true to your partner.
S: 1 + 1 is 2. That's kindergarten. → Two halves is the same as 1. → Half an apple + half an apple is 1 apple.
T: (Project a circle partitioned into 4 equal parts with 1 part shaded.) How many circles do you see?
S: 1.
T: How many equal parts does this circle have?
S: 4.
T: Write the fraction that is represented by the shaded part.
S: (Write $\frac{1}{4}$.)
T: (Write 1 fourth + 1 fourth + 1 fourth + 1 fourth = 4 fourths = 1.) True or false?
S: True.

Continue with the following possible fraction graphics: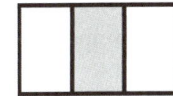

Application Problem (8 minutes)

Materials: (S) 1 index card with diagonals drawn, pair of scissors (1 per pair of students)

a. Discuss with your partner what you notice about the rectangle.

b. Use your scissors to cut your rectangle on the diagonal lines. Prove that you have cut the rectangle into 4 fourths. Include a drawing in your explanation.

Note: This Application Problem reviews and reinforces the concept that fractional parts have the same area. Many students may say that the diagonal lines do not create fourths because the triangles created by the diagonals do not look alike. Exploration helps students see that the areas are, in fact, equal and prepares them for the work with tape diagrams that is done in this lesson.

Lesson 1: Decompose fractions as a sum of unit fractions using tape diagrams.

A STORY OF UNITS   Lesson 1 4•5

Concept Development (33 minutes)

Materials: (T) 3 strips of paper, markers (S) 3 strips of paper, colored markers or colored pencils, personal white board

NOTES ON MULTIPLE MEANS OF ACTION AND EXPRESSION:

Folding paper strips into thirds may be challenging for some students. Offer the following guidance:

- Roll the paper strip into a loose tube. Lightly flatten the middle, and adjust so that the three parts are all approximately the same size. Align the edges of the paper strip. Then, flatten the tube.
- Provide prefolded strips.
- Depending on paper thickness, it may be desirable to score or line the paper to guide folds.

Problem 1: Fold a strip of paper to create thirds and sixths. Record the decompositions represented by the folded paper with addition.

T: The area of this strip of paper is my whole. What number represents this strip of paper?

S: 1.

T: Fold to decompose the whole into 3 equal parts. (Demonstrate.)

T: Draw lines on the creases you made. (Demonstrate.) Draw a number bond to represent the whole decomposed into 3 units of…?

S: 1 third.

T: (Allow students time to draw.) Tell me an addition number sentence to describe this decomposition, starting with "1 equals…" (Record the sentence as students speak.)

S: $1 = \frac{1}{3} + \frac{1}{3} + \frac{1}{3}$.

T: Let's show this decomposition in another way. (Shade 2 thirds of the paper strip.)

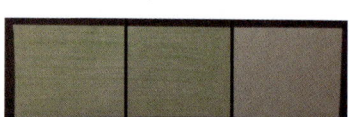

T: (Insert parentheses.) $1 = \left(\frac{1}{3} + \frac{1}{3}\right) + \frac{1}{3}$. Tell me a new addition sentence that matches the new groups starting with "1 equals…"

S: $1 = \frac{2}{3} + \frac{1}{3}$.

T: Decompose 5 sixths into 5 units of 1 sixth with a number bond. (Allow students time to work.)

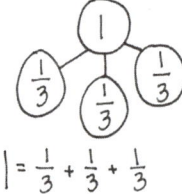

T: Give me an addition sentence representing this decomposition, starting with "5 sixths equals …" (Record the sentence as students speak.)

S: $\frac{5}{6} = \frac{1}{6} + \frac{1}{6} + \frac{1}{6} + \frac{1}{6} + \frac{1}{6}$.

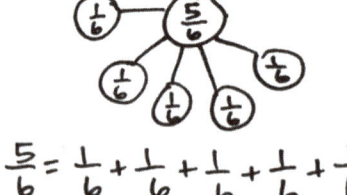

T: Let's double the number of units in our whole. Fold your strip on the creases. Fold one more time in half. Open up your strip. Into how many parts have we now decomposed the whole?

S: 6.

T: On the other side that has no lines, draw lines on the creases you made, and shade 5 sixths.

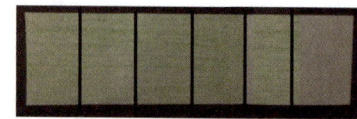

A STORY OF UNITS   Lesson 1   4•5

T: Show this decomposition in another way.

T: (Insert parentheses.) $\frac{5}{6} = \left(\frac{1}{6} + \frac{1}{6} + \frac{1}{6}\right) + \left(\frac{1}{6} + \frac{1}{6}\right)$. Tell me a new addition sentence that matches this new decomposition, starting with "5 sixths equals ..." (Record the sentence as students speak.)

S: $\frac{5}{6} = \frac{3}{6} + \frac{2}{6}$.

T: Draw a number bond and addition sentence to match.

S: (Draw a number bond and addition sentence.)

T: Use your paper strip to show your partner the units that match each part.

S: $\frac{5}{6} = \frac{3}{6} + \frac{2}{6}$.

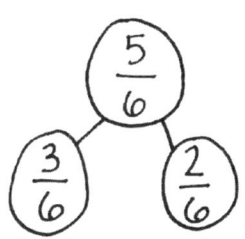

Problem 2: Fold two strips of paper into fourths. Shade $\frac{7}{4}$. Write the number sentence created.

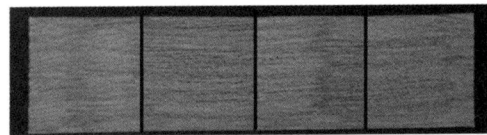   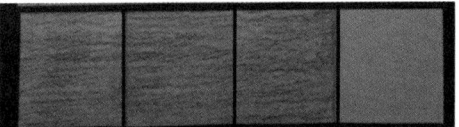

T: Take a new strip of paper. The area of this strip of paper is the whole. Fold this paper to create 4 equal parts. (Demonstrate creating fourths vertically.) Shade all 4 of the parts. Take one more strip of paper, fold it, and shade 3 of the 4 parts. How much is shaded?

S: The first strip of paper represents $\frac{4}{4}$. On the second strip of paper, we shaded $\frac{3}{4}$. $\frac{7}{4} = \frac{4}{4} + \frac{3}{4}$.

T: Draw a number bond to represent the 2 parts and their sum.

S: (Draw.)

T: Can $\frac{4}{4}$ be renamed?

S: Yes. $\frac{4}{4}$ is equal to 1.

T: Draw another number bond to rename $\frac{4}{4}$ with 1.

S: (Draw.)

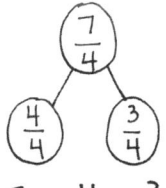

T: Write a number sentence that represents this number bond.

S: (Write $1\frac{3}{4} = 1 + \frac{3}{4}$.)

T: With a **fraction greater than 1**, like $\frac{7}{4}$, we can rename it. We say this is one and three-fourths. $1\frac{3}{4}$ is another way to record the decomposition of $\frac{7}{4}$ as $\frac{4}{4}$ and $\frac{3}{4}$. Compare and contrast $1\frac{3}{4}$ to $\frac{7}{4}$.

S: One has a whole number. The other has just a fraction. → They both represent the same area, so they are equivalent. → So, when a fraction is greater than 1, we can write it using a whole number and a fraction.

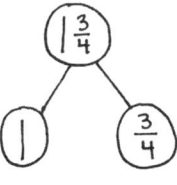

Lesson 1: Decompose fractions as a sum of unit fractions using tape diagrams.   21

A STORY OF UNITS   Lesson 1 4•5

Problem 3: Write decompositions of fractions represented by tape diagrams as number sentences.

Display the following tape diagram:

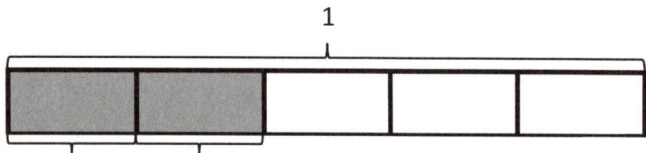

T: The rectangle represents 1. Name the unit fraction.
S: 1 fifth.
T: (Label $\frac{1}{5}$ underneath both shaded unit fractions.) Name the shaded fraction.
S: 2 fifths.
T: Decompose $\frac{2}{5}$ into unit fractions.
S: $\frac{2}{5} = \frac{1}{5} + \frac{1}{5}$.

Display the tape diagram pictured to the right.

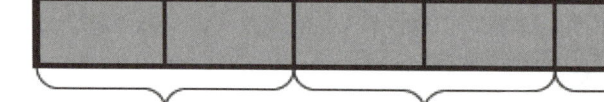

T: What is the unit fraction?
S: 1 fifth.
T: Use the model to write an addition sentence for the tape diagram showing the decomposition of 4 fifths indicated by the brackets.
S: (Write $\frac{4}{5} = \frac{1}{5} + \frac{1}{5} + \frac{2}{5}$.)

Display the tape diagram pictured to the right.

T: What is the unit fraction?
S: 1 sixth. → 1 fifth.
T: How do you know it is not 1 sixth?
S: This tape diagram shows 5 equal parts shaded as being 1. Then, there's another unit after that. → This tape diagram represents a number greater than 1. → This tape diagram is showing a whole number and a fraction.
T: Tell your partner the number this tape diagram represents.
S: $\frac{6}{5}$. → $1\frac{1}{5}$.

> **NOTES ON MULTIPLE MEANS OF ENGAGEMENT:**
>
> Challenge students working above grade level to use parentheses and what they understand about repeated addition to write as many number sentences as they can for the tape diagram of $1\frac{1}{5} = \frac{2}{5} + \frac{2}{5} + \frac{2}{5}$.

Lesson 1: Decompose fractions as a sum of unit fractions using tape diagrams.

A STORY OF UNITS   Lesson 1   4•5

T: On your personal white board, write the number sentence for the tape diagram showing a sum equal to 6 fifths.

S: $1\frac{1}{5} = \frac{2}{5} + \frac{2}{5} + \frac{2}{5}. \rightarrow \frac{6}{5} = \frac{3}{5} + \frac{3}{5}. \rightarrow \frac{6}{5} = \frac{1}{5} + \frac{5}{5}. \rightarrow \frac{6}{5} = 1 + \frac{1}{5}.$

Problem 4: Draw decompositions of fractions with tape diagrams from number sentences.

Display the number sentence $\frac{6}{6} = \frac{1}{6} + \frac{2}{6} + \frac{3}{6}.$

MP.3

T: Discuss with your partner how this number sentence can be modeled as a tape diagram.

S: Well, the sum is 1 because it is equal to $\frac{6}{6}$. → The unit fraction is 1 sixth, so we should partition the tape diagram into 6 equally sized pieces. → We can use brackets to label the sum and addends.

Allow partners to draw a tape diagram and share. Repeat with $\frac{8}{6} = \frac{4}{6} + \frac{4}{6}.$

Problem Set (10 minutes)

Students should do their personal best to complete the Problem Set within the allotted 10 minutes. Some problems do not specify a method for solving. This is an intentional reduction of scaffolding that invokes MP.5, Use Appropriate Tools Strategically. Students should solve these problems using the RDW approach used for Application Problems.

For some classes, it may be appropriate to modify the assignment by specifying which problems students should work on first. With this option, let the purposeful sequencing of the Problem Set guide the selections so that problems continue to be scaffolded. Balance word problems with other problem types to ensure a range of practice. Consider assigning incomplete problems for homework or at another time during the day.

Student Debrief (10 minutes)

Lesson Objective: Decompose fractions as a sum of unit fractions using tape diagrams.

The Student Debrief is intended to invite reflection and active processing of the total lesson experience.

Invite students to review their solutions for the Problem Set. They should check work by comparing answers with a partner before going over answers as a class. Look for misconceptions or misunderstandings that can be addressed in the Debrief. Guide students in a conversation to debrief the Problem Set and process the lesson.

Any combination of the questions below may be used to lead the discussion.

- How do Problems 1(f), 1(g), and 1(h) differ from Problems 1(a–e)? How do the tape diagrams model a **fraction greater than 1**?

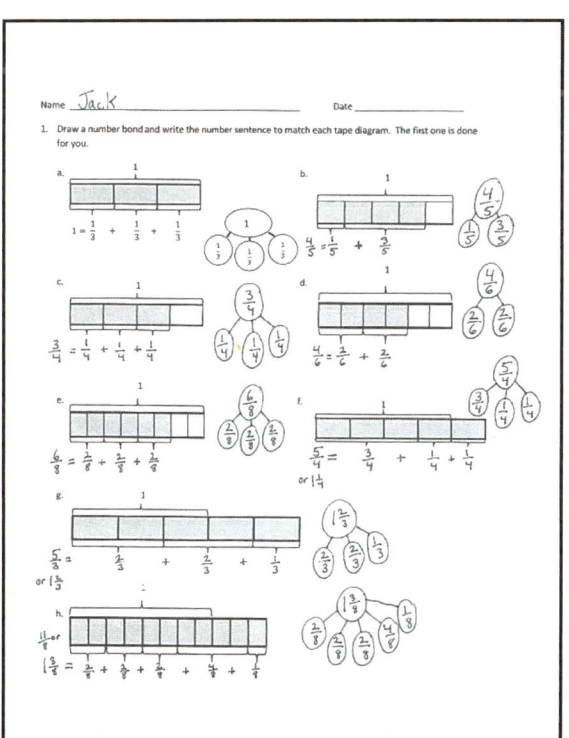

Lesson 1:   Decompose fractions as a sum of unit fractions using tape diagrams.   23

- Compare the size of the shaded fractions in Problems 1(c) and 1(e). Assume the wholes are equal. What can you infer about the two number sentences?
- How do the number bonds connect to the number sentences?
- How did using the paper strips during our lesson help you visualize the tape diagrams you had to draw in Problem 2?
- What relationship does the unit fraction have with the number of units in a whole?
- How did the Application Problem connect to today's lesson?

Exit Ticket (3 minutes)

After the Student Debrief, instruct students to complete the Exit Ticket. A review of their work will help with assessing students' understanding of the concepts that were presented in today's lesson and planning more effectively for future lessons. The questions may be read aloud to the students.

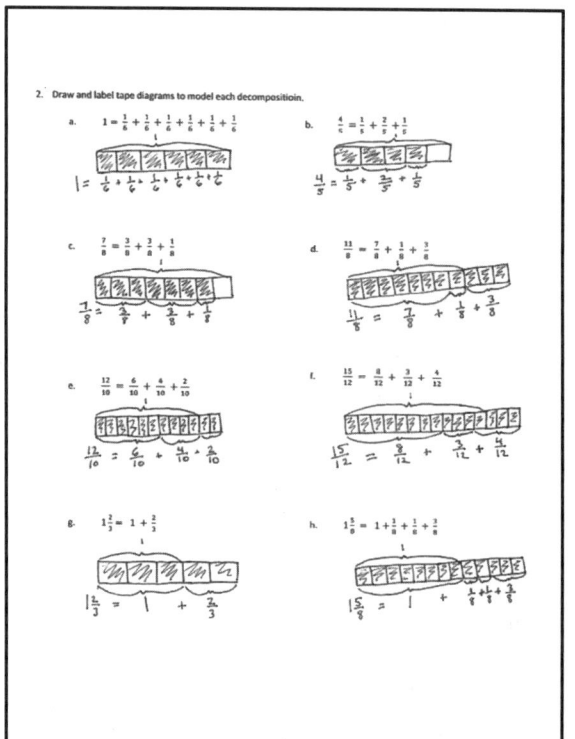

Lesson 1 Problem Set 4•5

Name _____ Date _____

1. Draw a number bond, and write the number sentence to match each tape diagram. The first one is done for you.

a.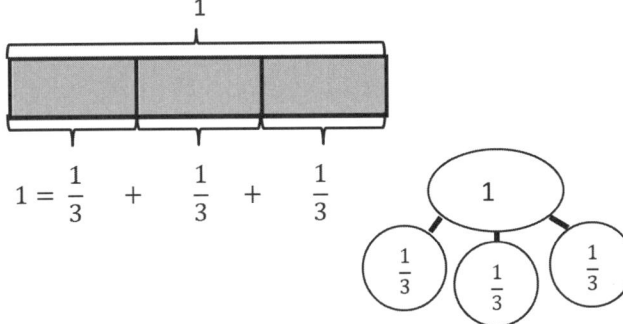

b.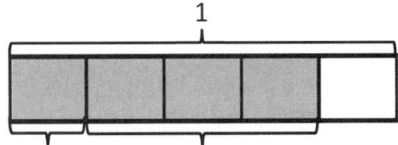

c.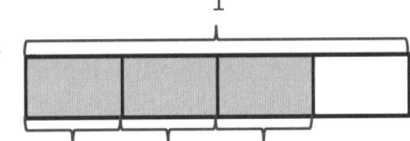

d.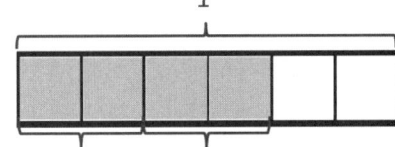

e.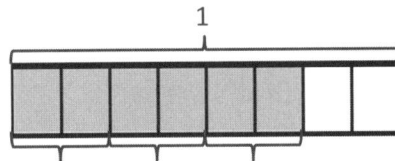

f.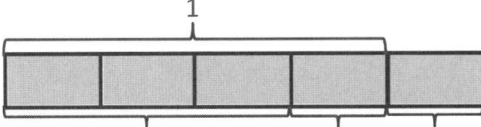

Lesson 1: Decompose fractions as a sum of unit fractions using tape diagrams.

25

g.

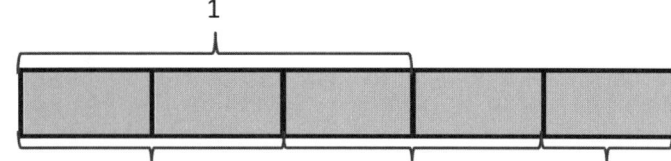

h.

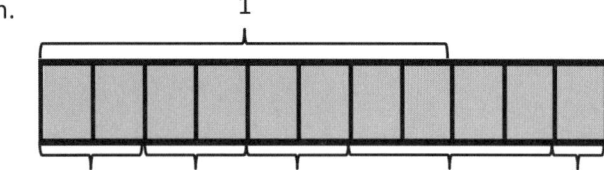

2. Draw and label tape diagrams to model each decomposition.

   a. $1 = \frac{1}{6} + \frac{1}{6} + \frac{1}{6} + \frac{1}{6} + \frac{1}{6} + \frac{1}{6}$

   b. $\frac{4}{5} = \frac{1}{5} + \frac{2}{5} + \frac{1}{5}$

   c. $\frac{7}{8} = \frac{3}{8} + \frac{3}{8} + \frac{1}{8}$

   d. $\frac{11}{8} = \frac{7}{8} + \frac{1}{8} + \frac{3}{8}$

e. $\dfrac{12}{10} = \dfrac{6}{10} + \dfrac{4}{10} + \dfrac{2}{10}$

f. $\dfrac{15}{12} = \dfrac{8}{12} + \dfrac{3}{12} + \dfrac{4}{12}$

g. $1\dfrac{2}{3} = 1 + \dfrac{2}{3}$

h. $1\dfrac{5}{8} = 1 + \dfrac{1}{8} + \dfrac{1}{8} + \dfrac{3}{8}$

Name _____ Date _____

1. Complete the number bond, and write the number sentence to match the tape diagram.

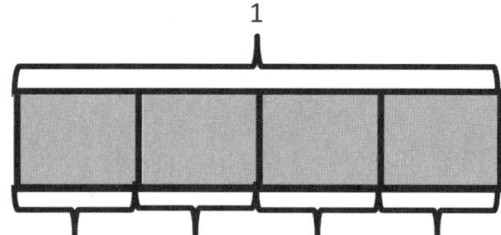 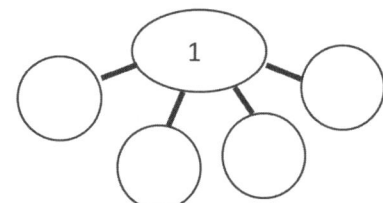

2. Draw and label tape diagrams to model each number sentence.

   a. $1 = \frac{1}{5} + \frac{1}{5} + \frac{1}{5} + \frac{1}{5} + \frac{1}{5}$

   b. $\frac{5}{6} = \frac{2}{6} + \frac{2}{6} + \frac{1}{6}$

Name _____ Date _____

1. Draw a number bond, and write the number sentence to match each tape diagram. The first one is done for you.

a.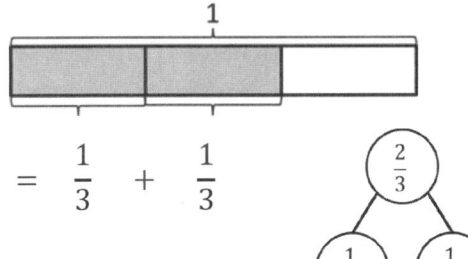

$$\frac{2}{3} = \frac{1}{3} + \frac{1}{3}$$

b.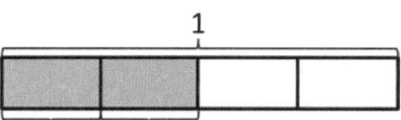

c.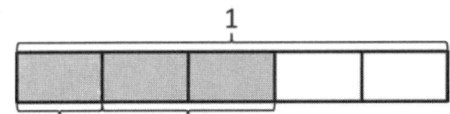

d.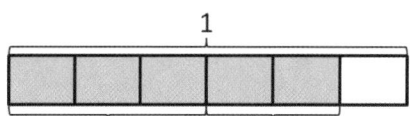

e.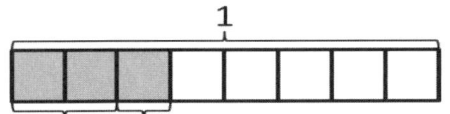

f.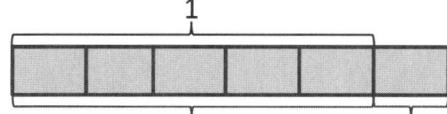

Lesson 1: Decompose fractions as a sum of unit fractions using tape diagrams.

g.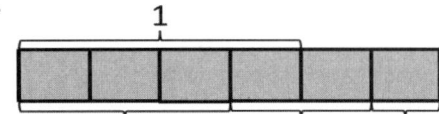

h.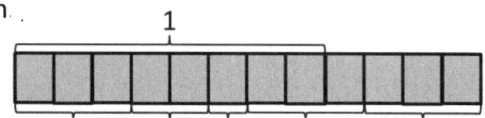

2. Draw and label tape diagrams to match each number sentence.

   a. $\frac{5}{8} = \frac{2}{8} + \frac{2}{8} + \frac{1}{8}$

   b. $\frac{12}{8} = \frac{6}{8} + \frac{2}{8} + \frac{4}{8}$

   c. $\frac{11}{10} = \frac{5}{10} + \frac{5}{10} + \frac{1}{10}$

   d. $\frac{13}{12} = \frac{7}{12} + \frac{1}{12} + \frac{5}{12}$

   e. $1\frac{1}{4} = A + \frac{1}{4}$

   f. $1\frac{2}{7} = A + \frac{2}{7}$

A STORY OF UNITS — Lesson 2 4•5

Lesson 2

Objective: Decompose fractions as a sum of unit fractions using tape diagrams.

Suggested Lesson Structure

- ■ Fluency Practice   (10 minutes)
- ■ Application Problem   (6 minutes)
- ■ Concept Development   (34 minutes)
- ■ Student Debrief   (10 minutes)
- **Total Time**   **(60 minutes)**

Fluency Practice (10 minutes)

- Read Tape Diagrams **3.OA.3**   (4 minutes)
- Break Apart Fractions **4.NF.3** (6 minutes)

Read Tape Diagrams (4 minutes)

Materials: (S) Personal white board

Note: This fluency activity prepares students for today's lesson.

T: (Project a tape diagram partitioned into 3 equal parts. Write 15 at the top.) Say the value of the whole.
S: 15.
T: Write the value of 1 unit as a division problem.
S: (Write 15 ÷ 3 = 5.)
T: (Write 5 in each unit.) Write the whole as a repeated addition sentence.
S: (Write 5 + 5 + 5 = 15.)
T: (Write 3 fives = 5 + 5 + 5 = 3 × ___.) Write the whole as a multiplication equation.
S: (Write 3 × 5 = 15.)

Continue with the following possible sequence: 8 ÷ 2, 20 ÷ 5, 12 ÷ 2, 8 ÷ 4, 21 ÷ 3, and 32 ÷ 4.

Lesson 2:   Decompose fractions as a sum of unit fractions using tape diagrams.   31

A STORY OF UNITS   Lesson 2 4•5

Break Apart Fractions (6 minutes)

Materials: (S) Personal white board

Note: This fluency activity reviews Lesson 1.

T: (Project a circle partitioned into 4 equal parts with 3 parts shaded.) How many circles do you see?
S: 1 circle.
T: How many equal parts does the circle have?
S: 4.
T: What fraction of the circle is shaded?
S: 3 fourths.
T: How many fourths are in 3 fourths?
S: 3.
T: (Write $\frac{3}{4}$ = ___ + ___ + ___.) On your personal white board, write $\frac{3}{4}$ as a repeated addition sentence.
S: (Write $\frac{3}{4} = \frac{1}{4} + \frac{1}{4} + \frac{1}{4}$.)
T: (Write $\frac{3}{4} = \frac{1}{4} + \frac{1}{4} + \frac{1}{4}$. Beneath it, write $\frac{3}{4} = \frac{2}{4} +$ ___.) Fill in the unknown fraction.
S: (Write $\frac{3}{4} = \frac{2}{4} + \frac{1}{4}$.)

Continue with the fraction graphics pictured to the right.

Application Problem (6 minutes)

Mrs. Salcido cut a small birthday cake into 6 equal pieces for 6 children. One child was not hungry, so she gave the birthday boy the extra piece. Draw a tape diagram to show how much cake each of the five children received.

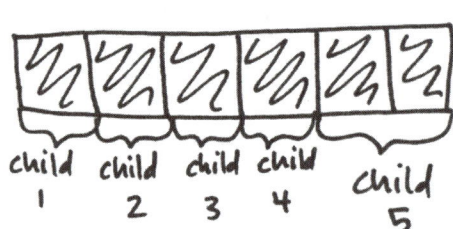
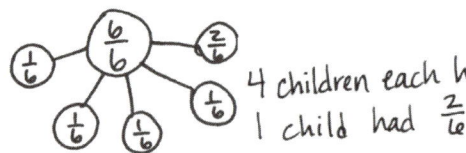

4 children each had $\frac{1}{6}$ of the cake.
1 child had $\frac{2}{6}$ of the cake.

$\frac{6}{6} = \frac{1}{6} + \frac{1}{6} + \frac{1}{6} + \frac{1}{6} + \frac{2}{6}$

Note: This Application Problem is a review of the material presented in Lesson 1 and prepares students for the more advanced portion of this lesson objective that they encounter in today's lesson.

Lesson 2: Decompose fractions as a sum of unit fractions using tape diagrams.

A STORY OF UNITS

Lesson 2   4•5

Concept Development (34 minutes)

Materials: (T) 2 strips of paper, markers (S) 2 strips of paper, markers or colored pencils, personal white board

Problem 1: Use a number bond to show how 1 can be decomposed into fourths and how fourths can be composed to make 1.

T: (Display a number bond to show 1 decomposed into 4 units of 1 fourth.) What does the number bond show?

S: 1 is the whole. The four 1 fourths are the parts. → 4 fourths make 1.

T: Let's say it as an addition sentence starting with "1 equals ..."

S: $1 = \frac{1}{4} + \frac{1}{4} + \frac{1}{4} + \frac{1}{4}$.

T: Fold a strip of paper to represent the same parts that our number bond shows. Work with a partner to see if there are any different number sentences we could create for decomposing 1 into fourths. Draw number bonds, and then write number sentences.

T: What number sentences did you create?

S: $1 = \frac{3}{4} + \frac{1}{4}$. → $1 = \frac{2}{4} + \frac{2}{4}$. → We could write $1 = \frac{1}{4} + \frac{1}{4} + \frac{2}{4}$. They all equal 1.

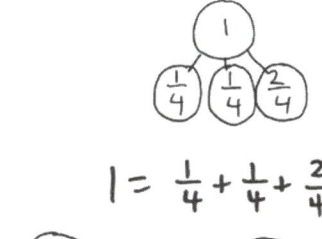

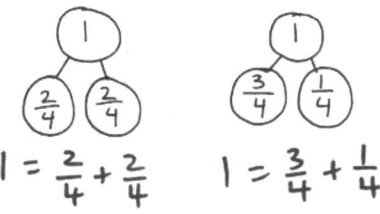

Problem 2: Fold a piece of paper to create eighths. Decompose fractions in different ways.

T: Turn over your strip of paper. The length of this strip of paper represents 1. Fold this paper to create 8 equal parts. (Demonstrate folding vertically.) Shade 7 of your 8 parts.

T: Give me one number sentence that shows the decomposition of 7 eighths into unit fractions.

S: (Write the sum of 7 units of 1 eighth.)

T: Use parentheses to decompose your sum of unit fractions into two parts.

S: (Write a possible answer such as $\frac{7}{8} = \left(\frac{1}{8} + \frac{1}{8} + \frac{1}{8}\right) + \left(\frac{1}{8} + \frac{1}{8} + \frac{1}{8} + \frac{1}{8}\right)$.)

T: On your board, record your decomposition of 7 eighths with 2 parts, and then look for other ways to decompose with 2 or more parts.

S: $\frac{7}{8} = \frac{3}{8} + \frac{4}{8}$. → $\frac{7}{8} = \frac{7}{8} + 0$. → $\frac{7}{8} = \frac{2}{8} + \frac{2}{8} + \frac{3}{8}$.

NOTES ON MULTIPLE MEANS OF ACTION AND EXPRESSION:

If paper strip eighths are difficult to color and manipulate in Problem 2 of the Concept Development, use concrete manipulatives such as fraction strips, Cuisenaire rods, or linking cubes. Alternatively, use larger fraction strips or tape diagram drawings.

Lesson 2: Decompose fractions as a sum of unit fractions using tape diagrams.

A STORY OF UNITS Lesson 2 4•5

T: What do all of the number sentences have in common? Discuss with a partner.

S: They all add up to $\frac{7}{8}$. → The parts are eighths in all of them.

Problem 3: Write number sentences to decompose $\frac{5}{6}$ as a sum of fractions with the same denominator.

MP.3

T: Form groups of three. Work on your personal white board. Each of you should write a number bond and sentence showing a decomposition of 5 sixths. If you have the same decomposition as someone else in your group, one of you must change your work. (Allow time for students to work.)

T: Let's share. What number bonds did you create? (Record number sentences.)

S: $\frac{5}{6} = \frac{1}{6} + \frac{1}{6} + \frac{1}{6} + \frac{1}{6} + \frac{1}{6}$. → $\frac{5}{6} = \frac{3}{6} + \frac{2}{6}$. → $\frac{5}{6} = \frac{4}{6} + \frac{1}{6}$. → (Other various answers.)

 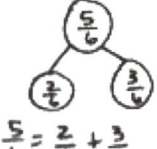 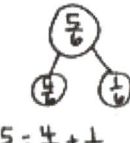 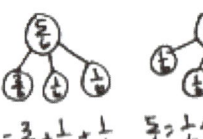 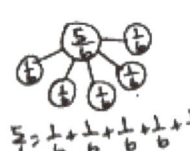 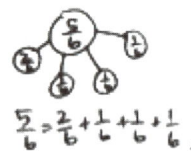

T: Now, on your board, instead of drawing number bonds, draw tape diagrams to show three different ways of decomposing the fraction $\frac{5}{4}$. Write the number sentence describing each tape diagram you drew next to the tape diagrams. What number sentences did you write?

S: $\frac{5}{4} = \frac{1}{4} + \frac{1}{4} + \frac{1}{4} + \frac{1}{4} + \frac{1}{4}$. → $\frac{5}{4} = \frac{2}{4} + \frac{1}{4} + \frac{2}{4}$. → $\frac{5}{4} = 1 + \frac{1}{4}$.

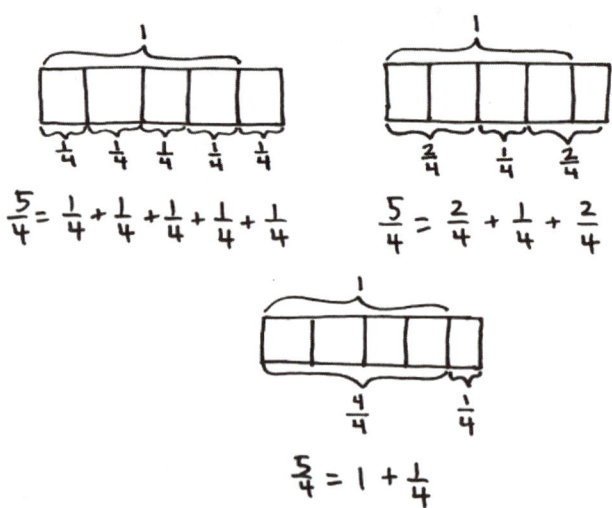

> **NOTES ON MULTIPLE MEANS OF REPRESENTATION:**
>
> Check that English language learners and others understand that they are to represent 5 fourths, not 5 *fifths*, with a tape diagram. Provide guidance as students model this improper fraction as a tape diagram. Ask, "What is the unit? How many units to make 1? How do you show 1 on your tape diagram?"

A STORY OF UNITS

Lesson 2 4•5

Problem Set (10 minutes)

Students should do their personal best to complete the Problem Set within the allotted 10 minutes. For some classes, it may be appropriate to modify the assignment by specifying which problems they work on first. Some problems do not specify a method for solving. Students should solve these problems using the RDW approach used for Application Problems.

Student Debrief (10 minutes)

Lesson Objective: Decompose fractions as a sum of unit fractions using tape diagrams.

The Student Debrief is intended to invite reflection and active processing of the total lesson experience.

Invite students to review their solutions for the Problem Set. They should check work by comparing answers with a partner before going over answers as a class. Look for misconceptions or misunderstandings that can be addressed in the Debrief. Guide students in a conversation to debrief the Problem Set and process the lesson.

Any combination of the questions below may be used to lead the discussion.

- Look at your answers for Problems 1(b) and 1(c). Problem 1(c) is a fraction greater than 1, but it has fewer ways to be decomposed. Why is that?
- In Problem 1(a), which was completed for you, the first number sentence was decomposed into the sum of unit fractions. The second number sentence bonded some of these unit fractions. Which ones? ($\frac{2}{8}$ bonded $\frac{1}{8} + \frac{1}{8}$.) Draw parentheses around the unit fractions in the first number sentence that match the second number sentence. Do the same for Problems 1(b) and 1(c). (Answers will vary.)

$$\frac{5}{8} = \left(\frac{1}{8} + \frac{1}{8}\right) + \left(\frac{1}{8} + \frac{1}{8}\right) + \frac{1}{8}. \rightarrow \frac{5}{8} = \frac{2}{8} + \frac{2}{8} + \frac{1}{8}.$$

- Give examples of when you decomposed numbers in earlier grades.
- How did the Application Problem connect to today's lesson?

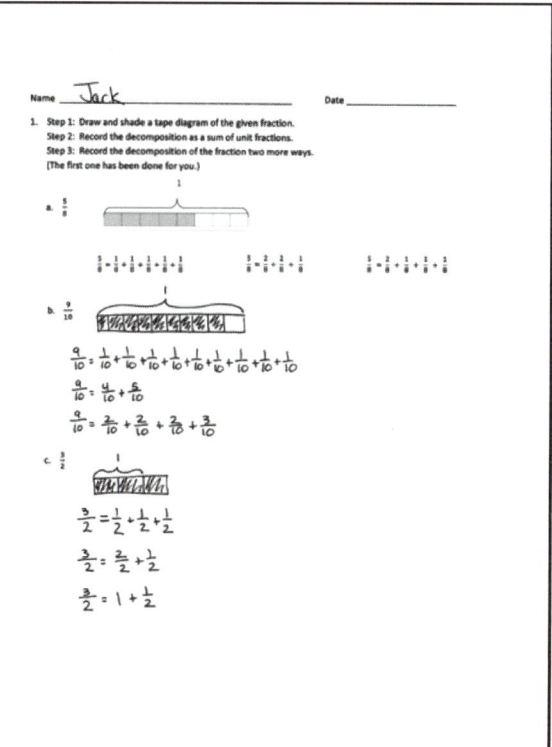

Lesson 2: Decompose fractions as a sum of unit fractions using tape diagrams.

Exit Ticket (3 minutes)

After the Student Debrief, instruct students to complete the Exit Ticket. A review of their work will help with assessing students' understanding of the concepts that were presented in today's lesson and planning more effectively for future lessons. The questions may be read aloud to the students.

A STORY OF UNITS — Lesson 2 Problem Set — 4•5

Name _____ Date _____

1. Step 1: Draw and shade a tape diagram of the given fraction.
   Step 2: Record the decomposition as a sum of unit fractions.
   Step 3: Record the decomposition of the fraction two more ways.
   (The first one has been done for you.)

   a. $\frac{5}{8}$

   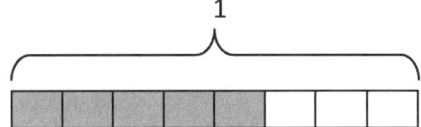

   $\frac{5}{8} = \frac{1}{8} + \frac{1}{8} + \frac{1}{8} + \frac{1}{8} + \frac{1}{8}$ $\quad\quad \frac{5}{8} = \frac{2}{8} + \frac{2}{8} + \frac{1}{8}$ $\quad\quad \frac{5}{8} = \frac{2}{8} + \frac{1}{8} + \frac{1}{8} + \frac{1}{8}$

   b. $\frac{9}{10}$

   c. $\frac{3}{2}$

Lesson 2: Decompose fractions as a sum of unit fractions using tape diagrams.

2. Step 1: Draw and shade a tape diagram of the given fraction.
Step 2: Record the decomposition of the fraction in three different ways using number sentences.

   a. $\dfrac{7}{8}$

   b. $\dfrac{5}{3}$

   c. $\dfrac{7}{5}$

   d. $1\dfrac{1}{3}$

A STORY OF UNITS	Lesson 2 Exit Ticket 4•5

Name _____ Date _____

Step 1: Draw and shade a tape diagram of the given fraction.

Step 2: Record the decomposition of the fraction in three different ways using number sentences.

$\frac{4}{7}$

Lesson 2: Decompose fractions as a sum of unit fractions using tape diagrams.

Name _____ Date _____

1. Step 1: Draw and shade a tape diagram of the given fraction.
   Step 2: Record the decomposition as a sum of unit fractions.
   Step 3: Record the decomposition of the fraction two more ways.
   (The first one has been done for you.)

   a. $\frac{5}{6}$

   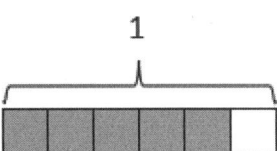

   $\frac{5}{6} = \frac{1}{6} + \frac{1}{6} + \frac{1}{6} + \frac{1}{6} + \frac{1}{6}$ $\frac{5}{6} = \frac{2}{6} + \frac{2}{6} + \frac{1}{6}$ $\frac{5}{6} = \frac{1}{6} + \frac{4}{6}$

   b. $\frac{6}{8}$

   c. $\frac{7}{10}$

A STORY OF UNITS

Lesson 2 Homework 4•5

2. Step 1: Draw and shade a tape diagram of the given fraction.
   Step 2: Record the decomposition of the fraction in three different ways using number sentences.

   a. $\frac{11}{12}$

   b. $\frac{5}{4}$

   c. $\frac{6}{5}$

   d. $1\frac{1}{4}$

Lesson 2: Decompose fractions as a sum of unit fractions using tape diagrams.

Lesson 3

Objective: Decompose non-unit fractions and represent them as a whole number times a unit fraction using tape diagrams.

Suggested Lesson Structure

- ■ Fluency Practice (12 minutes)
- ■ Application Problem (8 minutes)
- ■ Concept Development (30 minutes)
- ■ Student Debrief (10 minutes)

Total Time **(60 minutes)**

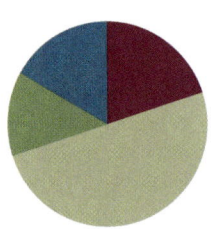

Fluency Practice (12 minutes)

- Multiply Mentally **4.OA.4** (4 minutes)
- Repeated Addition as Multiplication **4.OA.4** (4 minutes)
- Add Fractions **4.NF.3** (4 minutes)

Multiply Mentally (4 minutes)

Materials: (S) Personal white board

Note: This fluency activity reviews Module 3 content.

- T: (Write 34 × 2 = ___.) Say the multiplication sentence.
- S: 34 × 2 = 68.
- T: (Write 34 × 2 = 68. Below it, write 34 × 20 = ___.) Say the multiplication sentence.
- S: 34 × 20 = 680.
- T: (Write 34 × 20 = 680. Below it, write 34 × 22 = ___.) On your personal white board, solve 34 × 22.
- S: (Write 34 × 22 = 748.)

Continue with the following possible sequence: 23 × 3, 23 × 20, and 23 × 23; and 12 × 4, 12 × 30, and 12 × 34.

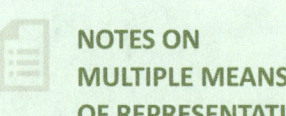

NOTES ON MULTIPLE MEANS OF REPRESENTATION:

Scaffold the Multiply Mentally fluency activity for students working below grade level and others having difficulty. Clarify that (34 × 2) + (34 × 20) is the same as 34 × 22, and so on. Ask students, "Why is this true?"

Repeated Addition as Multiplication (4 minutes)

Materials: (S) Personal white board

Note: This fluency activity reviews Module 3 content.

- T: (Write 2 + 2 + 2 = ___.) Say the addition sentence.
- S: 2 + 2 + 2 = 6.

A STORY OF UNITS   Lesson 3 4•5

T: (Write 2 + 2 + 2 = 6. Beneath it, write ___ × 2 = 6.) On your personal white board, fill in the unknown factor.

S: (Write 3 × 2 = 6.)

T: (Write 3 × 2 = 6. To the right, write 30 + 30 + 30 = ___.) Say the addition sentence.

2 + 2 + 2 = 6	30 + 30 + 30 = 90	32 + 32 + 32 = 96
3 × 2 = 6	3 × 30 = 90	3 × 32 = 96

S: 30 + 30 + 30 = 90.

T: (Write 30 + 30 + 30 = 90. Beneath it, write ___ × 30 = 90.) Fill in the unknown factor.

S: (Write 3 × 30 = 90.)

T: (Write 3 × 30 = 90. To the right, write 32 + 32 + 32 = ___.) On your board, write the repeated addition sentence. Then, beneath it, write a multiplication sentence to reflect the addition sentence.

S: (Write 32 + 32 + 32 = 96. Beneath it, write 3 × 32 = 96.)

Continue with the following possible sequence: 1 + 1 + 1 + 1, 4 × 1; 20 + 20 + 20 + 20, 4 × 20; 21 + 21 + 21 + 21, 4 × 21; and 23 + 23 + 23, 3 × 23.

Add Fractions (4 minutes)

Materials: (S) Personal white board

Note: This fluency activity reviews Lesson 2.

T: (Write $\frac{4}{5}$.) Say the fraction.

S: 4 fifths.

T: On your personal white board, draw a tape diagram representing 4 fifths.

S: (Draw a tape diagram partitioned into 5 equal units. Shade 4 units.)

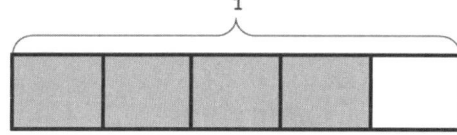

T: (Write $\frac{4}{5}$ = ___ + ___ + ___ + ___.) Write $\frac{4}{5}$ as the sum of unit fractions.

S: (Write $\frac{4}{5} = \frac{1}{5} + \frac{1}{5} + \frac{1}{5} + \frac{1}{5}$)

T: (Write $\frac{4}{5} = \frac{2}{5} + \frac{1}{5} + \frac{1}{5}$.) Bracket 2 fifths on your diagram, and complete this number sentence.

S: (Group $\frac{2}{5}$ on the diagram. Write $\frac{4}{5} = \frac{2}{5} + \frac{1}{5} + \frac{1}{5}$.)

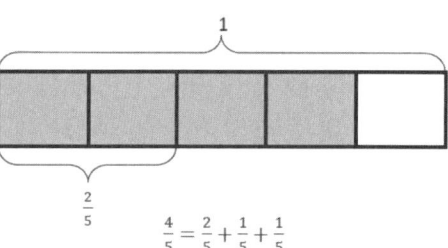

T: (Write $\frac{4}{5} = \frac{}{5} + \frac{}{5}$.) Bracket fifths again on your diagram, and write a number sentence to match. There's more than one correct answer.

S: (Group fifths on the diagram. Write $\frac{4}{5} = \frac{2}{5} + \frac{2}{5}, \frac{4}{5} = \frac{3}{5} + \frac{1}{5}$, or $\frac{4}{5} = \frac{1}{5} + \frac{3}{5}$.)

Continue with the following possible sequence: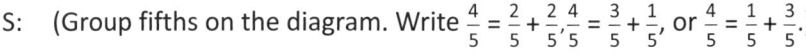

Lesson 3: Decompose non-unit fractions and represent them as a whole number times a unit fraction using tape diagrams.

A STORY OF UNITS   Lesson 3   4•5

Application Problem (8 minutes)

Mrs. Beach prepared copies for 4 reading groups. She made 6 copies for each group. How many copies did Mrs. Beach make?

a. Draw a tape diagram.

b. Write both an addition and a multiplication sentence to solve. Discuss with a partner why you are able to add or multiply to solve this problem.

c. What fraction of the copies is needed for 3 groups? To show that, shade the tape diagram.

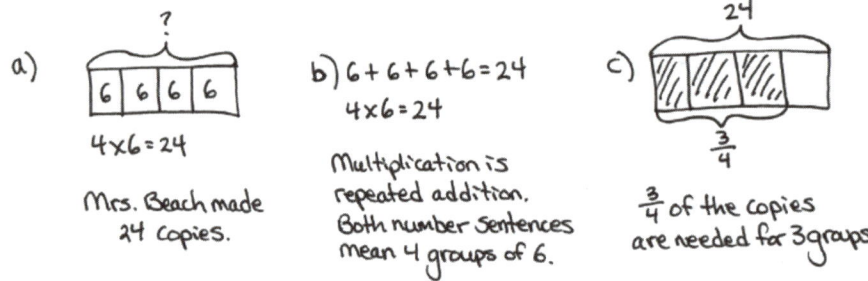

Note: This Application Problem builds from Grade 3 knowledge of interpreting products of whole numbers. This Application Problem bridges to today's lesson where students come to understand that a non-unit fraction can be decomposed and represented as a whole number times a unit fraction.

Concept Development (30 minutes)

Materials: (S) Personal white board

Problem 1: Express a non-unit fraction less than 1 as a whole number times a unit fraction using a tape diagram.

T: Look back at the tape diagram that we drew in the Application Problem. What fraction is represented by the shaded part?

S: $\frac{3}{4}$.

T: Say $\frac{3}{4}$ decomposed as the sum of unit fractions.

S: $\frac{3}{4} = \frac{1}{4} + \frac{1}{4} + \frac{1}{4}$.

T: How many fourths are there in $\frac{3}{4}$?

S: 3.

T: We know this because we count 1 fourth 3 times. Discuss with a partner. How might we express this using multiplication?

S: We have 3 fourths. That's $\frac{1}{4} + \frac{1}{4} + \frac{1}{4}$ or three groups of 1 fourth. Could we multiply $3 \times \frac{1}{4}$?

Lesson 3: Decompose non-unit fractions and represent them as a whole number times a unit fraction using tape diagrams.

T: Yes! If we want to add the same fraction of a certain amount many times, instead of adding, we can multiply. Just like you multiplied 6 copies 4 times, we can multiply 1 fourth 3 times. What is 3 copies of $\frac{1}{4}$?

S: It's $\frac{3}{4}$. My tape diagram proves it!

Repeat with $\frac{2}{3}$ and $\frac{7}{8}$. Instruct students to draw a tape diagram to represent each fraction (as on the previous page), to shade the given number of parts. Then, direct students to write an addition number sentence and a multiplication number sentence.

Problem 2: Determine the non-unit fraction greater than 1 that is represented by a tape diagram, and then write the fraction as a whole number times a unit fraction.

T: (Project the tape diagram of $\frac{10}{8}$ as shown below.) What fractional unit does the tape diagram show?

S: It shows tenths! → It shows eighths!

T: We first must identify 1. It's bracketed here. (Point.) How many units is 1 partitioned into?

S: 8.

T: The bracketed portion of the tape diagram shows 8 fractional units. What is the total number of eighths?

S: 10.

T: What is the fraction?

S: 10 eighths.

T: Say this as an addition number sentence. Use your fingers to keep track of the number of units as you say them.

S: $\frac{10}{8} = \frac{1}{8} + \frac{1}{8} + \frac{1}{8} + \frac{1}{8} + \frac{1}{8} + \frac{1}{8} + \frac{1}{8} + \frac{1}{8} + \frac{1}{8} + \frac{1}{8}$.

T: As a multiplication number sentence?

S: $\frac{10}{8} = 10 \times \frac{1}{8}$.

T: What are the advantages of multiplying fractions instead of adding?

S: It's easier to write. → It's faster. → It's more efficient.

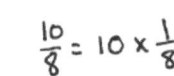

Problem 3: Express a non-unit fraction greater than 1 as a whole number times a unit fraction using a tape diagram.

T: Let's put parentheses around 8 eighths so that we can see 10 eighths can also be written to show 1 and 2 more eighths. (Write $\frac{10}{8} = \left(8 \times \frac{1}{8}\right) + \left(2 \times \frac{1}{8}\right)$.)

T: Discuss with your partner how to draw a tape diagram to show 5 thirds.

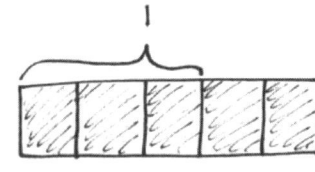

Lesson 3: Decompose non-unit fractions and represent them as a whole number times a unit fraction using tape diagrams.

A STORY OF UNITS   Lesson 3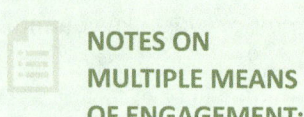

S: I can draw one unit at a time. The units are thirds, so I'll draw five small rectangles together. → I know 5 thirds is greater than 1, so I'll draw 1. That's 3 thirds. So, then I can draw another 1. I'll just shade 5 parts. → I will draw a long rectangle and break it into 5 equal parts. Each part represents 1 third. I'll bracket 3 thirds to show 1.

T: How can we express $\frac{5}{3}$ as a multiplication expression?

S: We have five thirds. That's $5 \times \frac{1}{3}$.

T: Is there another way we can express $\frac{5}{3}$ using multiplication?

S: Can we express the 1 as $3 \times \frac{1}{3}$ and then add $2 \times \frac{1}{3}$?

T: Yes! We can use multiplication and addition to decompose fractions.

> **NOTES ON MULTIPLE MEANS OF ENGAGEMENT:**
>
> Offer an alternative to Problem 2 on the Problem Set for students working above grade level. Challenge students to compose a word problem of their own to match one or more of the tape diagrams they construct for Problem 2. Always offer challenges and extensions to learners as alternatives, rather than additional *busy* work.

Problem Set (10 minutes)

Students should do their personal best to complete the Problem Set within the allotted 10 minutes. For some classes, it may be appropriate to modify the assignment by specifying which problems they work on first. Some problems do not specify a method for solving. Students should solve these problems using the RDW approach used for Application Problems.

Student Debrief (10 minutes)

Lesson Objective: Decompose non-unit fractions and represent them as a whole number times a unit fraction using tape diagrams.

The Student Debrief is intended to invite reflection and active processing of the total lesson experience.

Invite students to review their solutions for the Problem Set. They should check work by comparing answers with a partner before going over answers as a class. Look for misconceptions or misunderstandings that can be addressed in the Debrief. Guide students in a conversation to debrief the Problem Set and process the lesson.

Any combination of the questions below may be used to lead the discussion.

- In all of the problems, why do we need to label 1 on our tape diagrams? What would happen if we did not label 1?

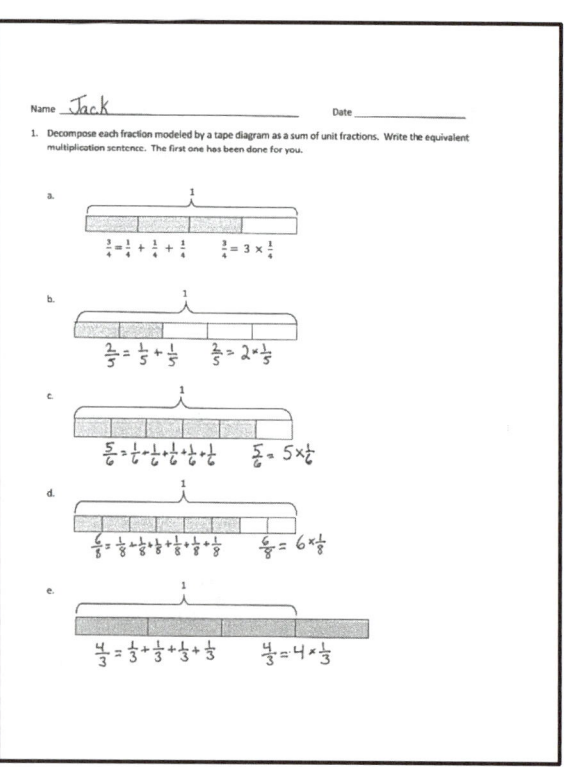

46   Lesson 3:   Decompose non-unit fractions and represent them as a whole number times a unit fraction using tape diagrams.

A STORY OF UNITS

Lesson 3 4•5

- What is an advantage to representing the fractions using multiplication?
- What is similar in Problems 3(c), 3(d), and 3(e)? Which fractions are greater than 1? Which is less than 1?
- Are you surprised to see multiplication sentences with products less than 1? Why?
- In our lesson, when we expressed $\frac{5}{3}$ as $\left(3 \times \frac{1}{3}\right) + \left(2 \times \frac{1}{3}\right)$, what property were we using?
- Consider the work we did in Lessons 1 and 2 where we decomposed a tape diagram multiple ways. Can we now rewrite those number sentences using addition and multiplication? Try it with this tape diagram (as shown below).

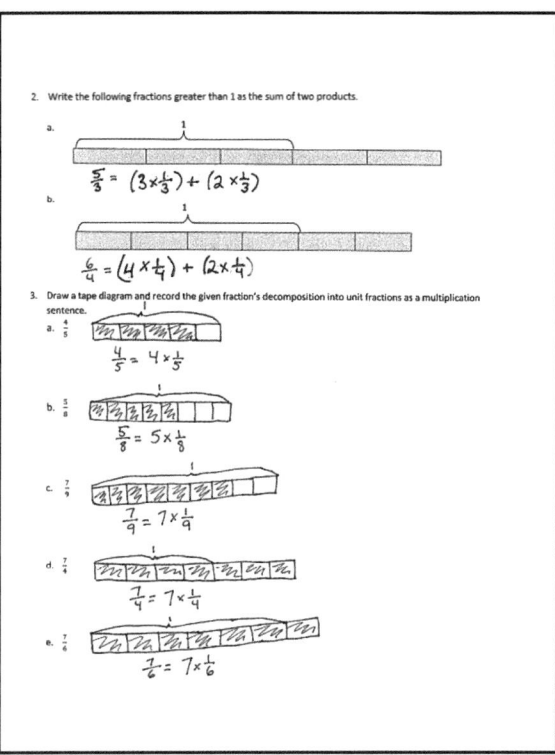

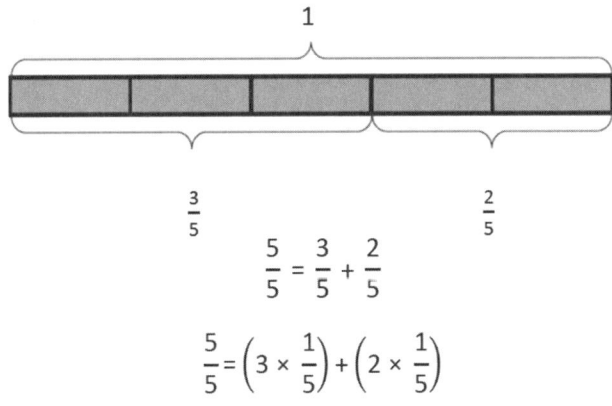

- How is multiplying fractions like multiplying whole numbers?
- How did the Application Problem connect to today's lesson?

Exit Ticket (3 minutes)

After the Student Debrief, instruct students to complete the Exit Ticket. A review of their work will help with assessing students' understanding of the concepts that were presented in today's lesson and planning more effectively for future lessons. The questions may be read aloud to the students.

Lesson 3: Decompose non-unit fractions and represent them as a whole number times a unit fraction using tape diagrams.

47

© 2015 Great Minds. eureka-math.org
G4-M5-TE-B4-1.3.1-01.2016

A STORY OF UNITS   Lesson 3 Problem Set   4•5

Name _____ Date _____

1. Decompose each fraction modeled by a tape diagram as a sum of unit fractions. Write the equivalent multiplication sentence. The first one has been done for you.

a.
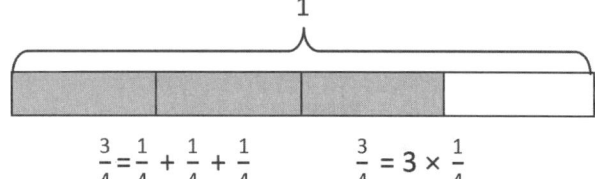

$\frac{3}{4} = \frac{1}{4} + \frac{1}{4} + \frac{1}{4}$ $\frac{3}{4} = 3 \times \frac{1}{4}$

b.
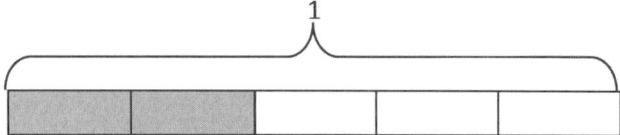

c.
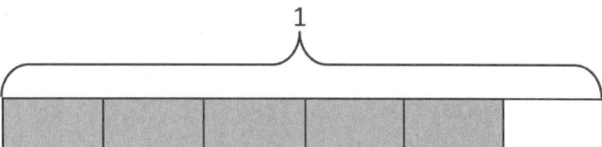

d.
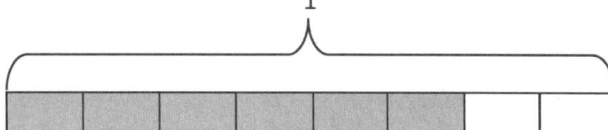

e.
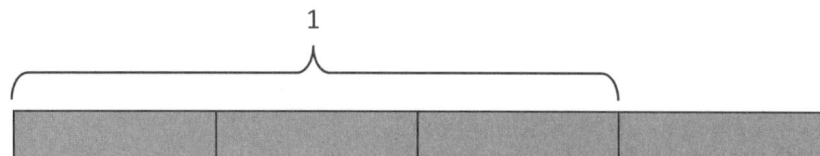

2. Write the following fractions greater than 1 as the sum of two products.

   a.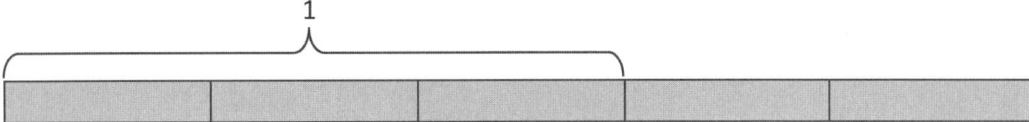

   b.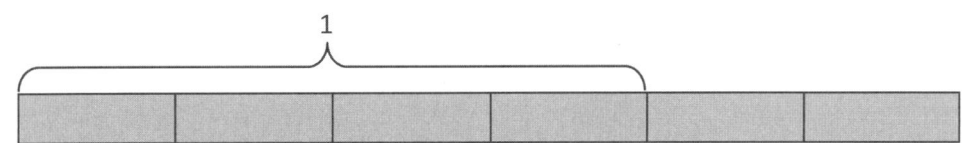

3. Draw a tape diagram, and record the given fraction's decomposition into unit fractions as a multiplication sentence.

   a. $\frac{4}{5}$

   b. $\frac{5}{8}$

   c. $\frac{7}{9}$

   d. $\frac{7}{4}$

   e. $\frac{7}{6}$

Name _____ Date _____

1. Decompose each fraction modeled by a tape diagram as a sum of unit fractions. Write the equivalent multiplication sentence.

   a.

   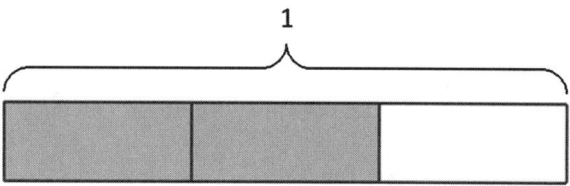

   b.

   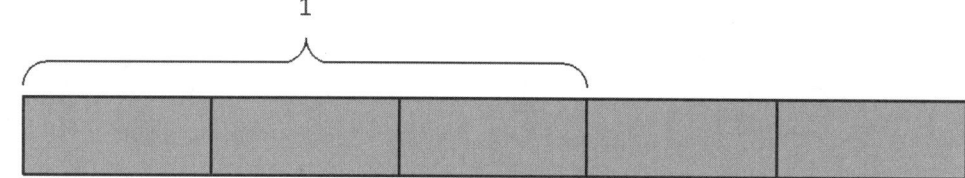

2. Draw a tape diagram, and record the given fraction's decomposition into unit fractions as a multiplication sentence.

   $$\frac{6}{9}$$

Name _____   Date _____

1. Decompose each fraction modeled by a tape diagram as a sum of unit fractions. Write the equivalent multiplication sentence. The first one has been done for you.

   a.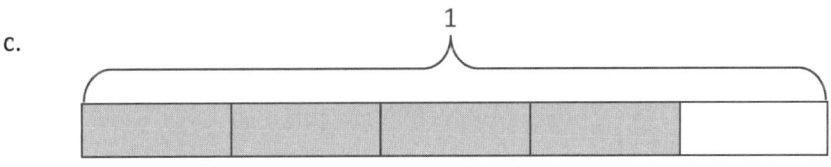

   $\frac{2}{3} = \frac{1}{3} + \frac{1}{3}$ $\frac{2}{3} = 2 \times \frac{1}{3}$

   b.

   c.

   d.

2. Write the following fractions greater than 1 as the sum of two products.

   a.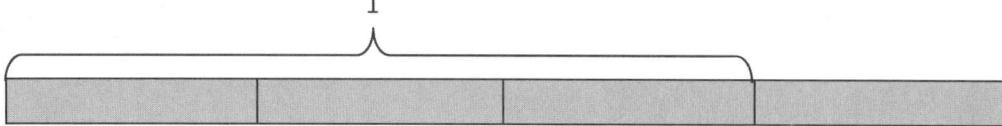

   b.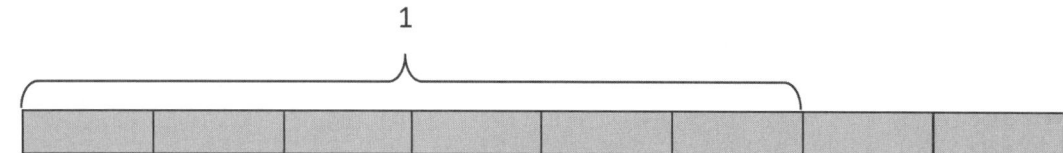

3. Draw a tape diagram, and record the given fraction's decomposition into unit fractions as a multiplication sentence.

   a. $\frac{3}{5}$

   b. $\frac{3}{8}$

   c. $\frac{5}{9}$

   d. $\frac{8}{5}$

   e. $\frac{12}{4}$

A STORY OF UNITS   Lesson 4   4•5

Lesson 4

Objective: Decompose fractions into sums of smaller unit fractions using tape diagrams.

Suggested Lesson Structure

- **Fluency Practice**   (12 minutes)
- **Application Problem**   (4 minutes)
- **Concept Development**   (34 minutes)
- **Student Debrief**   (10 minutes)
- **Total Time**   **(60 minutes)**

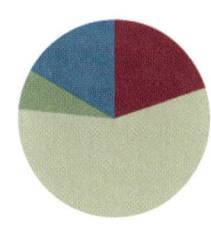

Fluency Practice (12 minutes)

- Break Apart Fractions **4.NF.3**   (7 minutes)
- Count by Equivalent Fractions **3.NF.3** (5 minutes)

Break Apart Fractions (7 minutes)

Materials: (S) Personal white board

Note: This fluency activity reviews Lesson 3.

T: (Project a tape diagram partitioned into 3 equal units. Write 1 above it. Shade 2 units.) How many equal parts does this 1 have?
S: 3 parts.
T: Say the value of 1 unit.
S: 1 third.
T: What fraction of 1 is shaded?
S: 2 thirds.
T: On your personal white board, write the value of the shaded part as a sum of unit fractions.
S: (Write $\frac{2}{3} = \frac{1}{3} + \frac{1}{3}$.)
T: (Write $\_\_ \times \frac{1}{3} = \frac{2}{3}$.) On your board, complete the number sentence.
S: (Write $2 \times \frac{1}{3} = \frac{2}{3}$.)

Continue with the following possible sequence: $\frac{3}{5}, \frac{5}{8}$, and $\frac{5}{4}$.

Lesson 4: Decompose fractions into sums of smaller unit fractions using tape diagrams.   53

A STORY OF UNITS

Lesson 4 4•5

T: (Write $\frac{3}{4}$.) Say the fraction.
S: 3 fourths.
T: On your board, draw a tape diagram of 3 fourths.
S: (Draw a tape diagram partitioned into 4 equal units. Shade 3 units.)
T: What's the value of each unit?
S: 1 fourth.
T: Express 3 fourths as a repeated addition sentence.
S: (Write $\frac{3}{4} = \frac{1}{4} + \frac{1}{4} + \frac{1}{4}$.)
T: (Write $\frac{3}{4} = \_ \times \frac{1}{4}$.) Fill in the unknown number.
S: (Write $\frac{3}{4} = 3 \times \frac{1}{4}$.)

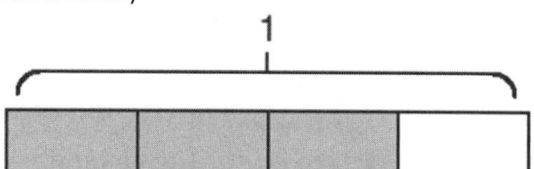

Continue with the following possible sequence: $\frac{4}{5}, \frac{8}{5}$, and $\frac{6}{3}$.

Count by Equivalent Fractions (5 minutes)

Note: This fluency activity prepares students for Lesson 4.

T: Count by ones to 6.
S: 1, 2, 3, 4, 5, 6.
T: Count by sixths to 6 sixths. Start at 0 sixths. (Write as students count.)
S: $\frac{0}{6}, \frac{1}{6}, \frac{2}{6}, \frac{3}{6}, \frac{4}{6}, \frac{5}{6}, \frac{6}{6}$.
T: 6 sixths is the same as one of what unit?
S: 1 one.
T: (Beneath $\frac{6}{6}$, write 1.) Count by sixths again. This time, say 1 one when you arrive at 6 sixths. Start at zero.
S: $0, \frac{1}{6}, \frac{2}{6}, \frac{3}{6}, \frac{4}{6}, \frac{5}{6}$, 1 one.
T: Let's count by thirds to 6 thirds. Start at 0 thirds. (Write as students count.)
S: $\frac{0}{3}, \frac{1}{3}, \frac{2}{3}, \frac{3}{3}, \frac{4}{3}, \frac{5}{3}, \frac{6}{3}$.
T: How many thirds are in 1?
S: 3 thirds.
T: (Beneath $\frac{3}{3}$, write 1.) How many thirds are in 2?
S: 6 thirds.
T: (Beneath $\frac{6}{3}$, write 2.) Let's count by thirds again. This time, when you arrive at 3 thirds and 6 thirds, say the whole number. Start at zero.
S: $0, \frac{1}{3}, \frac{2}{3}, 1, \frac{4}{3}, \frac{5}{3}, 2$.

Continue, counting by halves to 6 halves.

54 Lesson 4: Decompose fractions into sums of smaller unit fractions using tape diagrams.

A STORY OF UNITS — Lesson 4 4•5

Application Problem (4 minutes)

A recipe calls for $\frac{3}{4}$ cup of milk. Saisha only has a $\frac{1}{4}$-cup measuring cup. If she doubles the recipe, how many times will she need to fill the $\frac{1}{4}$ cup with milk? Draw a tape diagram, and record as a multiplication sentence.

Note: This Application Problem reviews students' knowledge of fractions from Lesson 3 and prepares them for today's objective of decomposing unit fractions into sums of smaller unit fractions.

$6 \times \frac{1}{4} = \frac{6}{4}$

She will need to fill her $\frac{1}{4}$ measuring cup 6 times.

Concept Development (34 minutes)

Materials: (S) Personal white board

Problem 1: Use tape diagrams to represent the decomposition of $\frac{1}{3}$ as the sum of unit fractions.

T: Draw a tape diagram that represents 1, and shade 1 third. Decompose each of the thirds in half. How many parts are there now?
S: 6.
T: What fraction of 1 does each part represent?
S: 1 sixth.
T: How many sixths are shaded?
S: 2 sixths.
T: What can we say about 1 third and 2 sixths?
S: They are the same.
T: How can you tell?
S: They both take up the same amount of space.
T: Let's write that as a number sentence: $\frac{1}{3} = \frac{1}{6} + \frac{1}{6} + \frac{2}{6}$.
T: Now, decompose each sixth into 2 equal parts on your tape diagram. How many parts are in 1 now?
S: 12.
T: What fractional part of 1 does each piece represent?
S: 1 twelfth.
T: How many twelfths equal $\frac{1}{6}$?

NOTES ON MULTIPLE MEANS OF ACTION AND EXPRESSION:

Cuisenaire rods can be used to model 1 whole (brown), 2 halves (pink), 4 fourths (red), and 8 eighths (white). If concrete Cuisenaire rods are unavailable or otherwise challenging, virtual rods can be found at the link below:
http://nrich.maths.org/4348.

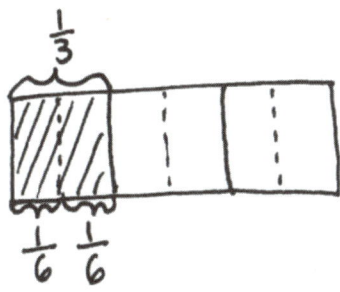

$\frac{1}{3} = \frac{1}{6} + \frac{1}{6} = \frac{2}{6}$

$\frac{1}{3} = 2 \times \frac{1}{6} = \frac{2}{6}$

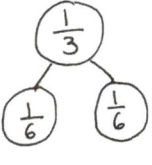

MP.2

Lesson 4: Decompose fractions into sums of smaller unit fractions using tape diagrams.

A STORY OF UNITS

Lesson 4 4•5

S: $\frac{2}{12}$ equals $\frac{1}{6}$.

T: Work with your partner to write a number sentence for how many twelfths equal $\frac{1}{3}$.

S: (Write $\frac{1}{3} = \frac{1}{6} + \frac{1}{6} = \left(\frac{1}{12} + \frac{1}{12}\right) + \left(\frac{1}{12} + \frac{1}{12}\right)$.)

T: We can put parentheses around two groups of 1 twelfth to show that each combines to make $\frac{1}{6}$.

$$\frac{1}{3} = \left(\frac{1}{12} + \frac{1}{12}\right) + \left(\frac{1}{12} + \frac{1}{12}\right)$$

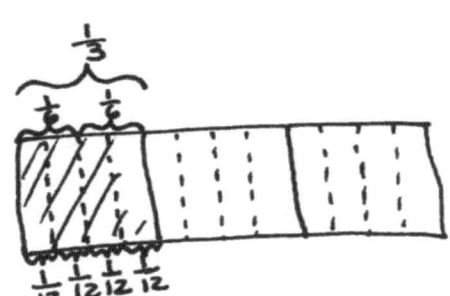

T: How can we represent this using multiplication?

S: (Write $\frac{1}{3} = \left(2 \times \frac{1}{12}\right) + \left(2 \times \frac{1}{12}\right) \rightarrow \frac{1}{3} = 4 \times \frac{1}{12} = \frac{4}{12}$).

Problem 2: Use tape diagrams to represent the decomposition of $\frac{1}{5}$ and $\frac{2}{5}$ as the sum of smaller unit fractions.

T: Draw a tape diagram, and shade $\frac{1}{5}$. Decompose each of the fifths into 3 equal parts. How many parts are there now?

S: There are 15 parts.

T: What fraction does each part represent?

S: $\frac{1}{15}$.

T: Write an addition sentence to show how many fifteenths it takes to make 1 fifth.

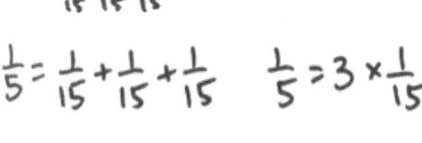 

S: (Write $\frac{1}{5} = \frac{1}{15} + \frac{1}{15} + \frac{1}{15} + \frac{3}{15}$).

T: What can we say about one-fifth and three-fifteenths?

S: They are equal.

T: With your partner, write a number sentence that represents $\frac{2}{5}$.

S: (Write $\frac{2}{5} = \frac{3}{15} + \frac{3}{15} + \frac{6}{15}. \rightarrow \frac{2}{5} = \left(3 \times \frac{1}{15}\right) + \left(3 \times \frac{1}{15}\right) = \frac{6}{15}. \rightarrow \frac{2}{5} = 2 \times \frac{1}{5} = 2 \times \frac{3}{15} = \frac{6}{15}$.)

Problem 3: Draw a tape diagram, and use addition to show that $\frac{2}{6}$ is the sum of 4 twelfths.

T: (Project $\frac{2}{6} = \frac{1}{12} + \frac{1}{12} + \frac{1}{12} + \frac{1}{12} = \frac{4}{12}$.) Using what you just learned, how can you model to show that $\frac{2}{6}$ is equal to $\frac{4}{12}$?

S: We can draw a tape diagram and shade $\frac{2}{6}$. Then, we can decompose it into twelfths.

T: How many twelfths are shaded?

S: 4.

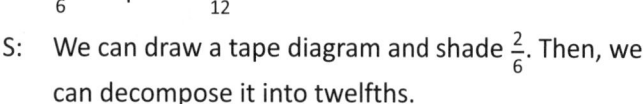
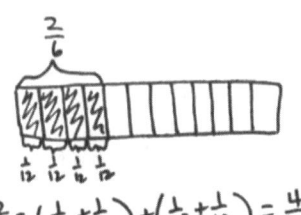
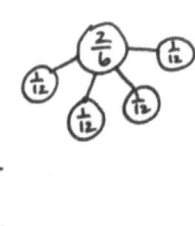
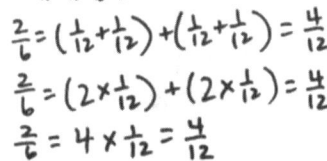

Lesson 4: Decompose fractions into sums of smaller unit fractions using tape diagrams.

EUREKA MATH

A STORY OF UNITS — Lesson 4 4•5

T: We have seen that 1 third is equal to 2 sixths. We have seen 1 sixth is equal to 2 twelfths. So, how many twelfths equal 1 third?

S: 4 twelfths!

T: So, 2 thirds is how many twelfths? Explain to your partner how you know using your diagrams.

S: 1 third is 4 twelfths, so 2 thirds is 8 twelfths. → It's just double. → It's twice the area on the tape diagram. → It's the same as 4 sixths. 1 third is 2 sixths. 2 thirds is 4 sixths. 1 sixth is the same as 2 twelfths, so 4 times 2 is 8. 8 twelfths.

Problem Set (10 minutes)

Students should do their personal best to complete the Problem Set within the allotted 10 minutes. For some classes, it may be appropriate to modify the assignment by specifying which problems they work on first. Some problems do not specify a method for solving. Students should solve these problems using the RDW approach used for Application Problems.

Student Debrief (10 minutes)

Lesson Objective: Decompose fractions into sums of smaller unit fractions using tape diagrams.

The Student Debrief is intended to invite reflection and active processing of the total lesson experience.

Invite students to review their solutions for the Problem Set. They should check work by comparing answers with a partner before going over answers as a class. Look for misconceptions or misunderstandings that can be addressed in the Debrief. Guide students in a conversation to debrief the Problem Set and process the lesson.

Any combination of the questions below may be used to lead the discussion.

- For Problem 1(a–d), what were some different ways that you decomposed the unit fraction?
- What is different about Problems 3(c) and 3(d)? Explain how fourths can be decomposed into both eighths and twelfths.
- For Problems 4, 5, and 6, explain the process you used to show equivalent fractions.
- Without using a tape diagram, what strategy would you use for decomposing a unit fraction?
- How did the Application Problem connect to today's lesson?

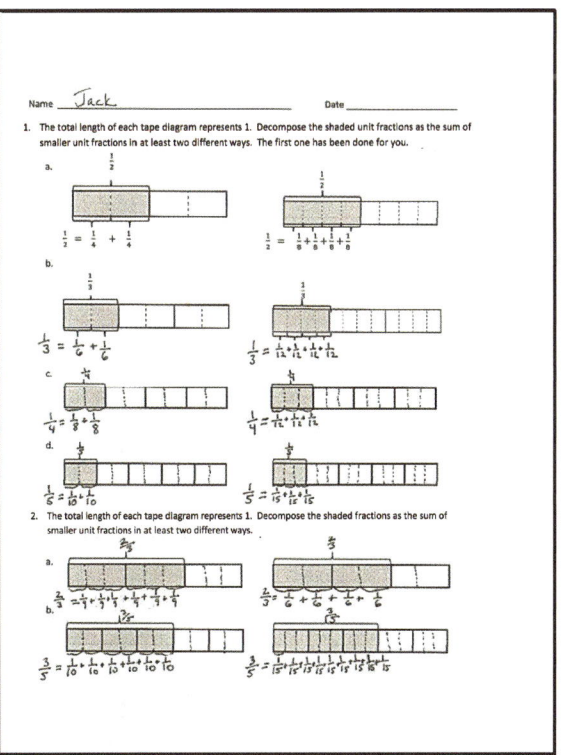

Lesson 4: Decompose fractions into sums of smaller unit fractions using tape diagrams.

A STORY OF UNITS
Lesson 4 4•5

Exit Ticket (3 minutes)

After the Student Debrief, instruct students to complete the Exit Ticket. A review of their work will help with assessing students' understanding of the concepts that were presented in today's lesson and planning more effectively for future lessons. The questions may be read aloud to the students.

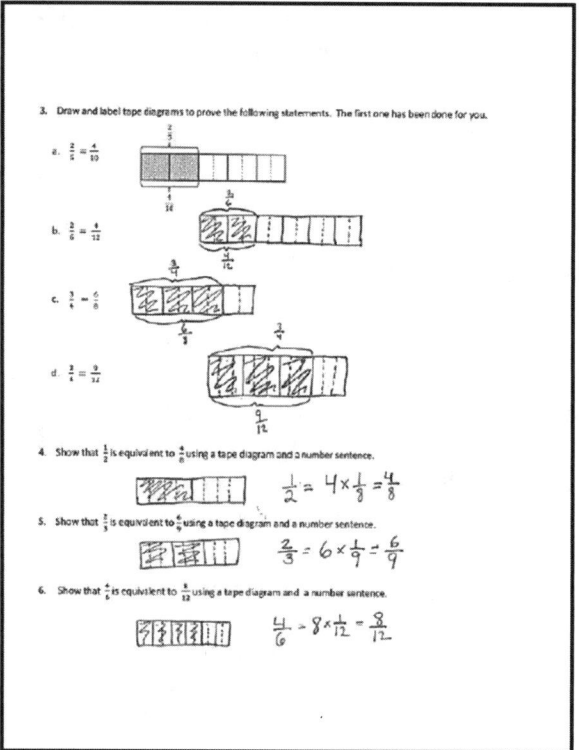

---

58   **Lesson 4:**   Decompose fractions into sums of smaller unit fractions using tape diagrams.

A STORY OF UNITS   Lesson 4 Problem Set 4•5

Name _____   Date _____

1. The total length of each tape diagram represents 1. Decompose the shaded unit fractions as the sum of smaller unit fractions in at least two different ways. The first one has been done for you.

a.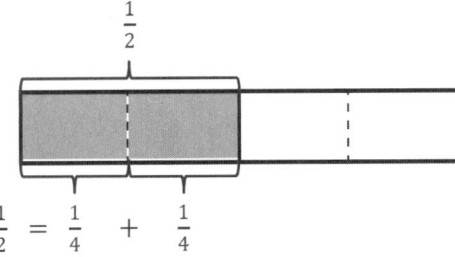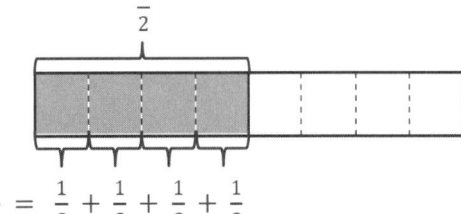

$\frac{1}{2} = \frac{1}{4} + \frac{1}{4}$ $\frac{1}{2} = \frac{1}{8} + \frac{1}{8} + \frac{1}{8} + \frac{1}{8}$

b.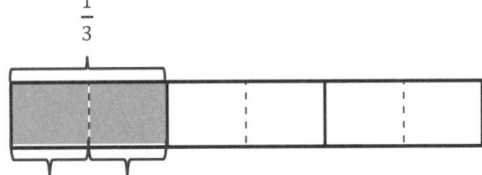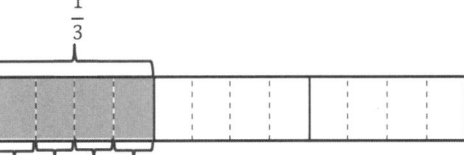

c.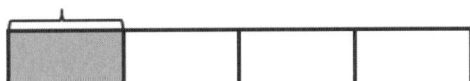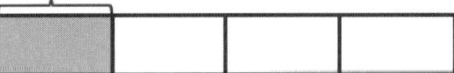

d.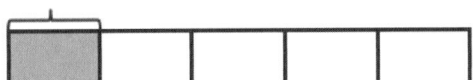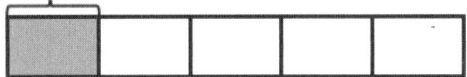

Lesson 4: Decompose fractions into sums of smaller unit fractions using tape diagrams.

2. The total length of each tape diagram represents 1. Decompose the shaded fractions as the sum of smaller unit fractions in at least two different ways.

   a.

   b.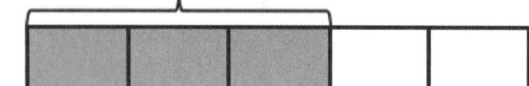

3. Draw and label tape diagrams to prove the following statements. The first one has been done for you.

   a. $\frac{2}{5} = \frac{4}{10}$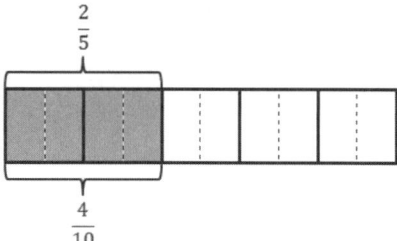

   b. $\frac{2}{6} = \frac{4}{12}$

A STORY OF UNITS Lesson 4 Problem Set 4•5

c. $\frac{3}{4} = \frac{6}{8}$

d. $\frac{3}{4} = \frac{9}{12}$

4. Show that $\frac{1}{2}$ is equivalent to $\frac{4}{8}$ using a tape diagram and a number sentence.

5. Show that $\frac{2}{3}$ is equivalent to $\frac{6}{9}$ using a tape diagram and a number sentence.

6. Show that $\frac{4}{6}$ is equivalent to $\frac{8}{12}$ using a tape diagram and a number sentence.

Lesson 4: Decompose fractions into sums of smaller unit fractions using tape diagrams.

Name _____ Date _____

1. The total length of the tape diagram represents 1. Decompose the shaded unit fraction as the sum of smaller unit fractions in at least two different ways.

   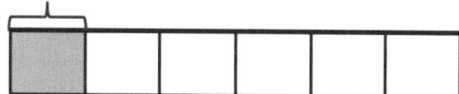

2. Draw a tape diagram to prove the following statement.

$$\frac{2}{3} = \frac{4}{6}$$

A STORY OF UNITS   Lesson 4 Homework 4•5

Name _____   Date _____

1. The total length of each tape diagram represents 1. Decompose the shaded unit fractions as the sum of smaller unit fractions in at least two different ways. The first one has been done for you.

   a.
   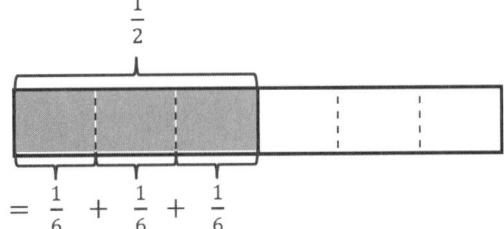
   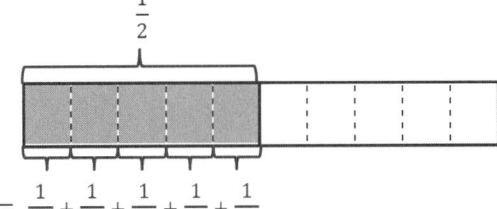

   $\frac{1}{2} = \frac{1}{6} + \frac{1}{6} + \frac{1}{6}$ $\frac{1}{2} = \frac{1}{10} + \frac{1}{10} + \frac{1}{10} + \frac{1}{10} + \frac{1}{10}$

   b.
   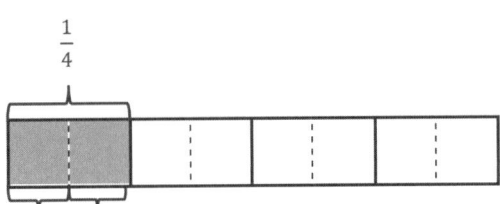

2. The total length of each tape diagram represents 1. Decompose the shaded fractions as the sum of smaller unit fractions in at least two different ways.

   a.
   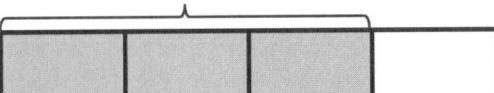
   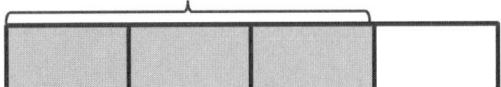

   b.
   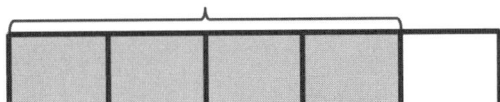

Lesson 4: Decompose fractions into sums of smaller unit fractions using tape diagrams.

c.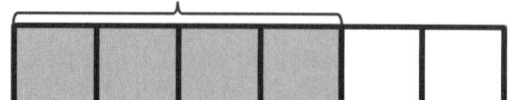

3. Draw tape diagrams to prove the following statements. The first one has been done for you.

   a. $\frac{2}{5} = \frac{4}{10}$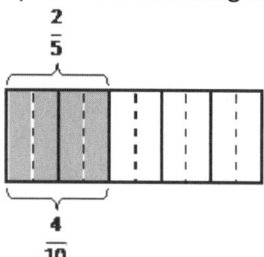

   b. $\frac{3}{6} = \frac{6}{12}$

   c. $\frac{2}{6} = \frac{6}{18}$

   d. $\frac{3}{4} = \frac{12}{16}$

4. Show that $\frac{1}{2}$ is equivalent to $\frac{6}{12}$ using a tape diagram and a number sentence.

5. Show that $\frac{2}{3}$ is equivalent to $\frac{8}{12}$ using a tape diagram and a number sentence.

6. Show that $\frac{4}{5}$ is equivalent to $\frac{12}{15}$ using a tape diagram and a number sentence.

Lesson 4: Decompose fractions into sums of smaller unit fractions using tape diagrams.

A STORY OF UNITS   Lesson 5   4•5

Lesson 5

Objective: Decompose unit fractions using area models to show equivalence.

Suggested Lesson Structure

■ Fluency Practice   (12 minutes)
■ Application Problem   (8 minutes)
■ Concept Development   (30 minutes)
■ Student Debrief   (10 minutes)
  Total Time   **(60 minutes)**

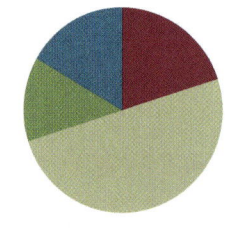

Fluency Practice (12 minutes)

- Count by Equivalent Fractions **3.NF.3**   (4 minutes)
- Add Fractions **4.NF.3**   (4 minutes)
- Break Apart the Unit Fraction **4.NF.3**   (4 minutes)

Count by Equivalent Fractions (4 minutes)

Note: This fluency activity reviews Lesson 4.

T: Count from 0 fourths to 4 fourths by 1 fourths. (Write as students count.)
S: $\frac{0}{4}, \frac{1}{4}, \frac{2}{4}, \frac{3}{4}, \frac{4}{4}$
T: 4 fourths is the same as one of what unit?
S: 1 one.
T: (Beneath $\frac{4}{4}$, write 1.) Count by fourths again. This time, say "1" when you arrive at 4 fourths. Start at zero.
S: $0, \frac{1}{4}, \frac{2}{4}, \frac{3}{4}, 1$.
T: Let's count by halves to 4 halves. (Write as students count.)
S: $\frac{0}{2}, \frac{1}{2}, \frac{2}{2}, \frac{3}{2}, \frac{4}{2}$.
T: How many halves are equal to 1?
S: 2 halves.
T: (Beneath $\frac{2}{2}$, write 1.) How many halves are equal to 2?
S: 4 halves.

$\frac{0}{4}$	$\frac{1}{4}$	$\frac{2}{4}$	$\frac{3}{4}$	$\frac{4}{4}$
0	$\frac{1}{4}$	$\frac{2}{4}$	$\frac{3}{4}$	1

$\frac{0}{2}$	$\frac{1}{2}$	$\frac{2}{2}$	$\frac{3}{2}$	$\frac{4}{2}$
0	$\frac{1}{2}$	1	$\frac{3}{2}$	2

Lesson 5:   Decompose unit fractions using area models to show equivalence.

A STORY OF UNITS   Lesson 5   4•5

T: (Beneath $\frac{4}{2}$, write 2.) Let's count by halves again. This time, when you arrive at 2 halves and 4 halves, say the whole number. Start at zero.

S: $0, \frac{1}{2}, 1\frac{3}{2}, 2$.

Repeat the process, counting by fourths to 12 fourths.

Add Fractions (4 minutes)

Materials: (S) Personal white board

Note: This fluency activity reviews Lesson 2.

T: (Write $\frac{4}{5}$.) Say the fraction.

S: 4 fifths.

T: On your personal white board, draw a tape diagram representing 4 fifths.

S: (Draw a tape diagram partitioned into 5 equal units. Shade 4 units.)

T: (Write $\frac{4}{5} = \_\_ + \_\_ + \_\_ + \_\_$.) On your board, fill in the unknown fractions.

S: (Write $\frac{4}{5} = \frac{1}{5} + \frac{1}{5} + \frac{1}{5} + \frac{1}{5}$.)

T: (Write $\frac{4}{5} = \frac{1}{5} + \frac{1}{5} + \frac{1}{5} + \frac{1}{5}$. Beneath it, write $\frac{4}{5} = \_\_ \times \frac{1}{5}$.) Fill in the unknown factor.

S: (Write $\frac{4}{5} = 4 \times \frac{1}{5}$.)

Continue the process with $\frac{5}{8}$ and $\frac{3}{7}$.

Break Apart the Unit Fraction (4 minutes)

Materials: (S) Personal white board

Note: This fluency activity reviews Lesson 4.

T: (Project a tape diagram partitioned into 2 equal units. Shade 1 unit.) Name the fraction of the diagram that is shaded.

S: 1 half.

T: (Write $\frac{1}{2}$ above the shaded unit. Decompose the shaded unit into 3 equal units.)

T: What fraction of the tape diagram is each smaller unit?

S: 1 sixth.

T: (Write $\frac{1}{2} = \_\_ + \_\_ + \_\_$.) On your personal white board, complete the number sentence.

S: (Write $\frac{1}{2} = \frac{1}{6} + \frac{1}{6} + \frac{1}{6}$.)

Repeat the process with $\frac{1}{3}$.

T: (Write $\frac{2}{3}$.) On your board, draw and shade a tape diagram to show $\frac{2}{3}$.

Lesson 5:   Decompose unit fractions using area models to show equivalence.   67

A STORY OF UNITS   Lesson 5   4•5

T: Decompose each third into 3 equal parts on your model with an addition sentence. (Pause.) Each third is the same as 3 of what unit?

S: 3 ninths.

T: (Write $\frac{2}{3} = \frac{}{9}$.) 2 thirds is the same as how many ninths? Write the answer on your board.

S: (Write $\frac{2}{3} = \frac{6}{9}$.)

Continue with the following possible sequence: $\frac{1}{2} = \frac{4}{8}, \frac{3}{4} = \frac{6}{8}, \frac{3}{4} = \frac{9}{12}$, and $\frac{5}{6} = \frac{10}{12}$.

Application Problem (8 minutes)

A loaf of bread was cut into 6 equal slices. Each of the 6 slices was cut in half to make thinner slices for sandwiches.

Mr. Beach used 4 slices. His daughter said, "Wow! You used $\frac{2}{6}$ of the loaf!" His son said, "No. He used $\frac{4}{12}$." Work with a partner to explain who was correct using a tape diagram.

$\frac{2}{6} = \frac{1}{12} + \frac{1}{12} + \frac{1}{12} + \frac{1}{12} = \frac{4}{12}$

$\frac{2}{6} = (2 \times \frac{1}{12}) + (2 \times \frac{1}{12})$

or

$\frac{2}{6} = 4 \times \frac{1}{12}$

Mr. Beach's son and daughter were both correct. $\frac{2}{6}$ represents the same amount as $\frac{4}{12}$.

Note: This Application Problem builds on Lesson 4's objective of decomposing a fraction as the sum of smaller fractions. It also bridges to today's lesson where students use the area model as another way to show both decomposition and equivalence.

A STORY OF UNITS

Lesson 5 4•5

Concept Development (30 minutes)

Materials: (S) Personal white board

Problem 1: Draw an area model to illustrate that $\frac{1}{5}$ is equal to $\frac{2}{10}$.

> **NOTES ON MULTIPLE MEANS OF ACTION AND EXPRESSION:**
>
> Drawing an area model representing fifths or other odd numbers may be challenging for some students. Slip grid paper into personal white boards to assist them, if beneficial. Students who find it easier may continue using folded paper strips to model fractions.

T: Draw an area model that is partitioned into 5 equal parts. Shade 1 of them. If the entire figure represents 1, what fractional part is shaded?

S: 1 fifth.

T: Draw a horizontal dotted line to decompose the whole into two equal rows. (Demonstrate.) What happened? Discuss with your partner.

S: There were 5 pieces, but now, there are 10. → We had fifths, but now, we have tenths. → We doubled the number of original units (fifths) to make a new unit (tenths). → We cut each fifth into 2 equal pieces to make tenths. → There are more parts, but they are smaller, so 2 times 1 tenth is the same as 1 fifth.

T: How many tenths are shaded?

S: 2 tenths.

T: Even though the parts changed, did the area covered by the shaded region change?

S: No.

T: What relationship does this show between $\frac{1}{5}$ and $\frac{2}{10}$? Say your answer as an addition sentence.

S: $\frac{1}{5} = \left(\frac{1}{10} + \frac{1}{10}\right) = \frac{2}{10}$. 1 fifth equals 2 tenths.

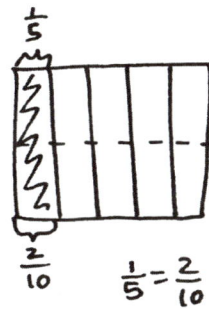

$\frac{1}{5} = \frac{2}{10}$

$\frac{1}{5} = \left(\frac{1}{10} + \frac{1}{10}\right) = 2 \times \frac{1}{10} = \frac{2}{10}$

Problem 2: Decompose $\frac{1}{3}$ as $\frac{4}{12}$ represented in an area model and as the sum and product of unit fractions.

T: Draw an area model that is partitioned into 3 equal parts. Shade 1 of them. If the entire figure represents 1, what fraction is shaded?

S: 1 third.

T: Discuss with your partner how to draw horizontal dotted lines to decompose 1 third to demonstrate that $\frac{1}{3} = \frac{4}{12}$.

MP.2

S: We can draw a horizontal line. → One line won't be enough. That will make sixths. Two lines will make ninths. Three lines!

T: How many parts do we have now?

S: 12.

T: How many twelfths are shaded?

S: $\frac{4}{12}$.

$\frac{1}{3} = \frac{4}{12}$

$\frac{1}{3} = \left(\frac{1}{12} + \frac{1}{12} + \frac{1}{12} + \frac{1}{12}\right) = 4 \times \frac{1}{12} = \frac{4}{12}$

Lesson 5: Decompose unit fractions using area models to show equivalence.

EUREKA MATH

A STORY OF UNITS

Lesson 5   4•5

MP.2

T: Represent the decomposition of $\frac{1}{3}$ as the sum of unit fractions.

S: $\frac{1}{3} = \frac{1}{12} + \frac{1}{12} + \frac{1}{12} + \frac{1}{12} = \frac{4}{12}$.

T: Now, like in the last lesson, represent this decomposition of $\frac{1}{3}$ using a multiplication sentence.

S: $\frac{1}{3} = \left(2 \times \frac{1}{12}\right) + \left(2 \times \frac{1}{12}\right) = \frac{4}{12}$. → $\frac{1}{3} = \left(4 \times \frac{1}{12}\right) = \frac{4}{12}$.

Problem 3: Model $\frac{1}{2} = \frac{5}{10}$, and represent the decomposition as the sum and product of unit fractions.

T: (Display $\frac{1}{2} = \frac{5}{10}$.) Discuss with your partner how to represent this equivalence using an area model.

S: We can partition an area model in half. We can draw lines across so that they make equal parts. → We need 10 parts. Since there are 2 halves, that would be 5 on each side.

T: Work with your partner to draw the model, and write a number sentence to represent the decomposition.

S: $\frac{1}{2} = \frac{1}{10} + \frac{1}{10} + \frac{1}{10} + \frac{1}{10} + \frac{1}{10} = \frac{5}{10}$. → $\frac{1}{2} = 5 \times \frac{1}{10} = \frac{5}{10}$.

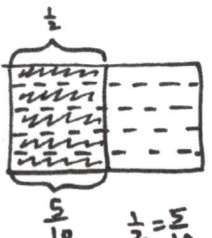

$\frac{1}{2} = 5 \times \frac{1}{10} = \frac{5}{10}$

Problem Set (10 minutes)

Students should do their personal best to complete the Problem Set within the allotted 10 minutes. For some classes, it may be appropriate to modify the assignment by specifying which problems they work on first. Some problems do not specify a method for solving. Students should solve these problems using the RDW approach used for Application Problems.

Student Debrief (10 minutes)

Lesson Objective: Decompose unit fractions using area models to show equivalence.

The Student Debrief is intended to invite reflection and active processing of the total lesson experience.

Invite students to review their solutions for the Problem Set. They should check work by comparing answers with a partner before going over answers as a class. Look for misconceptions or misunderstandings that can be addressed in the Debrief. Guide students in a conversation to debrief the Problem Set and process the lesson.

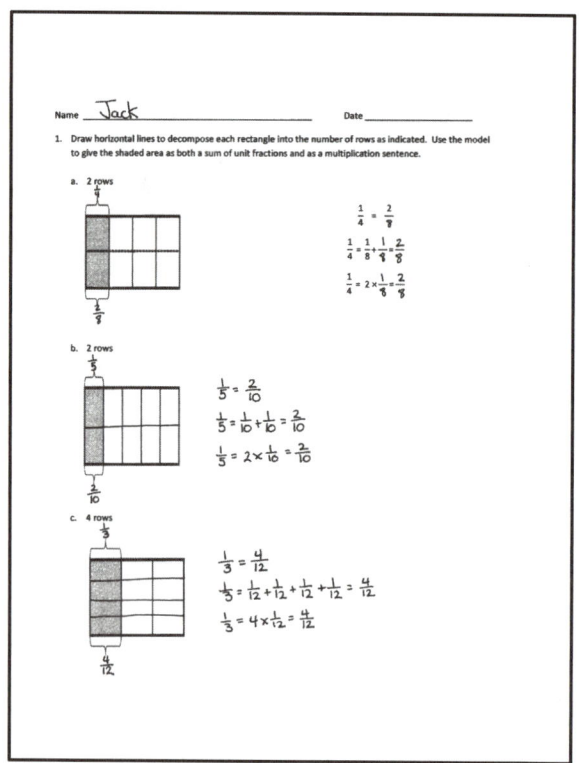

Lesson 5: Decompose unit fractions using area models to show equivalence.

A STORY OF UNITS Lesson 5 4•5

Any combination of the questions below may be used to lead the discussion.

- In Problem 1, why do you think the directions tell you how many rows to draw?
- How is Problem 2 more difficult than Problem 1?
- Problems 2(a), 2(b), and 2(c) all start with an area of 1 half. What does that tell you about the fractions 3 sixths, 4 eighths, and 5 tenths? What happens to the size and number of units as 1 half is decomposed into sixths, eighths, and tenths?
- Explain to your partner how you determined the answer for Problem 3.
- In Grade 3, we used tape diagrams to show equivalent fractions. In Grade 4, we are using area models, and we are including addition and multiplication statements. Why are these statements important?
- How did the Application Problem connect to today's lesson?

Exit Ticket (3 minutes)

After the Student Debrief, instruct students to complete the Exit Ticket. A review of their work will help with assessing students' understanding of the concepts that were presented in today's lesson and planning more effectively for future lessons. The questions may be read aloud to the students.

> **NOTES ON MULTIPLE MEANS OF REPRESENTATION:**
>
> While discussing sixths, eighths, tenths, and other fractional units that end in –*th*, check that English language learners are able to hear and say the ending digraph /th/. Help them distinguish the meaning and pronunciation of, for example, the whole number *six* and the fraction *sixths*.

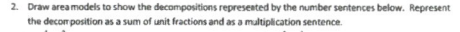

Lesson 5: Decompose unit fractions using area models to show equivalence.

A STORY OF UNITS   Lesson 5 Problem Set 4•5

Name _____   Date _____

1. Draw horizontal lines to decompose each rectangle into the number of rows as indicated. Use the model to give the shaded area as both a sum of unit fractions and as a multiplication sentence.

   a. 2 rows

   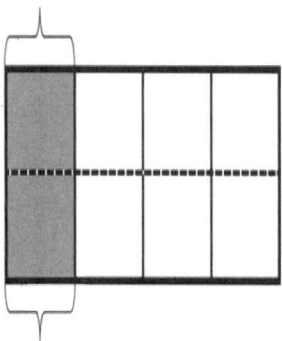

   $\dfrac{1}{4} = \dfrac{2}{\underline{}}$

   $\dfrac{1}{4} = \dfrac{1}{8} + \underline{} = \underline{}$

   $\dfrac{1}{4} = 2 \times \underline{} = \underline{}$

   b. 2 rows

   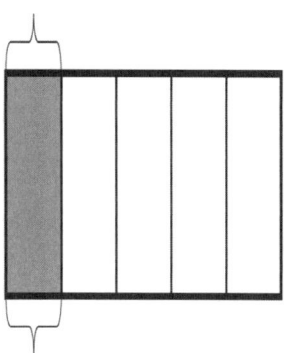

   c. 4 rows

   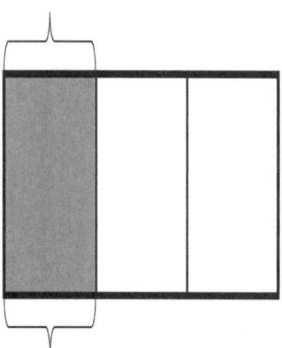

Lesson 5: Decompose unit fractions using area models to show equivalence.

A STORY OF UNITS

Lesson 5 Problem Set 4•5

2. Draw area models to show the decompositions represented by the number sentences below. Represent the decomposition as a sum of unit fractions and as a multiplication sentence.

   a. $\frac{1}{2} = \frac{3}{6}$

   b. $\frac{1}{2} = \frac{4}{8}$

   c. $\frac{1}{2} = \frac{5}{10}$

   d. $\frac{1}{3} = \frac{2}{6}$

   e. $\frac{1}{3} = \frac{4}{12}$

   f. $\frac{1}{4} = \frac{3}{12}$

3. Explain why $\frac{1}{12} + \frac{1}{12} + \frac{1}{12}$ is the same as $\frac{1}{4}$.

Lesson 5: Decompose unit fractions using area models to show equivalence.

73

A STORY OF UNITS   Lesson 5 Exit Ticket 4•5

Name _____ Date _____

1. Draw horizontal lines to decompose each rectangle into the number of rows as indicated. Use the model to give the shaded area as both a sum of unit fractions and as a multiplication sentence.

   a. 2 rows

   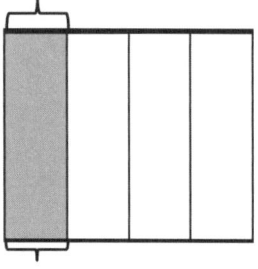

   b. 3 rows

   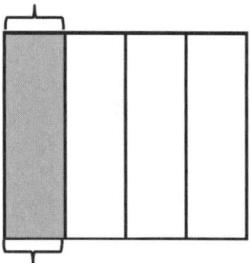

2. Draw an area model to show the decomposition represented by the number sentence below. Represent the decomposition as a sum of unit fractions and as a multiplication sentence.

   $\frac{3}{5} = \frac{6}{10}$

Lesson 5: Decompose unit fractions using area models to show equivalence.

A STORY OF UNITS

Lesson 5 Homework 4•5

Name _____ Date _____

1. Draw horizontal lines to decompose each rectangle into the number of rows as indicated. Use the model to give the shaded area as both a sum of unit fractions and as a multiplication sentence.

   a. 3 rows

   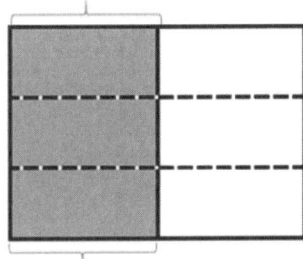

   $\dfrac{1}{2} = \dfrac{3}{\underline{}}$

   $\dfrac{1}{2} = \dfrac{1}{6} + \dfrac{}{} + \dfrac{}{} = \dfrac{3}{6}$

   $\dfrac{1}{2} = 3 \times \dfrac{}{} = \dfrac{3}{6}$

   b. 2 rows

   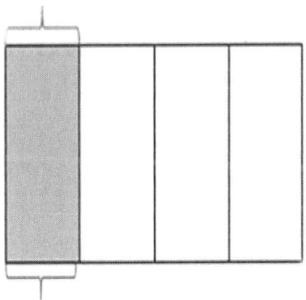

   c. 4 rows

   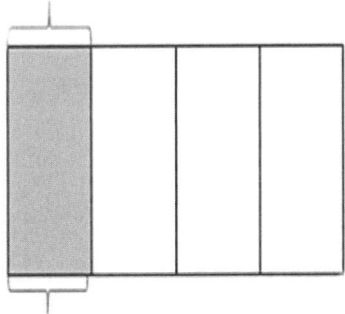

Lesson 5: Decompose unit fractions using area models to show equivalence.

2. Draw area models to show the decompositions represented by the number sentences below. Represent the decomposition as a sum of unit fractions and as a multiplication sentence.

   a. $\frac{1}{3} = \frac{2}{6}$

   b. $\frac{1}{3} = \frac{3}{9}$

   c. $\frac{1}{3} = \frac{4}{12}$

   d. $\frac{1}{3} = \frac{5}{15}$

   e. $\frac{1}{5} = \frac{2}{10}$

   f. $\frac{1}{5} = \frac{3}{15}$

3. Explain why $\frac{1}{12} + \frac{1}{12} + \frac{1}{12} + \frac{1}{12}$ is the same as $\frac{1}{3}$.

Lesson 6

Objective: Decompose fractions using area models to show equivalence.

Suggested Lesson Structure

- Fluency Practice   (12 minutes)
- Application Problem   (8 minutes)
- Concept Development   (30 minutes)
- Student Debrief   (10 minutes)

Total Time   **(60 minutes)**

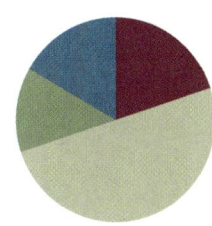

Fluency Practice (12 minutes)

- Sprint: Multiply Whole Numbers Times Fractions **4.NF.4**   (9 minutes)
- Find Equivalent Fractions **4.NF.1**   (3 minutes)

Sprint: Multiply Whole Numbers Times Fractions (9 minutes)

Materials: (S) Multiply Whole Numbers Times Fractions Sprint

Note: This fluency activity reviews Lesson 3.

Find Equivalent Fractions (3 minutes)

Materials: (S) Personal white board

Note: This fluency activity reviews Lesson 5.

T: (Write $\frac{1}{3}$.) Say the fraction.
S: 1 third.
T: On your personal white board, draw a model to show $\frac{1}{3}$.
S: (Draw a model partitioned into 3 equal units. Shade 1 unit.)
T: (Write $\frac{1}{3} = \frac{2}{}$.) Draw a dotted horizontal line to decompose 1 third into an equivalent fraction.
S: (Draw a dotted horizontal line, breaking 3 units into 6 smaller units. Write $\frac{1}{3} = \frac{2}{6}$.)

Continue with the following possible sequence: $\frac{1}{3} = \frac{3}{9}, \frac{1}{2} = \frac{2}{4}, \frac{1}{2} = \frac{4}{8}, \frac{1}{4} = \frac{2}{8}$, and $\frac{1}{5} = \frac{3}{15}$.

A STORY OF UNITS   Lesson 6 4•5

Application Problem (8 minutes)

Use area models to prove that $\frac{1}{2} = \frac{2}{4} = \frac{4}{8}, \frac{1}{2} = \frac{3}{6} = \frac{6}{12}$, and $\frac{1}{2} = \frac{5}{10}$. What conclusion can you make about $\frac{4}{8}, \frac{6}{12}$, and $\frac{5}{10}$? Explain.

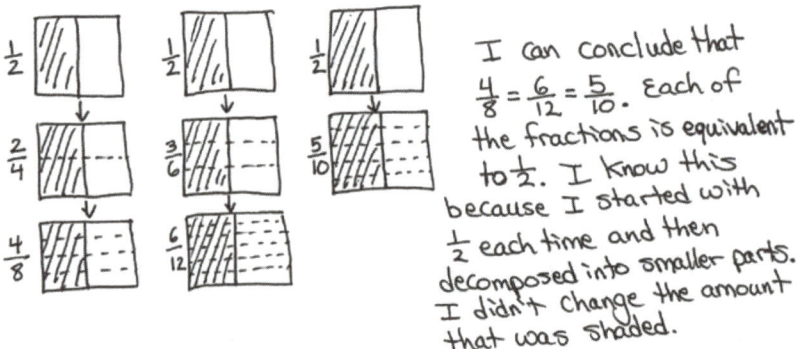

I can conclude that $\frac{4}{8} = \frac{6}{12} = \frac{5}{10}$. Each of the fractions is equivalent to $\frac{1}{2}$. I know this because I started with $\frac{1}{2}$ each time and then decomposed into smaller parts. I didn't change the amount that was shaded.

Note: This Application Problem builds from Lesson 5, where students decomposed unit fractions using area models to show equivalence. Consider leading a discussion with a question, such as "Why can you show $\frac{1}{2} = \frac{2}{4} = \frac{4}{8}$ on one model, $\frac{1}{2} = \frac{3}{6} = \frac{6}{12}$ on another, and $\frac{1}{2} = \frac{5}{10}$ on another?" Or perhaps lead with a question, such as "Why can't you show $\frac{1}{2} = \frac{2}{4} = \frac{5}{10}$ on the same area model?"

Concept Development (30 minutes)

Materials: (S) Personal white board

Problem 1: Use an area model to show that $\frac{3}{4} = \frac{6}{8}$.

T: Draw an area model representing 1, and then shade $\frac{3}{4}$.

T: Discuss with a partner how you can use this model to show the decomposition of 3 fourths into eighths.

S: We could draw a line so that each of the fourths is split into 2 equal parts. That would give us eighths. → Drawing a line will make each unit into 2 smaller units, which would be eighths.

T: How many eighths are shaded?

S: 6 eighths.

T: Work with a partner to write an addition and a multiplication sentence to describe the decomposition.

S: $\frac{3}{4} = \left(\frac{1}{8} + \frac{1}{8}\right) + \left(\frac{1}{8} + \frac{1}{8}\right) + \left(\frac{1}{8} + \frac{1}{8}\right) = \frac{6}{8}$. → $\frac{3}{4} = 3 \times \frac{2}{8} = 6 \times \frac{1}{8} = \frac{6}{8}$. → $\frac{3}{4}$ is equal to $\frac{6}{8}$.

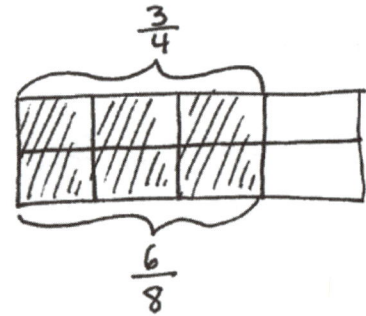

Lesson 6:   Decompose fractions using area models to show equivalence.

T: What do these addition and multiplication sentences tell you?

S: The shaded area didn't change. It's still the same amount. The number of pieces increased, but the size of the pieces got smaller. → Adding together all of the smaller units equals the total of the larger units shaded. → Multiplying also equals the total of the larger units shaded and is easier to write out!

Problem 2: Draw an area model to represent the equivalence of two fractions, and express the equivalence as the sum and product of unit fractions.

T: Let's draw an area model to show that $\frac{2}{3} = \frac{8}{12}$. What fraction will you model first, and why? Discuss with a partner.

S: I will represent $\frac{2}{3}$ first since thirds are the larger pieces. I can draw 1 divided into thirds and then shade 2 of them. Then, it's easy to split the thirds into parts to make twelfths. → We have to draw the larger units first and then decompose them into smaller ones, don't we?

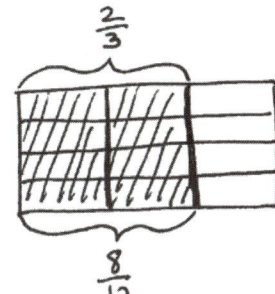

T: Draw an area model representing 2 thirds.

T: How can we show that $\frac{2}{3} = \frac{8}{12}$? Discuss.

S: We can split the thirds into parts until we have 12 of them. → Yes, but we need to make sure that they are equal parts. → We might have to erase our lines and then redraw to make them look equal. → We can draw three lines across the thirds. This will make 12 groups. → When I do that, the eight pieces are already shaded!

T: Express the equivalence as an addition sentence.

S: $\frac{2}{3} = \frac{1}{3} + \frac{1}{3} = \left(\frac{1}{12} + \frac{1}{12} + \frac{1}{12} + \frac{1}{12}\right) + \left(\frac{1}{12} + \frac{1}{12} + \frac{1}{12} + \frac{1}{12}\right) = \frac{8}{12}$.

T: Express the equivalence as a multiplication sentence.

S: $\frac{2}{3} = \left(8 \times \frac{1}{12}\right) = \frac{8}{12}$. → $\frac{2}{3} = \left(4 \times \frac{1}{12}\right) + \left(4 \times \frac{1}{12}\right) = \frac{8}{12}$.

Problem 3: Decompose to create equivalent fractions by drawing an area model and then dividing the area model into smaller parts.

T: Let's use what we know to model equivalent fractions.

1. Draw an area model. The entire figure is 1.
2. Choose a fraction, and partition the whole using vertical lines.
3. Shade your fraction.
4. Switch papers with a partner. Write down the fraction that your partner has represented.
5. Draw one to three horizontal lines. What equivalent fraction have you modeled?

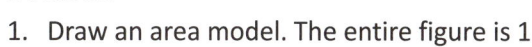
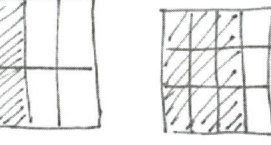

T: How could we model 5 thirds?

Lesson 6: Decompose fractions using area models to show equivalence.

A STORY OF UNITS   Lesson 6 4•5

S: We can draw an area model and partition it into 5 parts. Each part is 1 third. We have to label 1 after 3 units.

T: Draw one horizontal line to model an equivalent fraction. How many units are in 1?

S: 6.

T: What fraction is represented?

S: $\frac{10}{6}$.

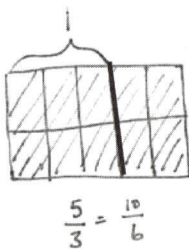

Problem Set (10 minutes)

Students should do their personal best to complete the Problem Set within the allotted 10 minutes. For some classes, it may be appropriate to modify the assignment by specifying which problems they work on first. Some problems do not specify a method for solving. Students should solve these problems using the RDW approach used for Application Problems.

> **NOTES ON MULTIPLE MEANS OF ENGAGEMENT:**
>
> Scaffold Problem 2 on the Problem Set for students working below grade level and others by providing number sentence frames such as the following:
>
> $\left(\frac{1}{-}+\frac{1}{-}\right)+\left(\frac{1}{-}+\frac{1}{-}\right)=$
>
> $(\ \times -)+(\ \times -)=-.$

Student Debrief (10 minutes)

Lesson Objective: Decompose fractions using area models to show equivalence.

The Student Debrief is intended to invite reflection and active processing of the total lesson experience.

Invite students to review their solutions for the Problem Set. They should check work by comparing answers with a partner before going over answers as a class. Look for misconceptions or misunderstandings that can be addressed in the Debrief. Guide students in a conversation to debrief the Problem Set and process the lesson.

Any combination of the questions below may be used to lead the discussion.

- Look at Problems 1(c) and 2(b). Compare the two problems. How can $\frac{3}{4}$ be equivalent to both fractions?
- Why do we use parentheses? What does it help show?

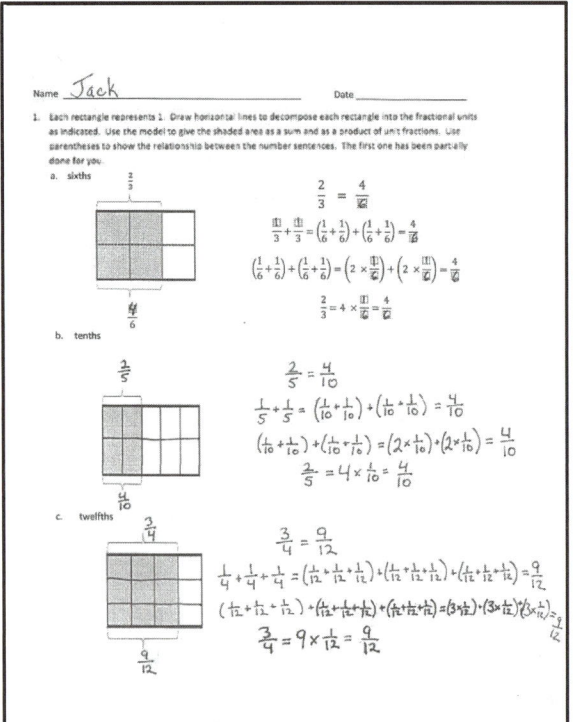

- In Problem 2 of the Concept Development, could you represent $\frac{8}{12}$ first and then show the equivalence to $\frac{2}{3}$? How would you show it?
- How can two different fractions represent the same portion of a whole?
- How did the Application Problem connect to today's lesson?

Exit Ticket (3 minutes)

After the Student Debrief, instruct students to complete the Exit Ticket. A review of their work will help with assessing students' understanding of the concepts that were presented in today's lesson and planning more effectively for future lessons. The questions may be read aloud to the students.

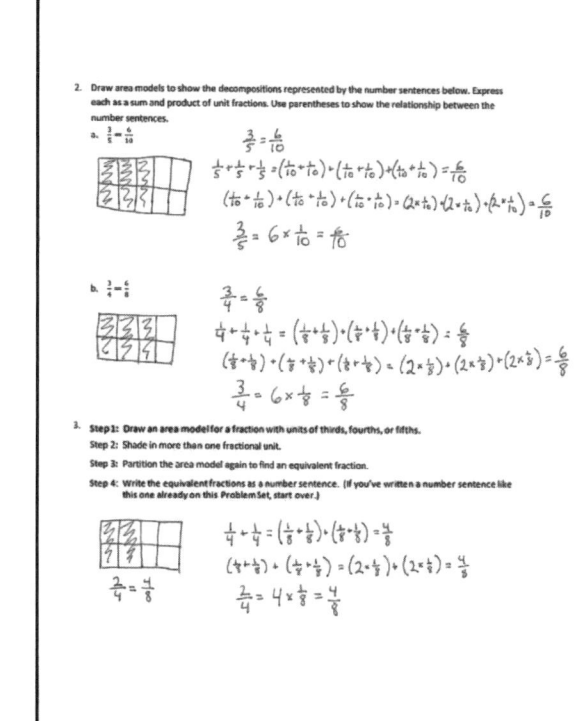

A STORY OF UNITS   Lesson 6 Sprint 4•5

A

Number Correct: _____

Multiply Whole Numbers Times Fractions

1.	$\frac{1}{3}+\frac{1}{3}=$		23.	$\frac{1}{3}+\frac{1}{3}+\frac{1}{3}+\frac{1}{3}=$	
2.	$2 \times \frac{1}{3}=$		24.	$4 \times \frac{1}{3}=$	
3.	$\frac{1}{4}+\frac{1}{4}+\frac{1}{4}=$		25.	$\frac{5}{6}=$	$\_\times \frac{1}{6}$
4.	$3 \times \frac{1}{4}=$		26.	$\frac{5}{6}=$	$5 \times \_\_$
5.	$\frac{1}{5}+\frac{1}{5}=$		27.	$\frac{5}{8}=$	$5 \times \_\_$
6.	$2 \times \frac{1}{5}=$		28.	$\frac{5}{8}=$	$\_\times \frac{1}{8}$
7.	$\frac{1}{5}+\frac{1}{5}+\frac{1}{5}=$		29.	$\frac{7}{8}=$	$7 \times \_\_$
8.	$3 \times \frac{1}{5}=$		30.	$\frac{7}{10}=$	$7 \times \_\_$
9.	$\frac{1}{5}+\frac{1}{5}+\frac{1}{5}+\frac{1}{5}=$		31.	$\frac{7}{8}=$	$\_\times \frac{1}{8}$
10.	$4 \times \frac{1}{5}=$		32.	$\frac{7}{10}=$	$\_\times \frac{1}{10}$
11.	$\frac{1}{10}+\frac{1}{10}+\frac{1}{10}=$		33.	$\frac{6}{6}=$	$6 \times \_\_$
12.	$3 \times \frac{1}{10}=$		34.	$1 =$	$6 \times \_\_$
13.	$\frac{1}{8}+\frac{1}{8}+\frac{1}{8}=$		35.	$\frac{8}{8}=$	$\_\times \frac{1}{8}$
14.	$3 \times \frac{1}{8}=$		36.	$1 =$	$\_\times \frac{1}{8}$
15.	$\frac{1}{2}+\frac{1}{2}=$		37.	$9 \times \frac{1}{10}=$	
16.	$2 \times \frac{1}{2}=$		38.	$7 \times \frac{1}{5}=$	
17.	$\frac{1}{3}+\frac{1}{3}+\frac{1}{3}=$		39.	$1 =$	$3 \times \_\_$
18.	$3 \times \frac{1}{3}=$		40.	$7 \times \frac{1}{12}=$	
19.	$\frac{1}{4}+\frac{1}{4}+\frac{1}{4}+\frac{1}{4}=$		41.	$1 =$	$\_\times \frac{1}{5}$
20.	$4 \times \frac{1}{4}=$		42.	$\frac{3}{5}=$	$\frac{1}{5}+\frac{1}{5}+\_\_$
21.	$\frac{1}{2}+\frac{1}{2}+\frac{1}{2}=$		43.	$3 \times \frac{1}{4}=$	$\_+\frac{1}{4}+\frac{1}{4}$
22.	$3 \times \frac{1}{2}=$		44.	$1 =$	$\_+\_+\_$

Lesson 6: Decompose fractions using area models to show equivalence.

B

Number Correct: _____

Improvement: _____

Multiply Whole Numbers Times Fractions

1.	$\frac{1}{5}+\frac{1}{5}=$		23.	$\frac{1}{2}+\frac{1}{2}+\frac{1}{2}=$	
2.	$2\times\frac{1}{5}=$		24.	$3\times\frac{1}{2}$	
3.	$\frac{1}{3}+\frac{1}{3}=$		25.	$\frac{5}{6}=$	$\_\times\frac{1}{6}$
4.	$2\times\frac{1}{3}=$		26.	$\frac{5}{6}=$	$5\times\_$
5.	$\frac{1}{4}+\frac{1}{4}+\frac{1}{4}=$		27.	$\frac{5}{8}=$	$5\times\_$
6.	$3\times\frac{1}{4}$		28.	$\frac{5}{8}=$	$\_\times\frac{1}{8}$
7.	$\frac{1}{5}+\frac{1}{5}+\frac{1}{5}=$		29.	$\frac{7}{8}=$	$7\times\_$
8.	$3\times\frac{1}{5}$		30.	$\frac{7}{10}=$	$7\times\_$
9.	$\frac{1}{5}+\frac{1}{5}+\frac{1}{5}+\frac{1}{5}=$		31.	$\frac{7}{8}=$	$\_\times\frac{1}{8}$
10.	$4\times\frac{1}{5}$		32.	$\frac{7}{10}=$	$\_\times\frac{1}{10}$
11.	$\frac{1}{8}+\frac{1}{8}+\frac{1}{8}=$		33.	$\frac{8}{8}=$	$8\times\_$
12.	$3\times\frac{1}{8}$		34.	$1=$	$8\times\_$
13.	$\frac{1}{10}+\frac{1}{10}+\frac{1}{10}=$		35.	$\frac{6}{6}=$	$\_\times\frac{1}{6}$
14.	$3\times\frac{1}{10}$		36.	$1=$	$\_\times\frac{1}{6}$
15.	$\frac{1}{3}+\frac{1}{3}+\frac{1}{3}=$		37.	$5\times\frac{1}{12}$	
16.	$3\times\frac{1}{3}$		38.	$6\times\frac{1}{5}$	
17.	$\frac{1}{4}+\frac{1}{4}+\frac{1}{4}+\frac{1}{4}=$		39.	$1=$	$4\times\_$
18.	$4\times\frac{1}{4}$		40.	$9\times\frac{1}{10}$	
19.	$\frac{1}{2}+\frac{1}{2}=$		41.	$1=$	$\_\times\frac{1}{3}$
20.	$2\times\frac{1}{2}$		42.	$\frac{3}{4}=$	$\frac{1}{4}+\frac{1}{4}+\_$
21.	$\frac{1}{3}+\frac{1}{3}+\frac{1}{3}+\frac{1}{3}=$		43.	$3\times\frac{1}{5}$	$\_+\frac{1}{5}+\frac{1}{5}$
22.	$4\times\frac{1}{3}$		44.	$1=$	$\_+\_+\_+\_$

Lesson 6: Decompose fractions using area models to show equivalence.

Lesson 6 Problem Set 4•5

Name _____ Date _____

1. Each rectangle represents 1. Draw horizontal lines to decompose each rectangle into the fractional units as indicated. Use the model to give the shaded area as a sum and as a product of unit fractions. Use parentheses to show the relationship between the number sentences. The first one has been partially done for you.

    a. Sixths $\frac{2}{3}$

    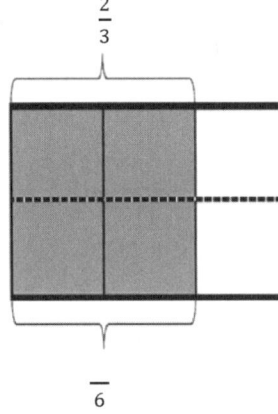

    $\frac{}{6}$

    $\frac{2}{3} = \frac{4}{-}$

    $\frac{1}{3} + \frac{1}{3} = \left(\frac{1}{6} + \frac{1}{6}\right) + \left(\frac{1}{6} + \frac{1}{6}\right) = \frac{4}{-}$

    $\left(\frac{1}{6} + \frac{1}{6}\right) + \left(\frac{1}{6} + \frac{1}{6}\right) = \left(2 \times \frac{}{-}\right) + \left(2 \times \frac{}{-}\right) = \frac{4}{-}$

    $\frac{2}{3} = 4 \times \frac{}{-} = \frac{4}{-}$

    b. Tenths

    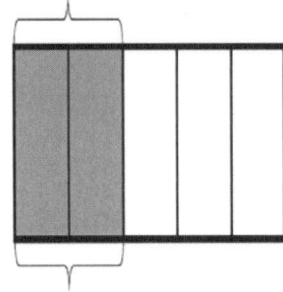

84   Lesson 6:   Decompose fractions using area models to show equivalence.

c. Twelfths

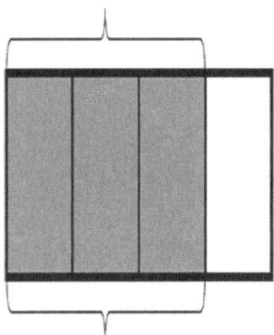

2. Draw area models to show the decompositions represented by the number sentences below. Express each as a sum and product of unit fractions. Use parentheses to show the relationship between the number sentences.

   a. $\dfrac{3}{5} = \dfrac{6}{1A}$

   b. $\dfrac{3}{4} = \dfrac{6}{8}$

Lesson 6: Decompose fractions using area models to show equivalence.

3. Step 1: Draw an area model for a fraction with units of thirds, fourths, or fifths.

   Step 2: Shade in more than one fractional unit.

   Step 3: Partition the area model again to find an equivalent fraction.

   Step 4: Write the equivalent fractions as a number sentence. (If you've written a number sentence like this one already on this Problem Set, start over.)

A STORY OF UNITS — Lesson 6 Exit Ticket 4•5

Name _____ Date _____

1. The rectangle below represents 1. Draw horizontal lines to decompose the rectangle into eighths. Use the model to give the shaded area as a sum and as a product of unit fractions. Use parentheses to show the relationship between the number sentences.

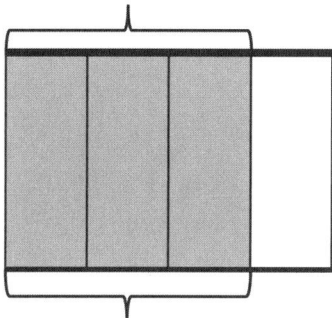

2. Draw an area model to show the decomposition represented by the number sentence below.

$$\frac{4}{5} = \frac{8}{10}$$

Lesson 6: Decompose fractions using area models to show equivalence.   87

Name _____ Date _____

1. Each rectangle represents 1. Draw horizontal lines to decompose each rectangle into the fractional units as indicated. Use the model to give the shaded area as a sum and as a product of unit fractions. Use parentheses to show the relationship between the number sentences. The first one has been partially done for you.

   a. Tenths $\frac{2}{5}$

   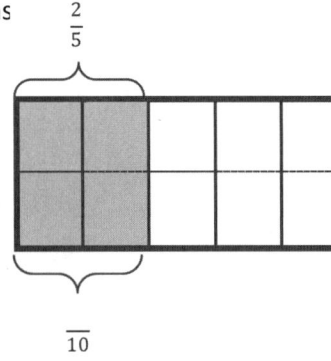

   $\frac{2}{5} = \frac{4}{-}$

   $\frac{1}{5} + \frac{1}{5} = \left(\frac{1}{10} + \frac{1}{10}\right) + \left(\frac{1}{10} + \frac{1}{10}\right) = \frac{4}{-}$

   $\left(\frac{1}{10} + \frac{1}{10}\right) + \left(\frac{1}{10} + \frac{1}{10}\right) = \left(2 \times \frac{}{-}\right) + \left(2 \times \frac{}{-}\right) = \frac{4}{-}$

   $\frac{2}{5} = 4 \times \frac{}{-} = \frac{4}{-}$

   b. Eighths

   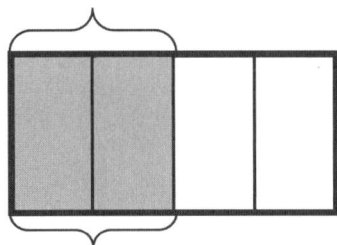

c. Fifteenths

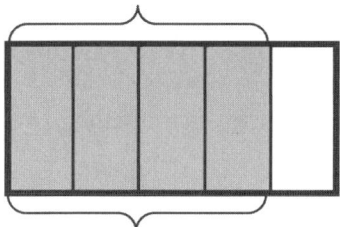

2. Draw area models to show the decompositions represented by the number sentences below. Express each as a sum and product of unit fractions. Use parentheses to show the relationship between the number sentences.

a. $\frac{2}{3} = \frac{4}{6}$

b. $\frac{4}{5} = \frac{8}{10}$

3. Step 1: Draw an area model for a fraction with units of thirds, fourths, or fifths.

   Step 2: Shade in more than one fractional unit.

   Step 3: Partition the area model again to find an equivalent fraction.

   Step 4: Write the equivalent fractions as a number sentence. (If you have written a number sentence like this one already in this Homework, start over.)

A STORY OF UNITS

Mathematics Curriculum

GRADE 4 • MODULE 5

Topic B
Fraction Equivalence Using Multiplication and Division

4.NF.1, 4.NF.3b

Focus Standard:	4.NF.1	Explain why a fraction *a/b* is equivalent to a fraction (*n* × *a*)/(*n* × *b*) by using visual fraction models, with attention to how the number and size of the parts differ even though the two fractions themselves are the same size. Use this principle to recognize and generate equivalent fractions.
Instructional Days:	5	
Coherence -Links from:	G3–M5	Fractions as Numbers on the Number Line
-Links to:	G5–M3	Addition and Subtraction of Fractions
	G5–M4	Multiplication and Division of Fractions and Decimal Fractions

In Topic B, students begin generalizing their work with fraction equivalence. In Lessons 7 and 8, students analyze their earlier work with tape diagrams and the area model in Lessons 3 through 5 to begin using multiplication to create an equivalent fraction that comprises smaller units, e.g., $\frac{2}{3} = \frac{2 \times 4}{3 \times 4} = \frac{8}{12}$. Conversely, students reason, in Lessons 9 and 10, that division can be used to create a fraction that comprises larger units (or a single unit) equivalent to a given fraction, e.g., $\frac{8}{12} = \frac{8 \div 4}{12 \div 4} = \frac{2}{3}$. The numerical work of Lessons 7 through 10 is introduced and supported using area models and tape diagrams.

In Lesson 11, students use tape diagrams to transition their knowledge of fraction equivalence to the number line. They see that any unit fraction length can be partitioned into *n* equal lengths. For example, each third in the interval from 0 to 1 may be partitioned into 4 equal parts. Doing so multiplies both the total number of fractional units (the denominator) and the number of selected units (the numerator) by 4. Conversely, students see that, in some cases, fractional units may be grouped together to form some number of larger fractional units. For example, when the interval from 0 to 1 is partitioned into twelfths, one may group 4 twelfths at a time to make thirds. By doing so, both the total number of fractional units and number of selected units are divided by 4.

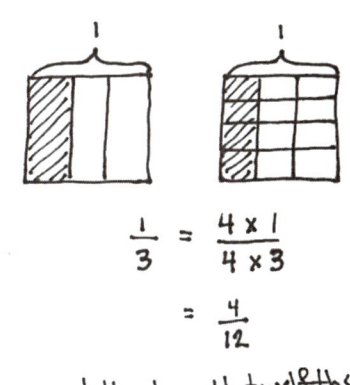

A Teaching Sequence Toward Mastery of Fraction Equivalence Using Multiplication and Division
Objective 1: Use the area model and multiplication to show the equivalence of two fractions. (Lessons 7–8)
Objective 2: Use the area model and division to show the equivalence of two fractions. (Lessons 9–10)
Objective 3: Explain fraction equivalence using a tape diagram and the number line, and relate that to the use of multiplication and division. (Lesson 11)

Lesson 7

Objective: Use the area model and multiplication to show the equivalence of two fractions.

Suggested Lesson Structure

■ Fluency Practice   (12 minutes)
■ Application Problem (4 minutes)
■ Concept Development (34 minutes)
■ Student Debrief   (10 minutes)
  Total Time   **(60 minutes)**

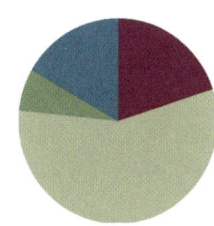

Fluency Practice (12 minutes)

- Break Apart Fractions **4.NF.3**   (4 minutes)
- Count by Equivalent Fractions **3.NF.3** (4 minutes)
- Draw Equivalent Fractions **4.NF.1**   (4 minutes)

Break Apart Fractions (4 minutes)

Materials: (S) Personal white board

Note: This fluency activity reviews Lessons 1–3.

T: (Project a tape diagram of 3 fifths with the whole labeled as 1.) Name the fraction that's shaded.
S: $\frac{3}{5}$.
T: (Write $\frac{3}{5}$ = ___.) Say the fraction.
S: 3 fifths.
T: On your personal white board, write $\frac{3}{5}$ as a repeated addition sentence using unit fractions.
S: (Write $\frac{3}{5} = \frac{1}{5} + \frac{1}{5} + \frac{1}{5}$.)
T: (Write $\frac{3}{5} = \frac{1}{5} + \frac{1}{5} + \frac{1}{5} = \underline{} \times \frac{1}{5}$.) On your board, complete the number sentence.
S: (Write $\frac{3}{5} = \frac{1}{5} + \frac{1}{5} + \frac{1}{5} = 3 \times \frac{1}{5}$.)

Continue with the following possible sequence: $\frac{5}{6} = \frac{1}{6} + \frac{1}{6} + \frac{1}{6} + \frac{1}{6} + \frac{1}{6} = 5 \times \frac{1}{6}$ and $\frac{5}{8} = \frac{1}{8} + \frac{1}{8} + \frac{1}{8} + \frac{1}{8} + \frac{1}{8} = 5 \times \frac{1}{8}$.

A STORY OF UNITS

Lesson 7 4•5

Count by Equivalent Fractions (4 minutes)

Materials: (S) Personal white board

Note: This fluency activity prepares students for lessons throughout this module.

T: Count from 0 to 10 by ones. Start at 0.
S: 0, 1, 2, 3, 4, 5, 6, 7, 8, 9, 10.
T: Count by 1 fourths to 10 fourths. Start at 0 fourths. (Write as students count.)

$\frac{0}{4}$	$\frac{1}{4}$	$\frac{2}{4}$	$\frac{3}{4}$	$\frac{4}{4}$	$\frac{5}{4}$	$\frac{6}{4}$	$\frac{7}{4}$	$\frac{8}{4}$	$\frac{9}{4}$	$\frac{10}{4}$
0	$\frac{1}{4}$	$\frac{2}{4}$	$\frac{3}{4}$	1	$\frac{5}{4}$	$\frac{6}{4}$	$\frac{7}{4}$	2	$\frac{9}{4}$	$\frac{10}{4}$

S: $\frac{0}{4}, \frac{1}{4}, \frac{2}{4}, \frac{3}{4}, \frac{4}{4}, \frac{5}{4}, \frac{6}{4}, \frac{7}{4}, \frac{8}{4}, \frac{9}{4}, \frac{10}{4}$.

T: 4 fourths is the same as 1 of what unit?
S: 1 one.
T: (Beneath 4 fourths, write 1.) 2 ones is the same as how many fourths?
S: 8 fourths.
T: (Beneath $\frac{8}{4}$, write 2.) Let's count to 10 fourths again, but this time, say the whole numbers when you come to a whole number. Start at 0.
S: $0, \frac{1}{4}, \frac{2}{4}, \frac{3}{4}, 1, \frac{5}{4}, \frac{6}{4}, \frac{7}{4}, 2, \frac{9}{4}, \frac{10}{4}$.

Repeat the process, counting by thirds to 10 thirds.

Draw Equivalent Fractions (4 minutes)

Materials: (S) Personal white board

Note: This fluency activity reviews Lesson 6.

T: (Write $\frac{2}{3}$.) Say the fraction.
S: 2 thirds.
T: On your personal white board, draw an area model to show $\frac{2}{3}$.
S: (Draw a model partitioned into 3 equal units. Shade 2 units.)
T: (Write $\frac{2}{3} = \frac{4}{6}$.) Draw a dotted horizontal line to find the equivalent fraction.
S: (Draw a dotted horizontal line, breaking 3 units into 6 smaller units. Write $\frac{2}{3} = \frac{4}{6}$.)

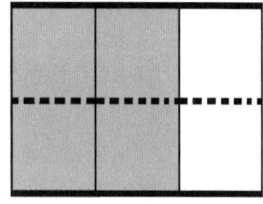

$\frac{2}{3} = \frac{4}{6}$

Continue with the following possible sequence: $\frac{2}{3} = \frac{6}{3}, \frac{3}{4} = \frac{6}{?}, \frac{2}{5} = \frac{4}{?}$, and $\frac{4}{5} = \frac{12}{?}$.

Lesson 7: Use the area model and multiplication to show the equivalence of two fractions.

A STORY OF UNITS   Lesson 7   4•5

Application Problem (4 minutes)

Model an equivalent fraction for $\frac{4}{7}$ using an area model.

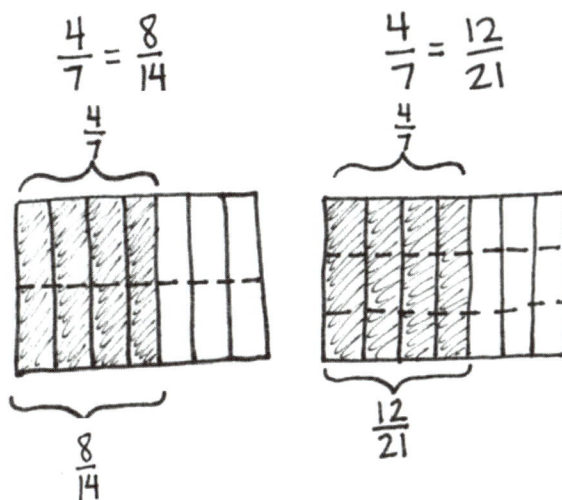

> **NOTES ON MULTIPLE MEANS OF REPRESENTATION:**
>
> Students working below grade level and others may benefit from explicit instruction as they decompose unit fractions. When doubling the number of units, instruct students to draw one horizontal dotted line. When tripling, draw two lines, and so forth.

Note: This Application Problem reviews Lesson 6 and leads into today's lesson as students find equivalent fractions using multiplication.

Concept Development (34 minutes)

Materials: (S) Personal white board

Problem 1: Determine that multiplying the numerator and denominator by n results in an equivalent fraction.

T: Draw an area model representing 1 whole partitioned into thirds. Shade and record $\frac{1}{3}$ below the area model. Draw 1 horizontal line across the area model.

S: (Draw, partition, and shade an area model.)

T: What happened to the size of the fractional units?

S: The units got smaller. → The unit became half the size.

T: What happened to the number of units in the whole?

S: There were 3; now there are 6. → We doubled the total number of units.

T: What happened to the number of selected units when we drew the dotted line?

S: There was 1 unit selected, and now there are 2. → It doubled, too!

T: That's right. We can record the doubling of units with multiplication: $\frac{1}{3} = \frac{1 \times 2}{3 \times 2} = \frac{2}{6}$.

S: Hey, I remember from third grade that $\frac{1}{3}$ is the same as $\frac{2}{6}$.

T: Yes. They are equivalent fractions.

T: Why didn't doubling the number of selected units make the fraction larger?

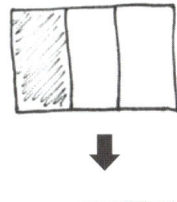

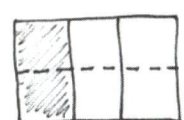

MP.7

Lesson 7: Use the area model and multiplication to show the equivalence of two fractions.

S: We didn't change the amount of the fraction, just the size of the units. → Yeah. So, the size of the units became half as big.

T: Draw an area model representing 1 partitioned with a vertical line into 2 halves.

T: Shade and record $\frac{1}{2}$ below the area model. If we want to rewrite $\frac{1}{2}$ using 4 times as many units, what should we do?

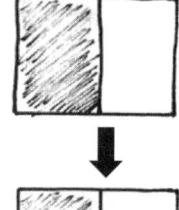

S: Draw horizontal dotted lines—three of them. → Then, we can write a number sentence using multiplication. → This time, it's 4 times as many, so we will multiply the top number and bottom number by 4.

T: Show me. (Allow time for students to partition the area model.) What happened to the size of the fractional unit?

S: The size of the fractional unit got smaller.

T: What happened to the number of units in the whole?

S: There are 4 times as many. → They quadrupled.

T: What happened to the number of selected units?

S: There was 1, and now there are 4. → The number of selected units quadrupled!

T: Has the size of the selected units changed?

S: There are more smaller-unit fractions instead of one bigger-unit fraction, but the area is still the same.

T: What can you conclude about $\frac{1}{2}$ and $\frac{4}{8}$?

S: They are equal!

T: Let's show that using multiplication: $\frac{1}{2} = \frac{1 \times 4}{2 \times 4} = \frac{4}{8}$. $\left(\frac{4 \text{ times as many selected units}}{4 \text{ times as many units in the whole}} \right)$

T: When we quadrupled the number of units, the number of selected units quadrupled. When we doubled the number of units, the number of selected units doubled. What do you predict would happen to the shaded fraction if we tripled the units?

S: The number of units within the shaded fraction would triple, too.

Problem 2: Given an area model, determine an equivalent fraction for the area selected.

T: (Display area model showing $\frac{1}{4}$.) Work with your partner to determine an equivalent fraction to $\frac{1}{4}$.

S: Let's draw one horizontal line. That will double the number of units. → We can draw two horizontal lines. That will triple the number of units and make them smaller, too. → If we multiply the top and bottom numbers by 4, we could quadruple the number of units. Each one will be a quarter of the size, too.

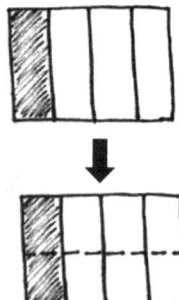

Circulate to listen for student understanding and monitor their work. Reconvene to examine one or more equivalent fractions.

T: Some groups drew one horizontal line. (Demonstrate.) Tell your partner what happened to the size of the units.

S: The units got smaller.

T: Tell your partner what happened to the number of units.

S: There are twice as many units.

T: Let's record that: $\frac{1}{4} = \frac{1 \times 2}{4 \times 2} = \frac{2}{8}$.

T: What is the relationship of the **denominators**, the fractional units, in the equivalent fractions?

S: The denominator in $\frac{2}{8}$ is double the denominator in $\frac{1}{4}$ because we doubled the number of units. → Since the size of the units is half as big, we doubled the denominator.

T: What is the relationship of the **numerators**, the number of fractional units selected, in the equivalent fractions?

S: The numerator in $\frac{2}{8}$ is double the numerator in $\frac{1}{4}$ because we doubled the number of selected units. → Since the size of the selected units is half as big, we doubled the numerator.

Problem 3: Express an equivalent fraction using multiplication, and verify by drawing an area model.

T: Discuss with your partner how to find another way to name $\frac{1}{3}$ without drawing an area model first.

S: Let's triple the number of units in the whole. → So, we have to multiply the numerator and denominator by 3. → Or we could triple the numerator and denominator.

T: Now, verify that the fraction you found is equivalent by drawing an area model.

S: (Work.)

$\frac{1}{3} = \frac{1 \times 3}{3 \times 3} = \frac{3}{9}$

My area model proves the equation. $\frac{1}{3} = \frac{3}{9}$.

Problem Set (10 minutes)

Students should do their personal best to complete the Problem Set within the allotted 10 minutes. For some classes, it may be appropriate to modify the assignment by specifying which problems they work on first. Some problems do not specify a method for solving. Students should solve these problems using the RDW approach used for Application Problems.

Student Debrief (10 minutes)

Lesson Objective: Use the area model and multiplication to show the equivalence of two fractions.

The Student Debrief is intended to invite reflection and active processing of the total lesson experience.

Invite students to review their solutions for the Problem Set. They should check work by comparing answers with a partner before going over answers as a class. Look for misconceptions or misunderstandings that can be addressed in the Debrief. Guide students in a conversation to debrief the Problem Set and process the lesson.

Any combination of the questions below may be used to lead the discussion.

- What pattern did you notice for Problem 1(a–d)?
- Discuss and compare with your partner your answers to Problems 2(e) and 2(f).
- In Problem 2, the unit fractions have different **denominators**. Discuss with your partner how the size of a unit fraction is related to the denominator.
- The **numerator** identifies the number of units selected. Can the numerator be larger than the denominator?

Exit Ticket (3 minutes)

After the Student Debrief, instruct students to complete the Exit Ticket. A review of their work will help with assessing students' understanding of the concepts that were presented in today's lesson and planning more effectively for future lessons. The questions may be read aloud to the students.

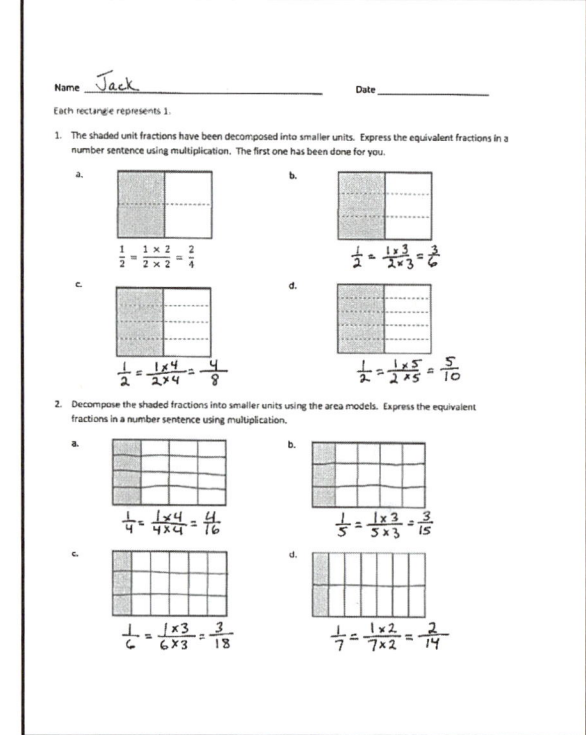

A STORY OF UNITS   Lesson 7 Problem Set 4•5

Name _____   Date _____

Each rectangle represents 1.

1. The shaded unit fractions have been decomposed into smaller units. Express the equivalent fractions in a number sentence using multiplication. The first one has been done for you.

   a.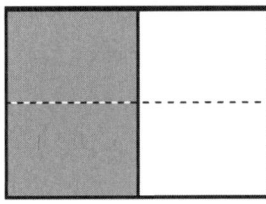

   $$\frac{1}{2} = \frac{1 \times 2}{2 \times 2} = \frac{2}{4}$$

   b.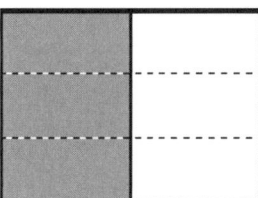

   c.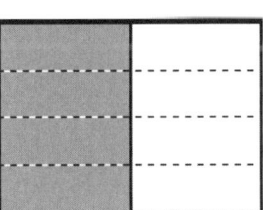

   d.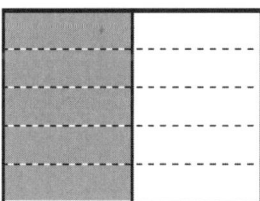

Lesson 7: Use the area model and multiplication to show the equivalence of two fractions.

2. Decompose the shaded fractions into smaller units using the area models. Express the equivalent fractions in a number sentence using multiplication.

a.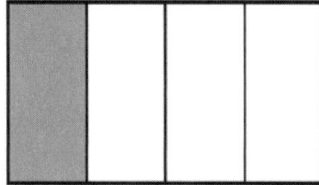

b.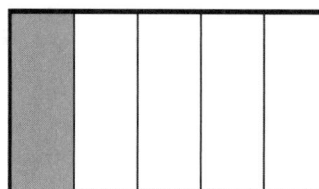

c.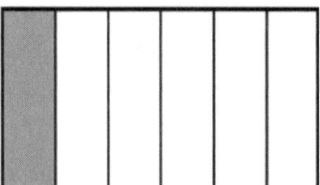

d.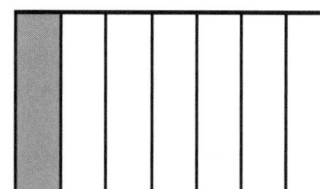

e. What happened to the size of the fractional units when you decomposed the fraction?

f. What happened to the total number of units in the whole when you decomposed the fraction?

3. Draw three different area models to represent 1 third by shading.
   Decompose the shaded fraction into (a) sixths, (b) ninths, and (c) twelfths.
   Use multiplication to show how each fraction is equivalent to 1 third.

   a.

   b.

   c.

A STORY OF UNITS

Lesson 7 Exit Ticket 4•5

Name _____ Date _____

Draw two different area models to represent 1 fourth by shading.
Decompose the shaded fraction into (a) eighths and (b) twelfths.
Use multiplication to show how each fraction is equivalent to 1 fourth.

a.

b.

Lesson 7: Use the area model and multiplication to show the equivalence of two fractions.

Name _____ Date _____

Each rectangle represents 1.

1. The shaded unit fractions have been decomposed into smaller units. Express the equivalent fractions in a number sentence using multiplication. The first one has been done for you.

   a.
   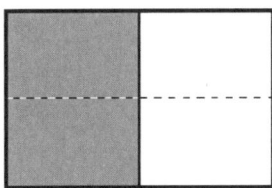
   $$\frac{1}{2} = \frac{1 \times 2}{2 \times 2} = \frac{2}{4}$$

   b.
   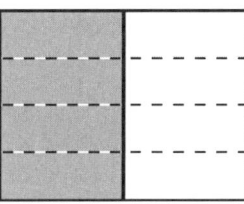

   c.
   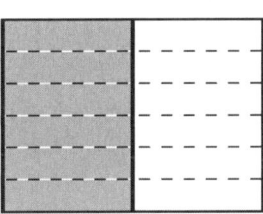

   d.
   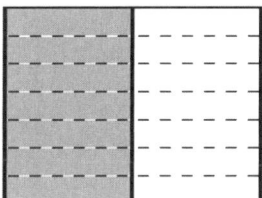

2. Decompose the shaded fractions into smaller units using the area models. Express the equivalent fractions in a number sentence using multiplication.

   a.

   b.

c. 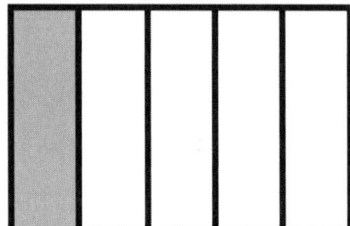   d.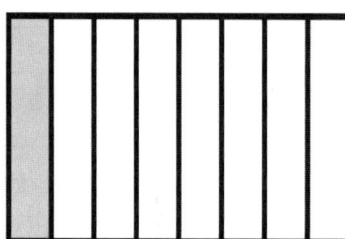

3. Draw three different area models to represent 1 fourth by shading.
   Decompose the shaded fraction into (a) eighths, (b) twelfths, and (c) sixteenths.
   Use multiplication to show how each fraction is equivalent to 1 fourth.

   a.

   b.

   c.

Lesson 8

Objective: Use the area model and multiplication to show the equivalence of two fractions.

Suggested Lesson Structure

- ■ Fluency Practice   (12 minutes)
- ■ Application Problem   (4 minutes)
- ■ Concept Development   (34 minutes)
- ■ Student Debrief   (10 minutes)

    Total Time   **(60 minutes)**

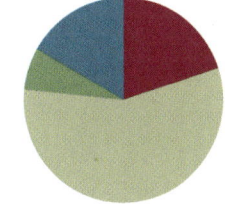

Fluency Practice (12 minutes)

- Multiply Mentally **4.OA.4**   (4 minutes)
- Count by Equivalent Fractions **3.NF.3**   (4 minutes)
- Draw Equivalent Fractions **4.NF.1**   (4 minutes)

Multiply Mentally (4 minutes)

Materials: (S) Personal white board

Note: This fluency activity reviews Module 3 content.

- T: (Write 32 × 3 = ___.) Say the multiplication sentence.
- S: 32 × 3 = 96.
- T: (Write 32 × 3 = 96. Below it, write 32 × 20 = ___.) Say the multiplication sentence.
- S: 32 × 20 = 640.
- T: (Write 32 × 20 = 640. Below it, write 32 × 23 = ___.) On your personal white board, solve 32 × 23.
- S: (Write 32 × 23 = 736.)

Continue with the following possible sequence: 42 × 2, 42 × 20, 42 × 22; and 21 × 4, 21 × 40, 21 × 44.

Count by Equivalent Fractions (4 minutes)

Note: This fluency activity reviews Lesson 4.

- T: Count by twos to 12. Start at 0.
- S: 0, 2, 4, 6, 8, 10, 12.

A STORY OF UNITS — Lesson 8 4•5

T: Count by 2 thirds to 12 thirds. Start at 0 thirds. (Write as students count.)

S: $\frac{0}{3}, \frac{2}{3}, \frac{4}{3}, \frac{6}{3}, \frac{8}{3}, \frac{10}{3}, \frac{12}{3}$.

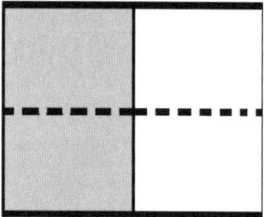

T: 1 is the same as how many thirds?

S: 3 thirds.

T: 2 is the same as how many thirds?

S: 6 thirds.

T: (Beneath $\frac{6}{3}$, write 2.) 3 is the same as how many thirds?

S: 9 thirds.

T: 4 is the same as how many thirds?

S: 12 thirds.

T: (Beneath $\frac{12}{3}$, write 4.) Count by 2 thirds again. This time, say the whole numbers when you arrive at them. Start at zero.

S: $0, \frac{2}{3}, \frac{4}{3}, 2, \frac{8}{3}, \frac{10}{3}, 4$.

Repeat the process, counting by 2 sixths to 18 sixths.

Draw Equivalent Fractions (4 minutes)

Materials: (S) Personal white board

Note: This fluency activity reviews Lesson 7.

T: (Write $\frac{1}{2}$.) Say the fraction.

S: 1 half.

T: On your personal white board, draw a model to show $\frac{1}{2}$.

S: (Draw a model partitioned into 2 equal units. Shade 1 unit.)

T: (Write $\frac{1}{2} = \frac{\times}{\times} = \frac{}{4}$.) Draw a dotted horizontal line to find the equivalent fraction. Then, complete the number sentence.

S: (Draw a dotted horizontal line, breaking 2 units into 4 smaller units. Write $\frac{1}{2} = \frac{1 \times 2}{2 \times 2} = \frac{2}{4}$.)

Continue with the following possible sequence: 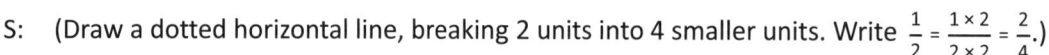 $\frac{1}{2} = \frac{}{8}, \frac{1}{3} = \frac{}{9}, \frac{1}{4} = \frac{}{8}, \frac{1}{5} = \frac{}{15}$, and $\frac{1}{7} = \frac{}{14}$.

Lesson 8: Use the area model and multiplication to show the equivalence of two fractions.

A STORY OF UNITS   Lesson 8   4•5

Application Problem (4 minutes)

Saisha gives some of her chocolate bar, pictured below, to her younger brother Lucas. He says, "Thanks for $\frac{3}{12}$ of the bar." Saisha responds, "No. I gave you $\frac{1}{4}$ of the bar." Explain why both Lucas and Saisha are correct.

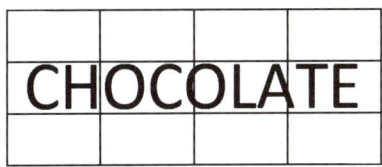
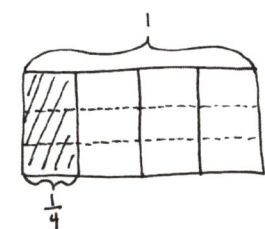

The smaller unit is twelfths. 3 twelfths is the same as 1 fourth.

Both Lucas and Saisha are correct because $\frac{3}{12} = \frac{1}{4}$.

$\frac{1}{4} = \frac{1 \times 3}{4 \times 3} = \frac{3}{12}$

Note: This Application Problem reviews content from Lesson 7. This bridges into today's lesson, where students determine equivalent fractions of non-unit fractions. Revisit this problem in the Student Debrief by asking students to write the remaining portion as two equivalent fractions.

Concept Development (34 minutes)

Materials: (S) Personal white board

Problem 1: Determine that multiplying both the numerator and denominator by *n* results in an equivalent fraction.

T: Draw an area model to represent 2 thirds. Draw three horizontal lines across the area model.
S: (Draw and partition the model.)
T: What happened to the size of the fractional units?
S: The units are 4 times as small because we divided each unit into 4 smaller units.
T: What happened to the number of units?
S: There were 3, and now there are 12. → There are 4 times as many units.
T: What happened to the number of selected units?
S: There were 2 units selected; now there are 8 units selected.
T: Discuss with your partner how to represent the equivalence of $\frac{2}{3}$ and $\frac{8}{12}$ using multiplication.
S: We can multiply the numerator and denominator by 4. We can write $\frac{2}{3} = \frac{2 \times 4}{3 \times 4} = \frac{8}{12}$.
T: How do you know the fraction is still representing the same amount?

$\frac{2}{3} = \frac{2 \times 4}{3 \times 4} = \frac{8}{12}$

There are more pieces and they are smaller, but it's still $\frac{2}{3}$!

Lesson 8: Use the area model and multiplication to show the equivalence of two fractions.

A STORY OF UNITS • Lesson 8 • 4•5

S: I know it's the same size because I didn't change how much is selected. → There are more smaller units instead of fewer bigger units, but the area of the selected fraction is still the same. → The fractions are equivalent.

T: What was different about this problem from the ones we did yesterday?

S: The fraction that we are starting with doesn't have 1 as the numerator.

T: We know any fraction can be decomposed into the sum of unit fractions. Yesterday, we saw that 1 third equals 4 twelfths. Today, we see that 2 thirds equals 4 twelfths plus 4 twelfths, or 8 twelfths.

MP.7

T: Draw an area model to represent $\frac{5}{6}$. Find an equivalent fraction with the denominator of 12. Explain to a partner how this is done.

S: We partition each of the 6 units into 2 parts so that we have 12 units. → We double the number of units to make twelfths. → There are twice as many units in the whole and twice as many units selected, but the parts are only half as big. → $\frac{5}{6} = \frac{5 \times 2}{6 \times 2} = \frac{10}{12}$.

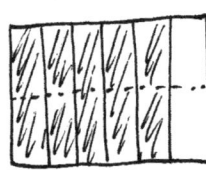

T: What have we discovered about finding equivalent fractions?

S: The area of the fraction stays the same, but the number and size of the units change. → The number of units increases. The size of the unit fraction decreases.

Problem 2: Determine that two fractions are equivalent using an area model and a number sentence.

T: (Project $\frac{3}{4} = \frac{6}{8}$.) If the whole is the same, is this statement true or false?

S: 3 times 2 is 6, and 4 times 2 is 8. Yes. It's true. → If we multiply both the numerator and denominator by 2, we get $\frac{6}{8}$. → Doubling the selected units and the number of units in the whole has the same area as $\frac{3}{4}$.

T: Represent the equivalence in a number sentence using multiplication, and draw an area model to show the equivalence.

S: (Do so as pictured to the right.)

T: (Project $\frac{3}{4} = \frac{6}{12}$.) If the wholes are the same, is this statement true or false? How do you know? Discuss with your partner.

S: Three times 2 is 6, and 4 times 3 is 12. It's false. We didn't multiply by the same number. → This is false. I drew a model for $\frac{3}{4}$ and then decomposed it into twelfths. There are 9 units shaded, not 6. → The numerator is being multiplied by 2, and the denominator is being multiplied by 3. They are not equivalent fractions.

T: With your partner, revise the right side of the equation to make a true number sentence.

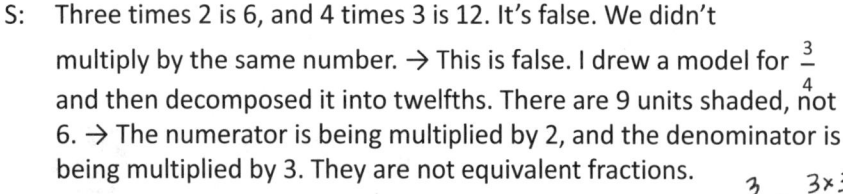

S: We could change $\frac{6}{12}$ to $\frac{9}{12}$. → Or we could change $\frac{6}{12}$ to $\frac{6}{8}$ because both the numerator and denominator would be multiplied by 2.

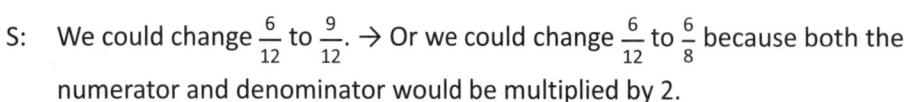

Lesson 8: Use the area model and multiplication to show the equivalence of two fractions.

A STORY OF UNITS   Lesson 8

Problem 3: Write a number sentence using multiplication to show the equivalence of two fractions. Draw the corresponding area model.

T: Find an equivalent fraction without drawing an area model first. Write $\frac{3}{5}$ on your personal white board. How have we found equivalent fractions?

S: We've doubled, tripled, or quadrupled the numerator and denominator. → We multiply the numerator and denominator by the same number.

T: Find an equivalent fraction to $\frac{3}{5}$ using multiplication.

S: When I multiply the numerator and denominator by 2, I get $\frac{6}{10}$.

T: Use an area model to confirm your number sentence.

S: (Do so, correcting any errors as necessary. Answers may vary.)

$$\frac{3}{5} = \frac{3 \times 2}{5 \times 2} = \frac{6}{10}$$

Problem Set (10 minutes)

Students should do their personal best to complete the Problem Set within the allotted 10 minutes. For some classes, it may be appropriate to modify the assignment by specifying which problems they work on first. Some problems do not specify a method for solving. Students should solve these problems using the RDW approach used for Application Problems.

> **NOTES ON MULTIPLE MEANS OF ENGAGEMENT**
>
> Invite students working above grade level and others to test their discoveries about multiplying fractions by partitioning shapes other than rectangles, such as circles and hexagons. This work may best be supported by means of concrete or virtual manipulatives.

Student Debrief (10 minutes)

Lesson Objective: Use the area model and multiplication to show the equivalence of two fractions.

The Student Debrief is intended to invite reflection and active processing of the total lesson experience.

Invite students to review their solutions for the Problem Set. They should check work by comparing answers with a partner before going over answers as a class. Look for misconceptions or misunderstandings that can be addressed in the Debrief. Guide students in a conversation to debrief the Problem Set and process the lesson.

Any combination of the questions below may be used to lead the discussion.

- For Problem 3(a–d), how did you determine the number of horizontal lines to draw in each area model?

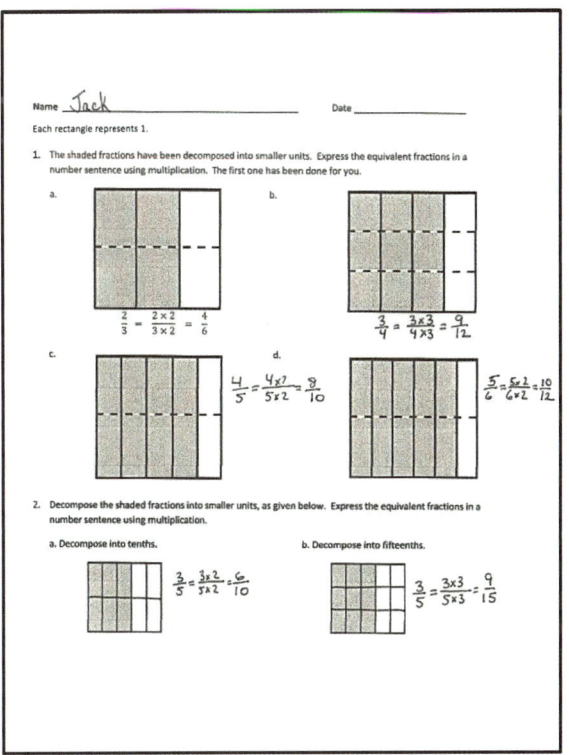

   Lesson 8: Use the area model and multiplication to show the equivalence of two fractions.

- For Problem 5(c), did you and your partner have the same answer? Explain why you might have different answers.
- Explain when someone might need to use equivalent fractions in daily life.
- How are we able to show equivalence without having to draw an area model?
- Think back to the Application Problem. What fraction of the bar did Saisha receive?

Exit Ticket (3 minutes)

After the Student Debrief, instruct students to complete the Exit Ticket. A review of their work will help with assessing students' understanding of the concepts that were presented in today's lesson and planning more effectively for future lessons. The questions may be read aloud to the students.

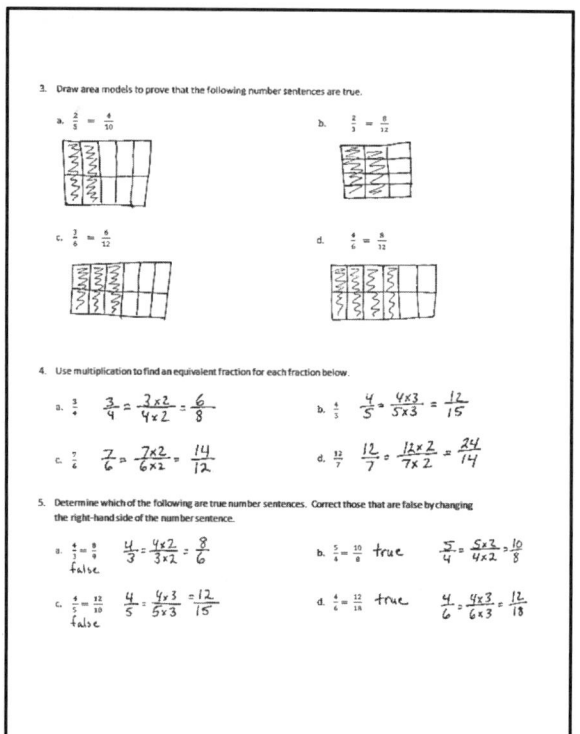

Lesson 8: Use the area model and multiplication to show the equivalence of two fractions.

A STORY OF UNITS　　　　　　　　　　　　　　　　　Lesson 8 Problem Set 4•5

Name _____   Date _____

Each rectangle represents 1.

1. The shaded fractions have been decomposed into smaller units. Express the equivalent fractions in a number sentence using multiplication. The first one has been done for you.

   a.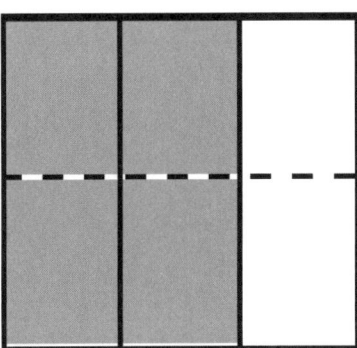

   $$\frac{2}{3} = \frac{2 \times 2}{3 \times 2} = \frac{4}{6}$$

   b.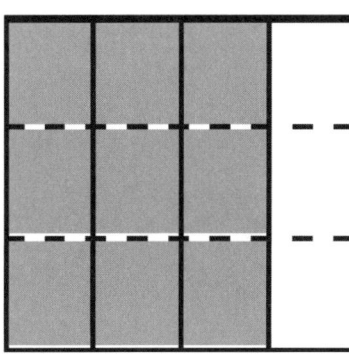

   c.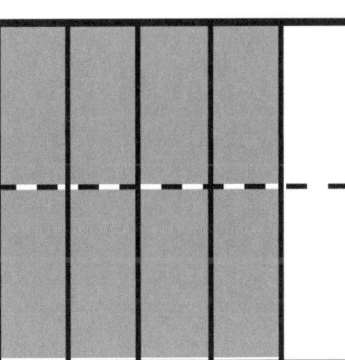

   d.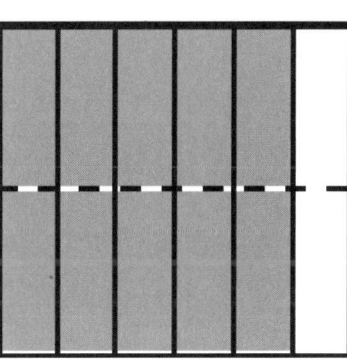

2. Decompose the shaded fractions into smaller units, as given below. Express the equivalent fractions in a number sentence using multiplication.

   a. Decompose into tenths.

   b. Decompose into fifteenths.

   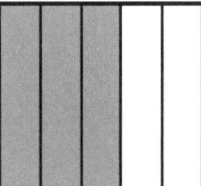

Lesson 8: Use the area model and multiplication to show the equivalence of two fractions.

A STORY OF UNITS Lesson 8 Problem Set 4•5

3. Draw area models to prove that the following number sentences are true.

   a. $\dfrac{2}{5} = \dfrac{4}{10}$

   b. $\dfrac{2}{3} = \dfrac{8}{12}$

   c. $\dfrac{3}{6} = \dfrac{6}{12}$

   d. $\dfrac{4}{6} = \dfrac{8}{12}$

4. Use multiplication to find an equivalent fraction for each fraction below.

   a. $\dfrac{3}{4}$

   b. $\dfrac{4}{5}$

   c. $\dfrac{7}{6}$

   d. $\dfrac{12}{7}$

5. Determine which of the following are true number sentences. Correct those that are false by changing the right-hand side of the number sentence.

   a. $\dfrac{4}{3} = \dfrac{8}{9}$

   b. $\dfrac{5}{4} = \dfrac{10}{8}$

   c. $\dfrac{4}{5} = \dfrac{12}{10}$

   d. $\dfrac{4}{6} = \dfrac{12}{18}$

Lesson 8: Use the area model and multiplication to show the equivalence of two fractions.

Name _____ Date _____

1. Use multiplication to create an equivalent fraction for the fraction below.

    $$\frac{2}{5}$$

2. Determine if the following is a true number sentence. If needed, correct the statement by changing the right-hand side of the number sentence.

    $$\frac{3}{4} = \frac{9}{8}$$

Name _____ Date _____

Each rectangle represents 1.

1. The shaded fractions have been decomposed into smaller units. Express the equivalent fractions in a number sentence using multiplication. The first one has been done for you.

a.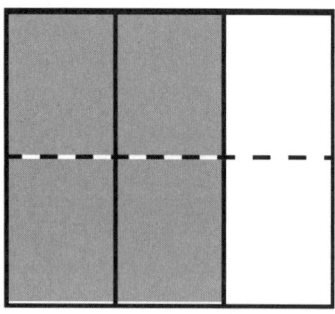

$$\frac{2}{3} = \frac{2 \times 2}{3 \times 2} = \frac{4}{6}$$

b.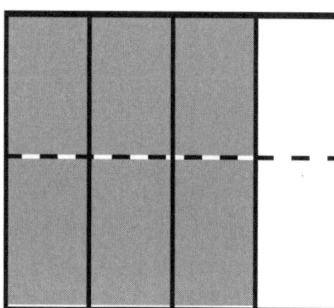

c.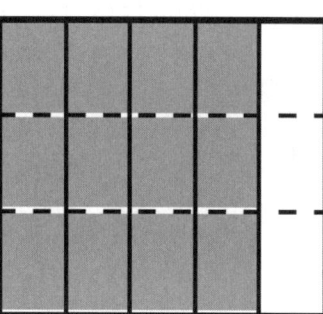

d.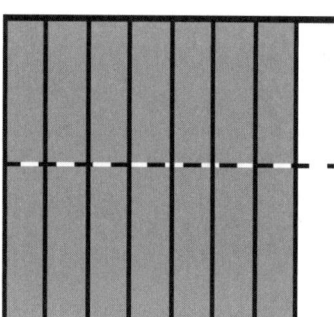

2. Decompose both shaded fractions into twelfths. Express the equivalent fractions in a number sentence using multiplication.

a.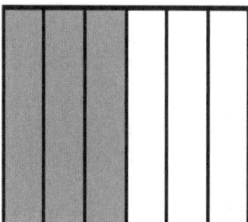

b.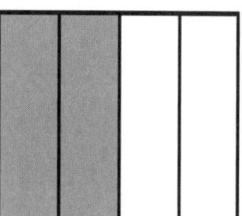

3. Draw area models to prove that the following number sentences are true.

   a. $\dfrac{1}{3} = \dfrac{2}{6}$

   b. $\dfrac{2}{5} = \dfrac{4}{10}$

   c. $\dfrac{5}{7} = \dfrac{10}{14}$

   d. $\dfrac{3}{6} = \dfrac{9}{18}$

4. Use multiplication to create an equivalent fraction for each fraction below.

   a. $\dfrac{2}{3}$

   b. $\dfrac{5}{6}$

   c. $\dfrac{6}{5}$

   d. $\dfrac{10}{8}$

5. Determine which of the following are true number sentences. Correct those that are false by changing the right-hand side of the number sentence.

   a. $\dfrac{2}{3} = \dfrac{4}{9}$

   b. $\dfrac{5}{6} = \dfrac{10}{12}$

   c. $\dfrac{3}{5} = \dfrac{6}{15}$

   d. $\dfrac{7}{4} = \dfrac{21}{12}$

Lesson 8: Use the area model and multiplication to show the equivalence of two fractions.

Lesson 9

Objective: Use the area model and division to show the equivalence of two fractions.

Suggested Lesson Structure

- Fluency Practice   (12 minutes)
- Application Problem   (6 minutes)
- Concept Development   (32 minutes)
- Student Debrief   (10 minutes)

Total Time   **(60 minutes)**

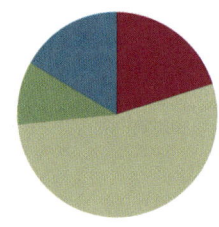

Fluency Practice (12 minutes)

- Add and Subtract **4.NBT.4**   (4 minutes)
- Find Equivalent Fractions **4.NF.1** (4 minutes)
- Draw Equivalent Fractions **4.NF.1** (4 minutes)

Add and Subtract (4 minutes)

Materials: (S) Personal white board

Note: This fluency activity reviews the year-long Grade 4 fluency standard for adding and subtracting using the standard algorithm.

   T: (Write 532 thousands 367 ones.) On your personal white board, write this number in standard form.
   S: (Write 532,367.)
   T: (Write 423 thousands 142 ones.) Add this number to 532,367 using the standard algorithm.
   S: (Write 532,367 + 423,142 = 955,509 using the standard algorithm.)

Continue the process for 671,526 + 264,756.

   T: (Write 916 thousands 450 ones.) On your board, write this number in standard form.
   S: (Write 916,450.)
   T: (Write 615 thousands 137 ones.) Subtract this number from 916,450 using the standard algorithm.
   S: (Write 916,450 – 615,137 = 301,313 using the standard algorithm.)

Continue with the following possible sequence: 762,162 – 335,616 and 500,000 – 358,219.

A STORY OF UNITS　　　　　　　　　　　　　　　　　　　　　　　　　　　Lesson 9　4•5

Find Equivalent Fractions (4 minutes)

Materials: (S) Personal white board

Note: This fluency activity reviews Lesson 7.

- T: (Write $\frac{1}{2} = \frac{\times}{\times} = \frac{2}{}$. Point to $\frac{1}{2}$.) Say the unit fraction.
- S: 1 half.
- T: On your personal white board, complete the number sentence to make an equivalent fraction.
- S: (Write $\frac{1}{2} = \frac{1 \times 2}{2 \times 2} = \frac{2}{4}$.)

Continue with the following possible suggestions: $\frac{1}{2} = \frac{4}{8}, \frac{1}{3} = \frac{2}{6}, \frac{2}{3} = \frac{3}{9}, \frac{1}{4} = \frac{4}{16}$, and $\frac{1}{5} = \frac{3}{15}$.

Draw Equivalent Fractions (4 minutes)

Materials: (S) Personal white board

Note: This fluency activity reviews Lesson 8.

- T: (Write $\frac{2}{3}$.) Say the fraction.
- S: 2 thirds.
- T: On your personal white board, draw a model to show $\frac{2}{3}$.
- T: (Write $\frac{2}{3} = \frac{\times}{\times} = \frac{}{6}$.) Draw a dotted horizontal line to find the equivalent fraction. Then, complete the number sentence.
- S: (Draw a dotted horizontal line, breaking 3 units into 6 smaller units. Write $\frac{2}{3} = \frac{2 \times 2}{3 \times 2} = \frac{4}{6}$.)

Continue with the following possible sequence: $\frac{2}{3} = \frac{}{9}, \frac{3}{4} = \frac{}{12}, \frac{3}{5} = \frac{}{10}$, and $\frac{4}{5} = \frac{}{15}$.

Application Problem (6 minutes)

What fraction of a foot is 1 inch? What fraction of a foot is 3 inches?
(Hint: 12 inches = 1 foot.) Draw a tape diagram to model your work.

Note: Students are asked to think about fractions within a context, such as measurement, that will be useful in upcoming word problems. This measurement work is developed more in Module 7.

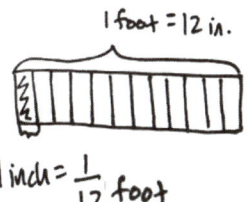

1 inch = $\frac{1}{12}$ foot

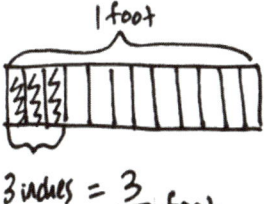

3 inches = $\frac{3}{12}$ foot

An inch is $\frac{1}{12}$ of a foot.　　3 inches is $\frac{3}{12}$ of a foot.

Lesson 9: Use the area model and division to show the equivalence of two fractions

117

A STORY OF UNITS   Lesson 9   4•5

Concept Development (32 minutes)

Materials: (S) Personal white board

Problem 1: Simplify $\frac{6}{12}$ by composing larger fractional units using division.

T: (Project area model showing $\frac{6}{12}$.) What fraction does the area model represent?

S: $\frac{6}{12}$.

T: Discuss with a partner. Do you see any fractions equivalent to $\frac{6}{12}$?

S: Half of the area model is shaded. The model shows $\frac{1}{2}$.

T: Which is the larger unit? Twelfths or halves?

S: Halves!

T: Circle the smaller units to make the larger units. Say the equivalent fractions.

S: $\frac{6}{12} = \frac{1}{2}$.

T: (Write $\frac{6 \div 6}{12 \div 6} = -$, and point to the denominator.) Twelve units were in the whole, and we made groups of 6 units. Say a division sentence to record that.

S: $12 \div 6 = 2$.

T: (Record the 2 in the denominator, and point to the numerator.) Six units were selected, and we made a group of 6 units. Say a division sentence to record that.

S: $6 \div 6 = 1$.

T: (Record the 1 in the numerator.) We write $\frac{6}{12} = \frac{6 \div 6}{12 \div 6} = \frac{1}{2}$, dividing both the numerator and denominator by 6 to find an equivalent fraction.

T: What happened to the size of the units and the total number of units?

S: The size of the units got larger. There are fewer units in the whole. → The units are 6 times as large, but the number of units is 6 times less. → The units got larger. The number of units got smaller.

> **NOTES ON MULTIPLE MEANS OF EXPRESSION:**
>
> As the conceptual foundation for simplification is being set, the word *simplify* is initially avoided with students as they compose higher-value units. The process is rather referred to as *composition,* the opposite of decomposition, which relates directly to their drawing, work throughout the last two lessons, and work with whole numbers. When working numerically, the process is referred to at times as *renaming,* again in an effort to relate to whole number work.

> **NOTES ON MULTIPLE MEANS OF REPRESENTATION:**
>
> English language learners may confuse the terms *decompose* and *compose.*
>
> - Demonstrate that the prefix *de* can be placed before some words to add an opposite meaning.
> - Use gestures to clarify the meanings: *Decompose* is to take apart, and *compose* is to put together.
> - Refresh students' memory of decomposition and composition in the context of the operations with whole numbers.

Lesson 9: Use the area model and division to show the equivalence of two fractions.

A STORY OF UNITS   Lesson 9   4•5

Problem 2: Simplify both $\frac{2}{8}$ and $\frac{3}{12}$ as $\frac{1}{4}$ by composing larger fractional units.

T: Draw an area model to represent $\frac{2}{8}$. Group two units to make larger units.

T: Write $\frac{2}{8} = \frac{2 \div 2}{8 \div 2} = $ —. How many groups of 2 are shaded?

S: 1.

T: How many groups of 2 are in the whole?

S: 4.

T: (Write $\frac{2}{8} = \frac{2 \div 2}{8 \div 2} = \frac{1}{4}$.) Talk to your partner about how we showed that 2 eighths is the same as 1 fourth. Discuss both the model and our use of division. (Allow students time to discuss.)

T: Draw an area model to represent $\frac{3}{12}$. Compose an equivalent fraction.

S: We can make groups of 2. → No. That won't work. Some of the groups could have shaded and unshaded units. → Groups of 3 will work. That's how many shaded units there are.

T: How many groups of 3 are shaded?

S: 1.

T: How many groups of 3 in all?

S: 4.

T: The new fraction is…?

S: $\frac{1}{4}$.

T: Write the number sentence to show that you composed groups of 3.

S: (Write $\frac{3}{12} = \frac{3 \div 3}{12 \div 3} = \frac{1}{4}$.)

T: Compare the area models for $\frac{3}{12}$ and $\frac{2}{8}$.

S: They both equal $\frac{1}{4}$.

Problem 3: Simplify both $\frac{2}{6}$ and $\frac{4}{12}$ as $\frac{1}{3}$ by composing larger fractional units.

T: When we composed fractions in the last two problems, what did you notice?

S: We divided to find equivalent fractions. → We made equal groups to make large units. → We composed a unit fraction from a non-unit fraction.

T: Draw area models to show $\frac{2}{6}$ and $\frac{4}{12}$. Rename both fractions as the same unit fraction.

S: I can make groups of 2 in both area models. I could make groups of 3, but I won't make equal groups of shaded and unshaded units. → Four is a factor of both 4 and 12, so I can make groups of 4. → First, I made groups of 2 when I was working with 4 twelfths, but then I noticed I could make groups of 2 again. → Hey, dividing by 2 twice is the same as dividing by 4.

Lesson 9:   Use the area model and division to show the equivalence of two fractions.

A STORY OF UNITS

Lesson 9 4•5

T: Circle the groups, and express each composition in a number sentence using division.

S: $\frac{2}{6} = \frac{2 \div 2}{6 \div 2} = \frac{1}{3}$. $\frac{4}{12} = \frac{4 \div 4}{12 \div 4} = \frac{1}{3}$.

T: How are $\frac{4}{12}$ and $\frac{2}{6}$ related?

S: When I model $\frac{4}{12}$ and $\frac{2}{6}$, I see that they both have the same area as $\frac{1}{3}$. → $\frac{1}{3} = \frac{4}{12} = \frac{2}{6}$. → The equivalent fraction for $\frac{4}{12}$ and $\frac{2}{6}$ with the largest units is $\frac{1}{3}$. → We composed $\frac{4}{12}$ and $\frac{2}{6}$ into the same unit fraction.

> **NOTES ON MULTIPLE MEANS OF ENGAGEMENT:**
>
> Challenge students working above grade level and others to couple the expressions of fraction composition with the related multiplication expression of decomposition (e.g., $\frac{4}{12} = \frac{4 \div 4}{12 \div 4} = \frac{1}{3}$ and $\frac{1}{3} = \frac{1 \times 4}{3 \times 4} = \frac{4}{12}$).

Problem Set (10 minutes)

Students should do their personal best to complete the Problem Set within the allotted 10 minutes. For some classes, it may be appropriate to modify the assignment by specifying which problems they work on first. Some problems do not specify a method for solving. Students should solve these problems using the RDW approach used for Application Problems.

Student Debrief (10 minutes)

Lesson Objective: Use the area model and division to show the equivalence of two fractions.

The Student Debrief is intended to invite reflection and active processing of the total lesson experience.

Invite students to review their solutions for the Problem Set. They should check work by comparing answers with a partner before going over answers as a class. Look for misconceptions or misunderstandings that can be addressed in the Debrief. Guide students in a conversation to debrief the Problem Set and process the lesson.

Any combination of the questions below may be used to lead the discussion.

- Look at Problem 1(a–d). Write some examples of fractions where the denominator is a multiple of the numerator. (Pause.) What do we know about these fractions?
- In Problems 3 and 4, does it matter how your area models are shaded? Will it still result in a correct answer?

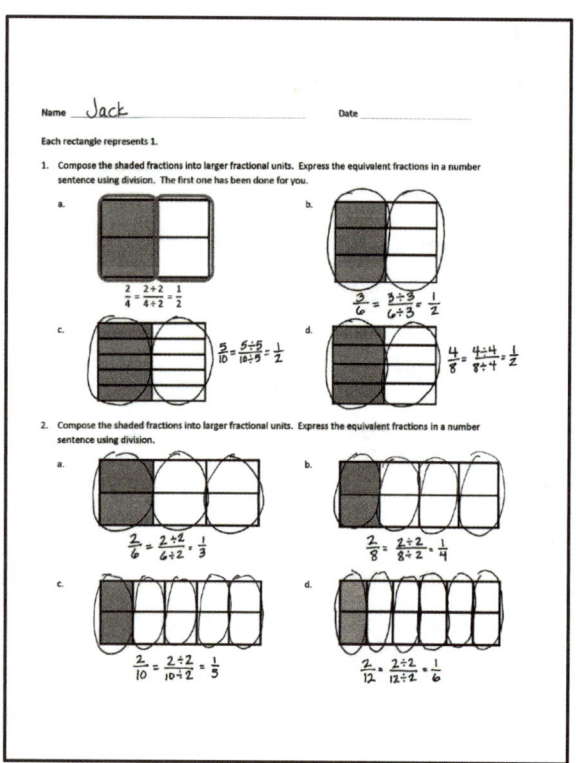

Lesson 9: Use the area model and division to show the equivalence of two fractions.

- Explain how two fractions can be composed into the same larger unit fraction.
- How can what you know about factors help rename a fraction in larger units?
- When we rename $\frac{3}{12}$ as $\frac{1}{4}$, why is it helpful to think about the factors of 3 and 12?
- Contrast the following: renaming fractions when you multiply versus when you divide and decomposing versus composing fractions. For each, discuss what happens to the size of the units and number of units.
- Use what you learned today to determine if $\frac{3}{8}$ can be renamed as a larger unit. Why or why not?

Exit Ticket (3 minutes)

After the Student Debrief, instruct students to complete the Exit Ticket. A review of their work will help with assessing students' understanding of the concepts that were presented in today's lesson and planning more effectively for future lessons. The questions may be read aloud to the students.

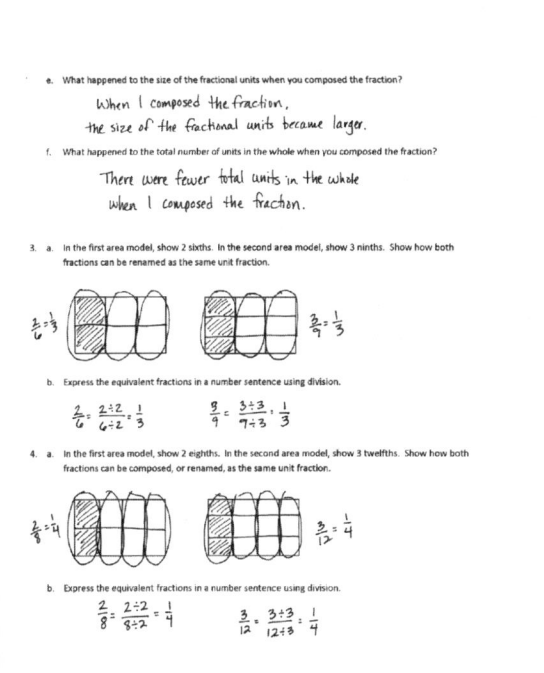

Name _____ Date _____

Each rectangle represents 1.

1. Compose the shaded fractions into larger fractional units. Express the equivalent fractions in a number sentence using division. The first one has been done for you.

a.

$$\frac{2}{4} = \frac{2 \div 2}{4 \div 2} = \frac{1}{2}$$

b.

c.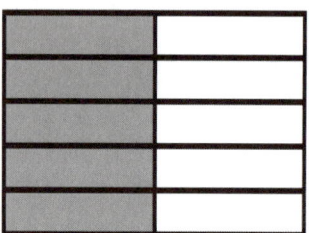

d.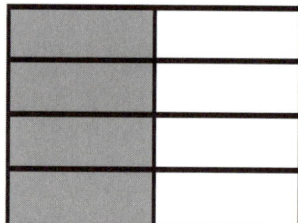

2. Compose the shaded fractions into larger fractional units. Express the equivalent fractions in a number sentence using division.

a.

b.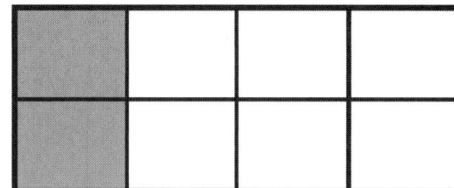

c.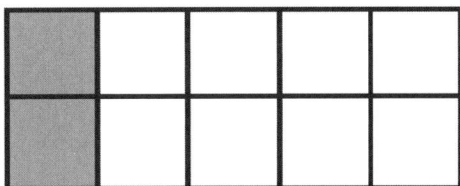

d.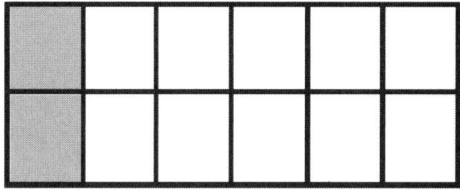

e. What happened to the size of the fractional units when you composed the fraction?

f. What happened to the total number of units in the whole when you composed the fraction?

3. a. In the first area model, show 2 sixths. In the second area model, show 3 ninths. Show how both fractions can be renamed as the same unit fraction.

b. Express the equivalent fractions in a number sentence using division.

4. a. In the first area model, show 2 eighths. In the second area model, show 3 twelfths. Show how both fractions can be composed, or renamed, as the same unit fraction.

b. Express the equivalent fractions in a number sentence using division.

A STORY OF UNITS

Lesson 9 Exit Ticket 4•5

Name _____ Date _____

a. In the first area model, show 2 sixths. In the second area model, show 4 twelfths. Show how both fractions can be composed, or renamed, as the same unit fraction.

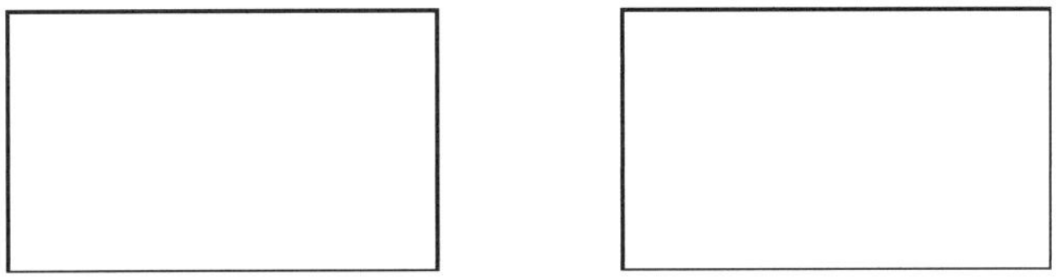

b. Express the equivalent fractions in a number sentence using division.

Name _____ Date _____

Each rectangle represents 1.

1. Compose the shaded fractions into larger fractional units. Express the equivalent fractions in a number sentence using division. The first one has been done for you.

   a.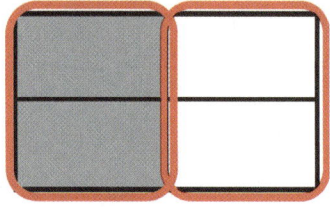

   $$\frac{2}{4} = \frac{2 \div 2}{4 \div 2} = \frac{1}{2}$$

   b.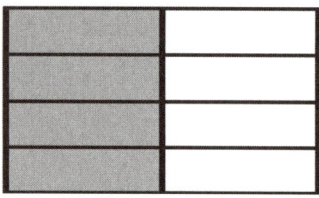

   c.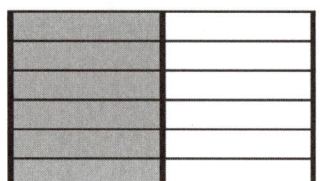

   d.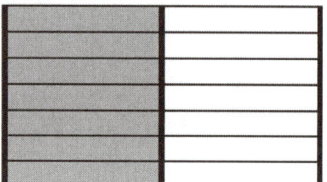

Lesson 9: Use the area model and division to show the equivalence of two fractions.

2. Compose the shaded fractions into larger fractional units. Express the equivalent fractions in a number sentence using division.

a.

b.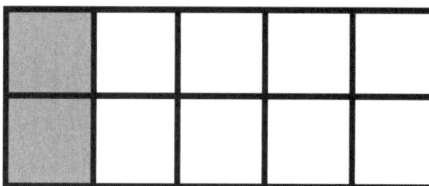

c.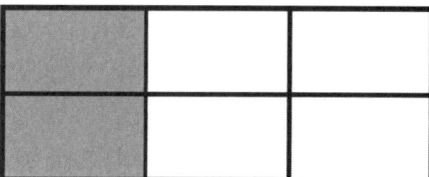

d.

e. What happened to the size of the fractional units when you composed the fraction?

f. What happened to the total number of units in the whole when you composed the fraction?

3. a. In the first area model, show 4 eighths. In the second area model, show 6 twelfths. Show how both fractions can be composed, or renamed, as the same unit fraction.

   b. Express the equivalent fractions in a number sentence using division.

4. a. In the first area model, show 4 eighths. In the second area model, show 8 sixteenths. Show how both fractions can be composed, or renamed, as the same unit fraction.

   b. Express the equivalent fractions in a number sentence using division.

A STORY OF UNITS

Lesson 10 4•5

Lesson 10

Objective: Use the area model and division to show the equivalence of two fractions.

Suggested Lesson Structure

- Fluency Practice (12 minutes)
- Application Problem (8 minutes)
- Concept Development (30 minutes)
- Student Debrief (10 minutes)

  Total Time **(60 minutes)**

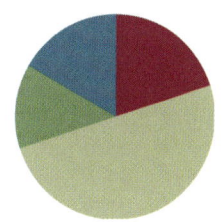

Fluency Practice (12 minutes)

- Count by Equivalent Fractions **3.NF.3** (4 minutes)
- Find Equivalent Fractions **4.NF.1** (4 minutes)
- Draw Equivalent Fractions **4.NF.1** (4 minutes)

Count by Equivalent Fractions (4 minutes)

Note: This fluency activity reinforces Module 5 fraction concepts.

T: Count by threes to 24. Start at zero.
S: 0, 3, 6, 9, 12, 15, 18, 21, 24.
T: Count by 3 fourths to 24 fourths. Start at 0 fourths. (Write as students count.)
S: $\frac{0}{4}, \frac{3}{4}, \frac{6}{4}, \frac{9}{4}, \frac{12}{4}, \frac{15}{4}, \frac{18}{4}, \frac{21}{4}, \frac{24}{4}$.
T: 1 is the same as how many fourths?
S: 4 fourths.
T: 2 is the same as how many fourths?
S: 8 fourths.
T: 3 is the same as how many fourths?
S: 12 fourths.
T: (Beneath $\frac{12}{4}$, write 3.) 4 is the same as how many fourths?

> **NOTES ON MULTIPLE MEANS OF REPRESENTATION:**
>
> While leading the Count by Equivalent Fractions fluency activity, enunciate the ending digraph /th/ of fraction names to help English language learners distinguish fractions from whole numbers (e.g., *fourths*, not *fours*).
>
> Couple numbers on the board with prepared visuals, if beneficial.

$\frac{0}{4}$	$\frac{3}{4}$	$\frac{6}{4}$	$\frac{9}{4}$	$\frac{12}{4}$	$\frac{15}{4}$	$\frac{18}{4}$	$\frac{21}{4}$	$\frac{24}{4}$
0	$\frac{3}{4}$	$\frac{6}{4}$	$\frac{9}{4}$	3	$\frac{15}{4}$	$\frac{18}{4}$	$\frac{21}{4}$	6

Lesson 10: Use the area model and division to show the equivalence of two fractions.

S: 16 fourths.

T: 5 is the same as how many fourths?

S: 20 fourths.

T: 6 is the same as how many fourths?

S: 24 fourths.

T: (Beneath $\frac{24}{4}$, write 6.) Count by 3 fourths again. This time, say the whole numbers when you arrive at them. Start at zero.

S: $0, \frac{3}{4}, \frac{6}{4}, \frac{9}{4}, 3, \frac{15}{4}, \frac{18}{4}, \frac{21}{4}, 6$.

Repeat the process, counting by 3 fifths to 30 fifths.

Find Equivalent Fractions (4 minutes)

Materials: (S) Personal white board

Note: This fluency activity reviews Lesson 8.

T: (Write $\frac{3}{4} = \frac{\times}{\times} = \frac{}{8}$. Point to $\frac{3}{4}$.) Say the fraction.

S: 3 fourths.

T: On your personal white board, complete the number sentence.

S: (Write $\frac{3}{4} = \frac{3 \times 2}{4 \times 2} = \frac{6}{8}$.)

Continue with the following possible suggestions: $\frac{3}{4} = \frac{9}{12}, \frac{2}{3} = \frac{4}{6}, \frac{2}{5} = \frac{4}{10}, \frac{4}{5} = \frac{8}{10}$, and $\frac{3}{5} = \frac{9}{15}$.

Draw Equivalent Fractions (4 minutes)

Materials: (S) Personal white board

Note: This fluency activity reviews Lesson 9.

T: (Project a model with 2 out of 4 equal units shaded.) Draw the model, and write the fraction that is shaded.

S: (Draw a model with 2 out of 4 equal units shaded. Write $\frac{2}{4}$.)

T: (Write $\frac{2}{4} = \frac{\div}{\div} = \frac{}{}$.) Compose the shaded units into 1 larger unit by circling. Then, complete the number sentence.

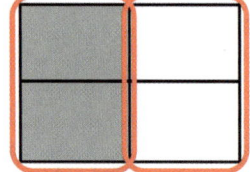

S: (Circle the shaded units into 1 larger unit. Write $\frac{2}{4} = \frac{2 \div 2}{4 \div 2} = \frac{1}{2}$.)

Continue with the following possible sequence: $\frac{3}{9} = \frac{1}{3}, \frac{4}{8} = \frac{1}{2}, \frac{2}{8} = \frac{1}{4}, \frac{5}{10} = \frac{1}{2}$, and $\frac{4}{12} = \frac{1}{3}$.

Lesson 10: Use the area model and division to show the equivalence of two fractions.

A STORY OF UNITS   Lesson 10   4•5

Application Problem (8 minutes)

Nuri spent $\frac{9}{12}$ of his money on a book and the rest of his money on a pencil.

  a. Express how much of his money he spent on the pencil in fourths.

  a. Nuri started with $1. How much did he spend on the pencil?

Note: This Application Problem connects Topic A and Lesson 9 by finding the other fractional part of the whole and expressing equivalent fractions. Using what students know about money, ask why it is preferable to answer in fourths rather than twelfths. Students connect fourths to quarters of a dollar. Revisit this problem in the Student Debrief to express how much money was spent on the book in fourths.

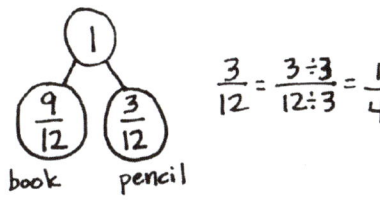

Concept Development (30 minutes)

Materials: (S) Personal white board

Problem 1: Simplify a fraction by drawing to find a common factor, and relate it to division.

T: Draw an area model that represents $\frac{10}{12}$.

T: If we want to compose an equivalent fraction, what do we do?

S: We make equal groups. → We divide the numerator and the denominator by the same number. → We should divide by 10. We divided by the same number that was in the numerator yesterday.

T: Can I divide both the numerator and denominator by 10?

S: No.

T: Discuss with your partner how to determine the largest possible unit.

S: We can try to make groups of 2, then 3, then 4, until we have the largest number of units in a group with no remainder. → We can only make equal groups of 2. The other numbers don't divide evenly into both the numerator and denominator.

T: Show me. (Allow time for students to compose an area model.) What happened to the number of shaded units?

S: There were 10 units shaded, and now there are 5 groups of 2 units shaded!

> **NOTES ON MULTIPLE MEANS OF REPRESENTATION:**
>
> There are multiple ways of showing a given fraction using an area model. Area models may, therefore, look different from student to student. Allow students to share how they have drawn different area models, and be accepting of those that are mathematically correct.

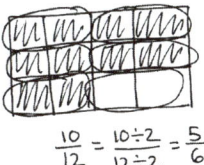

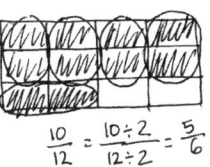

Lesson 10: Use the area model and division to show the equivalence of two fractions.   131

A STORY OF UNITS   Lesson 10   4•5

T: Consider the unit fractions $\frac{1}{12}$ and $\frac{1}{6}$. What do you notice about their denominators?

S: 6 is a factor of 12.

T: What about the numerators 10 and 5?

S: 5 is a factor of 10.

T: List the factors of 10 and 12.

S: The factors of 12 are 1, 2, 3, 4, 6, and 12. The factors of 10 are 1, 2, 5, and 10.

T: 1 and 2 are factors of both. We know then that we can make equal groups of 2. Equal groups of 1 bring us back to the original fraction.

Problem 2: Draw an area model of a number sentence that shows the simplification of a fraction.

T: (Project $\frac{6}{10} = \frac{6 \div 2}{10 \div 2} = \frac{3}{5}$.)

T: Draw an area model to show how this number sentence is true.

S: The numerator and denominator are both being divided by 2. I will circle groups of 2. → I know 2 is a factor of 6 and 10, so I could make groups of 2. → There are 3 shaded groups of 2 and 5 total groups of 2. → That's $\frac{3}{5}$.

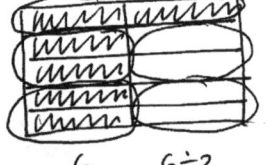
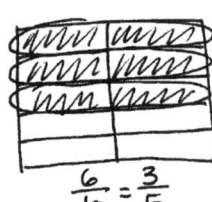
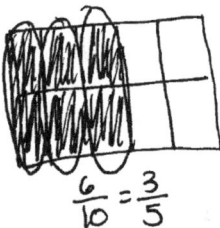

Problem 3: Simplify a fraction by drawing to find different common factors, and relate it to division.

T: With your partner, draw an area model to represent $\frac{8}{12}$. Rename $\frac{8}{12}$ using larger fractional units. You may talk as you work. (Circulate and listen.)

S: I can circle groups of 2 units. → 2 is a factor of 8 and 12. → There are 6 groups of 2 units. → Four groups are shaded. That's $\frac{4}{6}$.

T: What happens when I use 4 as a common factor instead of 2? Turn and talk.

S: Four is a factor of both 8 and 12. It works. → We can make larger units with groups of 4. → Thirds are larger than sixths. $\frac{8}{12} = \frac{2}{3}$. → We have fewer units, but they're bigger.

T: Express the equivalent fractions as two division number sentences.

S: (Write $\frac{8}{12} = \frac{8 \div 4}{12 \div 4} = \frac{2}{3}$ and $\frac{8}{12} = \frac{8 \div 2}{12 \div 2} = \frac{4}{6}$.)

T: What can you conclude about $\frac{2}{3}$ and $\frac{4}{6}$?

S: They are both equivalent to $\frac{8}{12}$.

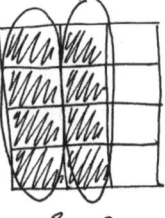
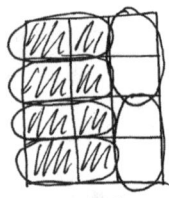

Lesson 10: Use the area model and division to show the equivalence of two fractions.

T: What is true about dividing the numerator and denominator in $\frac{8}{12}$ by 2 or 4?

S: Two and 4 are both factors of 8 and 12. → The larger the factor used, the larger the new fractional units will be.

T: Interesting. Discuss what your classmate said. "The larger the factor, the larger the new fractional units."

S: When we divided by 2, we got sixths, and when we divided by 4, we got thirds. Thirds are larger. Four is larger than 2. A larger factor gave a larger unit. → When the factor is larger, it means we can make fewer units but larger ones.

Problem 4: Simplify a fraction using the largest possible common factor.

T: Discuss with your partner how to rename $\frac{8}{12}$ with the largest units possible without using an area model.

S: Figure out the greatest number of units that can be placed in equal groups. → Divide the numerator and denominator by the same number, just like we've been doing. → Find a factor of both 8 and 12, and use it to divide the numerator and the denominator.

T: Express the equivalence using a division number sentence.

S: $\frac{8}{12} = \frac{8 \div 2}{12 \div 2} = \frac{4}{6}$. Four and 6 are still both even, so that wasn't the largest factor. → $\frac{8}{12} = \frac{8 \div 4}{12 \div 4} = \frac{2}{3}$. The only common factor 2 and 3 have is 1, so 4 must be the largest factor that 8 and 12 have in common.

T: How can we know we expressed an equivalent fraction with the largest units?

S: When we make equal groups, we need to see if we can make larger ones. → When we find the factors of the numerator and denominator, we have to pick the largest factor. Four is larger than 2, so dividing the numerator and denominator by 4 gets us the largest units. → When I found $\frac{4}{6}$, I realized 2 and 4 are both even, so I divided the numerator and denominator again by 2. Two and 3 only have a common factor of 1, so I knew I made the largest unit possible. → Dividing by 2 twice is the same as dividing by 4. Just get it over with faster, and divide by 4.

T: It's not wrong to say that $\frac{8}{12} = \frac{4}{6}$. It is true. It's just that, at times, it really is simpler to work with larger units because it means the denominator is a smaller number.

Problem Set (10 minutes)

Students should do their personal best to complete the Problem Set within the allotted 10 minutes. For some classes, it may be appropriate to modify the assignment by specifying which problems they work on first. Some problems do not specify a method for solving. Students should solve these problems using the RDW approach used for Application Problems.

Lesson 10: Use the area model and division to show the equivalence of two fractions.

A STORY OF UNITS

Lesson 10   4•5

Student Debrief (10 minutes)

Lesson Objective: Use the area model and division to show the equivalence of two fractions.

The Student Debrief is intended to invite reflection and active processing of the total lesson experience.

Invite students to review their solutions for the Problem Set. They should check work by comparing answers with a partner before going over answers as a class. Look for misconceptions or misunderstandings that can be addressed in the Debrief. Guide students in a conversation to debrief the Problem Set and process the lesson.

Any combination of the questions below may be used to lead the discussion.

- In Problem 2(b), did you compose the same units as your partner? Are both of your answers correct? Why?
- In Problem 4(a–d), how is it helpful to know the common factors for the numerators and denominators?
- In Problem 4, you were asked to use the largest common factor to rename the fraction: $\frac{4}{8} = \frac{1}{2}$. By doing so, you renamed $\frac{4}{8}$ using larger units. How is renaming fractions useful?
- Do fractions always need to be renamed to the largest unit? Explain.
- Why is it important to choose a common factor to make larger units?
- How can you tell that a fraction is composed of the largest possible fractional units?
- When you are drawing an area model and circling equal groups, do all of the groups have to appear the same in shape? How do you know that they still show the same amount?
- Explain how knowing the factors of the numerator and the factors of the denominator can be helpful in identifying equivalent fractions of a larger unit size.

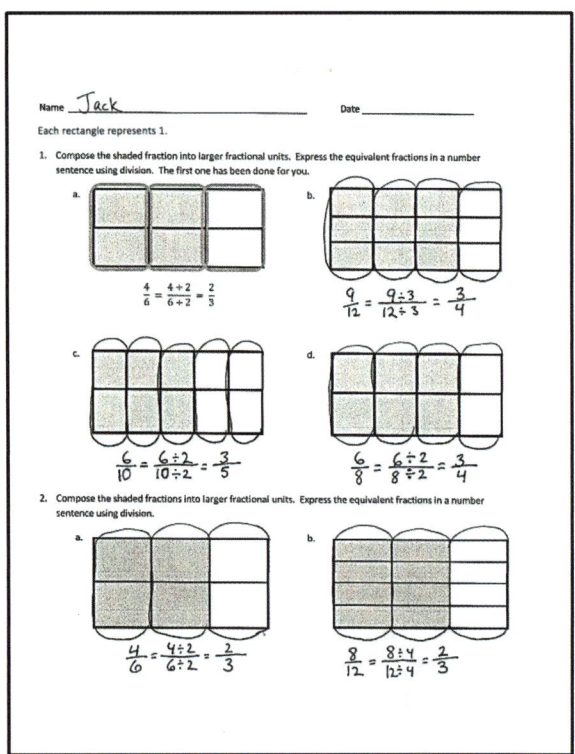

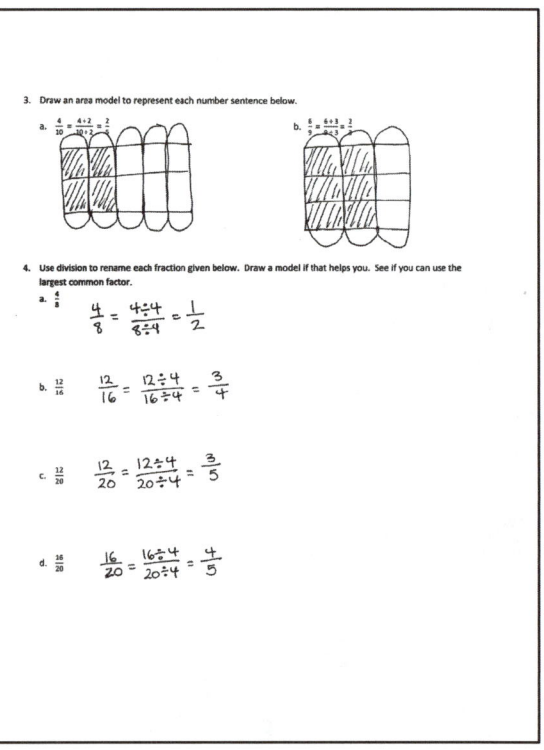

134   Lesson 10:   Use the area model and division to show the equivalence of two fractions.

Exit Ticket (3 minutes)

After the Student Debrief, instruct students to complete the Exit Ticket. A review of their work will help with assessing students' understanding of the concepts that were presented in today's lesson and planning more effectively for future lessons. The questions may be read aloud to the students.

Lesson 10: Use the area model and division to show the equivalence of two fractions.

A STORY OF UNITS   Lesson 10 Problem Set   4•5

Name _____   Date _____

Each rectangle represents 1.

1. Compose the shaded fraction into larger fractional units. Express the equivalent fractions in a number sentence using division. The first one has been done for you.

a.

$$\frac{4}{6} = \frac{4 \div 2}{6 \div 2} = \frac{2}{3}$$

b.

c.

d.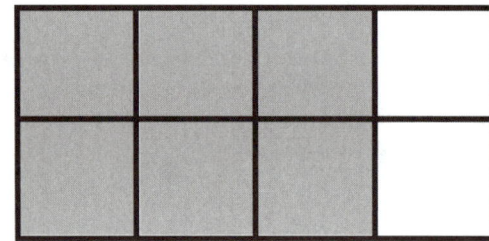

2. Compose the shaded fractions into larger fractional units. Express the equivalent fractions in a number sentence using division.

   a.

   b.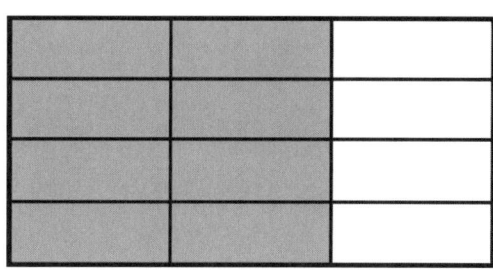

3. Draw an area model to represent each number sentence below.

   a. $\dfrac{4}{10} = \dfrac{4 \div 2}{10 \div 2} = \dfrac{2}{5}$

   b. $\dfrac{6}{9} = \dfrac{6 \div 3}{9 \div 3} = \dfrac{2}{3}$

4. Use division to rename each fraction given below. Draw a model if that helps you. See if you can use the largest common factor.

   a. $\dfrac{4}{8}$

   b. $\dfrac{12}{16}$

   c. $\dfrac{12}{20}$

   d. $\dfrac{16}{20}$

A STORY OF UNITS

Lesson 10 Exit Ticket 4•5

Name _____ Date _____

Draw an area model to show why the fractions are equivalent. Show the equivalence in a number sentence using division.

$$\frac{4}{10} = \frac{2}{5}$$

Lesson 10: Use the area model and division to show the equivalence of two fractions.

A STORY OF UNITS

Lesson 10 Homework 4•5

Name _____ Date _____

Each rectangle represents 1.

1. Compose the shaded fraction into larger fractional units. Express the equivalent fractions in a number sentence using division. The first one has been done for you.

a.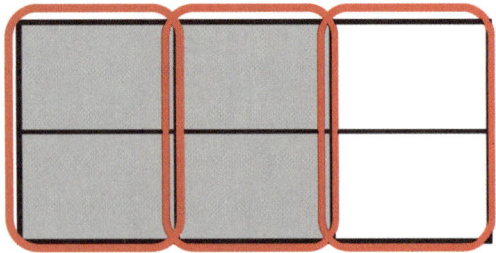

$$\frac{4}{6} = \frac{4 \div 2}{6 \div 2} = \frac{2}{3}$$

b.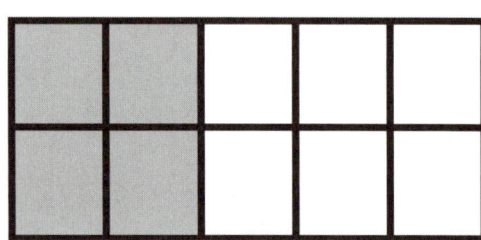

c.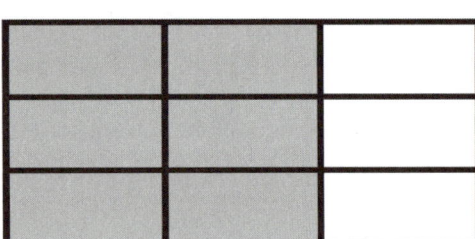

d.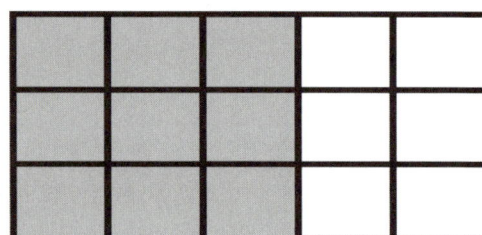

Lesson 10: Use the area model and division to show the equivalence of two fractions.

2. Compose the shaded fractions into larger fractional units. Express the equivalent fractions in a number sentence using division.

a.  b.

3. Draw an area model to represent each number sentence below.

   a. $\dfrac{6}{15} = \dfrac{6 \div 3}{15 \div 3} = \dfrac{2}{5}$

   b. $\dfrac{6}{18} = \dfrac{6 \div 3}{18 \div 3} = \dfrac{2}{6}$

4. Use division to rename each fraction given below. Draw a model if that helps you. See if you can use the largest common factor.

   a. $\frac{6}{12}$

   b. $\frac{4}{12}$

   c. $\frac{8}{12}$

   d. $\frac{12}{18}$

Lesson 11

Objective: Explain fraction equivalence using a tape diagram and the number line, and relate that to the use of multiplication and division.

Suggested Lesson Structure

■ Fluency Practice   (12 minutes)
■ Application Problem (5 minutes)
■ Concept Development (33 minutes)
■ Student Debrief   (10 minutes)
   Total Time   **(60 minutes)**

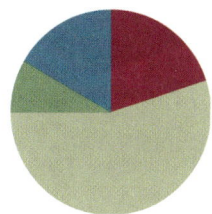

Fluency Practice (12 minutes)

- Find the Quotient and Remainder **4.NBT.6** (4 minutes)
- Find Equivalent Fractions **4.NF.1** (4 minutes)
- Draw Equivalent Fractions **4.NF.1** (4 minutes)

Find the Quotient and Remainder (4 minutes)

Materials: (S) Personal white board

Note: This fluency activity reviews Module 3 Lesson 28.

   T: (Write 6,765 ÷ 2.) On your personal white board, find the quotient and remainder.
   S: (Solve for and write 3,382 R1.)

Continue with the following possible sequence: 6,811 ÷ 5, 1,265 ÷ 4, and 1,736 ÷ 4.

Find Equivalent Fractions (4 minutes)

Materials: (S) Personal white board

Note: This fluency activity reviews Lesson 9.

   T: ($\frac{2}{10} = \frac{\div}{\div} = \frac{}{5}$. Point to $\frac{2}{10}$.) Say the fraction.
   S: 2 tenths.
   T: On your personal white board, fill in the unknown numbers to find the equivalent fraction.
   S: (Write $\frac{2}{10} = \frac{2 \div 2}{10 \div 2} = \frac{1}{5}$.)

Continue with the following possible sequence: $\frac{2}{4}, \frac{5}{10}, \frac{3}{6}$, and $\frac{4}{12}$.

Draw Equivalent Fractions (4 minutes)

Materials: (S) Personal white board

Note: This fluency activity reviews Lesson 10.

T: (Project a model with 4 out of 10 equal units shaded.) Draw the model, and write the fraction that is shaded.

S: (Draw a model with 4 out of 10 equal units shaded. Write $\frac{4}{10}$.)

T: (Write $\frac{4}{10} = \frac{\div}{\div} = \frac{}{}$.) Compose the shaded units into larger units by circling. Then, complete the number sentence.

S: (Circle the shaded units into 1 larger unit. Write $\frac{4}{10} = \frac{4 \div 2}{10 \div 2} = \frac{2}{5}$.)

Continue with the following possible sequence: $\frac{4}{6}, \frac{6}{9}, \frac{8}{10}$, and $\frac{9}{12}$.

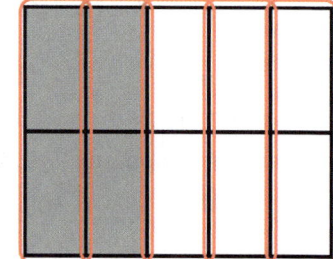

Application Problem (5 minutes)

Kelly was baking bread but could only find her $\frac{1}{8}$-cup measuring cup. She needs $\frac{1}{4}$ cup sugar, $\frac{3}{4}$ cup whole wheat flour, and $\frac{1}{2}$ cup all-purpose flour. How many $\frac{1}{8}$ cups will she need for each ingredient?

Solution 1

$\frac{1}{4}$ cup = $\frac{1 \times 2}{4 \times 2}$ = $\frac{2}{8}$ cup sugar

$\frac{3}{4}$ cup = $\frac{3 \times 2}{4 \times 2}$ = $\frac{6}{8}$ cup whole wheat flour

$\frac{1}{2}$ cup = $\frac{1 \times 4}{2 \times 4}$ = $\frac{4}{8}$ cup all purpose flour

Solution 2

Kelly needs 4 for the flour, 6 for the whole wheat, and 2 for the sugar.

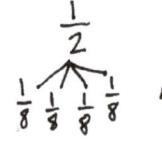 $4 \times \frac{1}{8} = \frac{4}{8}$

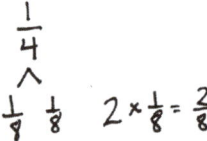 $2 \times \frac{1}{8} = \frac{2}{8}$

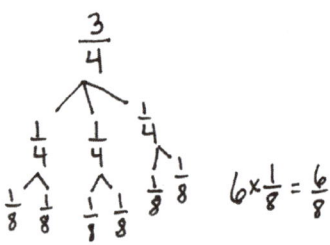 $6 \times \frac{1}{8} = \frac{6}{8}$

Note: This Application Problem places equivalent fractions into a context that may be familiar to students. Multiple solution strategies are possible. The first solution models the equivalency learned in Lessons 7 and 8. The second solution uses number bonds to find unit fractions, reviewing Topic A content.

A STORY OF UNITS

Lesson 11   4•5

Concept Development (33 minutes)

Materials: (S) Personal white board, ruler

Problem 1: Use a tape diagram and number line to find equivalent fractions for halves, fourths, and eighths.

T: Draw a tape diagram to show 1 partitioned into halves.

S: (Draw a tape diagram.)

T: Shade $\frac{1}{2}$. Now, decompose halves to make fourths. How many fourths are shaded?

S: 2 fourths.

T: On your personal white board, write what we did as a multiplication number sentence.

S: (Write $\frac{1}{2} = \frac{1 \times 2}{2 \times 2} = \frac{2}{4}$.)

T: Decompose fourths to make eighths. How many eighths are shaded?

S: 4 eighths.

T: Write a multiplication number sentence to show that 2 fourths and 4 eighths are equal.

S: (Write $\frac{2}{4} = \frac{2 \times 2}{4 \times 2} = \frac{4}{8}$.)

T: Now, use a ruler to draw a number line slightly longer than the tape diagram. Label points 0 and 1 so that they align with the ends of the tape diagram.

S: (Draw a number line.)

T: Label $\frac{1}{2}$ on the number line. Decompose the number line into fourths. What is equivalent to $\frac{2}{4}$ on the number line?

S: $\frac{1}{2} = \frac{2}{4}$. We showed that on the tape diagram.

T: Decompose the number line into eighths.

S: (Label the eighths.)

T: What is $\frac{4}{8}$ equal to on the number line?

S: $\frac{1}{2} = \frac{4}{8}$. → $\frac{2}{4} = \frac{4}{8}$. → That also means $\frac{1}{2} = \frac{2}{4} = \frac{4}{8}$.

T: Explain what happened on the number line as you decomposed the half.

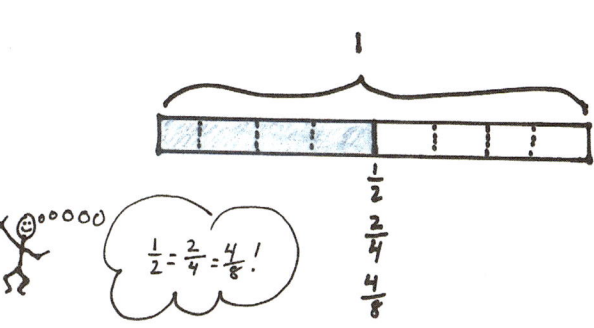

> **NOTES ON MULTIPLE MEANS OF REPRESENTATION:**
>
> To preserve the pace of the lesson, provide a tape diagram and number line Template (see Lesson 12) for some learners. Students may also choose to transform the tape diagram into a number line by erasing the top line, labeling points, and extending the endpoints.

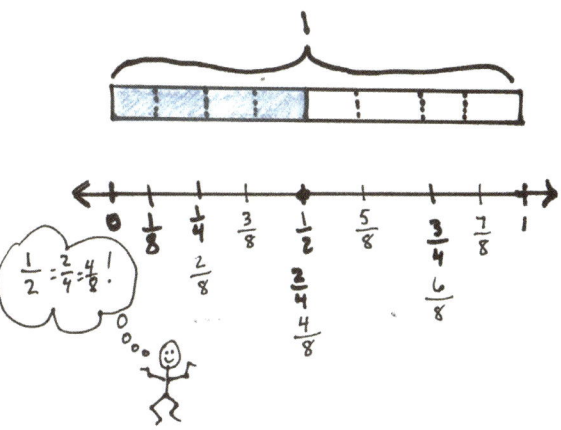

Lesson 11: Explain fraction equivalence using a tape diagram and the number line, and relate that to the use of multiplication and division.

145

A STORY OF UNITS   Lesson 11   4•5

S: When we decomposed the half into fourths, it was like sharing a licorice strip with four people instead of two. → We got 4 smaller parts instead of 2 larger parts. → There are 4 smaller segments in the whole instead of 2 larger segments. → We doubled the number of parts but made smaller parts, just like with the area model. → It made 2 lengths that were the same length as 1 half.

Problem 2: Use a number line, multiplication, and division to decompose and compose fractions.

T: Partition a number line into thirds. Decompose 1 third into 4 equal parts.

T: Write a number sentence using multiplication to show what fraction is equivalent to 1 third on this number line.

S: (Write $\frac{1}{3} = \frac{1 \times 4}{3 \times 4} = \frac{4}{12}$.)

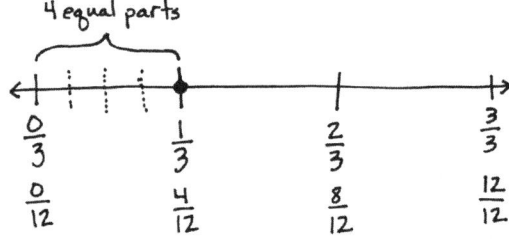

T: Explain to your partner why that is true.

S: It's just like the area model. We made more smaller units, but the lengths stayed the same instead of the area staying the same. → If we multiply a numerator and a denominator by the same number, we find an equivalent fraction. → 1 third was decomposed into fourths, so we multiplied the number of units in the whole and the number of selected units by 4.

T: Write the equivalence as a number sentence using division.

S: (Write $\frac{4}{12} = \frac{4 \div 4}{12 \div 4} = \frac{1}{3}$.)

T: Explain to your partner why that is true.

S: We can join four smaller segments to make one longer one that is the same as 1 third. → We can group the twelfths together to make thirds. → Four copies of $\frac{1}{12}$ equals $\frac{1}{3}$. → Just like the area model, we are composing units to make a larger unit.

Problem 3: Decompose a non-unit fraction using a number line and division.

T: Draw a number line. Partition it into fifths, label it, and locate $\frac{2}{5}$.

S: (Draw.)

T: Decompose $\frac{2}{5}$ into 6 equal parts. First, discuss your strategy with your partner.

S: I will make each fifth into 6 parts. → No. We have to decompose 2 units, not 1 unit. Each unit will be decomposed into 3 equal parts. → Two units are becoming 6 units. We are multiplying the numerator and denominator by 3.

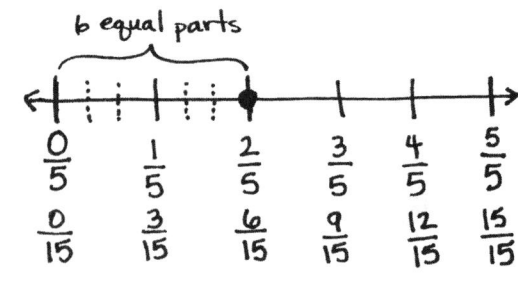

T: Write a number sentence to express the equivalent fractions.

S: (Write $\frac{2}{5} = \frac{2 \times 3}{5 \times 3} = \frac{6}{15}$.)

Lesson 11: Explain fraction equivalence using a tape diagram and the number line, and relate that to the use of multiplication and division.

A STORY OF UNITS

Lesson 11 4•5

Problem Set (10 minutes)

Students should do their personal best to complete the Problem Set within the allotted 10 minutes. For some classes, it may be appropriate to modify the assignment by specifying which problems they work on first. Some problems do not specify a method for solving. Students should solve these problems using the RDW approach used for Application Problems.

Student Debrief (10 minutes)

Lesson Objective: Explain fraction equivalence using a tape diagram and the number line, and relate that to the use of multiplication and division.

The Student Debrief is intended to invite reflection and active processing of the total lesson experience.

Invite students to review their solutions for the Problem Set. They should check work by comparing answers with a partner before going over answers as a class. Look for misconceptions or misunderstandings that can be addressed in the Debrief. Guide students in a conversation to debrief the Problem Set and process the lesson.

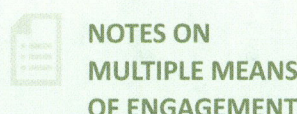

NOTES ON MULTIPLE MEANS OF ENGAGEMENT:

Challenge students working above grade level and others to discuss or journal about the three models used for finding equivalent fractions. Ask, "How do the tape diagram and number line relate to one another? When might you choose to use a number line rather than an area model? Why?"

Any combination of the questions below may be used to lead the discussion.

- In Problem 1, compare the distance from 0 to each point on the number line you circled. What do you notice?
- In Problem 1, does the unshaded portion of the tape diagram represent the same length from the point to 1 on every number line? How do you know?
- Compare your number sentences in Problem 2. Could they be rewritten using division?
- In Problem 5, what new units were created when 2 fifths was decomposed into 4 equal parts?
- How is modeling with a number line similar to modeling with an area model? How is it different?
- In Grade 3, you found equivalent fractions by locating them on a number line. Do you now require a number line to find equivalent fractions? What other ways can you determine equivalent fractions?

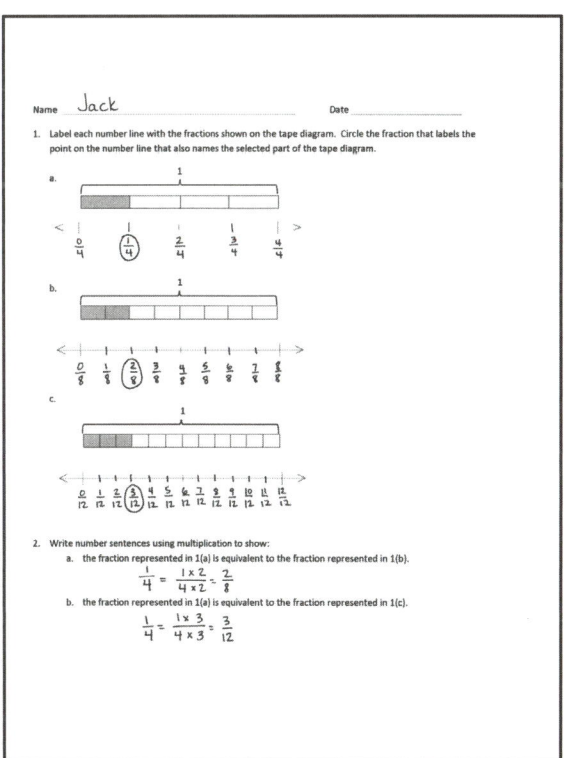

Lesson 11: Explain fraction equivalence using a tape diagram and the number line, and relate that to the use of multiplication and division.

147

A STORY OF UNITS

Lesson 11 4•5

Exit Ticket (3 minutes)

After the Student Debrief, instruct students to complete the Exit Ticket. A review of their work will help with assessing students' understanding of the concepts that were presented in today's lesson and planning more effectively for future lessons. The questions may be read aloud to the students.

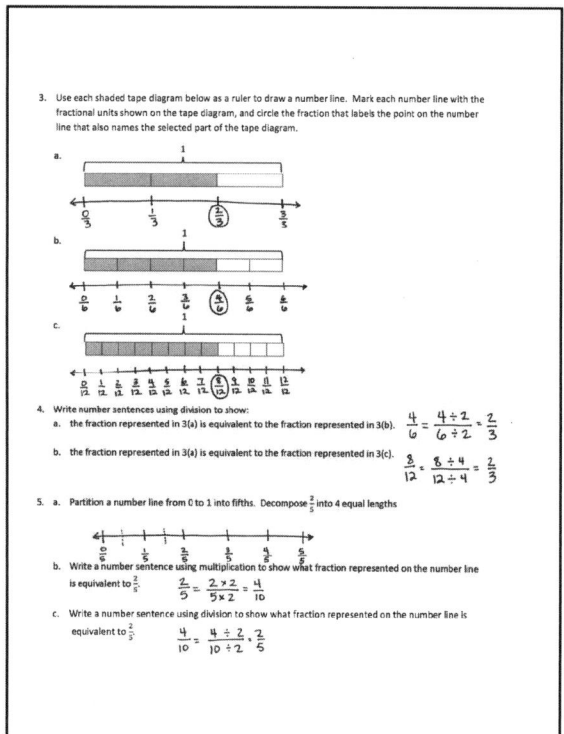

Lesson 11: Explain fraction equivalence using a tape diagram and the number line, and relate that to the use of multiplication and division.

A STORY OF UNITS   Lesson 11 Problem Set 4•5

Name _____   Date _____

1. Label each number line with the fractions shown on the tape diagram. Circle the fraction that labels the point on the number line that also names the shaded part of the tape diagram.

   a.
   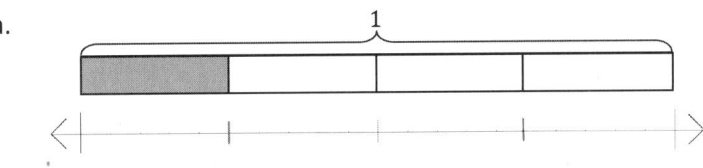

   b.
   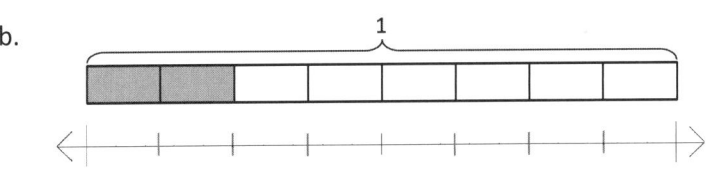

   c.
   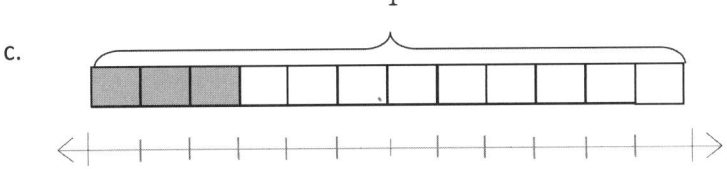

Lesson 11: Explain fraction equivalence using a tape diagram and the number line, and relate that to the use of multiplication and division.

2. Write number sentences using multiplication to show:

   a. The fraction represented in 1(a) is equivalent to the fraction represented in 1(b).

   b. The fraction represented in 1(a) is equivalent to the fraction represented in 1(c).

3. Use each shaded tape diagram below as a ruler to draw a number line. Mark each number line with the fractional units shown on the tape diagram, and circle the fraction that labels the point on the number line that also names the shaded part of the tape diagram.

   a.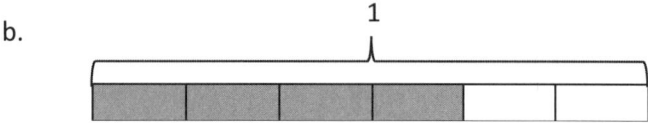

   b.

   c.

4. Write number sentences using division to show:

   a. The fraction represented in 3(a) is equivalent to the fraction represented in 3(b).

   b. The fraction represented in 3(a) is equivalent to the fraction represented in 3(c).

5. a. Partition a number line from 0 to 1 into fifths. Decompose $\frac{2}{5}$ into 4 equal lengths.

   b. Write a number sentence using multiplication to show what fraction represented on the number line is equivalent to $\frac{2}{5}$.

   c. Write a number sentence using division to show what fraction represented on the number line is equivalent to $\frac{2}{5}$.

Lesson 11: Explain fraction equivalence using a tape diagram and the number line, and relate that to the use of multiplication and division.

151

Name _____ Date _____

1. Partition a number line from 0 to 1 into sixths. Decompose $\frac{2}{6}$ into 4 equal lengths.

2. Write a number sentence using multiplication to show what fraction represented on the number line is equivalent to $\frac{2}{6}$.

3. Write a number sentence using division to show what fraction represented on the number line is equivalent to $\frac{2}{6}$.

A STORY OF UNITS   Lesson 11 Homework 4•5

Name _____   Date _____

1. Label each number line with the fractions shown on the tape diagram. Circle the fraction that labels the point on the number line that also names the shaded part of the tape diagram.

   a.

   b.

   c.

Lesson 11: Explain fraction equivalence using a tape diagram and the number line, and relate that to the use of multiplication and division.

2. Write number sentences using multiplication to show:

   a. The fraction represented in 1(a) is equivalent to the fraction represented in 1(b).

   b. The fraction represented in 1(a) is equivalent to the fraction represented in 1(c).

3. Use each shaded tape diagram below as a ruler to draw a number line. Mark each number line with the fractional units shown on the tape diagram, and circle the fraction that labels the point on the number line that also names the shaded part of the tape diagram.

   a.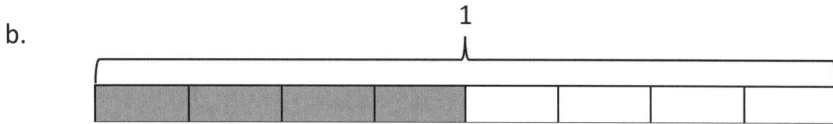

   b.

   c.

4. Write a number sentence using division to show the fraction represented in 3(a) is equivalent to the fraction represented in 3(b).

5. a. Partition a number line from 0 to 1 into fourths. Decompose $\frac{3}{4}$ into 6 equal lengths.

   b. Write a number sentence using multiplication to show what fraction represented on the number line is equivalent to $\frac{3}{4}$.

   c. Write a number sentence using division to show what fraction represented on the number line is equivalent to $\frac{3}{4}$.

A STORY OF UNITS

4
GRADE

Mathematics Curriculum

GRADE 4 • MODULE 5

Topic C
Fraction Comparison

4.NF.2

Focus Standard:	4.NF.2	Compare two fractions with different numerators and different denominators, e.g., by creating common denominators or numerators, or by comparing to a benchmark fraction such as 1/2. Recognize that comparisons are valid only when the two fractions refer to the same whole. Record the results of comparisons with symbols >, =, or <, and justify the conclusions, e.g., by using a visual fraction model.
Instructional Days:	4	
Coherence -Links from:	G3–M5	Fractions as Numbers on the Number Line
-Links to:	G5–M3	Addition and Subtraction of Fractions

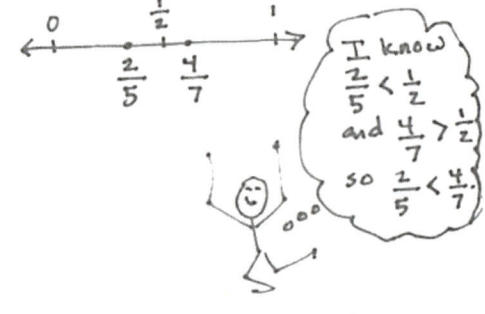

In Topic C, students use benchmarks and common units to compare fractions with different numerators and different denominators. The use of benchmarks is the focus of Lessons 12 and 13 and is modeled using a number line. Students use the relationship between the numerator and denominator of a fraction to compare to a known benchmark (e.g., 0, $\frac{1}{2}$, or 1) and then use that information to compare the given fractions. For example, when comparing $\frac{4}{7}$ and $\frac{2}{5}$, students reason that 4 sevenths is more than 1 half, while 2 fifths is less than 1 half. They then conclude that 4 sevenths is greater than 2 fifths.

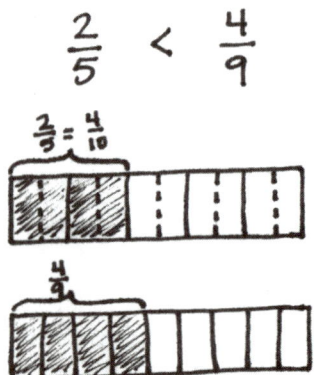

In Lesson 14, students reason that they can also use like numerators based on what they know about the size of the fractional units. They begin at a simple level by reasoning, for example, that 3 fifths is less than 3 fourths because fifths are smaller than fourths. They then see, too, that it is easy to make like numerators at times to compare, e.g., $\frac{2}{5} < \frac{4}{9}$ because $\frac{2}{5} = \frac{4}{10}$, and $\frac{4}{10} < \frac{4}{9}$ because $\frac{1}{10} < \frac{1}{9}$. Using their experience with fractions in Grade 3, they know the larger the denominator of a unit fraction, the smaller the size of the fractional unit.

156   Topic C: Fraction Comparison

EUREKA MATH

A STORY OF UNITS

Topic C 4•5

Like numerators are modeled using tape diagrams directly above each other, where one fractional unit is partitioned into smaller unit fractions. The lesson then moves to comparing fractions with related denominators, such as $\frac{2}{3}$ and $\frac{5}{6}$, wherein one denominator is a factor of the other, using both tape diagrams and the number line. In Lesson 15, students compare fractions by using an area model to express two fractions, wherein one denominator is not a factor of the other, in terms of the same unit using multiplication, e.g., $\frac{2}{3} < \frac{3}{4}$ because $\frac{2}{3} = \frac{2 \times 4}{3 \times 4} = \frac{8}{12}$ and $\frac{3}{4} = \frac{3 \times 3}{4 \times 3} = \frac{9}{12}$ and $\frac{8}{12} < \frac{9}{12}$. The area for $\frac{2}{3}$ is partitioned vertically, and the area for $\frac{3}{4}$ is partitioned horizontally.

To find the equivalent fraction and create the same size units, the areas are decomposed horizontally and vertically, respectively. Now the unit fractions are the same in each model or equation, and students can easily compare. The topic culminates with students comparing pairs of fractions and, by doing so, deciding which strategy is either necessary or efficient: reasoning using benchmarks and what they know about units, drawing a model (such as a number line, a tape diagram, or an area model), or the general method of finding like denominators through multiplication.

$$\frac{2}{3} < \frac{3}{4}$$

$$\frac{2}{3} = \frac{8}{12} \qquad \frac{3}{4} = \frac{9}{12}$$

A Teaching Sequence Toward Mastery of Fraction Comparison
Objective 1: Reason using benchmarks to compare two fractions on the number line. (Lessons 12–13)
Objective 2: Find common units or number of units to compare two fractions. (Lessons 14–15)

Topic C: Fraction Comparison

Lesson 12

Objective: Reason using benchmarks to compare two fractions on the number line.

Suggested Lesson Structure

- ■ Fluency Practice (12 minutes)
- ■ Application Problem (8 minutes)
- ■ Concept Development (30 minutes)
- ■ Student Debrief (10 minutes)

Total Time **(60 minutes)**

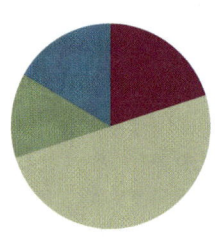

Fluency Practice (12 minutes)

- Add and Subtract **4.NBT.4** (4 minutes)
- Find Equivalent Fractions **4.NF.1** (4 minutes)
- Construct a Number Line with Fractions **4.NF.3** (4 minutes)

Add and Subtract (4 minutes)

Materials: (S) Personal white board

Note: This fluency activity reviews adding and subtracting using the standard algorithm.

- T: (Write 547 thousands 686 ones.) On your personal white board, write this number in standard form.
- S: (Write 547,686.)
- T: (Write 294 thousands 453 ones.) Add this number to 547,686 using the standard algorithm.
- S: (Write 547,686 + 294,453 = 842,139 using the standard algorithm.)

Continue the process with 645,838 + 284,567.

- T: (Write 800 thousands.) On your board, write this number in standard form.
- S: (Write 800,000.)
- T: (Write 648 thousands 745 ones.) Subtract this number from 800,000 using the standard algorithm.
- S: (Write 800,000 − 648,745 = 151,255 using the standard algorithm.)

Continue the process with 754,912 − 154,189.

A STORY OF UNITS   Lesson 12   4•5

Find Equivalent Fractions (4 minutes)

Materials: (S) Personal white board

Note: This fluency activity reviews Lesson 9.

T: (Write $\frac{6}{8} = \frac{÷}{÷} = \frac{6}{4}$. Point to $\frac{6}{8}$.) Say the fraction.
S: 6 eighths.
T: On your personal white board, complete the number sentence to find the equivalent fraction.
S: (Write $\frac{6}{8} = \frac{6 ÷ 2}{8 ÷ 2} = \frac{3}{4}$.)

Continue with the following possible sequence: $\frac{4}{6} = \frac{2}{3}$, $\frac{4}{10} = \frac{2}{5}$, $\frac{8}{10} = \frac{4}{5}$, $\frac{8}{12} = \frac{2}{3}$, and $\frac{9}{12} = \frac{3}{4}$.

Construct a Number Line with Fractions (4 minutes)

Materials: (S) Personal white board

Note: This fluency activity reviews Lesson 11.

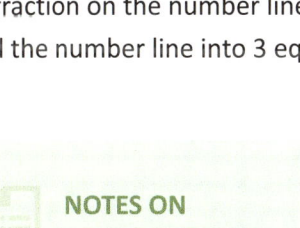

T: (Write $\frac{2}{3}$.) Say the fraction.
S: 2 thirds.
T: On your personal white board, draw a tape diagram. Label the whole diagram 1, and then shade in units to show $\frac{2}{3}$.
S: (Draw a tape diagram partitioned into 3 equal units. Write 1 at the top. Shade 2 units.)
T: Beneath your tape diagram, draw a number line. Then, label each fraction on the number line.
S: (Beneath the tape diagram, draw a number line. Partition and label the number line into 3 equal intervals.)

Continue with the following possible sequence: $\frac{2}{5}, \frac{3}{4}, \frac{3}{6}$, and $\frac{6}{9}$.

Application Problem (8 minutes)

Materials: (S) Number line (Template)

Plot $\frac{1}{4}, \frac{4}{5}$, and $\frac{5}{8}$ on a number line, and compare the three points.

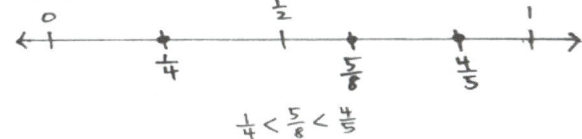

NOTES ON MULTIPLE MEANS OF ACTION AND EXPRESSION:

If students find various fractional units on one number line frustrating, give them the option of plotting $\frac{1}{4}, \frac{4}{5}, \frac{5}{8}$ on two number lines placed parallel for comparison.

Note: This Application Problem reviews equivalent fractions and bridges to today's lesson, in which students use reasoning and benchmarks to compare fractions.

Lesson 12:   Reason using benchmarks to compare two fractions on the number line.   159

A STORY OF UNITS   Lesson 12   4•5

Concept Development (30 minutes)

Materials: (S) Personal white board, number line (Template)

Problem 1: Reason about the size of a fraction compared to $\frac{1}{2}$.

T: How many sixths equal 1 whole? Say the unit.
S: 6 sixths.
T: How many sixths equal 1 half?
S: 3 sixths. → $\frac{3}{6} = \frac{1}{2}$. We already know that!
T: Is $\frac{2}{6}$ greater than or less than $\frac{3}{6}$?
S: Less than.
T: Is $\frac{2}{6}$ less than $\frac{1}{2}$ or greater than $\frac{1}{2}$?
S: Less than.

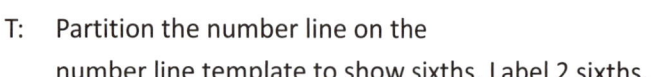

> **NOTES ON MULTIPLE MEANS OF REPRESENTATION:**
>
> Clarify for English language learners math language such as *greater than* and *less than*. Explain that *greater* has multiple meanings. Here, it means *larger* or *bigger*. Offer explanations in students' first language, if possible. If students are more comfortable, allow them to use *bigger*, *larger*, *smaller*, etc.

T: Partition the number line on the number line template to show sixths. Label 2 sixths.
T: Write a number sentence comparing 2 sixths and 1 half.
S: (Write $\frac{2}{6} < \frac{1}{2}$.)

Repeat the process with $\frac{5}{8}$.

T: (Write $\frac{2}{3}$.) Talk to your partner. Is $\frac{2}{3}$ greater than $\frac{1}{2}$ or less than $\frac{1}{2}$?
S: There are no thirds that are equal to $\frac{1}{2}$. → $\frac{1}{2}$ is between $\frac{1}{3}$ and $\frac{2}{3}$, so $\frac{2}{3}$ must be greater. → I can draw a model to prove that. → $\frac{2}{3}$ is almost 1. → 1 third is less than 1 half, and 2 thirds is greater than 1 half. → We can see on our other number line that $\frac{3}{6}$ is right between $\frac{1}{3}$ and $\frac{2}{3}$, and $\frac{2}{3}$ is equal to $\frac{4}{6}$. So, $\frac{2}{3}$ is greater than $\frac{1}{2}$.

T: (Write $\frac{2}{5}$.) Talk to your partner. Is $\frac{2}{5}$ greater than $\frac{1}{2}$ or less than $\frac{1}{2}$?
S: Five is an odd number, so it doesn't divide evenly by 2. Halfway between 0 fifths and 5 fifths should be somewhere between 2 fifths and 3 fifths. So, 2 fifths must be less than 1 half. → $\frac{1}{2}$ is halfway between $\frac{2}{5}$ and $\frac{3}{5}$, so $\frac{2}{5}$ is less than $\frac{1}{2}$.

Lesson 12: Reason using benchmarks to compare two fractions on the number line.

A STORY OF UNITS   Lesson 12   4•5

T: Model $\frac{2}{5}$ on a number line. Then, compare $\frac{2}{5}$ and $\frac{1}{2}$.

S: I can partition the fifths in half on the number line. That makes tenths. → $\frac{2}{5} = \frac{2 \times 2}{5 \times 2} = \frac{4}{10}$. → I know $\frac{5}{10}$ is the same as $\frac{1}{2}$.

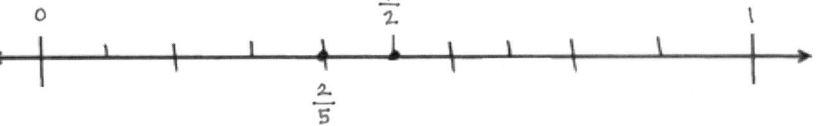

T: Write your conclusion on your board.

S: (Write $\frac{2}{5} < \frac{1}{2}$.)

Problem 2: Plot points on a number line by thinking about fractions in relation to 0, $\frac{1}{2}$, or 1. Compare the fractions.

T: (Write $\frac{5}{12}$.) What do we know about $\frac{5}{12}$ in relation to 0, $\frac{1}{2}$, and 1?

S: $\frac{5}{12}$ is greater than 0. → It's less than 1. → $\frac{5}{12}$ is about halfway between 0 and 1. → $\frac{5}{12}$ is less than $\frac{1}{2}$. I know because $\frac{6}{12}$ is equal to $\frac{1}{2}$, and $\frac{5}{12}$ is less than $\frac{6}{12}$.

T: Plot and label $\frac{5}{12}$ on a number line. Is $\frac{5}{12}$ closer to 0 or $\frac{1}{2}$?

S: It looks closer to $\frac{1}{2}$.

T: How close? Count the twelfths.

S: $\frac{5}{12}$ is just $\frac{1}{12}$ away from $\frac{6}{12}$. It's $\frac{5}{12}$ away from 0, so it's closer to $\frac{1}{2}$.

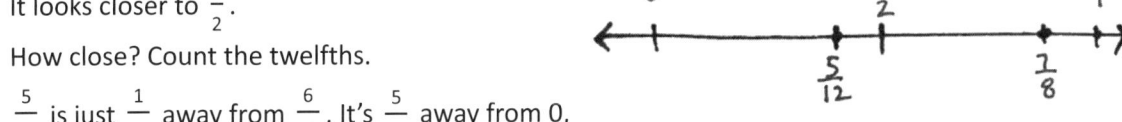

T: (Write $\frac{7}{8}$) What do we know about $\frac{7}{8}$ in relation to 0, $\frac{1}{2}$, and 1?

S: It's greater than 0. → It's less than 1. → It's greater than $\frac{1}{2}$. $\frac{4}{8}$ is equal to $\frac{1}{2}$, so $\frac{7}{8}$ is definitely more.

T: Discuss with your partner. Is $\frac{7}{8}$ closer to $\frac{1}{2}$ or to 1?

S: It is closer to 1, just 1 eighth away from 1. → $\frac{7}{8}$ is 3 eighths greater than $\frac{1}{2}$ and only $\frac{1}{8}$ less than 1.

T: Plot and label $\frac{7}{8}$ on the same number line as you labeled $\frac{5}{12}$. Write a number sentence comparing $\frac{7}{8}$ and $\frac{5}{12}$.

S: (Write $\frac{7}{8} > \frac{5}{12}$. → $\frac{5}{12} < \frac{7}{8}$.)

T: (Write $\frac{2}{6}$.) Here is a challenge! Plot $\frac{2}{6}$ on the same number line. Discuss with your partner the relationship $\frac{2}{6}$ has to the other points on the number line. Consider the size of each unit.

S: $\frac{2}{6}$ is really close to $\frac{5}{12}$. → I know $\frac{2}{6}$ is less than $\frac{7}{8}$ and less than $\frac{1}{2}$. → $\frac{2}{6}$ is 1 sixth away from 1 half, and $\frac{5}{12}$ is 1 twelfth away from 1 half. → So, if sixths are larger units than twelfths, then $\frac{2}{6}$ is farther away from $\frac{1}{2}$ than $\frac{5}{12}$ is. → I know that 2 sixths is equal to 4 twelfths, so 5 twelfths is greater.

Lesson 12: Reason using benchmarks to compare two fractions on the number line.   161

© 2015 Great Minds. eureka-math.org
G4-M5-TE-B4-1.3.1-01.2016

A STORY OF UNITS

Lesson 12 4•5

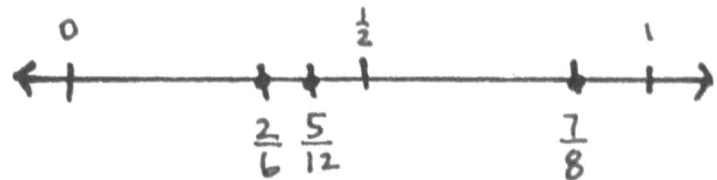

T: Excellent thinking. We can compare the distance of a point from $\frac{1}{2}$ based on the size of the fractional units. We can use these important locations on the number line as **benchmarks** to help us compare fractions.

Problem 3: Use the benchmarks 0, $\frac{1}{2}$, and 1 to compare two fractions without using a number line.

T: Talk to your partner. Compare $\frac{5}{8}$ and $\frac{4}{5}$. Consider the relationship $\frac{5}{8}$ has to 0, $\frac{1}{2}$, and 1.

S: $\frac{5}{12}$ is greater than $\frac{1}{2}$ since $\frac{1}{2} = \frac{4}{8}$. → It's close to $\frac{1}{2}$ since it's only a little more than $\frac{4}{8}$. → $\frac{5}{8}$ is $\frac{1}{8}$ more than $\frac{1}{2}$ but $\frac{3}{8}$ from 1.

T: What about $\frac{4}{5}$?

S: $\frac{4}{5}$ is greater than $\frac{1}{2}$. → It's close to 1. → It's only $\frac{1}{5}$ away. → If you have 4 fifths of something, you have most of it.

T: What can we conclude about $\frac{5}{8}$ and $\frac{4}{5}$? Think about the size of the units.

S: Eighths are smaller than fifths, so $\frac{5}{8}$ is closer to $\frac{1}{2}$ than $\frac{4}{5}$ is. → $\frac{5}{8}$ is less than $\frac{4}{5}$. → 5 eighths is a little more than half, but 4 fifths is a little less than 1.

T: Compare $\frac{2}{5}$ and $\frac{6}{10}$. Again, consider the relationship $\frac{2}{5}$ has to 0, $\frac{1}{2}$, and 1.

S: I know that $\frac{1}{2}$ is between $\frac{2}{5}$ and $\frac{3}{5}$, so $\frac{2}{5}$ is a little less than $\frac{1}{2}$. I know that $\frac{5}{10}$ is the same as $\frac{1}{2}$, so $\frac{6}{10}$ is greater than $\frac{1}{2}$. → $\frac{2}{5}$ is less than $\frac{6}{10}$.

T: Talk to your partner, and compare $\frac{33}{100}$ and $\frac{2}{3}$.

S: $\frac{50}{100}$ is equal to half, so $\frac{33}{100}$ is less than $\frac{1}{2}$. → $\frac{2}{3}$ is greater than $\frac{1}{2}$. → $\frac{33}{100}$ is less than $\frac{2}{3}$.

Problem Set (10 minutes)

Students should do their personal best to complete the Problem Set within the allotted 10 minutes. For some classes, it may be appropriate to modify the assignment by specifying which problems they work on first. Some problems do not specify a method for solving. Students should solve these problems using the RDW approach used for Application Problems.

A STORY OF UNITS   Lesson 12   4•5

Student Debrief (10 minutes)

Lesson Objective: Reason using benchmarks to compare two fractions on the number line.

The Student Debrief is intended to invite reflection and active processing of the total lesson experience.

Invite students to review their solutions for the Problem Set. They should check work by comparing answers with a partner before going over answers as a class. Look for misconceptions or misunderstandings that can be addressed in the Debrief. Guide students in a conversation to debrief the Problem Set and process the lesson.

Any combinations of the questions below may be used to lead the discussion.

- How was the number line helpful as we compared the fractions in Problem 1(b)?
- For Problem 3(a–j), explain how you used the benchmarks 0, $\frac{1}{2}$, and 1 to compare the fractions. When both fractions were greater than $\frac{1}{2}$, how did you know which one was greater?
- Will the strategy of using the benchmarks 0, $\frac{1}{2}$, and 1 always help us to compare two fractions? Explain.
- How did the Application Problem connect to today's lesson?

Exit Ticket (3 minutes)

After the Student Debrief, instruct students to complete the Exit Ticket. A review of their work will help with assessing students' understanding of the concepts that were presented in today's lesson and planning more effectively for future lessons. The questions may be read aloud to the students.

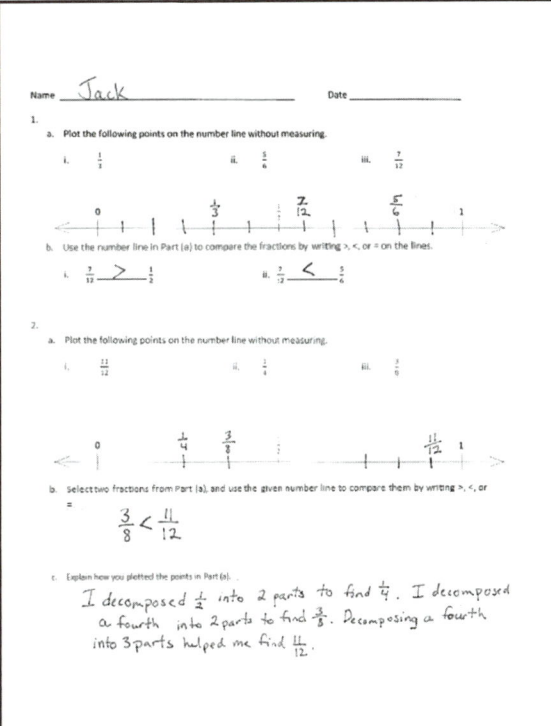

Lesson 12: Reason using benchmarks to compare two fractions on the number line.

163

A STORY OF UNITS   Lesson 12 Problem Set   4•5

Name _____   Date _____

1. a. Plot the following points on the number line without measuring.

   i. $\frac{1}{3}$ ii. $\frac{5}{6}$ iii. $\frac{7}{12}$

   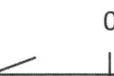

   b. Use the number line in Part (a) to compare the fractions by writing >, <, or = on the lines.

   i. $\frac{7}{12}$ _____ $\frac{1}{2}$ ii. $\frac{7}{12}$ _____ $\frac{5}{6}$

2. a. Plot the following points on the number line without measuring.

   i. $\frac{11}{12}$ ii. $\frac{1}{4}$ iii. $\frac{3}{8}$

   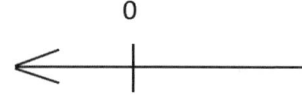 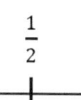

   b. Select two fractions from Part (a), and use the given number line to compare them by writing >, <, or =.

   c. Explain how you plotted the points in Part (a).

164   Lesson 12:   Reason using benchmarks to compare two fractions on the number line.

A STORY OF UNITS　　　　　　　　　　　　　　　　　　　　Lesson 12 Problem Set　4•5

3. Compare the fractions given below by writing > or < on the lines.
   Give a brief explanation for each answer referring to the benchmarks 0, $\frac{1}{2}$, and 1.

   a. $\frac{1}{2}$ _____ $\frac{3}{4}$ b. $\frac{1}{2}$ _____ $\frac{7}{8}$

   c. $\frac{2}{3}$ _____ $\frac{2}{5}$ d. $\frac{9}{10}$ _____ $\frac{3}{5}$

   e. $\frac{2}{3}$ _____ $\frac{7}{8}$ f. $\frac{1}{3}$ _____ $\frac{2}{4}$

   g. $\frac{2}{3}$ _____ $\frac{5}{10}$ h. $\frac{11}{12}$ _____ $\frac{2}{5}$

   i. $\frac{40}{100}$ _____ $\frac{51}{100}$ j. $\frac{7}{16}$ _____ $\frac{51}{100}$

Lesson 12: Reason using benchmarks to compare two fractions on the number line.

A STORY OF UNITS   Lesson 12 Exit Ticket   4•5

Name _____   Date _____

1. Plot the following points on the number line without measuring.

   a. $\frac{8}{10}$ b. $\frac{3}{5}$ c. $\frac{1}{4}$

   <-----|----------------|----------------|----->
         0               $\frac{1}{2}$ 1

2. Use the number line in Problem 1 to compare the fractions by writing >, <, or = on the lines.

   a. $\frac{1}{4}$ _____ $\frac{1}{2}$

   b. $\frac{8}{10}$ _____ $\frac{3}{5}$

   c. $\frac{1}{2}$ _____ $\frac{3}{5}$

   d. $\frac{1}{4}$ _____ $\frac{8}{10}$

Lesson 12: Reason using benchmarks to compare two fractions on the number line.

A STORY OF UNITS   Lesson 12 Homework 4•5

Name _____   Date _____

1. a. Plot the following points on the number line without measuring.

   i. $\frac{2}{3}$ ii. $\frac{1}{6}$ iii. $\frac{4}{10}$

   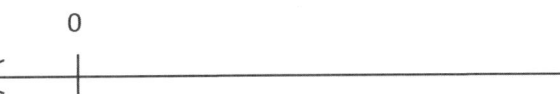

   b. Use the number line in Part (a) to compare the fractions by writing >, <, or = on the lines.

   i. $\frac{2}{3}$ _____ $\frac{1}{2}$ ii. $\frac{4}{10}$ _____ $\frac{1}{6}$

2. a. Plot the following points on the number line without measuring.

   i. $\frac{5}{12}$ ii. $\frac{3}{4}$ iii. $\frac{2}{6}$

   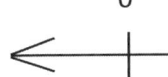

   b. Select two fractions from Part (a), and use the given number line to compare them by writing >, <, or =.

   c. Explain how you plotted the points in Part (a).

Lesson 12: Reason using benchmarks to compare two fractions on the number line.

A STORY OF UNITS   Lesson 12 Homework 4•5

3. Compare the fractions given below by writing > or < on the lines.
   Give a brief explanation for each answer referring to the benchmark of 0, $\frac{1}{2}$, and 1.

   a. $\frac{1}{2}$ _____ $\frac{1}{4}$ b. $\frac{6}{8}$ _____ $\frac{1}{2}$

   c. $\frac{3}{4}$ _____ $\frac{3}{5}$ d. $\frac{4}{6}$ _____ $\frac{9}{12}$

   e. $\frac{2}{3}$ _____ $\frac{1}{4}$ f. $\frac{4}{5}$ _____ $\frac{8}{12}$

   g. $\frac{1}{3}$ _____ $\frac{3}{6}$ h. $\frac{7}{8}$ _____ $\frac{3}{5}$

   i. $\frac{51}{100}$ _____ $\frac{5}{10}$ j. $\frac{8}{14}$ _____ $\frac{40}{100}$

Lesson 12: Reason using benchmarks to compare two fractions on the number line.

Application Problem

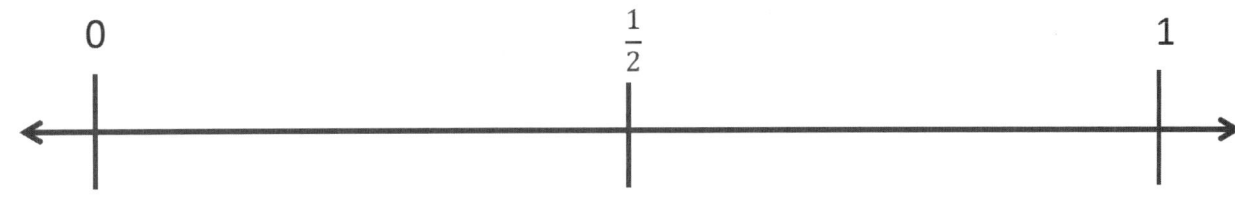

1.

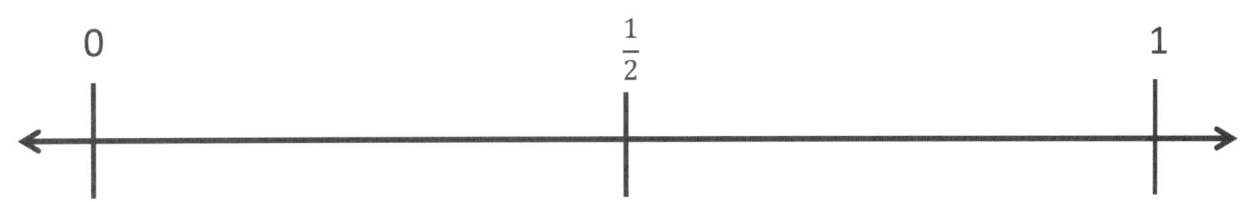

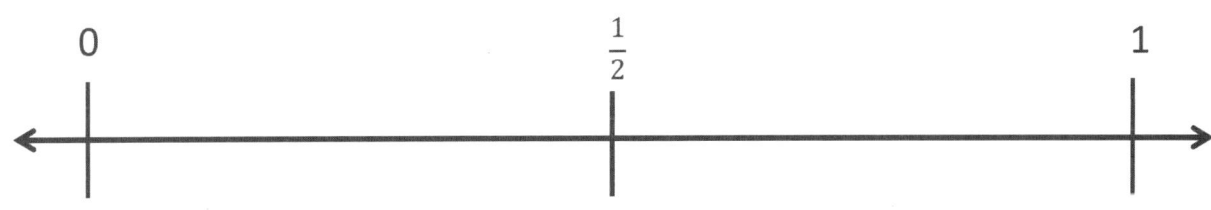

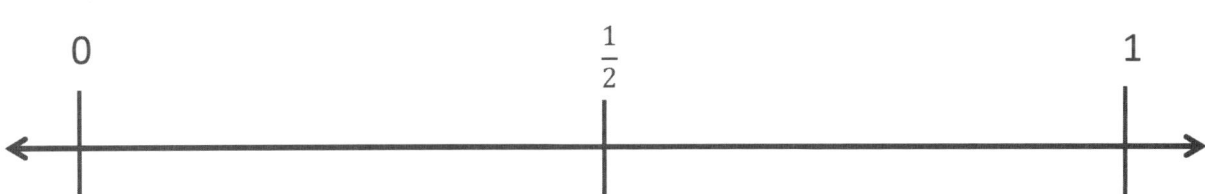

2.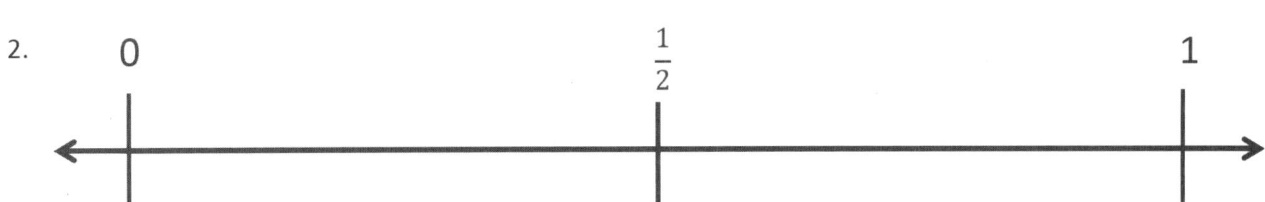

number line

Lesson 13

Objective: Reason using benchmarks to compare two fractions on the number line.

Suggested Lesson Structure

- **■** Fluency Practice (12 minutes)
- **■** Application Problem (5 minutes)
- **■** Concept Development (33 minutes)
- **■** Student Debrief (10 minutes)
- **Total Time** **(60 minutes)**

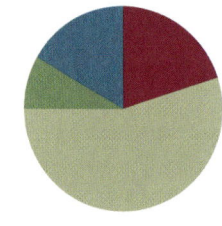

Fluency Practice (12 minutes)

- Divide 3 Different Ways **4.NBT.6**   (4 minutes)
- Count by Equivalent Fractions **3.NF.3**   (4 minutes)
- Plot Fractions on a Number Line **4.NF.3**   (4 minutes)

Divide 3 Different Ways (4 minutes)

Materials: (S) Personal white board

Note: This fluency activity reviews concepts covered in Module 3. Alternately, have students choose to solve the division problem using one of the three methods.

- T: (Write 435 ÷ 3.) Solve this problem by drawing place value disks.
- S: (Solve.)
- T: Solve 435 ÷ 3 using the area model.
- S: (Solve.)
- T: Solve 435 ÷ 3 using the standard algorithm.
- S: (Solve.)

Continue with 184 ÷ 4.

Count by Equivalent Fractions (4 minutes)

Note: This fluency activity reinforces Module 5 fraction concepts and prepares students for today's lesson.

- T: Count by fours to 40. Start at zero.
- S: 0, 4, 8, 12, 16, 20, 24, 28, 32, 36, 40.

T: Count by 4 fifths from 0 fifths to 40 fifths. (Write as students count.)

S: $\frac{0}{5}, \frac{4}{5}, \frac{8}{5}, \frac{12}{5}, \frac{16}{5}, \frac{20}{5}, \frac{24}{5}, \frac{28}{5}, \frac{32}{5}, \frac{36}{5}, \frac{40}{5}$.

$\frac{0}{5}$	$\frac{4}{5}$	$\frac{8}{5}$	$\frac{12}{5}$	$\frac{16}{5}$	$\frac{20}{5}$	$\frac{24}{5}$	$\frac{28}{5}$	$\frac{32}{5}$	$\frac{36}{5}$	$\frac{40}{5}$
0	$\frac{4}{5}$	$\frac{8}{5}$	$\frac{12}{5}$	$\frac{16}{5}$	4	$\frac{24}{5}$	$\frac{28}{5}$	$\frac{32}{5}$	$\frac{36}{5}$	8

T: 1 one is the same as how many fifths?
S: 5 fifths.
T: 2 ones is the same as how many fifths?
S: 10 fifths.
T: 3 ones is the same as how many fifths?
S: 15 fifths.

Continue asking through 8 ones.

T: (Beneath $\frac{40}{5}$, write 8.) Count by 4 fifths again. This time, say the whole numbers when you arrive at them. Start at zero.

S: $0, \frac{4}{5}, \frac{8}{5}, \frac{12}{5}, \frac{16}{5}, 4, \frac{24}{5}, \frac{28}{5}, \frac{32}{5}, \frac{36}{5}, 8$.

Plot Fractions on a Number Line (4 minutes)

Materials: (S) Personal white board

Note: This fluency activity reviews Lesson 12.

T: (Project a blank number line, partitioned into 2 equal parts.) Draw a number line on your personal white board, and then partition it into 2 equal parts.
S: (Draw a number line partitioned into 2 equal parts.)
T: (Write 0 below the left endpoint. Write 1 below the right endpoint.) Fill in the endpoints, and write the fraction that belongs at the halfway point.
S: (Write 0 below the left endpoint, 1 below the right endpoint, and $\frac{1}{2}$ below the halfway point.)
T: (Write $\frac{1}{5}$.) Label 1 fifth on your number line.
S: (Write $\frac{1}{5}$ between 0 and $\frac{1}{2}$ on the number line.)
T: (Write $\frac{1}{5} \_\_ \frac{1}{2}$.) On your board, fill in the blank with a greater than or less than symbol.
S: (Write $\frac{1}{5} < \frac{1}{2}$.)

Continue with the following possible sequence: Compare $\frac{1}{2}$ and $\frac{7}{10}$, $\frac{1}{5}$ and $\frac{7}{10}$, $\frac{1}{2}$ and $\frac{4}{5}$, and $\frac{4}{5}$ and $\frac{7}{10}$.

Lesson 13: Reason using benchmarks to compare two fractions on the number line.

A STORY OF UNITS   Lesson 13   4•5

Application Problem (5 minutes)

Mr. and Mrs. Reynolds went for a run. Mr. Reynolds ran for $\frac{6}{10}$ mile. Mrs. Reynolds ran for $\frac{2}{5}$ mile. Who ran farther? Explain how you know. Use the benchmarks 0, $\frac{1}{2}$, and 1 to explain your answer.

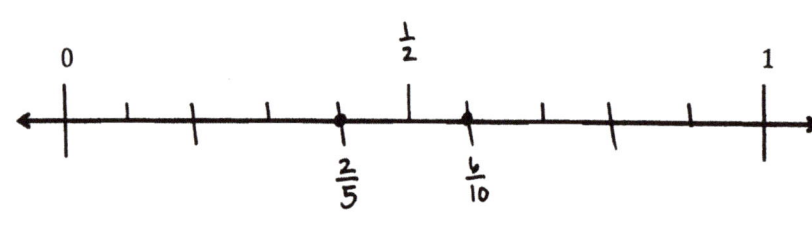

Mr. Reynolds ran farther than Mrs. Reynolds. I know this because $\frac{2}{5}$ is less than $\frac{1}{2}$ and $\frac{6}{10}$ is greater than $\frac{1}{2}$. $\frac{6}{10} = \frac{3}{5}$ so $\frac{3}{5} > \frac{2}{5}$.

Note: This Application Problem builds on Lesson 12 in which students learned to use benchmarks to compare two fractions. This Application Problem bridges to today's lesson in which students once again compare fractions using benchmarks.

Concept Development (33 minutes)

Materials: (S) Personal white board, blank number lines with midpoint (Template)

NOTES ON MULTIPLE MEANS OF ENGAGEMENT:

Some students may benefit from a review of how to change an improper fraction to a mixed number by drawing a number bond. Before the lesson, instruct students to draw a number bond for an improper fraction in which one addend has a value of 1 whole.

Problem 1: Reason to compare fractions between 1 and 2.

T: Compare $\frac{7}{8}$ and $\frac{6}{4}$ with your partner.

S: $\frac{7}{8}$ is less than 1. $\frac{6}{4}$ is greater than 1 because 1 is equal to $\frac{4}{4}$.

T: Draw a number bond for $\frac{6}{4}$ partitioning the whole and parts.

S: (Draw.)

T: We can use the bond to help us locate $\frac{6}{4}$ on the number line. Label a number line with endpoints 0 to 2, and locate $\frac{4}{4}$.

S: (Put pencils on $\frac{4}{4}$.)

T: $\frac{6}{4}$ is $\frac{2}{4}$ more. Imagine partitioning the line into fourths between 1 and 2. Where would you plot $\frac{6}{4}$?

S: $\frac{6}{4}$ is halfway between 1 and 2. → That's because $\frac{6}{4} = 1\frac{1}{2}$. → 6 fourths is 2 more fourths than 1. 2 fourths is the same as a half.

172   Lesson 13:   Reason using benchmarks to compare two fractions on the number line.

T: Plot $\frac{6}{4}$ and $\frac{7}{8}$. Write a statement to compare the two fractions.

S: $\frac{7}{8} < \frac{6}{4}$. → $\frac{6}{4} > \frac{7}{8}$.

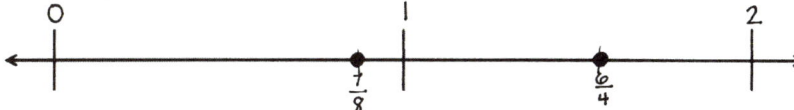

T: Next, compare $\frac{5}{3}$ and $\frac{9}{5}$. Discuss their relationship to 1.

S: Both are greater than 1 because $\frac{3}{3}$ and $\frac{5}{5}$ equal 1. → Neither is very close to 1, because $\frac{4}{3}$ and $\frac{6}{5}$ would be the fractions just a little bigger than 1.

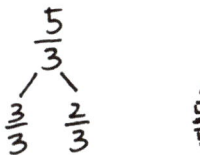

T: Write a number bond to show $\frac{5}{3}$ and $\frac{9}{5}$ as a whole and some parts.

S: (Draw bonds.)

T: Use the number bond to write each fraction as 1 and some more fractional units.

S: $\frac{5}{3} = 1\frac{2}{3}$. → $\frac{9}{5} = 1\frac{4}{5}$.

T: Label 0, 1, and 2 on another number line. We are plotting two points. One point is $\frac{2}{3}$ greater than 1. The other is $\frac{4}{5}$ greater than 1. Discuss with your partner how to plot these two points. Consider their placement in relation to 2.

> **NOTES ON MULTIPLE MEANS OF ENGAGEMENT:**
>
> Define the term *comparison symbol* for English language learners. Students may well be proficient at using *greater than* and *less than* symbols but may not recognize the term.

S: $\frac{2}{3}$ is 1 third less than 1. $\frac{4}{5}$ is 1 fifth less than 1. Thirds are greater than fifths, so $\frac{2}{3}$ is farther from 1 than $\frac{4}{5}$. → $1\frac{2}{3}$ is farther from 2 than $1\frac{4}{5}$. → The number bond lets me see that both fractions have 1 and some parts. The whole is the same, so I can compare just the parts and plot them between 1 and 2.

T: Plot the points. Compare $\frac{5}{3}$ and $\frac{9}{5}$. Write your statement using a comparison symbol.

S: (Write $\frac{5}{3} < \frac{9}{5}$ → $1\frac{2}{3} < 1\frac{4}{5}$.)

Continue the process with $\frac{7}{4}$ and $\frac{9}{5}$.

Lesson 13: Reason using benchmarks to compare two fractions on the number line.

A STORY OF UNITS

Lesson 13 4•5

Problem 2: Reason about the size of fractions as compared to $1\frac{1}{2}$.

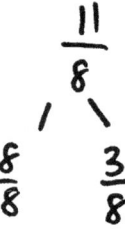

T: Is $\frac{11}{8}$ less than 1 or greater than 1? Create a number bond to guide you in your thinking.

S: $\frac{11}{8}$ is greater than 1 because $\frac{11}{8} = \frac{8}{8} + \frac{3}{8}$. $\frac{8}{8}$ is equal to 1, so $\frac{11}{8}$ must be greater than 1.

T: Is $\frac{11}{8}$ less than $1\frac{1}{2}$ or greater than $1\frac{1}{2}$?

S: $1\frac{1}{2} = \frac{8}{8} + \frac{4}{8}$, and $\frac{3}{8}$ is less than $\frac{4}{8}$, so $\frac{11}{8}$ is less than $1\frac{1}{2}$. → $1\frac{1}{2}$ is the same as $\frac{12}{8}$. $\frac{11}{8}$ is less than $\frac{12}{8}$, so $\frac{11}{8}$ is less than $1\frac{1}{2}$.

T: Discuss with your partner if $\frac{5}{4}$ is greater than or less than 1.

S: (Discuss.)

T: Plot $\frac{11}{8}$ and $\frac{5}{4}$ on another number line. You reasoned that both are between 1 and 2. Let's determine their placement using the benchmark $1\frac{1}{2}$. Label the number line with 1, $1\frac{1}{2}$, and 2. Talk it over with your partner before plotting.

S: $\frac{5}{4}$ is the same as $1\frac{1}{4}$. That's halfway between 1 and $1\frac{1}{2}$. → There are 2 fourths in a half, so $\frac{5}{4}$ is one unit away from $1\frac{1}{2}$, and $\frac{11}{8}$ is one unit away from $1\frac{1}{2}$. → Eighths are smaller than fourths, so $\frac{11}{8}$ is closer to $1\frac{1}{2}$.

T: Compare $\frac{11}{8}$ and $\frac{5}{4}$. Write your statement using a comparison symbol.

S: (Write $\frac{11}{8} > \frac{5}{4}$ or $1\frac{3}{8} > 1\frac{1}{4}$.)

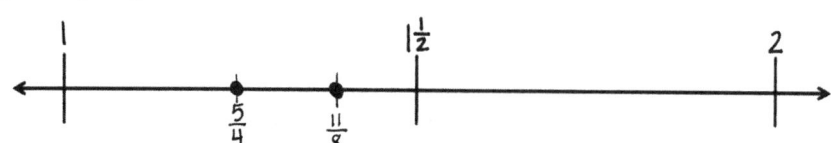

T: Compare $\frac{11}{8}$ and $\frac{10}{6}$. Discuss with a partner using benchmarks to help explain.

S: Both fractions are greater than a whole but less than 2. → $\frac{12}{8} = 1\frac{1}{2}$. So, $\frac{11}{8}$ is one unit less than $1\frac{1}{2}$. → $\frac{9}{6} = 1\frac{1}{2}$, so $\frac{10}{6}$ is one unit more than $1\frac{1}{2}$. → I drew number bonds. Both numbers have a whole, so I just compared the parts. I thought of $\frac{3}{8}$ and $\frac{4}{6}$ compared to $\frac{1}{2}$. I know $\frac{4}{6}$ is more than $\frac{1}{2}$, so I know $1\frac{4}{6} > 1\frac{3}{8}$. → $\frac{11}{8} < \frac{10}{6}$.

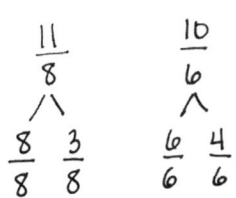
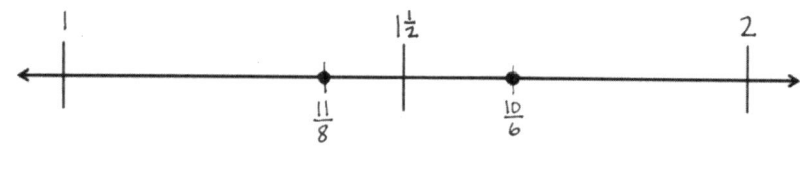

Lesson 13: Reason using benchmarks to compare two fractions on the number line.

A STORY OF UNITS　　　　　　　　　　　　　　　　　　　　　　　　　　　Lesson 13　4•5

Problem 3: Reason using benchmarks to compare two fractions.

T: Which is greater: $\frac{14}{10}$ or $\frac{7}{5}$? Discuss with a partner. Use the benchmarks to help explain.

S: I used number bonds. Since both have 1 whole, I compared the parts: $\frac{4}{10}$ and $\frac{2}{5}$ are both less than 1 half. $\frac{4}{10}$ is one unit away from 1 half. But there are no fifths equal to 1 half. → $\frac{4}{10}$ is 4 units from zero. $\frac{2}{5}$ is 2 units from zero. Fifths are half of tenths. I think they are equal! → I can make an equivalent fraction to compare. $\frac{7}{5} = \frac{7 \times 2}{5 \times 2} = \frac{14}{10}$. $\frac{14}{10}$ is equal to $\frac{7}{5}$. → $\frac{14}{10} = \frac{7}{5}$.

T: Compare $\frac{6}{4}$ and $\frac{11}{10}$.

S: $\frac{11}{10}$ is $\frac{1}{10}$ past 1. $\frac{6}{4} = 1\frac{1}{2}$. → $\frac{6}{4} > \frac{11}{10}$.

T: Compare $\frac{10}{8}$ and $\frac{8}{4}$.

S: $1\frac{2}{8}$ is halfway between 1 and $1\frac{1}{2}$. → $\frac{8}{4} = 2$. → $\frac{10}{8} < \frac{8}{4}$.

Problem Set (10 minutes)

Students should do their personal best to complete the Problem Set within the allotted 10 minutes. For some classes, it may be appropriate to modify the assignment by specifying which problems they work on first. Some problems do not specify a method for solving. Students should solve these problems using the RDW approach used for Application Problems.

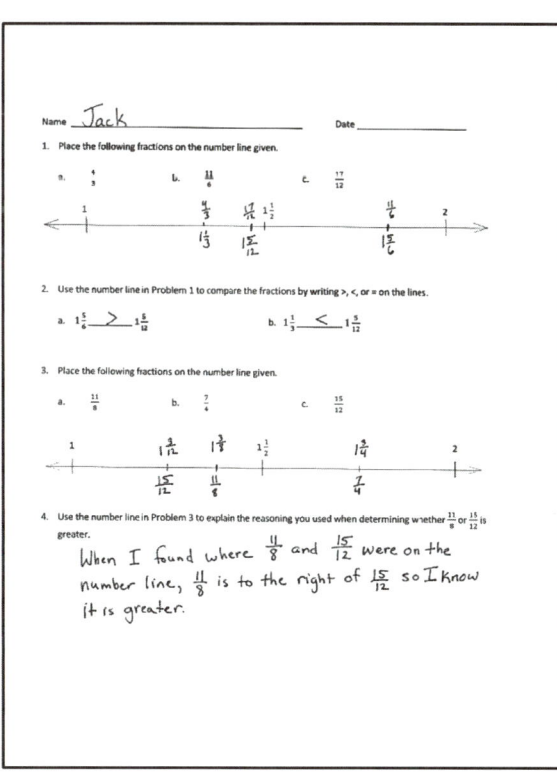

Student Debrief (10 minutes)

Lesson Objective: Reason using benchmarks to compare two fractions on the number line.

The Student Debrief is intended to invite reflection and active processing of the total lesson experience.

Invite students to review their solutions for the Problem Set. They should check work by comparing answers with a partner before going over answers as a class. Look for misconceptions or misunderstandings that can be addressed in the Debrief. Guide students in a conversation to debrief the Problem Set and process the lesson.

Lesson 13: Reason using benchmarks to compare two fractions on the number line.

175

Any combination of the questions below may be used to lead the discussion.

- When were number bonds helpful in solving some of the problems on the Problem Set? Explain.
- Explain your thinking in comparing the fractions when you solved Problem 5(a–j). Were benchmarks always helpful?
- How did you solve Problem 5(h)?
- What other benchmarks could you use when comparing fractions? Why are benchmarks helpful?
- How did the Application Problem connect to today's lesson?

Exit Ticket (3 minutes)

After the Student Debrief, instruct students to complete the Exit Ticket. A review of their work will help with assessing students' understanding of the concepts that were presented in today's lesson and planning more effectively for future lessons. The questions may be read aloud to the students.

A STORY OF UNITS

Lesson 13 Problem Set 4•5

Name _____ Date _____

1. Place the following fractions on the number line given.

   a. $\frac{4}{3}$ b. $\frac{11}{6}$ c. $\frac{17}{12}$

2. Use the number line in Problem 1 to compare the fractions by writing >, <, or = on the lines.

   a. $1\frac{5}{6}$ _____ $1\frac{5}{12}$ b. $1\frac{1}{3}$ _____ $1\frac{5}{12}$

3. Place the following fractions on the number line given.

   a. $\frac{11}{8}$ b. $\frac{7}{4}$ c. $\frac{15}{12}$

4. Use the number line in Problem 3 to explain the reasoning you used when determining whether $\frac{11}{8}$ or $\frac{15}{12}$ is greater.

Lesson 13: Reason using benchmarks to compare two fractions on the number line.

A STORY OF UNITS   Lesson 13 Problem Set   4•5

5. Compare the fractions given below by writing > or < on the lines. Give a brief explanation for each answer referring to benchmarks.

a. $\dfrac{3}{8}$ _____ $\dfrac{7}{12}$ b. $\dfrac{5}{12}$ _____ $\dfrac{7}{8}$

c. $\dfrac{8}{6}$ _____ $\dfrac{11}{12}$ d. $\dfrac{5}{12}$ _____ $\dfrac{1}{3}$

e. $\dfrac{7}{5}$ _____ $\dfrac{11}{10}$ f. $\dfrac{5}{4}$ _____ $\dfrac{7}{8}$

g. $\dfrac{13}{12}$ _____ $\dfrac{9}{10}$ h. $\dfrac{6}{8}$ _____ $\dfrac{5}{4}$

i. $\dfrac{8}{12}$ _____ $\dfrac{8}{4}$ j. $\dfrac{7}{5}$ _____ $\dfrac{16}{10}$

Lesson 13:   Reason using benchmarks to compare two fractions on the number line.

Name _____   Date _____

1. Place the following fractions on the number line given.

   a. $\dfrac{5}{4}$ b. $\dfrac{10}{7}$ c. $\dfrac{16}{9}$

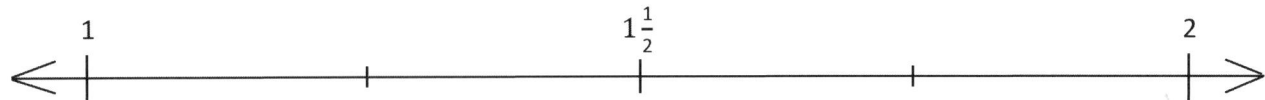

2. Compare the fractions using >, <, or =.

   a. $\dfrac{5}{4}$ _____ $\dfrac{10}{7}$ b. $\dfrac{5}{4}$ _____ $\dfrac{16}{9}$ c. $\dfrac{16}{9}$ _____ $\dfrac{10}{7}$

A STORY OF UNITS Lesson 13 Homework 4•5

Name _____ Date _____

1. Place the following fractions on the number line given.

   a. $\frac{3}{2}$ b. $\frac{9}{5}$ c. $\frac{14}{10}$

   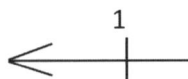

2. Use the number line in Problem 1 to compare the fractions by writing >, <, or = on the lines.

   a. $1\frac{1}{6}$ _____ $1\frac{4}{12}$ b. $1\frac{1}{2}$ _____ $1\frac{4}{5}$

3. Place the following fractions on the number line given.

   a. $\frac{12}{9}$ b. $\frac{6}{5}$ c. $\frac{18}{15}$

4. Use the number line in Problem 3 to explain the reasoning you used when determining whether $\frac{12}{9}$ or $\frac{18}{15}$ was greater.

Lesson 13: Reason using benchmarks to compare two fractions on the number line.

A STORY OF UNITS　　　　　　　　　　　　　　　　　　　Lesson 13 Homework 4•5

5. Compare the fractions given below by writing > or < on the lines. Give a brief explanation for each answer referring to benchmarks.

   a. $\dfrac{2}{5}$ _____ $\dfrac{6}{8}$

   b. $\dfrac{6}{10}$ _____ $\dfrac{5}{6}$

   c. $\dfrac{6}{4}$ _____ $\dfrac{7}{8}$

   d. $\dfrac{1}{4}$ _____ $\dfrac{8}{12}$

   e. $\dfrac{14}{12}$ _____ $\dfrac{11}{6}$

   f. $\dfrac{8}{9}$ _____ $\dfrac{3}{2}$

   g. $\dfrac{7}{8}$ _____ $\dfrac{11}{10}$

   h. $\dfrac{3}{4}$ _____ $\dfrac{4}{3}$

   i. $\dfrac{3}{8}$ _____ $\dfrac{3}{2}$

   j. $\dfrac{9}{6}$ _____ $\dfrac{16}{12}$

Lesson 13: Reason using benchmarks to compare two fractions on the number line.

A STORY OF UNITS

Lesson 13 Template 4•5

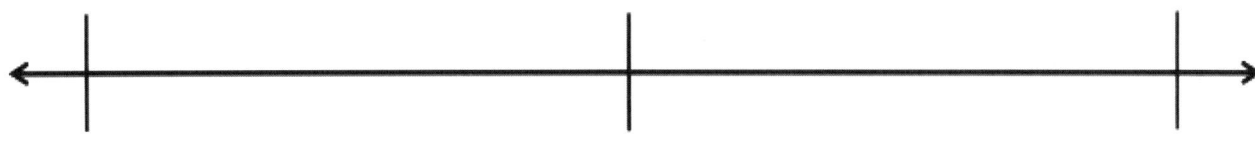

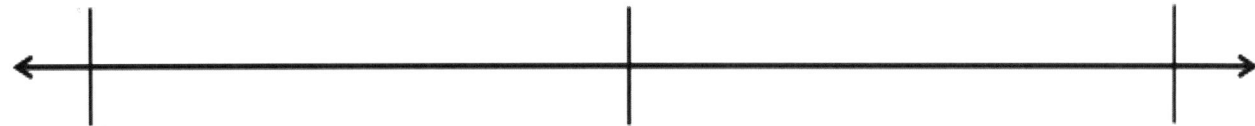

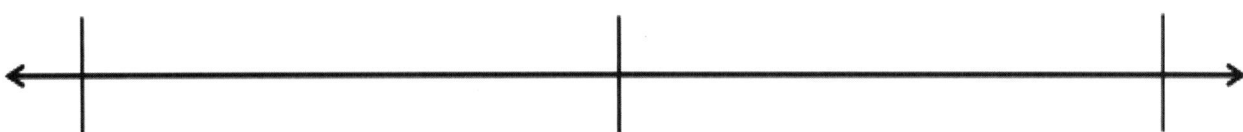

blank number lines with midpoint

Lesson 13: Reason using benchmarks to compare two fractions on the number line.

A STORY OF UNITS

Lesson 14

Objective: Find common units or number of units to compare two fractions.

Suggested Lesson Structure

■ Fluency Practice   (12 minutes)
■ Application Problem   (5 minutes)
■ Concept Development (33 minutes)
■ Student Debrief   (10 minutes)
   Total Time   **(60 minutes)**

Fluency Practice (12 minutes)

- Add and Subtract **4.NBT.4**   (4 minutes)
- Compare Fractions **4.NF.2**   (4 minutes)
- Construct a Number Line with Fractions **4.NF.2**   (4 minutes)

Add and Subtract (4 minutes)

Materials: (S) Personal white board

Note: This fluency activity reviews adding and subtracting using the standard algorithm.

   T: (Write 458 thousands 397 ones.) On your personal white board, write this number in standard form.
   S: (Write 458,397.)
   T: (Write 281 thousands 563 ones.) Add this number to 458,397 using the standard algorithm.
   S: (Write 458,397 + 281,563 = 739,960 using the standard algorithm.)

Continue the process with 456,919 + 292,689.

   T: (Write 900 thousands.) On your board, write this number in standard form.
   S: (Write 900,000.)
   T: (Write 523 thousands 536 ones.) Subtract this number from 900,000 using the standard algorithm.
   S: (Write 900,000 − 523,536 = 376,464 using the standard algorithm.)

Continue the process with 512,807 − 255,258.

Lesson 14:   Find common units or number of units to compare two fractions.

A STORY OF UNITS **Lesson 14 4•5**

Compare Fractions (4 minutes)

Materials: (S) Personal white board

Note: This fluency activity reviews Lesson 12.

- T: (Project a blank number line, partitioned into 2 equal parts.) Draw a number line on your personal white board, and then partition it into 2 equal parts.
- S: (Draw and partition a number line.)
- T: (Write 0 below the left endpoint. Write 1 below the right endpoint.) Fill in the endpoints, and write the fraction that belongs at the halfway point.
- S: (Label 0, $\frac{1}{2}$, and 1.)
- T: (Write $\frac{7}{8}$.) Plot 7 eighths on your number line.
- S: (Plot $\frac{7}{8}$.)
- T: (Write $\frac{7}{8}$ __ $\frac{1}{2}$.) On your board, fill in the blank with a greater than or less than symbol.
- S: (Write $\frac{7}{8} > \frac{1}{2}$.)
- T: (Write $\frac{1}{2}$ __ $\frac{3}{4}$.) On your board, fill in the blank with a greater than or less than symbol. Use your number line if you need to.
- S: (Write $\frac{1}{2} < \frac{3}{4}$.)

Continue with the following possible sequence: Compare $\frac{3}{4}$ and $\frac{7}{8}$, $\frac{5}{6}$ and $\frac{1}{2}$, and $\frac{5}{6}$ and $\frac{2}{3}$.

Construct a Number Line with Fractions (4 minutes)

Materials: (S) Personal white board

Note: This fluency activity reviews Lesson 13.

- T: (Project a blank number line, partitioned into 2 equal parts.) Draw a number line on your personal white board, and then partition it into 2 equal parts.
- S: (Draw a number line partitioned into 2 equal parts.)
- T: (Write 1 below the left endpoint. Write 2 below the right endpoint.) Fill in the endpoints, and write the fraction that belongs at the halfway point.
- S: (Write 1 below the left endpoint, 2 below the right endpoint, and $1\frac{1}{2}$ below the halfway point.)
- T: (Write $\frac{6}{5}$.) Plot 6 fifths on your number line.
- T: (Write $\frac{6}{5}$ __ $1\frac{1}{2}$.) On your board, fill in the blank with a greater than or less than symbol.
- S: (Write $\frac{6}{5} < 1\frac{1}{2}$.)

Continue with the following possible sequence: Compare $\frac{17}{10}$ and $1\frac{1}{2}$, $\frac{17}{10}$ and $\frac{6}{5}$, and $\frac{19}{12}$ and $\frac{7}{4}$.

Lesson 14:   Find common units or number of units to compare two fractions.

A STORY OF UNITS   Lesson 14   4•5

Application Problem (5 minutes)

Compare $\frac{4}{5}$, $\frac{3}{4}$, and $\frac{9}{10}$ using <, >, or =. Explain your reasoning using a benchmark.

Note: This Application Problem reviews all of Topic C and bridges to today's lesson in which students compare fractions with unrelated denominators using area models.

$\frac{9}{10}$ is 1 tenth from 1.
$\frac{4}{5}$ is 1 fifth from 1.
$\frac{3}{4}$ is 1 fourth from 1.
$\frac{1}{10} < \frac{1}{5} < \frac{1}{4}$ so $\frac{9}{10}$ is closest to 1, then $\frac{4}{5}$, then $\frac{3}{4}$. So $\frac{9}{10} > \frac{4}{5} > \frac{3}{4}$.

Concept Development (33 minutes)

Materials: (S) Personal white board

Problem 1: Reason about fraction size using unit language.

- T: Which is greater—1 apple or 3 apples?
- S: 3 apples!
- T: (Write 3 apples > 1 apple.)
- T: Which is greater—1 fourth or 3 fourths?
- S: 3 fourths!
- T: (Write 3 fourths > 1 fourth.)
- T: What do you notice about these two statements?
  3 apples > 1 apple.
  3 fourths > 1 fourth.
- S: The units are the same in each. One is apples, and the other is fourths. → We can compare the number of fourths like we compare the number of apples. → It is easy to compare when the units are the same!
- T: Which is greater—1 fourth or 1 fifth?
- S: 1 fourth.
- T: (Write 1 fourth > 1 fifth.)
- T: How do you know?
- S: I can draw two tape diagrams to compare. I can partition a whole into fourths on one tape diagram and into fifths on the other. There are more fifths than fourths, so each fourth is going to be bigger than a fifth. → $\frac{1}{5}$ is less than $\frac{1}{4}$ because fifths are smaller than fourths.
- T: (Write $\frac{1}{4} > \frac{1}{5}$.)
- T: Which is greater—2 fourths or 2 sixths?
- S: 2 fourths is greater than 2 sixths.

NOTES ON MULTIPLE MEANS OF ACTION AND EXPRESSION:

To accurately compare two fractions using a tape diagram, both tape diagrams must be the same length and aligned precisely. Providing a template of two blank parallel tape diagrams of equal length may be helpful in assisting students.

MP.7

Lesson 14: Find common units or number of units to compare two fractions.   185

A STORY OF UNITS — Lesson 14   4•5

T: (Write $\frac{2}{4} > \frac{2}{6}$.)

T: What do you notice about these statements?

$\frac{1}{4} > \frac{1}{5}$ $\frac{2}{4} > \frac{2}{6}$

S: Fourths are greater than fifths and sixths. → In each comparison, the numerators are the same.

T: Which would be greater—2 inches or 2 feet?

S: 2 feet! I know feet are greater than inches.

T: In the same way, 2 fourths is greater than 2 sixths because fourths are greater than sixths.

T: When the numerator is the same, we look at the denominator to reason about which fraction is greater. The greater the denominator, the smaller the fractional unit. Explain why $\frac{5}{7}$ is greater than $\frac{5}{12}$ of the same whole.

S: Sevenths are greater fractional units than twelfths. 5 sevenths are greater than 5 twelfths because 1 seventh is greater than 1 twelfth. → The sum of 5 larger units is going to be greater than the sum of 5 smaller units.

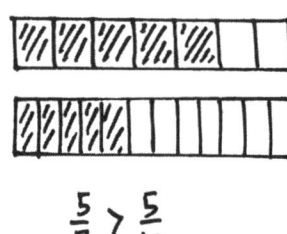

Problem 2: Compare fractions with related numerators.

T: (Display $\frac{2}{8}$ and $\frac{4}{10}$.) Draw a tape diagram to show each.

T: Partition the eighths in half. What fraction is now shown?

S: $\frac{4}{16}$. The numerators are the same! → The number of shaded units is the same.

T: Compare $\frac{4}{16}$ and $\frac{4}{10}$.

S: $\frac{4}{16}$ is less than $\frac{4}{10}$ since sixteenths are smaller units than tenths. I can compare the size of the units because the numerators are the same.

T: Compare $\frac{2}{8}$ and $\frac{4}{10}$.

S:  is less than $\frac{4}{10}$.

T: (Display $\frac{9}{10}$ and $\frac{3}{4}$.) Discuss a strategy for comparing these two fractions with your partner.

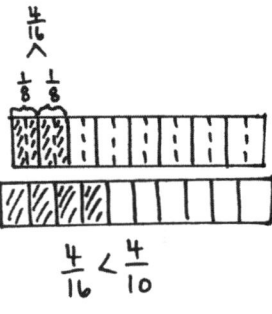

S: Let's make a common numerator of 9. $\frac{3}{4} = \frac{3 \times 3}{4 \times 3} = \frac{9}{12}$. → $\frac{9}{10}$ is greater than $\frac{9}{12}$. → $\frac{9}{10}$ is greater than $\frac{3}{4}$. → $\frac{9}{10} + \frac{1}{10} = 1$, and $\frac{3}{4} + \frac{1}{4} = 1$. 1 tenth is less than 1 fourth, so 9 tenths is greater.

Lesson 14: Find common units or number of units to compare two fractions.

A STORY OF UNITS — Lesson 14 4•5

Problem 3: Compare fractions having related denominators where one denominator is a factor of the other.

T: (Display $\frac{7}{10}$ and $\frac{3}{5}$.) Model each fraction using a tape diagram. Can we make a common numerator?

S: No. We can't multiply 3 by a number to get 7. → We could make them both numerators of 21.

T: Finding a common numerator does not work easily here. Consider the denominators. Can we make like units, or **common denominators**?

S: Yes. We can partition each fifth in half to make tenths. → $\frac{3}{5} = \frac{3 \times 2}{5 \times 2} = \frac{6}{10}$.

T: Compare $\frac{6}{10}$ and $\frac{7}{10}$.

S: $\frac{6}{10}$ is less than $\frac{7}{10}$. → That means that $\frac{3}{5}$ is less than $\frac{7}{10}$.

T: Draw a number line to show 3 fifths. Decompose the line into tenths to show 7 tenths. $\frac{3}{5}$ is equal to how many tenths?

S: $\frac{6}{10}$.

T: Compare $\frac{6}{10}$ and $\frac{7}{10}$.

S: $\frac{6}{10}$ is less than $\frac{7}{10}$, so $\frac{3}{5} < \frac{7}{10}$.

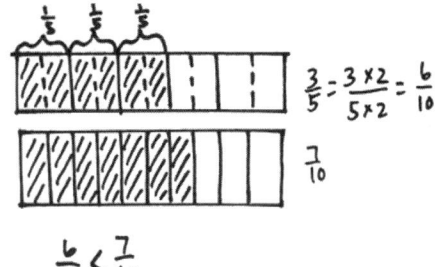

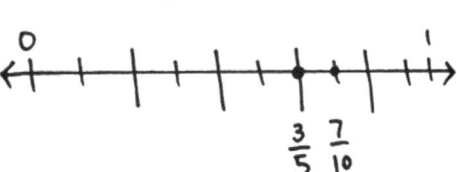

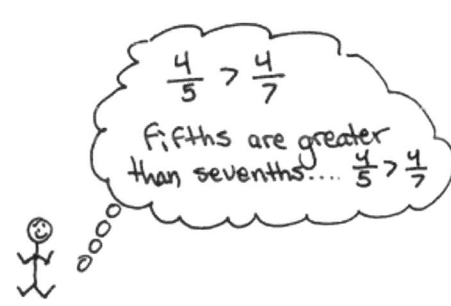

Problem 4: Compare fractions using different methods of reasoning.

T: Think about the strategies that we have learned. What strategy would you use to compare $\frac{4}{5}$ and $\frac{4}{7}$? Discuss with your partner. Defend your reasoning.

S: The numerators are the same. $\frac{4}{5}$ is greater than $\frac{4}{7}$. → There are 4 fifths and 4 sevenths. Since fifths are greater than sevenths, $\frac{4}{5}$ is greater than $\frac{4}{7}$. → 4 fifths is a lot more than 1 half. 4 sevenths is a little more than 1 half.

T: Compare $\frac{8}{10}$ and $\frac{4}{6}$.

S: It looks like we can make numerators that are the same because 8 is a multiple of 4. $\frac{4}{6}$ is the same as $\frac{8}{12}$. $\frac{8}{12}$ is less than $\frac{8}{10}$. So, $\frac{4}{6}$ is less than $\frac{8}{10}$. → $\frac{8}{10} + \frac{2}{10} = 1$, and $\frac{4}{6} + \frac{2}{6} = 1$. I know that 2 tenths is less than 2 sixths, so 8 tenths is greater.

T: Compare $\frac{5}{12}$ and $\frac{2}{3}$.

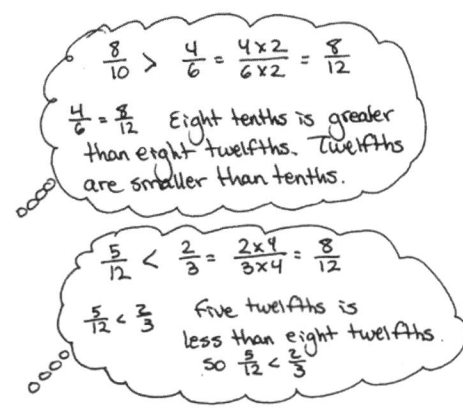

Lesson 14: Find common units or number of units to compare two fractions. 187

A STORY OF UNITS   Lesson 14 4•5

S: The units are different! Twelfths are not thirds, but we can decompose thirds to make twelfths! We can make like denominators. $\frac{2}{3}$ is the same as $\frac{8}{12}$. $\frac{8}{12}$ is more than $\frac{5}{12}$, $\frac{2}{3} > \frac{5}{12}$. → I wouldn't try to make the same number of units, because 5 is not a multiple of 2, but it might be possible. → 5 twelfths is less than a half, and 2 thirds is more than a half.

T: How might we use what we know to compare $1\frac{2}{5}$ and $1\frac{6}{8}$? Share your thoughts with your partner.

S: I see that the whole numbers are the same, so we can just compare the fractions. Let's compare $\frac{2}{5}$ and $\frac{6}{8}$. The numerators are related. 6 is a multiple of 2, so we can make fractions that have equal numerators. $\frac{2}{5}$ is the same as $\frac{6}{15}$, which is smaller than $\frac{6}{8}$. So, $1\frac{2}{5}$ is less than $1\frac{6}{8}$. → 2 fifths is less than half. 6 eighths is greater than half, so $1\frac{6}{8}$ is greater.

Problem Set (10 minutes)

Students should do their personal best to complete the Problem Set within the allotted 10 minutes. For some classes, it may be appropriate to modify the assignment by specifying which problems they work on first. Some problems do not specify a method for solving. Students should solve these problems using the RDW approach used for Application Problems.

Student Debrief (10 minutes)

Lesson Objective: Find common units or number of units to compare two fractions.

The Student Debrief is intended to invite reflection and active processing of the total lesson experience.

Invite students to review their solutions for the Problem Set. They should check work by comparing answers with a partner before going over answers as a class. Look for misconceptions or misunderstandings that can be addressed in the Debrief. Guide students in a conversation to debrief the Problem Set and process the lesson.

Any combination of the questions below may be used to lead the discussion.

- Why were the fractions in Problem 1 easier to compare than in Problem 2?
- Problems 5(a), 5(d), and 5(f) can be compared using different types of reasoning. Explain the reasoning you used for each.
- How can you determine whether you can make common numerators or **common denominators** when comparing fractions?
- How are tape diagrams and number lines helpful in comparing fractions?
- What new (or significant) math vocabulary did we use today to communicate precisely?

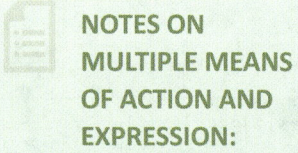

NOTES ON
MULTIPLE MEANS
OF ACTION AND
EXPRESSION:

Support English language learners as they explain their reasoning for Problems 5(a), 5(d), and 5(f) on the Problem Set. Provide a word bank with corresponding pictures.

Possible words for the word bank are listed below:

fourth   seventh   third   fifteenth
whole   ninth   one   closer
greater than   less than   almost
tape diagram

188   Lesson 14:   Find common units or number of units to compare two fractions.

A STORY OF UNITS   Lesson 14 4•5

- How did the Application Problem connect to today's lesson?

Exit Ticket (3 minutes)

After the Student Debrief, instruct students to complete the Exit Ticket. A review of their work will help with assessing students' understanding of the concepts that were presented in today's lesson and planning more effectively for future lessons. The questions may be read aloud to the students.

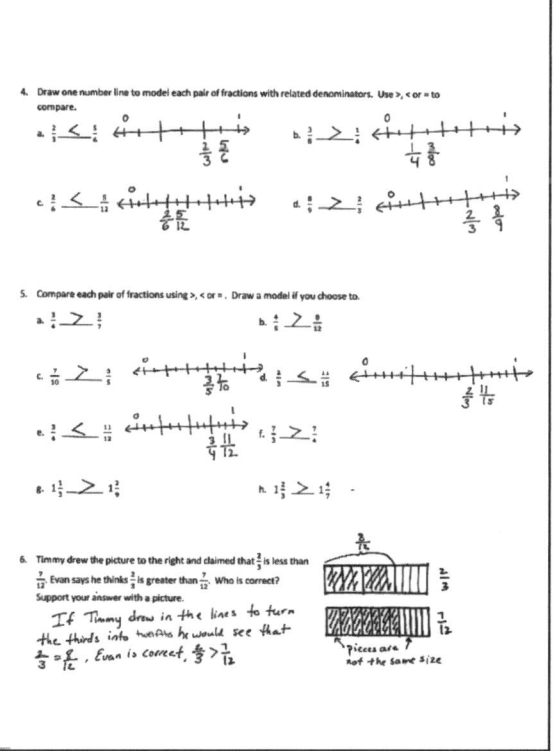

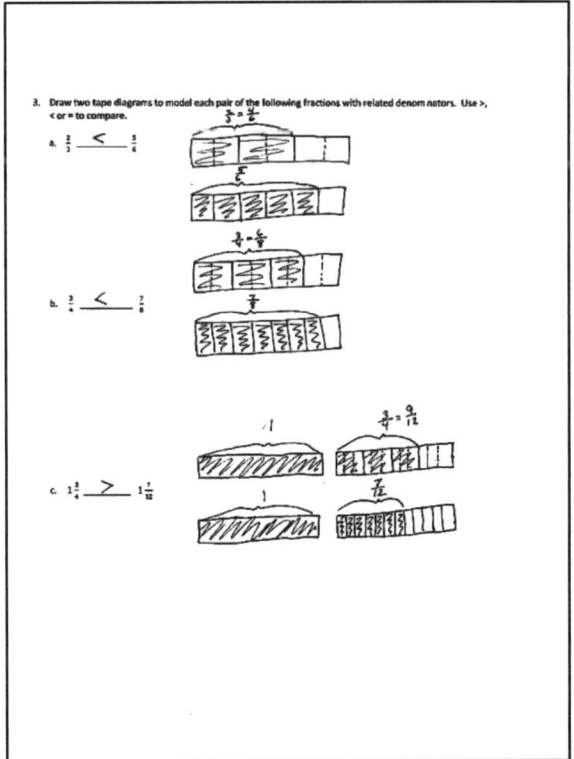

Lesson 14: Find common units or number of units to compare two fractions.

A STORY OF UNITS   Lesson 14 Problem Set   4•5

Name _____   Date _____

1. Compare the pairs of fractions by reasoning about the size of the units. Use >, <, or =.

   a. 1 fourth _____ 1 fifth   b. 3 fourths _____ 3 fifths

   c. 1 tenth _____ 1 twelfth   d. 7 tenths _____ 7 twelfths

2. Compare by reasoning about the following pairs of fractions with the same or related numerators. Use >, <, or =. Explain your thinking using words, pictures, or numbers. Problem 2(b) has been done for you.

   a. $\frac{3}{5}$ _____ $\frac{3}{4}$

   b. $\frac{2}{5} < \frac{4}{9}$
   because $\frac{2}{5} = \frac{4}{10}$
   4 tenths is less than 4 ninths because tenths are smaller than ninths.

   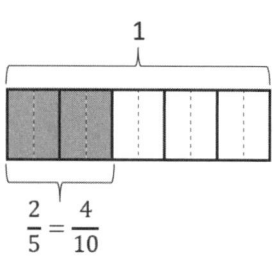
   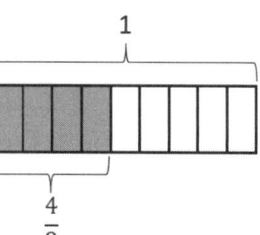

   c. $\frac{7}{11}$ _____ $\frac{7}{13}$ d. $\frac{6}{7}$ _____ $\frac{12}{15}$

Lesson 14: Find common units or number of units to compare two fractions.

3. Draw two tape diagrams to model each pair of the following fractions with related denominators. Use >, <, or = to compare.

   a. $\dfrac{2}{3}$ _____ $\dfrac{5}{6}$

   b. $\dfrac{3}{4}$ _____ $\dfrac{7}{8}$

   c. $1\dfrac{3}{4}$ _____ $1\dfrac{7}{12}$

A STORY OF UNITS Lesson 14 Problem Set 4•5

4. Draw one number line to model each pair of fractions with related denominators. Use >, <, or = to compare.

   a. $\dfrac{2}{3}$ _____ $\dfrac{5}{6}$

   b. $\dfrac{3}{8}$ _____ $\dfrac{1}{4}$

   c. $\dfrac{2}{6}$ _____ $\dfrac{5}{12}$

   d. $\dfrac{8}{9}$ _____ $\dfrac{2}{3}$

5. Compare each pair of fractions using >, <, or =. Draw a model if you choose to.

   a. $\dfrac{3}{4}$ _____ $\dfrac{3}{7}$

   b. $\dfrac{4}{5}$ _____ $\dfrac{8}{12}$

   c. $\dfrac{7}{10}$ _____ $\dfrac{3}{5}$

   d. $\dfrac{2}{3}$ _____ $\dfrac{11}{15}$

   e. $\dfrac{3}{4}$ _____ $\dfrac{11}{12}$

   f. $\dfrac{7}{3}$ _____ $\dfrac{7}{4}$

   g. $1\dfrac{1}{3}$ _____ $1\dfrac{2}{9}$

   h. $1\dfrac{2}{3}$ _____ $1\dfrac{4}{7}$

Lesson 14: Find common units or number of units to compare two fractions.

6. Timmy drew the picture to the right and claimed that $\frac{2}{3}$ is less than $\frac{7}{12}$. Evan says he thinks $\frac{2}{3}$ is greater than $\frac{7}{12}$. Who is correct? Support your answer with a picture.

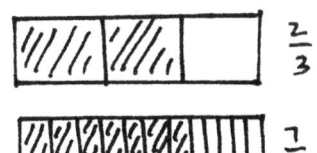

Name _____ Date _____

1. Draw tape diagrams to compare the following fractions:

   $\frac{2}{5}$ _____ $\frac{3}{10}$

2. Use a number line to compare the following fractions:

   $\frac{4}{3}$ _____ $\frac{7}{6}$

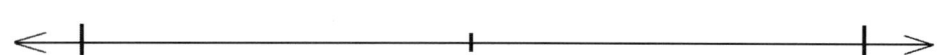

A STORY OF UNITS   Lesson 14 Homework   4•5

Name _____   Date _____

1. Compare the pairs of fractions by reasoning about the size of the units. Use >, <, or =.

   a. 1 third _____ 1 sixth

   b. 2 halves _____ 2 thirds

   c. 2 fourths _____ 2 sixths

   d. 5 eighths _____ 5 tenths

2. Compare by reasoning about the following pairs of fractions with the same or related numerators. Use >, <, or =. Explain your thinking using words, pictures, or numbers. Problem 2(b) has been done for you.

   a. $\dfrac{3}{6}$ _____ $\dfrac{3}{7}$

   b. $\dfrac{2}{5} < \dfrac{4}{9}$

   because $\dfrac{2}{5} = \dfrac{4}{10}$

   4 tenths is less than 4 ninths because tenths are smaller than ninths.

   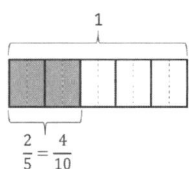

   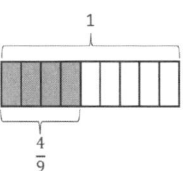

   c. $\dfrac{3}{11}$ _____ $\dfrac{3}{13}$

   d. $\dfrac{5}{7}$ _____ $\dfrac{10}{13}$

Lesson 14: Find common units or number of units to compare two fractions.

A STORY OF UNITS

Lesson 14 Homework 4•5

3. Draw two tape diagrams to model each pair of the following fractions with related denominators. Use >, <, or = to compare.

   a. $\frac{3}{4}$ _____ $\frac{7}{12}$

   b. $\frac{2}{4}$ _____ $\frac{1}{8}$

   c. $1\frac{4}{10}$ _____ $1\frac{3}{5}$

4. Draw one number line to model each pair of fractions with related denominators. Use >, <, or = to compare.

   a. $\dfrac{3}{4}$ _____ $\dfrac{5}{8}$

   b. $\dfrac{11}{12}$ _____ $\dfrac{3}{4}$

   c. $\dfrac{4}{5}$ _____ $\dfrac{7}{10}$

   d. $\dfrac{8}{9}$ _____ $\dfrac{2}{3}$

5. Compare each pair of fractions using >, <, or =. Draw a model if you choose to.

   a. $\dfrac{1}{7}$ _____ $\dfrac{2}{7}$

   b. $\dfrac{5}{7}$ _____ $\dfrac{11}{14}$

   c. $\dfrac{7}{10}$ _____ $\dfrac{3}{5}$

   d. $\dfrac{2}{3}$ _____ $\dfrac{9}{15}$

   e. $\dfrac{3}{4}$ _____ $\dfrac{9}{12}$

   f. $\dfrac{5}{3}$ _____ $\dfrac{5}{2}$

Lesson 14: Find common units or number of units to compare two fractions.

6. Simon claims $\frac{4}{9}$ is greater than $\frac{1}{3}$. Ted thinks $\frac{4}{9}$ is less than $\frac{1}{3}$. Who is correct? Support your answer with a picture.

Lesson 15

Objective: Find common units or number of units to compare two fractions.

Suggested Lesson Structure

- **Fluency Practice** (12 minutes)
- **Application Problem** (5 minutes)
- **Concept Development** (33 minutes)
- **Student Debrief** (10 minutes)

Total Time **(60 minutes)**

Fluency Practice (12 minutes)

- Count by Equivalent Fractions **4.NF.1**   (4 minutes)
- Find Equivalent Fractions **4.NF.1**   (4 minutes)
- Compare Fractions **4.NF.2**   (4 minutes)

Count by Equivalent Fractions (4 minutes)

Note: This activity builds fluency with equivalent fractions. The progression builds in complexity. Work students up to the highest level of complexity at which they can confidently participate.

- T: Count by ones to 4, starting at 0.
- S: 0, 1, 2, 3, 4.
- T: Count by fourths to 4 fourths. Start at 0 fourths. (Write as students count.)
- S: $\frac{0}{4}, \frac{1}{4}, \frac{2}{4}, \frac{3}{4}, \frac{4}{4}$.
- T: (Point to $\frac{4}{4}$.) 4 fourths is the same as 1 of what unit?
- S: 1 one.
- T: (Beneath $\frac{4}{4}$, write 1.) Count by fourths again. This time, when you come to 1, say 1. Start at zero. Try not to look at the board.
- S: $0, \frac{1}{4}, \frac{2}{4}, \frac{3}{4}, 1$.
- T: (Point to $\frac{2}{4}$.) 2 fourths is the same as 1 of what unit?
- S: 1 half.

$\frac{0}{4}$	$\frac{1}{4}$	$\frac{2}{4}$	$\frac{3}{4}$	$\frac{4}{4}$
0	$\frac{1}{4}$	$\frac{2}{4}$	$\frac{3}{4}$	1
0	$\frac{1}{4}$	$\frac{1}{2}$	$\frac{3}{4}$	1

Lesson 15: Find common units or number of units to compare two fractions.

T: (Beneath $\frac{2}{4}$, write $\frac{1}{2}$.) Count by fourths again. This time, convert to halves and whole numbers. Try not to look at the board.

S: $0, \frac{1}{4}, \frac{1}{2}, \frac{3}{4}, 1$.

Direct students to count forward and backward by fourths from 0 to 1, occasionally changing directions.

Find Equivalent Fractions (4 minutes)

Materials: (S) Personal white board

Note: This fluency activity reviews skills applied in Lesson 14.

T: (Write $\frac{1}{2} = \frac{\times}{\times} = \frac{2}{\quad}$. Point to $\frac{1}{2}$.) Say the unit fraction.

S: 1 half.

T: On your personal white board, fill in the unknown numbers to make an equivalent fraction.

S: (Write $\frac{1}{2} = \frac{1 \times 2}{2 \times 2} = \frac{2}{4}$.)

Continue with the following possible sequence: $\frac{1}{2} = \frac{4}{8}, \frac{1}{3} = \frac{2}{6}, \frac{1}{3} = \frac{3}{9}, \frac{1}{4} = \frac{4}{16}, \frac{1}{5} = \frac{3}{15}$.

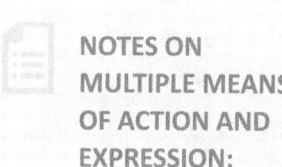

NOTES ON MULTIPLE MEANS OF ACTION AND EXPRESSION:

Fluency drills are fun, fast-paced math games. Be careful not to leave English language learners behind. Make sure to clarify that *common unit, common denominator, like unit,* and *like denominator* are terms that refer to the same thing and are often used in math class interchangeably.

Compare Fractions (4 minutes)

Materials: (S) Personal white board

Note: This fluency activity reviews Lesson 14.

T: (Write $\frac{1}{2}$ ___ $\frac{3}{4}$.) Write this on your personal white board. Then, find a common denominator, and write the greater than or less than sign.

S: (Write $\frac{1}{2}$ ___ $\frac{3}{4}$. Beneath it, write $\frac{2}{4} < \frac{3}{4}$.)

Continue with the following possible sequence: $\frac{1}{2}$—$\frac{3}{8}$, $\frac{1}{4}$—$\frac{3}{8}$, $\frac{5}{6}$—$\frac{1}{3}$, $\frac{1}{4}$—$\frac{5}{12}$, and $\frac{1}{3}$—$\frac{2}{9}$.

A STORY OF UNITS

Lesson 15 4•5

Application Problem (5 minutes)

Jamal ran $\frac{2}{3}$ mile. Ming ran $\frac{2}{4}$ mile. Laina ran $\frac{7}{12}$ mile. Who ran the farthest? What do you think is the easiest way to determine the answer to this question? Talk with a partner about your ideas.

Jamal ran the farthest.
It is easiest to form equivalent fractions since all 3 fractions have different denominators.

$\frac{2}{4} = \frac{2 \times 3}{4 \times 3} = \frac{6}{12}$

$\frac{2}{3} = \frac{2 \times 4}{3 \times 4} = \frac{8}{12}$

$\frac{7}{12}$

$\frac{6}{12} < \frac{7}{12} < \frac{8}{12}$

$\frac{2}{4} < \frac{7}{12} < \frac{2}{3}$

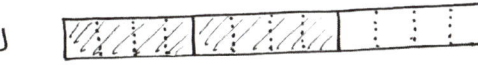
J

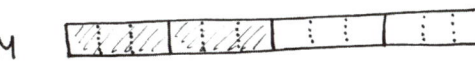
M

L

Note: This Application Problem reviews skills learned in Topic B to compare fractions and anticipates finding common units in this lesson. Be ready for conversations centered around comparing the fractions in other ways. Such conversations might include area models, tape diagrams, and finding equivalent fractions.

Concept Development (33 minutes)

Materials: (S) Personal white board

Problem 1: Compare two fractions with unrelated denominators using area models.

- T: (Display $\frac{3}{4}$ and $\frac{4}{5}$.) We have compared fractions by using benchmarks to help us reason. Another way to compare fractions is to find like units.
- T: Draw two almost-square rectangles that are the same size. Each model is 1. Partition the left area model into fourths by drawing vertical lines. (Model.)
- S: (Draw two almost-square rectangles, partitioning the left area model into fourths.)
- T: Shade $\frac{3}{4}$ of the left area model. Partition the right area model into fifths by drawing horizontal lines. Shade $\frac{4}{5}$. (Demonstrate.)
- S: (Shade $\frac{3}{4}$ of the left area model. Partition the right area model into fifths by drawing horizontal lines. Shade $\frac{4}{5}$.)

> **NOTES ON MULTIPLE MEANS OF REPRESENTATION:**
>
> When comparing fractions, we seek to make common units. This can be modeled by representing $\frac{3}{4}$ vertically, while representing $\frac{4}{5}$ horizontally. Then, each model is decomposed to make twentieths. Both models then show common units of the same size and shape, even if the whole units are not drawn perfectly square.

MP.2

Lesson 15: Find common units or number of units to compare two fractions.

A STORY OF UNITS

Lesson 15 4•5

T: Do we have like denominators?
S: No.
T: Partition each fourth into 5 equal pieces. (Demonstrate drawing horizontal lines.)
T: How many units are in the whole now?
S: 20.
T: What is the value of one of the new units?
S: 1 twentieth.
T: How many twentieths are shaded?
S: 15.
T: Now, let's decompose $\frac{4}{5}$. Partition each fifth into 4 equal pieces. (Model the decomposition drawing vertical lines.) How many twentieths are the same as $\frac{4}{5}$?
S: $\frac{16}{20}$ is the same as $\frac{4}{5}$.
T: Now that we have common units, can you compare the fractions?
S: Yes! $\frac{15}{20}$ is less than $\frac{16}{20}$, so $\frac{3}{4}$ is less than $\frac{4}{5}$.
T: How did we decompose $\frac{4}{5}$ and $\frac{3}{4}$ to compare?
S: We made common units so that we would be able to compare the fractions. First, we drew area models to show each fraction. We partitioned one using vertical lines and the other using horizontal lines. Then, we partitioned each model again to create like units. Once we had like units, it was easy to compare the fractions. We compared $\frac{15}{20}$ and $\frac{16}{20}$. Then, we knew that $\frac{3}{4} < \frac{4}{5}$.

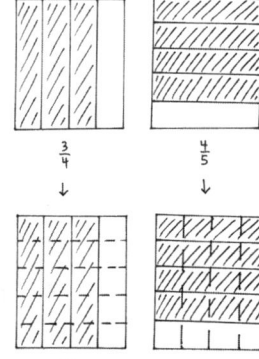

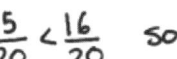 so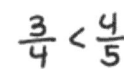

Repeat with $\frac{2}{3}$ and $\frac{3}{5}$, drawing thirds vertically and fifths horizontally. Then, partition the thirds into fifths and the fifths into thirds.

Problem 2: Compare two fractions greater than one with unrelated denominators using number bonds and area models.

T: (Display $\frac{5}{3}$ and $\frac{7}{4}$.) These fractions are greater than 1. Draw number bonds to show how $\frac{5}{3}$ and $\frac{7}{4}$ can be expressed as the sum of a whole number and a fraction.
S: $\frac{5}{3} = \frac{3}{3} + \frac{2}{3}$ and $\frac{7}{4} = \frac{4}{4} + \frac{3}{4}$.
T: Since the wholes are the same, we can just compare $\frac{2}{3}$ and $\frac{3}{4}$. Draw area models once again to help.
S: $\frac{2}{3}$ is less than $\frac{3}{4}$. → Since $\frac{2}{3}$ is less than $\frac{3}{4}$, $1\frac{2}{3}$ is less than $1\frac{3}{4}$. → $\frac{5}{3}$ is less than $\frac{7}{4}$.

Repeat with $\frac{6}{4}$ and $\frac{7}{5}$.

Lesson 15: Find common units or number of units to compare two fractions.

A STORY OF UNITS

Lesson 15 4•5

Problem 3: Compare two fractions with unrelated denominators without an area model.

T: We modeled common units to compare $\frac{4}{5}$ and $\frac{3}{4}$. What was the common unit?

S: Twentieths!

T: Use multiplication to show that $\frac{4}{5}$ is the same as $\frac{16}{20}$.

S: $\frac{4}{5} = \frac{4 \times 4}{5 \times 4} = \frac{16}{20}$.

T: Use multiplication to show that $\frac{3}{4}$ is the same as $\frac{15}{20}$.

S: $\frac{3}{4} = \frac{3 \times 5}{4 \times 5} = \frac{15}{20}$.

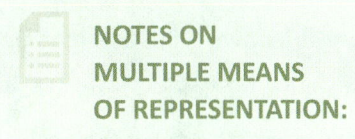

NOTES ON MULTIPLE MEANS OF REPRESENTATION:

For students who struggle to represent fractions precisely, provide a template of equally sized rectangles that can be partitioned as area models. This helps them to compare fractions more easily.

T: We decomposed by multiplying by the denominator of the other fraction.

T: Let's compare $\frac{3}{5}$ and $\frac{8}{12}$ by multiplying the denominators. We could use area models, but that would be a lot of little boxes!

T: (Write $\frac{3}{5} = \frac{3 \times 12}{5 \times 12} = \frac{}{60}$.) How many sixtieths are the same as 3 fifths? Write your answer as a multiplication sentence.

S: $\frac{3}{5} = \frac{3 \times 12}{5 \times 12} = \frac{36}{60}$.

T: (Write $\frac{8}{12} = \frac{8 \times 5}{12 \times 5} = \frac{}{60}$.) How many sixtieths are the same as 8 twelfths? Write your answer as a multiplication sentence.

S: $\frac{8}{12} = \frac{8 \times 5}{12 \times 5} = \frac{40}{60}$.

T: Compare $\frac{3}{5}$ and $\frac{8}{12}$.

S: $\frac{36}{60} < \frac{40}{60}$, so $\frac{3}{5} < \frac{8}{12}$.

T: Write $\frac{9}{5}$ and $\frac{10}{8}$. Express each as an equivalent fraction using multiplication.

S: $\frac{9}{5} = \frac{9 \times 8}{5 \times 8} = \frac{72}{40}$.

$\frac{10}{8} = \frac{10 \times 5}{8 \times 5} = \frac{50}{40}$.

T: $\frac{72}{40} > \frac{50}{40}$. That means $\frac{9}{5} > \frac{10}{8}$.

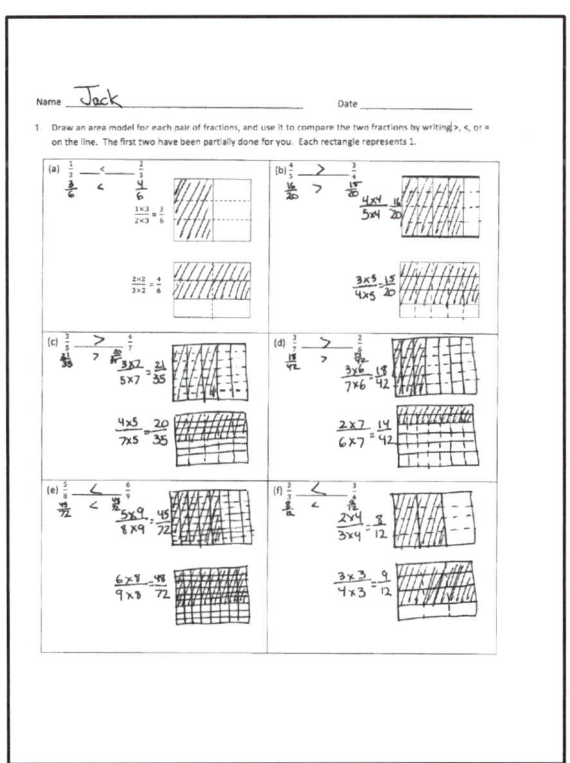

Problem Set (10 minutes)

Students should do their personal best to complete the Problem Set within the allotted 10 minutes. For some classes, it may be appropriate to modify the assignment by specifying which problems they work on first. Some problems do not specify a method for solving. Students should solve these problems using the RDW approach used for Application Problems.

Lesson 15: Find common units or number of units to compare two fractions.

203

© 2015 Great Minds. eureka-math.org
G4-M5-TE-B4-1.3.1-01.2016

Student Debrief (10 minutes)

Lesson Objective: Find common units or number of units to compare two fractions.

The Student Debrief is intended to invite reflection and active processing of the total lesson experience.

Invite students to review their solutions for the Problem Set. They should check work by comparing answers with a partner before going over answers as a class. Look for misconceptions or misunderstandings that can be addressed in the Debrief. Guide students in a conversation to debrief the Problem Set and process the lesson.

Any combination of the questions below can be used to lead the discussion.

- In Problem 2, did you need to use multiplication for every part? Why or why not? When is multiplication not needed, even with different denominators?
- In Problem 2(b), did everyone use forty-eighths? Did anyone use twenty-fourths?
- In Problem 3, how did you compare the fractions? Why?
- Do we always need to multiply the denominators to make like units?
- If fractions are hard to compare, we can always get like units by multiplying denominators—a method that always works. Why is it sometimes not the best way to compare fractions?
- What new or significant math vocabulary did we use today to communicate precisely?
- How did the Application Problem connect to today's lesson?

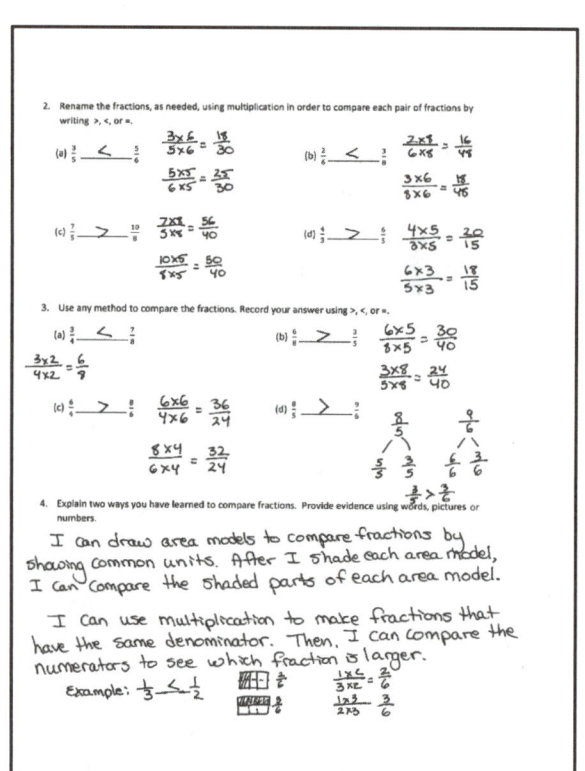

Exit Ticket (3 minutes)

After the Student Debrief, instruct students to complete the Exit Ticket. A review of their work will help with assessing students' understanding of the concepts that were presented in today's lesson and planning more effectively for future lessons. The questions may be read aloud to the students.

A STORY OF UNITS Lesson 15 Problem Set 4•5

Name _____ Date _____

1. Draw an area model for each pair of fractions, and use it to compare the two fractions by writing >, <, or = on the line. The first two have been partially done for you. Each rectangle represents 1.

a. $\dfrac{1}{2}$ ____<____ $\dfrac{2}{3}$

$\dfrac{1 \times 3}{2 \times 3} = \dfrac{3}{6}$

$\dfrac{2 \times 2}{3 \times 2} = \dfrac{4}{6}$

b. $\dfrac{4}{5}$ _____ $\dfrac{3}{4}$

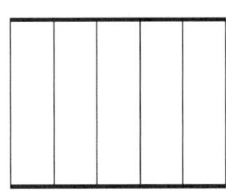

c. $\dfrac{3}{5}$ _____ $\dfrac{4}{7}$

d. $\dfrac{3}{7}$ _____ $\dfrac{2}{6}$

e. $\dfrac{5}{8}$ _____ $\dfrac{6}{9}$

f. $\dfrac{2}{3}$ _____ $\dfrac{3}{4}$

Lesson 15: Find common units or number of units to compare two fractions.

A STORY OF UNITS   Lesson 15 Problem Set   4•5

2. Rename the fractions, as needed, using multiplication in order to compare each pair of fractions by writing >, <, or =.

    a. $\frac{3}{5}$ _____ $\frac{5}{6}$ b. $\frac{2}{6}$ _____ $\frac{3}{8}$

    c. $\frac{7}{5}$ _____ $\frac{10}{8}$ d. $\frac{4}{3}$ _____ $\frac{6}{5}$

3. Use any method to compare the fractions. Record your answer using >, <, or =.

    a. $\frac{3}{4}$ _____ $\frac{7}{8}$ b. $\frac{6}{8}$ _____ $\frac{3}{5}$

    c. $\frac{6}{4}$ _____ $\frac{8}{6}$ d. $\frac{8}{5}$ _____ $\frac{9}{6}$

Lesson 15:   Find common units or number of units to compare two fractions.

EUREKA MATH

4. Explain two ways you have learned to compare fractions. Provide evidence using words, pictures, or numbers.

A STORY OF UNITS

Lesson 15 Exit Ticket 4•5

Name _____ Date _____

Draw an area model for each pair of fractions, and use it to compare the two fractions by writing >, <, or = on the line.

1. $\dfrac{3}{4}$ _____ $\dfrac{4}{5}$

2. $\dfrac{2}{6}$ _____ $\dfrac{3}{5}$

Lesson 15: Find common units or number of units to compare two fractions.

A STORY OF UNITS

Lesson 15 Homework 4•5

Name _____ Date _____

1. Draw an area model for each pair of fractions, and use it to compare the two fractions by writing >, <, or = on the line. The first two have been partially done for you. Each rectangle represents 1.

   a. $\frac{1}{2}$ ____<____ $\frac{3}{5}$

   $\frac{1\times5}{2\times5}=\frac{5}{10}$ $\frac{3\times2}{5\times2}=\frac{6}{10}$

   $\frac{5}{10}<\frac{6}{10}$, so $\frac{1}{2}<\frac{3}{5}$

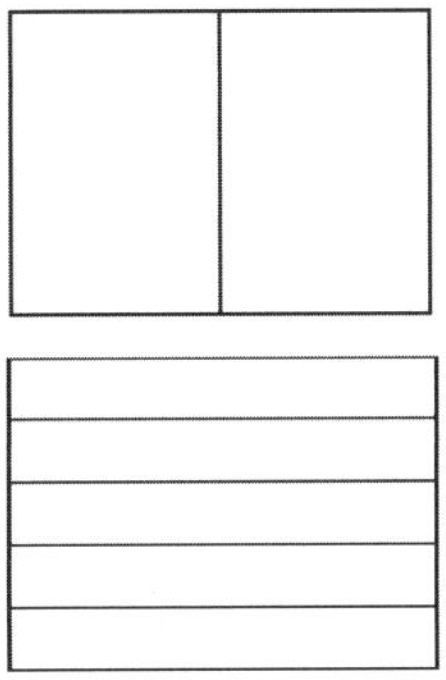

   b. $\frac{2}{3}$ _____ $\frac{3}{4}$

   c. $\frac{4}{6}$ _____ $\frac{5}{8}$

   d. $\frac{2}{7}$ _____ $\frac{3}{5}$

   e. $\frac{4}{6}$ _____ $\frac{6}{9}$

   f. $\frac{4}{5}$ _____ $\frac{5}{6}$

Lesson 15: Find common units or number of units to compare two fractions.

209

2. Rename the fractions, as needed, using multiplication in order to compare each pair of fractions by writing >, <, or =.

   a. $\dfrac{2}{3}$ _____ $\dfrac{2}{4}$

   b. $\dfrac{4}{7}$ _____ $\dfrac{1}{2}$

   c. $\dfrac{5}{4}$ _____ $\dfrac{9}{8}$

   d. $\dfrac{8}{12}$ _____ $\dfrac{5}{8}$

3. Use any method to compare the fractions. Record your answer using >, <, or =.

   a. $\dfrac{8}{9}$ _____ $\dfrac{2}{3}$

   b. $\dfrac{4}{7}$ _____ $\dfrac{4}{5}$

   c. $\dfrac{3}{2}$ _____ $\dfrac{9}{6}$

   d. $\dfrac{11}{7}$ _____ $\dfrac{5}{3}$

4. Explain which method you prefer using to compare fractions. Provide an example using words, pictures, or numbers.

A STORY OF UNITS

GRADE 4

Mathematics Curriculum

GRADE 4 • MODULE 5

Topic D
Fraction Addition and Subtraction

4.NF.3ad, 4.NF.1, 4.MD.2

Focus Standards:	4.NF.3ad	Understand a fraction *a/b* with *a* > 1 as a sum of fractions 1/*b*.
		a. Understand addition and subtraction of fractions as joining and separating parts referring to the same whole.
		d. Solve word problems involving addition and subtraction of fractions referring to the same whole and having like denominators, e.g., by using visual fraction models and equations to represent the problem.
Instructional Days:	6	
Coherence -Links from:	G3–M5	Fractions as Numbers on the Number Line
-Links to:	G5–M3	Addition and Subtraction of Fractions

Topic D bridges students' understanding of whole number addition and subtraction to fractions. Everything that they know to be true of addition and subtraction with whole numbers now applies to fractions. Addition is finding a total by combining like units. Subtraction is finding an unknown part. Implicit in the equations 3 + 2 = 5 and 2 = 5 – 3 is the assumption that the numbers are referring to the *same* units.

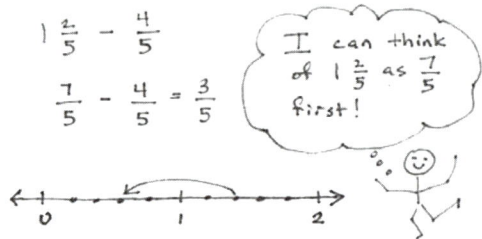

In Lessons 16 and 17, students generalize familiar facts about whole number addition and subtraction to work with fractions. Just as 3 apples – 2 apples = 1 apple, students note that 3 fourths – 2 fourths = 1 fourth. Just as 6 days + 3 days = 9 days = 1 week 2 days, students note that $\frac{6}{7} + \frac{3}{7} = \frac{9}{7} = \frac{7}{7} + \frac{2}{7} = 1\frac{2}{7}$. In Lesson 17, students decompose a whole into a fraction having the same denominator as the subtrahend. For example, 1 – 4 fifths becomes 5 fifths – 4 fifths = 1 fifth, connecting with Topic B skills. They then see that, when solving $1\frac{2}{5} - \frac{4}{5}$, they have a choice of subtracting $\frac{4}{5}$ from $\frac{7}{5}$ or from 1 (as pictured to the right). Students model with tape diagrams and number lines to understand and then verify their numerical work.

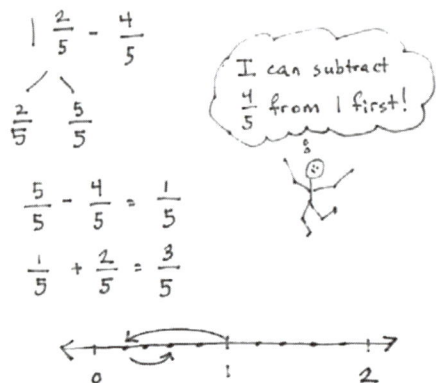

212   Topic D: Fraction Addition and Subtraction

A STORY OF UNITS

Topic D 4•5

In Lesson 18, students add more than two fractions and see sums of more than one whole, such as $\frac{2}{8} + \frac{5}{8} + \frac{7}{8} = \frac{14}{8}$. As students move into problem solving in Lesson 19, they create tape diagrams or number lines to represent and solve fraction addition and subtraction word problems (see the example below). These problems bridge students into work with mixed numbers, which follows the Mid-Module Assessment.

Mary mixed $\frac{3}{4}$ cup of wheat flour, $\frac{2}{4}$ cup of rice flour, and $\frac{1}{4}$ cup of oat flour for her bread dough. How many cups of flour did she put in her bread in all?

$\frac{3}{4} + \frac{2}{4} + \frac{1}{4} = \frac{6}{4}$

$\frac{6}{4} = \frac{4}{4} + \frac{2}{4} = 1 + \frac{2}{4} = 1\frac{2}{4}$

Mary used $\frac{6}{4}$ or $1\frac{2}{4}$ cups flour.

In Lessons 20 and 21, students add fractions with related units, where one denominator is a multiple (or factor) of the other. To add such fractions, a decomposition is necessary. Decomposing one unit into another is familiar territory: Students have had ample practice composing and decomposing in Topics A and B when working with place value units, converting units of measurement, and using the distributive property. For example, they have converted between equivalent measurement units (e.g., 100 cm = 1 m), and they have used such conversions to do arithmetic (e.g., 1 meter – 54 centimeters). With fractions, the concept is the same. To find the sum of $\frac{1}{2}$ and $\frac{1}{4}$, one simply renames (converts, decomposes) $\frac{1}{2}$ as $\frac{2}{4}$ and adds: $\frac{2}{4} + \frac{1}{4} = \frac{3}{4}$.

All numerical work is accompanied by visual models that allow students to use and apply their known skills and understandings. The addition of fractions with related units is also foundational to decimal work when adding tenths and hundredths in Module 6. Please note that addition of fractions with related denominators is not assessed.

Topic D: Fraction Addition and Subtraction

213

| A STORY OF UNITS | Topic D | 4•5 |

A Teaching Sequence Toward Mastery of Fraction Addition and Subtraction

Objective 1: Use visual models to add and subtract two fractions with the same units.
(Lesson 16)

Objective 2: Use visual models to add and subtract two fractions with the same units, including subtracting from one whole.
(Lesson 17)

Objective 3: Add and subtract more than two fractions.
(Lesson 18)

Objective 4: Solve word problems involving addition and subtraction of fractions.
(Lesson 19)

Objective 5: Use visual models to add two fractions with related units using the denominators 2, 3, 4, 5, 6, 8, 10, and 12.
(Lessons 20–21)

A STORY OF UNITS   Lesson 16   4•5

Lesson 16

Objective: Use visual models to add and subtract two fractions with the same units.

Suggested Lesson Structure

- ■ Fluency Practice   (12 minutes)
- ■ Application Problem   (5 minutes)
- ■ Concept Development   (33 minutes)
- ■ Student Debrief   (10 minutes)
-    **Total Time**   **(60 minutes)**

Fluency Practice (12 minutes)

- Count by Equivalent Fractions **4.NF.1**   (6 minutes)
- Compare Fractions **4.NF.2**   (6 minutes)

Count by Equivalent Fractions (6 minutes)

Note: This activity builds fluency with equivalent fractions. The progression builds in complexity. Work students up to the highest level of complexity in which they can confidently participate.

$\frac{0}{8}$	$\frac{1}{8}$	$\frac{2}{8}$	$\frac{3}{8}$	$\frac{4}{8}$	$\frac{5}{8}$	$\frac{6}{8}$	$\frac{7}{8}$	$\frac{8}{8}$
0	$\frac{1}{8}$	$\frac{2}{8}$	$\frac{3}{8}$	$\frac{4}{8}$	$\frac{5}{8}$	$\frac{6}{8}$	$\frac{7}{8}$	1
0	$\frac{1}{8}$	$\frac{2}{8}$	$\frac{3}{8}$	$\frac{1}{2}$	$\frac{5}{8}$	$\frac{6}{8}$	$\frac{7}{8}$	1
0	$\frac{1}{8}$	$\frac{1}{4}$	$\frac{3}{8}$	$\frac{1}{2}$	$\frac{5}{8}$	$\frac{3}{4}$	$\frac{7}{8}$	1

T: Starting at 0, count by ones to 8.
S: 0, 1, 2, 3, 4, 5, 6, 7, 8.
T: Starting at 0 eighths, count by 1 eighths to 8 eighths. (Write as students count.)
S: $\frac{0}{8}, \frac{1}{8}, \frac{2}{8}, \frac{3}{8}, \frac{4}{8}, \frac{5}{8}, \frac{6}{8}, \frac{7}{8}, \frac{8}{8}$.
T: (Point to $\frac{8}{8}$.) 8 eighths is the same as 1 of what unit?
S: 1 one.

Lesson 16: Use visual models to add and subtract two fractions with the same units.   215

T: (Beneath $\frac{8}{8}$ write 1.) Count by 1 eighths from zero to 1. This time, when you come to 1, say "1." Try not to look at the board.

S: 0, $\frac{1}{8}$, $\frac{2}{8}$, $\frac{3}{8}$, $\frac{4}{8}$, $\frac{5}{8}$, $\frac{6}{8}$, $\frac{7}{8}$, 1.

T: (Point to $\frac{4}{8}$.) 4 eighths is the same as 1 of what unit?

S: 1 half.

T: (Beneath $\frac{4}{8}$, write $\frac{1}{2}$.) Count by 1 eighths again. This time, convert to $\frac{1}{2}$ and 1. Try not to look at the board.

S: 0, $\frac{1}{8}$, $\frac{2}{8}$, $\frac{3}{8}$, $\frac{1}{2}$, $\frac{5}{8}$, $\frac{6}{8}$, $\frac{7}{8}$, 1.

T: What other fractions can we rename to make smaller-sized units?

S: $\frac{2}{8}$ and $\frac{6}{8}$.

T: (Point to $\frac{2}{8}$.) What's 2 eighths renamed?

S: $\frac{1}{4}$.

T: (Beneath $\frac{2}{8}$, write $\frac{1}{4}$. Point to $\frac{6}{8}$.) What's $\frac{6}{8}$ renamed?

S: $\frac{3}{4}$.

T: (Beneath $\frac{6}{8}$, write $\frac{3}{4}$.) Count by 1 eighths again. This time, convert to $\frac{1}{4}$ and $\frac{3}{4}$. Try not to look at the board.

S: 0, $\frac{1}{8}$, $\frac{1}{4}$, $\frac{3}{8}$, $\frac{1}{2}$, $\frac{5}{8}$, $\frac{3}{4}$, $\frac{7}{8}$, 1.

Direct students to count by eighths back and forth from 0 to 1, occasionally changing directions.

Compare Fractions (6 minutes)

Materials: (S) Personal white board

Note: This fluency activity reviews Lesson 15.

T: On your personal white board, draw two area models. (Allow students time to draw.)

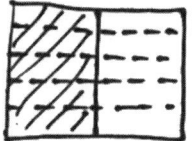

T: (Write $\frac{1}{2}$.) Partition and shade your first area model to show $\frac{1}{2}$. Then, write $\frac{1}{2}$ beneath it.

S: (Partition the first area model into 2 equal units. Shade one unit. Write $\frac{1}{2}$ beneath it.)

T: (Write $\frac{1}{2}$ ___ $\frac{2}{5}$.) Partition and shade your second area model to show $\frac{2}{5}$. Then, write $\frac{2}{5}$ beneath it.

S: (Partition the second area model into 5 equal units. Shade 2 units. Write $\frac{2}{5}$ beneath the shaded area.)

T: Partition the area models so that both fractions have common denominators.

S: (Draw dotted lines through the area models.)

A STORY OF UNITS   Lesson 16 4•5

T: Write a greater than, less than, or equal sign to compare the fractions.

S: (Write $\frac{1}{2} > \frac{2}{5}$.)

Continue with the following possible sequence: Compare $\frac{1}{5}$ and $\frac{3}{10}$, $\frac{1}{4}$ and $\frac{5}{8}$, and $\frac{1}{3}$ and $\frac{3}{4}$.

Application Problem (5 minutes)

Keisha ran $\frac{5}{6}$ mile in the morning and $\frac{2}{3}$ mile in the afternoon. Did Keisha run farther in the morning or in the afternoon? Solve independently. Share your solution with your partner. Did your partner solve the problem in the same way or a different way? Explain.

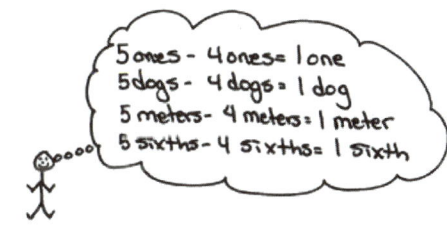

Note: This Application Problem builds on the Concept Developments of Lessons 14 and 15 where students learned to compare fractions with unrelated denominators by finding common units.

Concept Development (33 minutes)

Materials: (S) Personal white board, blank number lines (Template)

Problem 1: Solve for the difference using unit language and a number line.

T: (Project 5 – 4.) Solve. Say the number sentence using units of ones.

S: 5 ones – 4 ones = 1 one.

T: Say the number sentence if the unit is dogs.

S: 5 dogs – 4 dogs = 1 dog.

T: Say the number sentence if the unit is meters.

S: 5 meters – 4 meters = 1 meter.

T: Say the number sentence if the unit is sixths.

S: 5 sixths – 4 sixths = 1 sixth.

T: Let's show that 5 sixths – 4 sixths = 1 sixth.

T: (Project a number line with endpoints 0 and 1, partitioned into sixths.) Make tick marks on the first number line on your Template to make a number line with endpoints 0 and 1 above the number line. Partition the number line into sixths. (See the illustration on the next page.)

> **NOTES ON MULTIPLE MEANS OF REPRESENTATION:**
>
> Be sure to articulate the ending digraph /th/ to distinguish *six* from *sixth* for English language learners. Coupling spoken expressions with words or models may also improve student comprehension. For example, write out 5 *sixths* – 4 *sixths* = 1 *sixth*.

Lesson 16: Use visual models to add and subtract two fractions with the same units.

217

A STORY OF UNITS   Lesson 16   4•5

T: Draw a point at 5 sixths. Put the tip of your pencil on the point. Count backward to subtract 4 sixths.

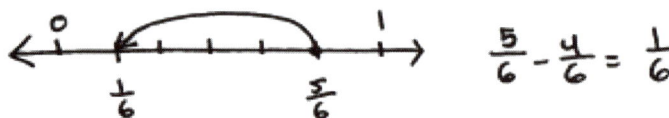

T: Move your pencil and count back with me as we subtract.
S: 4 sixths, 3 sixths, 2 sixths, 1 sixth!
T: Draw one arrow above the number line to model $\frac{5}{6} - \frac{4}{6}$. (Demonstrate.) Tell me the subtraction sentence.
S: $\frac{5}{6} - \frac{4}{6} = \frac{1}{6}$.

Repeat with $\frac{7}{8} - \frac{3}{8}$.

T: Solve for 7 sixths – 2 sixths. Work with a partner. Use the language of units and subtraction.
S: 7 sixths – 2 sixths = 5 sixths. → I know 7 ones minus 2 ones is 5 ones. I can subtract sixths like I subtract ones. $\frac{7}{6} - \frac{2}{6} = \frac{5}{6}$.
T: Discuss with your partner how to draw a number line to represent this problem.
S: We partition it like the first problem and draw the arrow to subtract. → But $\frac{7}{6}$ is more than 1 whole. 6 sixths is equal to 1. We have 7 sixths. → Let's make the number line with endpoints 0 and 2.
T: Label the endpoints 0 and 2. Partition the number line into sixths. Subtract.
S: On the number line, we started at 7 sixths and then went back 2 sixths. The answer is 5 sixths. → $\frac{7}{6} - \frac{2}{6} = \frac{5}{6}$.

> **NOTES ON MULTIPLE MEANS OF ENGAGEMENT:**
>
> Students working above grade level and others may present alternative subtraction strategies, such as counting up rather than counting down to solve $\frac{7}{6} - \frac{5}{6}$. Though not introduced in this lesson, the appropriate use of these strategies is desirable and is introduced later in the module.

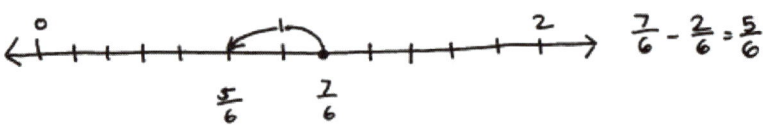

Repeat with $\frac{7}{4} - \frac{5}{4}$.

Problem 2: Decompose to record a difference greater than 1 as a mixed number.

T: (Display 10 sixths – 2 sixths.) Solve in unit form, and write a number sentence using fractions.
S: (Write 10 sixths – 2 sixths = 8 sixths and $\frac{10}{6} - \frac{2}{6} = \frac{8}{6}$.)
T: Use a number bond to decompose $\frac{8}{6}$ into the whole and fractional parts.
S: (Draw a number bond as pictured to the right.)
T: $\frac{6}{6}$ is the same as...?
S: 1 whole.

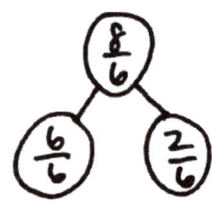

A STORY OF UNITS　　　　　　　　　　　　　　　　　　　　　Lesson 16　4•5

T: We can rename $\frac{8}{6}$ as a **mixed number**, $1\frac{2}{6}$, using a whole number and fractional parts.

$\frac{9}{5} - \frac{3}{5} = \frac{6}{5} = 1\frac{1}{5}$

$\wedge$
$\frac{5}{5}\ \frac{1}{5}$

Repeat with 9 fifths – 3 fifths.

Problem 3: Solve for the sum using unit language and a number line.

T: Look back at the first example. (Point to the number line representing 5 sixths – 4 sixths.) Put your finger on 1 sixth. To 1 sixth, let's add the 4 sixths that we took away.

T: Count as we add. 1 sixth, 2 sixths, 3 sixths, 4 sixths. Where are we now?

S: 5 sixths.

T: What is 1 sixth plus 4 sixths?

S: 5 sixths.

T: Let's show that on the number line. (Model with students as shown to the right.)

T: 1 one plus 4 ones is…?

S: 5 ones.

T: 1 apple plus 4 apples is…?

S: 5 apples.

T: 1 sixth plus 4 sixths equals…?

S: 5 sixths.

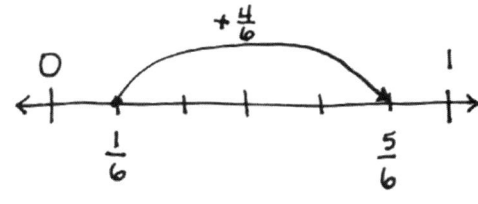

Repeat with $\frac{2}{8} + \frac{3}{8} = \frac{5}{8}$.

Problem 4: Decompose to record a sum greater than 1 as a mixed number.

T: (Display 5 fourths + 2 fourths.) Solve in unit form, and write a number sentence using fractions.

S: (Write 5 fourths + 2 fourths = 7 fourths and $\frac{5}{4} + \frac{2}{4} = \frac{7}{4}$.)

T: Use a number bond to decompose $\frac{7}{4}$ into the whole and some parts.

S: (Draw a number bond as pictured to the right.)

T: $\frac{4}{4}$ is the same as…?

S: 1 whole.

T: We can rename $\frac{7}{4}$ as a mixed number, $1\frac{3}{4}$.

5 fourths + 2 fourths = 7 fourths

$\frac{5}{4} + \frac{2}{4} = \frac{7}{4} = 1\frac{3}{4}$

$\wedge$
$\frac{4}{4}\ \frac{3}{4}$

Repeat with 6 sixths + 4 sixths.

Problem Set (10 minutes)

Students should do their personal best to complete the Problem Set within the allotted 10 minutes. For some classes, it may be appropriate to modify the assignment by specifying which problems they work on first. Some problems do not specify a method for solving. Students should solve these problems using the RDW approach used for Application Problems.

Lesson 16: Use visual models to add and subtract two fractions with the same units.

Student Debrief (10 minutes)

Lesson Objective: Use visual models to add and subtract two fractions with the same units.

The Student Debrief is intended to invite reflection and active processing of the total lesson experience.

Invite students to review their solutions for the Problem Set. They should check work by comparing answers with a partner before going over answers as a class. Look for misconceptions or misunderstandings that can be addressed in the Debrief. Guide students in a conversation to debrief the Problem Set and process the lesson.

Any combination of the questions below may be used to lead the discussion.

- How do Problems 1(a–d), 4(a), and 4(b) help you understand how to subtract or add fractions?
- In Problems 3 and 6 on the Problem Set, how do the number bonds help to decompose the fraction into a **mixed number**?
- Why would we want to name a fraction greater than 1 using a mixed number?
- How is the number line helpful in showing how we can subtract and add fractions with like units?
- How are number bonds helpful in showing how we can rename fractions greater than 1 as 1 whole and a fraction?
- How would you describe to a friend how to subtract and add fractions with like units?

Exit Ticket (3 minutes)

After the Student Debrief, instruct students to complete the Exit Ticket. A review of their work will help with assessing students' understanding of the concepts that were presented in today's lesson and planning more effectively for future lessons. The questions may be read aloud to the students.

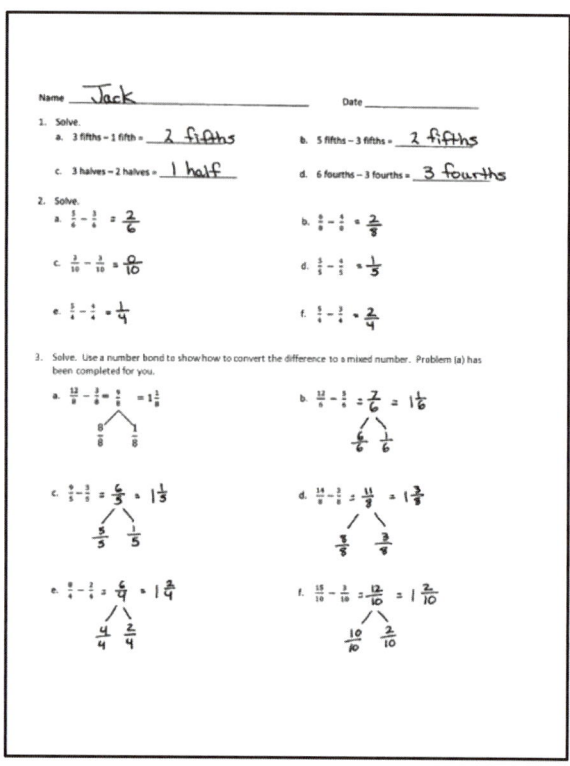

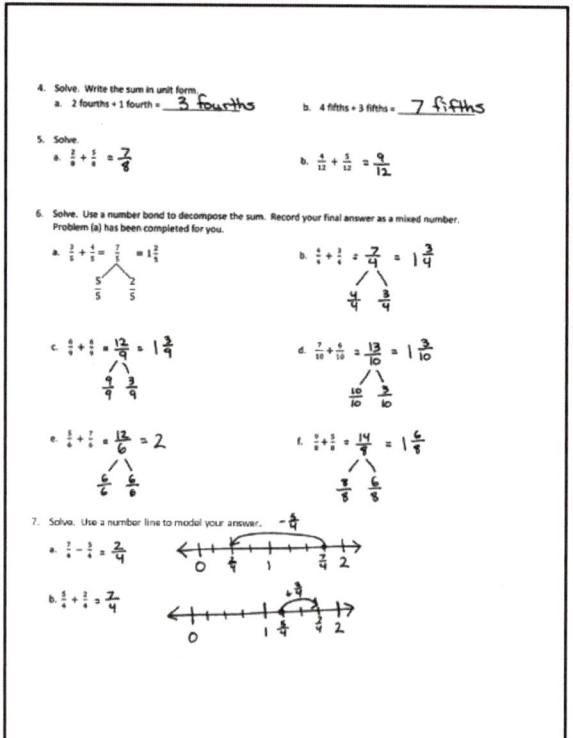

A STORY OF UNITS   **Lesson 16 Problem Set**   **4•5**

Name _____ Date _____

1. Solve.

   a. 3 fifths − 1 fifth = _____

   b. 5 fifths − 3 fifths = _____

   c. 3 halves − 2 halves = _____

   d. 6 fourths − 3 fourths = _____

2. Solve.

   a. $\dfrac{5}{6} - \dfrac{3}{6}$

   b. $\dfrac{6}{8} - \dfrac{4}{8}$

   c. $\dfrac{3}{10} - \dfrac{3}{10}$

   d. $\dfrac{5}{5} - \dfrac{4}{5}$

   e. $\dfrac{5}{4} - \dfrac{4}{4}$

   f. $\dfrac{5}{4} - \dfrac{3}{4}$

3. Solve. Use a number bond to show how to convert the difference to a mixed number. Problem (a) has been completed for you.

   a. $\dfrac{12}{8} - \dfrac{3}{8} = \dfrac{9}{8} = 1\dfrac{1}{8}$

   b. $\dfrac{12}{6} - \dfrac{5}{6}$

   c. $\dfrac{9}{5} - \dfrac{3}{5}$

   d. $\dfrac{14}{8} - \dfrac{3}{8}$

   e. $\dfrac{8}{4} - \dfrac{2}{4}$

   f. $\dfrac{15}{10} - \dfrac{3}{10}$

Lesson 16: Use visual models to add and subtract two fractions with the same units.

4. Solve. Write the sum in unit form.

   a. 2 fourths + 1 fourth = _____

   b. 4 fifths + 3 fifths = _____

5. Solve.

   a. $\frac{2}{8} + \frac{5}{8}$

   b. $\frac{4}{12} + \frac{5}{12}$

6. Solve. Use a number bond to decompose the sum. Record your final answer as a mixed number. Problem (a) has been completed for you.

   a. $\frac{3}{5} + \frac{4}{5} = \frac{7}{5} = 1\frac{2}{5}$

   number bond: $\frac{7}{5}$ decomposed into $\frac{5}{5}$ and $\frac{2}{5}$

   b. $\frac{4}{4} + \frac{3}{4}$

   c. $\frac{6}{9} + \frac{6}{9}$

   d. $\frac{7}{10} + \frac{6}{10}$

   e. $\frac{5}{6} + \frac{7}{6}$

   f. $\frac{9}{8} + \frac{5}{8}$

7. Solve. Use a number line to model your answer.

   a. $\frac{7}{4} - \frac{5}{4}$

   b. $\frac{5}{4} + \frac{2}{4}$

Name _____   Date _____

1. Solve. Use a number bond to decompose the difference. Record your final answer as a mixed number.

   $\dfrac{16}{9} - \dfrac{5}{9}$

2. Solve. Use a number bond to decompose the sum. Record your final answer as a mixed number.

   $\dfrac{5}{12} + \dfrac{11}{12}$

A STORY OF UNITS   Lesson 16 Homework 4•5

Name _____   Date _____

1. Solve.

   a. 3 sixths − 2 sixths = _____   b. 5 tenths − 3 tenths = _____

   c. 3 fourths − 2 fourths = _____ d. 5 thirds − 2 thirds = _____

2. Solve.

   a. $\frac{3}{5} - \frac{2}{5}$ b. $\frac{7}{9} - \frac{3}{9}$

   c. $\frac{7}{12} - \frac{3}{12}$ d. $\frac{6}{6} - \frac{4}{6}$

   e. $\frac{5}{3} - \frac{2}{3}$ f. $\frac{7}{4} - \frac{5}{4}$

3. Solve. Use a number bond to decompose the difference. Record your final answer as a mixed number. Problem (a) has been completed for you.

   a. $\frac{12}{6} - \frac{3}{6} = \frac{9}{6} = 1\frac{3}{6}$ b. $\frac{10}{8} - \frac{6}{8}$

      (number bond: $\frac{9}{6}$ decomposed into $\frac{6}{6}$ and $\frac{3}{6}$)

   c. $\frac{9}{5} - \frac{3}{5}$ d. $\frac{11}{4} - \frac{6}{4}$

   e. $\frac{10}{7} - \frac{2}{7}$ f. $\frac{21}{10} - \frac{9}{10}$

224   Lesson 16:   Use visual models to add and subtract two fractions with the same units.

A STORY OF UNITS　　　　　　　　　　　　　　　　　　　　　　Lesson 16 Homework　4•5

4. Solve. Write the sum in unit form.

   a. 4 fifths + 2 fifths = _____

   b. 5 eighths + 2 eighths = _____

5. Solve.

   a. $\frac{3}{11} + \frac{6}{11}$

   b. $\frac{3}{12} + \frac{6}{12}$

6. Solve. Use a number bond to decompose the sum. Record your final answer as a mixed number.

   a. $\frac{3}{4} + \frac{3}{4}$

   b. $\frac{8}{12} + \frac{6}{12}$

   c. $\frac{5}{8} + \frac{7}{8}$

   d. $\frac{8}{12} + \frac{5}{12}$

   e. $\frac{3}{5} + \frac{6}{5}$

   f. $\frac{4}{3} + \frac{2}{3}$

7. Solve. Use a number line to model your answer.

   a. $\frac{11}{9} - \frac{5}{9}$

   b. $\frac{13}{12} + \frac{4}{12}$

A STORY OF UNITS   **Lesson 16 Template**   4•5

Name _____ Date _____

blank number lines

A STORY OF UNITS

Lesson 17 4•5

Lesson 17

Objective: Use visual models to add and subtract two fractions with the same units, including subtracting from one whole.

Suggested Lesson Structure

- ■ Fluency Practice   (12 minutes)
- ■ Application Problem   (5 minutes)
- ■ Concept Development (33 minutes)
- ■ Student Debrief   (10 minutes)

   Total Time   **(60 minutes)**

Fluency Practice (12 minutes)

- Count by Equivalent Fractions **4.NF.1**   (4 minutes)
- Take Out the Whole Number **4.NF.3**   (4 minutes)
- Draw Tape Diagrams **4.NF.3**   (4 minutes)

Count by Equivalent Fractions (4 minutes)

Note: This activity builds fluency with equivalent fractions. The progression builds in complexity. Work students up to the highest level of complexity in which they can confidently participate.

T: Starting at 0, count by ones to 6.

S: 0, 1, 2, 3, 4, 5, 6.

T: Count by sixths from 0 sixths to 6 sixths. (Write as students count.)

S: $\frac{0}{6}, \frac{1}{6}, \frac{2}{6}, \frac{3}{6}, \frac{4}{6}, \frac{5}{6}, \frac{6}{6}$.

$\frac{0}{6}$	$\frac{1}{6}$	$\frac{2}{6}$	$\frac{3}{6}$	$\frac{4}{6}$	$\frac{5}{6}$	$\frac{6}{6}$
0	$\frac{1}{6}$	$\frac{2}{6}$	$\frac{3}{6}$	$\frac{4}{6}$	$\frac{5}{6}$	1
0	$\frac{1}{6}$	$\frac{2}{6}$	$\frac{1}{2}$	$\frac{4}{6}$	$\frac{5}{6}$	1
0	$\frac{1}{6}$	$\frac{1}{3}$	$\frac{1}{2}$	$\frac{2}{3}$	$\frac{5}{6}$	1

T: (Point to $\frac{6}{6}$.) 6 sixths is the same as 1 of what unit?

S: 1 one.

Lesson 17: Use visual models to add and subtract two fractions with the same units, including subtracting from one whole.

227

T: (Beneath $\frac{6}{6}$, write 1.) Count by 1 sixths again from 0 to 1. This time, when you come to $\frac{6}{6}$, say 1. Try not to look at the board. (Write as students count.)

S: 0, $\frac{1}{6}$, $\frac{2}{6}$, $\frac{3}{6}$, $\frac{4}{6}$, $\frac{5}{6}$, 1.

T: (Point to $\frac{3}{6}$.) 3 sixths is the same as 1 of what unit?

S: 1 half.

T: (Beneath $\frac{3}{6}$, write $\frac{1}{2}$.) Count by 1 sixths again. This time, include 1 half and 1. Try not to look at the board.

S: 0, $\frac{1}{6}$, $\frac{2}{6}$, $\frac{1}{2}$, $\frac{4}{6}$, $\frac{5}{6}$, 1.

T: What other fractions can we convert to larger units?

S: $\frac{2}{6}$ and $\frac{4}{6}$.

T: (Point to $\frac{2}{6}$.) 2 sixths is the same as what unit fraction?

S: $\frac{1}{3}$.

T: (Beneath $\frac{2}{6}$, write $\frac{1}{3}$. Point to $\frac{4}{6}$.) $\frac{4}{6}$ is the same as how many thirds?

S: $\frac{2}{3}$.

T: (Beneath $\frac{4}{6}$, write $\frac{2}{3}$.) Count by 1 sixths again. This time, include $\frac{1}{3}$ and $\frac{2}{3}$. Try not to look at the board.

S: 0, $\frac{1}{6}$, $\frac{1}{3}$, $\frac{1}{2}$, $\frac{2}{3}$, $\frac{5}{6}$, 1.

> **NOTES ON MULTIPLE MEANS OF ENGAGEMENT:**
>
> If students appear challenged beyond their comfort level, scaffold converting smaller units to larger units in the Count by Equivalent Fractions fluency activity. Before students count by sixths up to 1, have them count by sixths up to 1 half, converting one fraction at a time.

Direct students to count by sixths forward and backward from 0 to 1, occasionally changing directions.

Take Out the Whole Number (4 minutes)

Materials: (S) Personal white board

Note: This fluency activity prepares students for today's lesson.

T: How many halves are in 1?
S: 2 halves.
T: How many thirds are in 1?
S: 3 thirds.
T: How many fifths are in 1?
S: 5 fifths.
T: (Write $1\frac{2}{5}$. Beneath it, write a number bond. Write $\frac{2}{5}$ as one of the parts. Write $\frac{5}{5}$ for the other part.) On your personal white board, write the completed number bond.
S: (Write the completed number bond.)

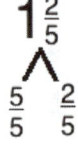

Continue with the following possible sequence of mixed numbers: $1\frac{3}{4}$, $1\frac{2}{3}$, $1\frac{3}{10}$, and $1\frac{5}{8}$.

A STORY OF UNITS   Lesson 17 4•5

Draw Tape Diagrams (4 minutes)

Materials: (S) Personal white board

Note: This fluency activity reviews Lesson 16.

T: (Write $\frac{2}{3} + \frac{2}{3} = \underline{\quad}$.) Complete the addition sentence.

S: $\frac{2}{3} + \frac{2}{3} = \frac{4}{3}$.

T: (Write $\frac{2}{3} + \frac{2}{3} = \frac{4}{3}$.) Draw a tape diagram to show $\frac{2}{3} + \frac{2}{3} = \frac{4}{3}$.

S: (Draw tape diagram showing $\frac{2}{3} + \frac{2}{3} = \frac{4}{3}$.)

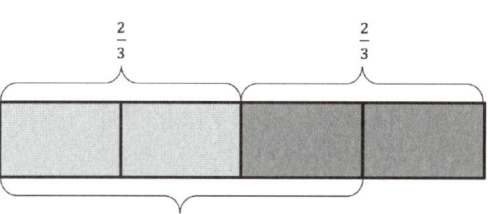

T: How many thirds are in 1?

S: 3 thirds.

T: (Write $\frac{2}{3} + \frac{2}{3} = \frac{4}{3} = 1\frac{\ }{3}$.) On your personal white board, fill in the unknown numerator.

S: (Write $\frac{2}{3} + \frac{2}{3} = \frac{4}{3} = 1\frac{1}{3}$.)

Continue with the following possible sequence: $\frac{3}{4} + \frac{2}{4}$ and $\frac{4}{5} + \frac{3}{5}$.

Application Problem (5 minutes)

Use a number bond to show the relationship between $\frac{2}{6}$, $\frac{3}{6}$, and $\frac{5}{6}$. Then, use the fractions to write two addition and two subtraction sentences.

Note: This Application Problem reviews work from earlier grades using related facts. The number sentences could also be written with the sum or difference on the left. The process of creating number bonds to show the relationship between addition and subtraction helps to bridge to the beginning of today's lesson where students identify related fraction facts when 1 one is the whole.

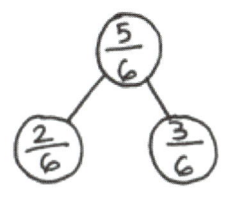

$\frac{2}{6} + \frac{3}{6} = \frac{5}{6}$

$\frac{3}{6} + \frac{2}{6} = \frac{5}{6}$

$\frac{5}{6} - \frac{3}{6} = \frac{2}{6}$

$\frac{5}{6} - \frac{2}{6} = \frac{3}{6}$

> **NOTES ON MULTIPLE MEANS OF REPRESENTATION:**
>
> Students working below grade level may benefit from drawing a tape diagram or another pictorial model of $\frac{5}{6}$, $\frac{2}{6}$, and $\frac{3}{6}$ in order to meaningfully derive two addition and two subtraction sentences from the number bond.

Lesson 17: Use visual models to add and subtract two fractions with the same units, including subtracting from one whole.

229

A STORY OF UNITS   Lesson 17   4•5

Concept Development (33 minutes)

Materials: (S) Personal white board

Problem 1: Subtract a fraction from 1.

T: Let's find the value of $1 - \frac{3}{8}$. Are the units the same?

S: No. There are ones and eighths.

T: Rename 1 one as eighths.

S: 8 eighths.

T: 8 eighths minus 3 eighths is...?

S: 5 eighths.

T: Model the subtraction using a number line. To simplify our number lines, use hash marks to show the eighths. Label 0, 1, and the numbers used to solve.

T: Record your work from the number line as a number sentence.

S: (Write $\frac{8}{8} - \frac{3}{8} = \frac{5}{8}$ or $1 - \frac{3}{8} = \frac{5}{8}$.)

T: (Display $1 - \frac{2}{5}$.) Discuss with your partner how to solve.

S: We have to make like units. 1 one is equal to 5 fifths.
→ 5 fifths minus 2 fifths equals 3 fifths. → $\frac{5}{5} - \frac{2}{5} = \frac{3}{5}$
→ $1 - \frac{2}{5} = \frac{3}{5}$.

T: Work with a partner to show $1 - \frac{2}{5}$ is the same as $\frac{5}{5} - \frac{2}{5}$ using a number line.

T: (Display $1 - \frac{2}{3} = x$.) Draw a number bond to show $\frac{2}{3}$, x, and 1.

T: Write two subtraction and two addition sentences using $\frac{2}{3}$, x, and 1.

S: $\frac{2}{3} + x = 1$. $x + \frac{2}{3} = 1$. $1 - \frac{2}{3} = x$. $1 - x = \frac{2}{3}$.

T: Draw a number line with endpoints 0 and 1. Partition and label thirds.

T: $\frac{2}{3} + x = 1$. Draw a point at $\frac{2}{3}$. How many more thirds does it take to make 1?

S: 1 third.

T: We can think of subtraction as an *unknown addend* problem and count up.

Repeat with $1 - \frac{7}{12}$.

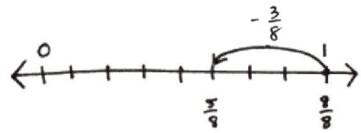

$\frac{8}{8} - \frac{3}{8} = \frac{5}{8}$

$1 - \frac{3}{8} = \frac{5}{8}$

> **NOTES ON MULTIPLE MEANS OF REPRESENTATION:**
>
> Student modeling of subtraction and addition on the number line may vary slightly depending on how students solve. For example, students working below grade level may model counting down with an arrow representing a series of hops. Encourage part–whole thinking and modeling by means of modeling with the number bond before the number line, if beneficial.

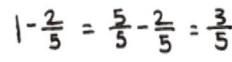

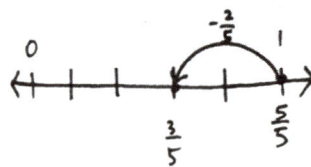

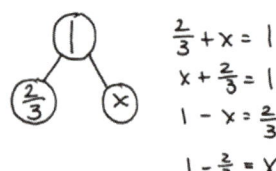

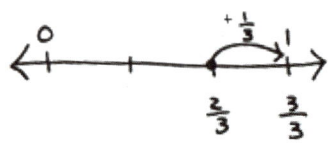

Lesson 17: Use visual models to add and subtract two fractions with the same units, including subtracting from one whole.

A STORY OF UNITS　　　　　　　　　　　　　　　　　　　　　　　　　　　　　　Lesson 17　4•5

Problem 2: Subtract a fraction from a number between 1 and 2.

T: Let's solve $1\frac{1}{5} - \frac{2}{5}$. First, draw a number bond to decompose $1\frac{1}{5}$ into a whole number and fractional parts. Show the whole number as fifths.

S: (Show $1\frac{1}{5}$ decomposed as $\frac{5}{5}$ and $\frac{1}{5}$.)

T: I'm going to draw two tape diagrams to show the whole of $1\frac{1}{5}$ with $\frac{2}{5}$ subtracted in different ways. (Draw two tape diagrams side by side. Cross off 2 fifths, as shown below, and write the related number sentences. See the illustration below.) Compare the methods with your partner.

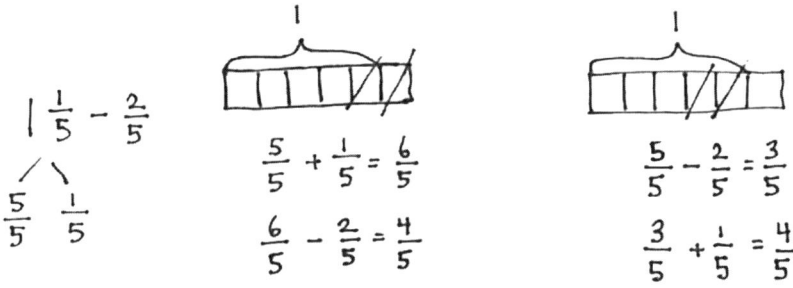

S: The solution on the left added 5 fifths and 1 fifth to get 6 fifths and then subtracted 2 fifths. → The second solution subtracted 2 fifths from 5 fifths and added that to 1 fifth. → That's how we learned how to subtract in Grades 1 and 2. When I subtract 8 from 13, I take it from the ten and add back 3.

T: Did both methods give the same answer?

S: Yes.

T: We can subtract from the total number of fifths, or we can subtract from 1, or take from 1, and add back the extra fifth.

T: Practice both methods using $1\frac{1}{4} - \frac{3}{4}$. Start by showing our number bond. Partner A, subtract from the total. Partner B, take from 1. Draw a tape diagram if it helps you.

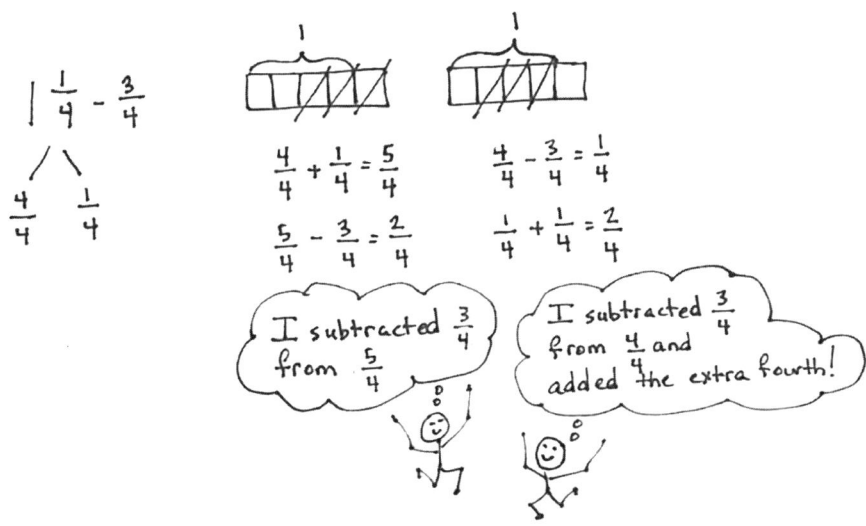

Lesson 17: Use visual models to add and subtract two fractions with the same units, including subtracting from one whole.

A STORY OF UNITS   Lesson 17   4•5

T: Try $1\frac{3}{8} - \frac{5}{8}$, switching strategies with your partner.

S: (Solve.)

T: By the way, 13 – 8 can also be solved by thinking 8 + ____ = 13 and counting up. What number sentence shows counting up as a strategy for solving $1\frac{1}{5} - \frac{2}{5}$? Talk to your partner.

S: $\frac{2}{5}$ + ____ = $\frac{6}{5}$. → $\frac{2}{5}$ + ____ = $1\frac{1}{5}$. → It's an unknown addend. → An unknown part.

T: Let's show it on the number line. (Draw the image to the right.)

T: I could also jump up to the whole number and add on.

T: The number line is a nice way to show counting up where the tape diagram was better for showing taking from the total and taking from 1. I chose to use the models that I thought would help you best understand. With your partner, take a moment to think about what subtracting from the total and subtracting from the whole number would look like on the number line.

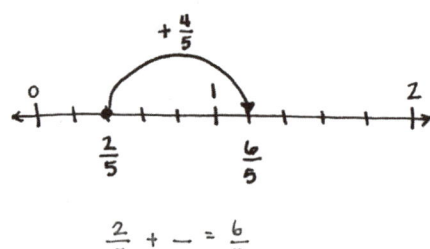

Problem Set (10 minutes)

Students should do their personal best to complete the Problem Set within the allotted 10 minutes. For some classes, it may be appropriate to modify the assignment by specifying which problems they work on first. Some problems do not specify a method for solving. Students should solve these problems using the RDW approach used for Application Problems.

Student Debrief (10 minutes)

Lesson Objective: Use visual models to add and subtract two fractions with the same units, including subtracting from one whole.

The Student Debrief is intended to invite reflection and active processing of the total lesson experience.

Invite students to review their solutions for the Problem Set. They should check work by comparing answers with a partner before going over answers as a class. Look for misconceptions or misunderstandings that can be addressed in the Debrief. Guide students in a conversation to debrief the Problem Set and process the lesson.

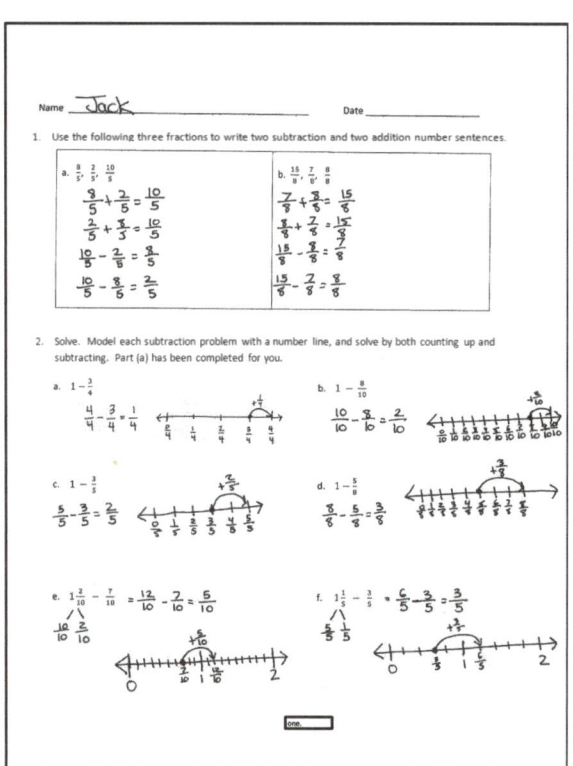

232   Lesson 17:   Use visual models to add and subtract two fractions with the same units, including subtracting from one whole.

Any combination of the questions below may be used to lead the discussion.

- For Problems 1(a) and (b), how did you determine the two addition and subtraction number sentences?
- Which strategy did you prefer for Problem 2(a–f)?
- What support does the number line offer you when solving problems such as these?
- Is the counting up strategy useful when solving subtraction problems? Explain.
- What extra step is there in solving when the fraction is written as a whole or mixed number instead of as a fraction?
- How is subtract from 1, or take from 1, similar to the take from 10 strategy?
- What role do fact families play in fractions? How are fraction fact families similar to whole number fact families?
- How did the Application Problem connect to today's lesson?

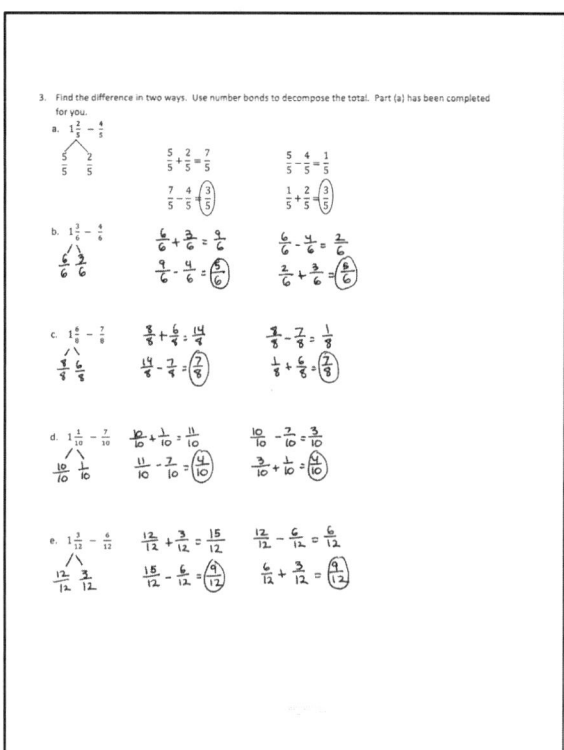

Exit Ticket (3 minutes)

After the Student Debrief, instruct students to complete the Exit Ticket. A review of their work will help with assessing students' understanding of the concepts that were presented in today's lesson and planning more effectively for future lessons. The questions may be read aloud to the students.

Lesson 17: Use visual models to add and subtract two fractions with the same units, including subtracting from one whole.

Name _____ Date _____

1. Use the following three fractions to write two subtraction and two addition number sentences.

   a. $\frac{8}{5}, \frac{2}{5}, \frac{10}{5}$

   b. $\frac{15}{8}, \frac{7}{8}, \frac{8}{8}$

2. Solve. Model each subtraction problem with a number line, and solve by both counting up and subtracting. Part (a) has been completed for you.

   a. $1 - \frac{3}{4}$

   $\frac{4}{4} - \frac{3}{4} = \frac{1}{4}$

   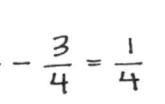

   b. $1 - \frac{8}{10}$

   c. $1 - \frac{3}{5}$

   d. $1 - \frac{5}{8}$

   e. $1\frac{2}{10} - \frac{7}{10}$

   f. $1\frac{1}{5} - \frac{3}{5}$

3. Find the difference in two ways. Use number bonds to decompose the total. Part (a) has been completed for you.

a. $1\frac{2}{5} - \frac{4}{5}$

   $\frac{5}{5} \quad \frac{2}{5}$

   $\frac{5}{5} + \frac{2}{5} = \frac{7}{5}$

   $\frac{7}{5} - \frac{4}{5} = \boxed{\frac{3}{5}}$

   $\frac{5}{5} - \frac{4}{5} = \frac{1}{5}$

   $\frac{1}{5} + \frac{2}{5} = \boxed{\frac{3}{5}}$

b. $1\frac{3}{6} - \frac{4}{6}$

c. $1\frac{6}{8} - \frac{7}{8}$

d. $1\frac{1}{10} - \frac{7}{10}$

e. $1\frac{3}{12} - \frac{6}{12}$

A STORY OF UNITS　　　　　　　　　　　　　　　　　　　　Lesson 17 Exit Ticket　4•5

Name _____   Date _____

1. Solve. Model the problem with a number line, and solve by both counting up and subtracting.

   $1 - \frac{2}{5}$

2. Find the difference in two ways. Use a number bond to show the decomposition.

   $1\frac{2}{7} - \frac{5}{7}$

Name _____ Date _____

1. Use the following three fractions to write two subtraction and two addition number sentences.

   a. $\frac{5}{6}, \frac{4}{6}, \frac{9}{6}$

   b. $\frac{5}{9}, \frac{13}{9}, \frac{8}{9}$

2. Solve. Model each subtraction problem with a number line, and solve by both counting up and subtracting.

   a. $1 - \frac{5}{8}$

   b. $1 - \frac{2}{5}$

   c. $1\frac{3}{6} - \frac{5}{6}$

   d. $1 - \frac{1}{4}$

   e. $1\frac{1}{3} - \frac{2}{3}$

   f. $1\frac{1}{5} - \frac{2}{5}$

Lesson 17: Use visual models to add and subtract two fractions with the same units, including subtracting from one whole.

3. Find the difference in two ways. Use number bonds to decompose the total. Part (a) has been completed for you.

a. $1\frac{2}{5} - \frac{4}{5}$

   $\frac{5}{5} \quad \frac{2}{5}$

   $\frac{5}{5} + \frac{2}{5} = \frac{7}{5}$

   $\frac{7}{5} - \frac{4}{5} = \boxed{\frac{3}{5}}$

   $\frac{5}{5} - \frac{4}{5} = \frac{1}{5}$

   $\frac{1}{5} + \frac{2}{5} = \boxed{\frac{3}{5}}$

b. $1\frac{3}{8} - \frac{7}{8}$

c. $1\frac{1}{4} - \frac{3}{4}$

d. $1\frac{2}{7} - \frac{5}{7}$

e. $1\frac{3}{10} - \frac{7}{10}$

A STORY OF UNITS Lesson 18 4•5

Lesson 18

Objective: Add and subtract more than two fractions.

Suggested Lesson Structure

- **Fluency Practice**   (12 minutes)
- **Concept Development** (38 minutes)
- **Student Debrief**   (10 minutes)
- **Total Time**   **(60 minutes)**

Fluency Practice (12 minutes)

- Count by Equivalent Fractions **4.NF.1**   (6 minutes)
- Subtract Fractions **4.NF.2**   (6 minutes)

Count by Equivalent Fractions (6 minutes)

Note: This activity builds fluency with equivalent fractions. The progression builds in complexity. Work students up to the highest level of complexity in which they can confidently participate.

T: Starting at 0, count by ones to 10.
S: 0, 1, 2, 3, 4, 5, 6, 7, 8, 9, 10.
T: Count by 1 tenths to 10 tenths starting at 0 tenths. (Write as students count.)
S: $\frac{0}{10}, \frac{1}{10}, \frac{2}{10}, \frac{3}{10}, \frac{4}{10}, \frac{5}{10}, \frac{6}{10}, \frac{7}{10}, \frac{8}{10}, \frac{9}{10}, \frac{10}{10}$.
T: (Point to $\frac{10}{10}$.) 10 tenths is the same as 1 of what unit?

$\frac{0}{10}$	$\frac{1}{10}$	$\frac{2}{10}$	$\frac{3}{10}$	$\frac{4}{10}$	$\frac{5}{10}$	$\frac{6}{10}$	$\frac{7}{10}$	$\frac{8}{10}$	$\frac{9}{10}$	$\frac{10}{10}$
0	$\frac{1}{10}$	$\frac{2}{10}$	$\frac{3}{10}$	$\frac{4}{10}$	$\frac{5}{10}$	$\frac{6}{10}$	$\frac{7}{10}$	$\frac{8}{10}$	$\frac{9}{10}$	1
0	$\frac{1}{10}$	$\frac{2}{10}$	$\frac{3}{10}$	$\frac{4}{10}$	$\frac{1}{2}$	$\frac{6}{10}$	$\frac{7}{10}$	$\frac{8}{10}$	$\frac{9}{10}$	1
0	$\frac{1}{10}$	$\frac{1}{5}$	$\frac{3}{10}$	$\frac{2}{5}$	$\frac{1}{2}$	$\frac{3}{5}$	$\frac{7}{10}$	$\frac{4}{5}$	$\frac{9}{10}$	1

S: 1 one.
T: (Beneath $\frac{10}{10}$, write 1.) Count by 1 tenths from 0 to 1 again. This time, when you come to $\frac{10}{10}$ say 1. Try not to look at the board.

Lesson 18: Add and subtract more than two fractions. 239

A STORY OF UNITS   Lesson 18 4•5

S: $0, \frac{1}{10}, \frac{2}{10}, \frac{3}{10}, \frac{4}{10}, \frac{5}{10}, \frac{6}{10}, \frac{7}{10}, \frac{8}{10}, \frac{9}{10}, 1$.

T: (Point to $\frac{5}{10}$.) 5 tenths is the same as 1 of what unit?

S: 1 half.

T: (Beneath $\frac{5}{10}$, write $\frac{1}{2}$.) Count by 1 tenths again. This time, convert to $\frac{1}{2}$ and 1. Try not to look at the board.

S: $0, \frac{1}{10}, \frac{2}{10}, \frac{3}{10}, \frac{4}{10}, \frac{1}{2}, \frac{6}{10}, \frac{7}{10}, \frac{8}{10}, \frac{9}{10}, 1$.

T: (Point to $\frac{2}{10}$.) What larger unit is $\frac{2}{10}$ equivalent to?

S: $\frac{1}{5}$.

> **NOTES ON MULTIPLE MEANS OF ENGAGEMENT:**
>
> One way to differentiate the Counting by Equivalent Fractions fluency activity for students working above grade level is to grant them more autonomy. Students may enjoy this as a partner activity in which students take turns leading and counting. Students can make individualized choices about when to convert larger units, counting forward and backward, and speed.

Repeat the process, replacing even numbers of tenths with fifths.

T: (Beneath $\frac{6}{10}$, write $\frac{3}{5}$.) Count by 1 tenths again. This time, count in the largest unit for each.

S: $0, \frac{1}{10}, \frac{1}{5}, \frac{3}{10}, \frac{2}{5}, \frac{1}{2}, \frac{3}{5}, \frac{7}{10}, \frac{4}{5}, \frac{9}{10}, 1$.

Direct students to count by tenths back and forth from 0 to 1, occasionally changing directions.

Subtract Fractions (6 minutes)

Materials: (S) Personal white board

Note: This fluency activity reviews Lesson 17.

T: (Write $1 - \frac{1}{3} = $ __.) How many thirds are in 1?

S: 3 thirds.

T: Write the subtraction sentence. Beneath it, rewrite the subtraction sentence, renaming the whole number in thirds. (Allow students time to work.)

T: Say the subtraction sentence with 1 renamed as thirds.

S: $\frac{3}{3} - \frac{1}{3} = \frac{2}{3}$.

Continue with the following possible sequence: $1 - \frac{1}{4}, 1 - \frac{3}{4}$, and $1 - \frac{4}{5}$.

T: (Write $1\frac{1}{3} - \frac{2}{3} = $ __.) Write the subtraction sentence on your personal white board.

S: (Write $1\frac{1}{3} - \frac{2}{3} = $ __.)

T: Can we take $\frac{2}{3}$ from $\frac{1}{3}$?

S: No.

T: (Break apart $1\frac{1}{3}$, writing $\frac{3}{3}$ as one of the parts.) Take $\frac{2}{3}$ from $\frac{3}{3}$, and solve using an addition sentence.

240   Lesson 18:   Add and subtract more than two fractions.

S: (Break apart $1\frac{1}{3}$ into $\frac{1}{3}$ and $\frac{3}{3}$. Take $\frac{2}{3}$ from $\frac{3}{3}$. Write $\frac{1}{3} + \frac{1}{3} = \frac{2}{3}$ to show the part of the whole number that remains plus the fractional part of the mixed number.)

Continue with the following possible sequence: $1\frac{1}{4} - \frac{3}{4}$, $1\frac{3}{5} - \frac{4}{5}$, and $1\frac{3}{10} - \frac{9}{10}$.

Concept Development (38 minutes)

Materials: (S) Adding and subtracting fractions (Practice Sheet)

Note: Arrange students in groups of three to solve and critique each other's work.

Exploration:

- Problems are sequenced from simple to complex and comprise addition and subtraction problems.
- All begin solving Problem A in the first rectangle.
- Students switch papers clockwise in their groups. Students analyze the solution in the first rectangle and critique it by discussing the solution with the writer. Then, students consider a different method to solve and record it in the second rectangle for Problem A.
- Students switch papers clockwise again for the second round of critiquing and third round of solving.
- Switching papers for the last time of the round, the original owner of the paper analyzes the three different methods used to solve the problem. A brief discussion may ensue as more than three methods could have been used within the group.
- The process continues as students solve Problems B through F.
- Some groups may not finish all problems during the time allotted, but the varied problems allow students to analyze and solve a wide variety of problems to prepare them for the Problem Set.
- Use the last five minutes of the Concept Development, prior to handing out the Problem Set, to review the many different solutions. The teacher may select one solution from three problems or three solutions from one problem to debrief. Identify common methods for solving addition and subtraction problems when there are more than two fractions.

NOTES ON MULTIPLE MEANS OF REPRESENTATION:

Exploration stations are sequenced from simple (Problem A) to complex (Problem F). To best guide student understanding, consider giving students working below grade level additional time to solve Problems A, B, and C, and then advance in order.

Lesson 18: Add and subtract more than two fractions.

A STORY OF UNITS

Lesson 18 4•5

Below are possible solutions for each problem. Students are encouraged to solve using computation through decomposition or other strategies.

Problem A: $\frac{1}{8} + \frac{3}{8} + \frac{4}{8}$

| $\frac{1}{8} + \frac{3}{8} + \frac{4}{8}$ $\frac{4}{8} + \frac{4}{8} = \frac{8}{8}$ | $\frac{1}{8} + \frac{3}{8} + \frac{4}{8}$ $\frac{1}{8} + \frac{7}{8} = \frac{8}{8}$ | $\frac{1}{8} + \frac{3}{8} + \frac{4}{8} = \frac{8}{8} = 1$ |

Problem B: $\frac{1}{6} + \frac{4}{6} + \frac{2}{6}$

| $\frac{1}{6} + \frac{4}{6} + \frac{2}{6}$ $\frac{6}{6} + \frac{1}{6} = \frac{7}{6}$ | $\frac{1}{6} + \frac{4}{6} + \frac{2}{6}$ $\frac{6}{6} = 1$ $1 + \frac{1}{6} = 1\frac{1}{6}$ | $\frac{1}{6} + \frac{2}{6} + \frac{4}{6}$ $\frac{3}{6} \quad \frac{3}{6} \quad \frac{1}{6}$ $\frac{3}{6} + \frac{3}{6} + \frac{1}{6} = \frac{6}{6} + \frac{1}{6} = 1\frac{1}{6}$ |

Problem C: $\frac{11}{10} - \frac{4}{10} - \frac{1}{10}$

| $\frac{11}{10} - \frac{4}{10} - \frac{1}{10}$ $\frac{11}{10} - \frac{5}{10} = \frac{6}{10}$ | $\frac{11}{10} - \frac{4}{10} = \frac{7}{10}$ $\frac{7}{10} - \frac{1}{10} = \frac{6}{10}$ $\frac{6}{10} = \frac{3}{5}$ | $\frac{11}{10} - \frac{1}{10} = \frac{10}{10}$ $\frac{10}{10} - \frac{4}{10} = \frac{6}{10}$ |

Lesson 18: Add and subtract more than two fractions.

A STORY OF UNITS — Lesson 18 4•5

Problem D:	$1 - \frac{3}{12} - \frac{5}{12}$	
$1 - \frac{3}{12} - \frac{5}{12}$ $\vee$ $\frac{12}{12} - \frac{8}{12} = \frac{4}{12}$	$\frac{12}{12} - \frac{3}{12} = \frac{9}{12}$ $\frac{9}{12} - \frac{5}{12} = \frac{4}{12}$	$\frac{3}{12} + \frac{5}{12} = \frac{8}{12}$ $\frac{8}{12} + \frac{4}{12} = \frac{12}{12}$ $\frac{4}{12} = \frac{1}{3}$

Problem E:	$\frac{5}{8} + \frac{4}{8} + \frac{1}{8}$	
$\frac{5}{8} + \frac{4}{8} + \frac{1}{8} = \frac{10}{8}$ $\wedge$ $\frac{8}{8}\;\;\frac{2}{8}$ $= 1\frac{2}{8}$	$\frac{5}{8} + \frac{4}{8} + \frac{1}{8}$ $\wedge$ $\frac{3}{8}\;\;\frac{1}{8}$ $\frac{8}{8} + \frac{2}{8} = 1 + \frac{2}{8} = 1\frac{2}{8}$	$\frac{1}{8} + \frac{5}{8} + \frac{4}{8}$ $\wedge$ $\frac{2}{8}\;\;\frac{2}{8}$ $\frac{8}{8} + \frac{2}{8} = 1\frac{2}{8} = 1\frac{1}{4}$

Problem F:	$1\frac{1}{5} - \frac{2}{5} - \frac{3}{5}$	
$1\frac{1}{5} - \frac{2}{5} - \frac{3}{5}$ $\vee$ $1 - \frac{5}{5} = 0$ $0 + \frac{1}{5} = \frac{1}{5}$	$\frac{6}{5} - \frac{2}{5} = \frac{4}{5}$ $\frac{4}{5} - \frac{3}{5} = \frac{1}{5}$	$1\frac{1}{5} - \frac{2}{5}$ $\wedge$ $\frac{1}{5}\;\;\frac{1}{5}$ $= 1 - \frac{4}{5}$ $= \frac{1}{5}$

Problem Set (10 minutes)

Students should do their personal best to complete the Problem Set within the allotted 10 minutes. For some classes, it may be appropriate to modify the assignment by specifying which problems they work on first. Some problems do not specify a method for solving. Students should solve these problems using the RDW approach used for Application Problems.

Student Debrief (10 minutes)

Lesson Objective: Add and subtract more than two fractions.

The Student Debrief is intended to invite reflection and active processing of the total lesson experience.

Invite students to review their solutions for the Problem Set. They should check work by comparing answers with a partner before going over answers as a class. Look for misconceptions or misunderstandings that can be addressed in the Debrief. Guide students in a conversation to debrief the Problem Set and process the lesson.

Lesson 18: Add and subtract more than two fractions.

Any combination of the questions below may be used to lead the discussion.

- In Problem 1(h), the total is a mixed number. Was it necessary to change the mixed number to a fraction in this case? Explain.
- Discuss your solution strategy for Problem 1(i). Grouping fractions to make 1 is a strategy that can help in solving problems mentally. Solving for $\frac{2}{12} + \frac{10}{12}$ and $\frac{5}{12} + \frac{7}{12}$ can lead to the solution more rapidly.
- For Problem 2, did you agree with Monica or Stuart? Explain why you chose that strategy. Do you see a different method?
- Consider how you solved Problem 1(c) and the other solution for it in Problem 3. Would this solution be accurate? (Display $\frac{5+7+2}{7} = \frac{14}{7} = 2$.) Explain why this representation for addition of fractions is correct.
- Observe your solution to Problem 1(d). Is my solution correct? Why? Explain. (Display $\frac{7-3-1}{8} = \frac{3}{8}$.)
- Explain in words how we add or subtract more than two fractions with like units.
- When is it necessary to decompose the total in a subtraction problem into fractions? Give an example.

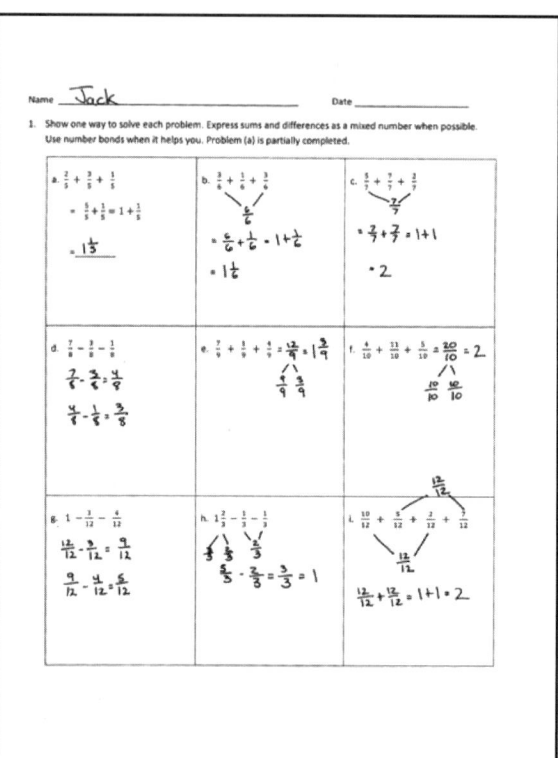

Exit Ticket (3 minutes)

After the Student Debrief, instruct students to complete the Exit Ticket. A review of their work will help with assessing students' understanding of the concepts that were presented in today's lesson and planning more effectively for future lessons. The questions may be read aloud to the students.

Lesson 18 Practice Sheet 4•5

Name _____ Date _____

Problem A:	$\frac{1}{8} + \frac{3}{8} + \frac{4}{8}$	

Problem B:	$\frac{1}{6} + \frac{4}{6} + \frac{2}{6}$	

Problem C:	$\frac{11}{1A} - \frac{4}{1A} - \frac{1}{1A}$	

adding and subtracting fractions

Lesson 18: Add and subtract more than two fractions.

A STORY OF UNITS   Lesson 18 Practice Sheet   4•5

Problem D: $\quad 1 - \dfrac{3}{12} - \dfrac{5}{12}$

Problem E: $\quad \dfrac{5}{8} + \dfrac{4}{8} + \dfrac{1}{8}$

Problem F: $\quad 1\dfrac{1}{5} - \dfrac{2}{5} - \dfrac{3}{5}$

adding and subtracting fractions

Lesson 18: Add and subtract more than two fractions.

A STORY OF UNITS — Lesson 18 Problem Set 4•5

Name _____ Date _____

1. Show one way to solve each problem. Express sums and differences as a mixed number when possible. Use number bonds when it helps you. Part (a) is partially completed.

a. $\frac{2}{5} + \frac{3}{5} + \frac{1}{5}$ $= \frac{5}{5} + \frac{1}{5} = 1 + \frac{1}{5}$ = _____	b. $\frac{3}{6} + \frac{1}{6} + \frac{3}{6}$	c. $\frac{5}{7} + \frac{7}{7} + \frac{2}{7}$
d. $\frac{7}{8} - \frac{3}{8} - \frac{1}{8}$	e. $\frac{7}{9} + \frac{1}{9} + \frac{4}{9}$	f. $\frac{4}{10} + \frac{11}{10} + \frac{5}{10}$
g. $1 - \frac{3}{12} - \frac{4}{12}$	h. $1\frac{2}{3} - \frac{1}{3} - \frac{1}{3}$	i. $\frac{10}{12} + \frac{5}{12} + \frac{2}{12} + \frac{7}{12}$

Lesson 18: Add and subtract more than two fractions.

A STORY OF UNITS   Lesson 18 Problem Set   4•5

2. Monica and Stuart used different strategies to solve $\frac{5}{8} + \frac{2}{8} + \frac{5}{8}$.

   Monica's Way

   $\frac{5}{8} + \frac{2}{8} + \frac{5}{8} = \frac{7}{8} + \frac{5}{8} = \frac{8}{8} + \frac{4}{8} = 1\frac{4}{8}$

   $\frac{1}{8} \quad \frac{4}{8}$

   Stuart's Way

   $\frac{5}{8} + \frac{2}{8} + \frac{5}{8} = \frac{12}{8} = 1 + \frac{4}{8} = 1\frac{4}{8}$

   $\frac{8}{8} \quad \frac{4}{8}$

   Whose strategy do you like best? Why?

3. You gave one solution for each part of Problem 1. Now, for each problem indicated below, give a different solution method.

   1(c) $\frac{5}{7} + \frac{7}{7} + \frac{2}{7}$

   1(f) $\frac{4}{1A} + \frac{11}{1A} + \frac{5}{1A}$

   1(g) $1 - \frac{3}{12} - \frac{4}{12}$

248   Lesson 18:   Add and subtract more than two fractions.

A STORY OF UNITS — Lesson 18 Exit Ticket 4•5

Name _____ Date _____

Solve the following problems. Use number bonds to help you.

1. $\frac{5}{9} + \frac{2}{9} + \frac{4}{9}$

2. $1 - \frac{5}{8} - \frac{1}{8}$

Lesson 18: Add and subtract more than two fractions.

Name _____ Date _____

1. Show one way to solve each problem. Express sums and differences as a mixed number when possible. Use number bonds when it helps you. Part (a) is partially completed.

a. $\frac{1}{3} + \frac{2}{3} + \frac{1}{3}$ $= \frac{3}{3} + \frac{1}{3} = 1 + \frac{1}{3}$ = _____	b. $\frac{5}{8} + \frac{5}{8} + \frac{3}{8}$	c. $\frac{4}{6} + \frac{6}{6} + \frac{1}{6}$
d. $1\frac{2}{12} - \frac{2}{12} - \frac{1}{12}$	e. $\frac{5}{7} + \frac{1}{7} + \frac{4}{7}$	f. $\frac{4}{10} + \frac{7}{10} + \frac{9}{10}$
g. $1 - \frac{3}{10} - \frac{1}{10}$	h. $1\frac{3}{5} - \frac{4}{5} - \frac{1}{5}$	i. $\frac{10}{15} + \frac{7}{15} + \frac{12}{15} + \frac{1}{15}$

Lesson 18: Add and subtract more than two fractions.

Lesson 18 Homework 4•5

2. Bonnie used two different strategies to solve $\frac{5}{10} + \frac{4}{10} + \frac{3}{10}$.

Bonnie's First Strategy

$$\frac{5}{10} + \frac{4}{10} + \frac{3}{10} = \frac{9}{10} + \frac{3}{10} = \frac{10}{10} + \frac{2}{10} = 1\frac{2}{10}$$

$$\underset{\frac{1}{10}\quad\frac{2}{10}}{\wedge}$$

Bonnie's Second Strategy

$$\frac{5}{10} + \frac{4}{10} + \frac{3}{10} = \frac{12}{10} = 1 + \frac{2}{10} = 1\frac{2}{10}$$

$$\underset{\frac{10}{10}\quad\frac{2}{10}}{\wedge}$$

Which strategy do you like best? Why?

3. You gave one solution for each part of Problem 1. Now, for each problem indicated below, give a different solution method.

1(b) $\frac{5}{8} + \frac{5}{8} + \frac{3}{8}$

1(e) $\frac{5}{7} + \frac{1}{7} + \frac{4}{7}$

1(h) $1\frac{3}{5} - \frac{4}{5} - \frac{1}{5}$

Lesson 18: Add and subtract more than two fractions.

Lesson 19

Objective: Solve word problems involving addition and subtraction of fractions.

Suggested Lesson Structure

- **Fluency Practice** (12 minutes)
- **Application Problem** (6 minutes)
- **Concept Development** (32 minutes)
- **Student Debrief** (10 minutes)

Total Time **(60 minutes)**

Fluency Practice (12 minutes)

- Count by Equivalent Fractions **4.NF.1** (6 minutes)
- Add and Subtract Fractions **4.NF.3** (6 minutes)

Count by Equivalent Fractions (6 minutes)

Note: This activity builds fluency with equivalent fractions. The progression builds in complexity. Work students up to the highest level of complexity in which they can confidently participate.

T: Starting at 0, count by twos to 12.
S: 0, 2, 4, 6, 8, 10, 12.
T: Count by 2 twelfths from 0 twelfths to 12 twelfths. (Write as students count.)
S: $\frac{0}{12}, \frac{2}{12}, \frac{4}{12}, \frac{6}{12}, \frac{8}{12}, \frac{10}{12}, \frac{12}{12}$.
T: (Point to $\frac{12}{12}$.) 12 twelfths is the same as 1 of what unit?
S: 1 one.
T: (Beneath $\frac{12}{12}$, write 1.) Count by 2 twelfths again from 0 to 1. Try not to look at the board.
S: 0, $\frac{2}{12}, \frac{4}{12}, \frac{6}{12}, \frac{8}{12}, \frac{10}{12}$, 1.
T: (Point to $\frac{6}{12}$.) 6 twelfths is the same as what unit fraction?

S: 1 half.

T: (Beneath $\frac{6}{12}$, write $\frac{1}{2}$.) Count by 2 twelfths again. This time, convert $\frac{6}{12}$ to $\frac{1}{2}$ and $\frac{12}{12}$ to 1. Try not to look at the board.

S: $0, \frac{2}{12}, \frac{4}{12}, \frac{1}{2}, \frac{8}{12}, \frac{10}{12}, 1$.

T: (Point to $\frac{2}{12}$.) What's $\frac{2}{12}$ renamed to a larger unit?

S: 1 sixth.

T: (Beneath $\frac{2}{12}$, write $\frac{1}{6}$. Point to $\frac{4}{12}$.) What's $\frac{4}{12}$ renamed as sixths?

S: 2 sixths.

Continue, renaming $\frac{8}{12}$ and $\frac{10}{12}$ as sixths.

S: $0, \frac{1}{6}, \frac{2}{6}, \frac{1}{2}, \frac{4}{6}, \frac{5}{6}, 1$.

Continue, renaming $\frac{2}{6}$ and $\frac{4}{6}$ as thirds.

S: $0, \frac{1}{6}, \frac{1}{3}, \frac{1}{2}, \frac{2}{3}, \frac{5}{6}, 1$.

Direct students to count back and forth by 2 twelves from 0 to 1, occasionally changing directions.

Add and Subtract Fractions (6 minutes)

Materials: (S) Personal white board

Note: This fluency activity reviews Lesson 18.

T: (Write $\frac{3}{6} + \frac{1}{6} + \frac{1}{6} = $ ___.) Write the complete number sentence on your personal white board.

S: (Write $\frac{3}{6} + \frac{1}{6} + \frac{1}{6} = \frac{5}{6}$.)

T: (Write $\frac{7}{8} - \frac{2}{8} = $ ___.) Write the complete number sentence on your personal white board.

S: (Write $\frac{7}{8} - \frac{2}{8} = \frac{5}{8}$.)

Continue with the following possible sequence: $\frac{5}{10} + \frac{2}{10} + \frac{2}{10}$, $\frac{2}{5} + \frac{2}{5} + \frac{2}{5}$, $\frac{5}{6} - \frac{1}{6} - \frac{3}{6}$, $1 - \frac{1}{5}$, $1 - \frac{3}{5}$, $1 - \frac{3}{8}$, $1 - \frac{7}{10}$, $1\frac{9}{10} - \frac{3}{10} - \frac{5}{10}$, and $1\frac{7}{12} - \frac{5}{12} - \frac{11}{12}$.

Lesson 19: Solve word problems involving addition and subtraction of fractions.

A STORY OF UNITS   Lesson 19   4•5

Application Problem (6 minutes)

Fractions are all around us! Make a list of times that you have used fractions, heard fractions, or seen fractions. Be ready to share your ideas.

Note: This Application Problem encourages students to think of real life examples of fractions. Additionally, this Application Problem contextualizes previously learned skills in the module and prepares students for today's problem-solving lesson involving fractions. Have students spend a few minutes brainstorming together in small groups and then share ideas with the whole group.

Concept Development (32 minutes)

Materials: (S) Problem Set

Suggested Delivery of Instruction for Solving Lesson 19's Word Problems

1. Model the problem.

Have two pairs of students, who can successfully model the problem, work at the board while the others work independently or in pairs at their seats. Review the following questions before beginning the first problem.

- Can you draw something?
- What can you draw?
- What conclusions can you make from your drawing?

As students work, circulate. Reiterate the questions above. After two minutes, have the two pairs of students share only their labeled diagrams. For about one minute, have the demonstrating students receive and respond to feedback and questions from their peers.

2. Calculate to solve and write a statement.

Give everyone two minutes to finish work on that question, sharing their work and thinking with a peer. All should then write their equations and statements of the answer.

3. Assess the solution for reasonableness.

Give students one to two minutes to assess and explain the reasonableness of their solutions.

Note: Problems 1–4 of the Problem Set are used during the Concept Development portion of the lesson.

254   Lesson 19:   Solve word problems involving addition and subtraction of fractions.

A STORY OF UNITS

Lesson 19 4•5

Problem 1: Use the RDW process to solve a word problem involving the addition of fractions.

Sue ran $\frac{9}{10}$ mile on Monday and $\frac{7}{10}$ mile on Tuesday. How many miles did Sue run in the 2 days?

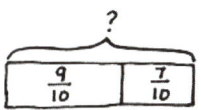

Sue ran $1\frac{6}{10}$ miles.

> **NOTES ON MULTIPLE MEANS OF ENGAGEMENT:**
>
> Differentiate the difficulty of Problem 1 by adjusting the numbers. Students working above grade level may enjoy the challenge of adding three addends, for example $\frac{37}{100} + \frac{18}{100} + \frac{65}{100}$. Grade 4 expectations in this domain are limited to fractions with like denominators 2, 3, 4, 5, 6, 8, 10, 12, and 100.

Solution 1

$\frac{9}{10} + \frac{7}{10} = \frac{16}{10} = 1\frac{6}{10}$

$\wedge$
$\frac{10}{10}\ \frac{6}{10}$

Solution 2

$\frac{9}{10} + \frac{7}{10} = \frac{10}{10} + \frac{6}{10} = 1\frac{3}{5}$

$\wedge$
$\frac{1}{10}\ \frac{6}{10}$ $\frac{6}{10} = \frac{3}{5}$

Students may initially represent the problem by drawing number bonds or number lines as they did in the previous lessons to model addition. Assist students in finding the parts and wholes. In Problem 1, the 2 parts, $\frac{9}{10}$ and $\frac{7}{10}$, make the whole, $1\frac{3}{5}$. Encourage students to represent this relationship as a tape diagram to model, as done with whole number addition. In contrast to their previous solutions, students are not drawing the fractional units to count. Instead, they are seeing the relationship the two fractions have with each other and calculating based on what they know about whole number and fraction addition. Possible strategies and solutions to this problem may include, but are not limited to, those shown above.

Problem 2: Use the RDW process to solve a word problem involving the addition and subtraction of fractions.

Mr. Salazar cut his son's birthday cake into 8 equal pieces. Mr. Salazar, Mrs. Salazar, and the birthday boy each ate 1 piece of cake. What fraction of the cake was left?

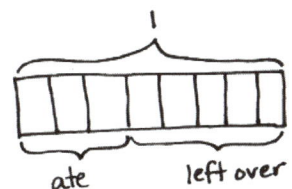

$\frac{5}{8}$ of the cake is left.

Solution 1

$1 - \frac{3}{8} = \frac{8}{8} - \frac{3}{8} = \frac{5}{8}$

Solution 2

$\frac{1}{8} + \frac{1}{8} + \frac{1}{8} + x = \frac{8}{8}$

$\frac{3}{8} + x = \frac{8}{8}$

$\frac{3}{8} + \frac{5}{8} = \frac{8}{8}$ $x = \frac{5}{8}$

Lesson 19: Solve word problems involving addition and subtraction of fractions.

A STORY OF UNITS Lesson 19 4•5

Although each person had 1 piece of cake, students must consider the 1 piece as a fractional unit of the whole. The whole is represented as 1, and students may choose to take from or add to the whole. Again, encourage students to think about the parts and the whole when drawing a picture to represent the problem. A tape diagram is a good way to connect the part–whole relationship, with which they are familiar in whole number addition and subtraction, to fraction computation. The parts can be taken or added one at a time, or students may group them as $\frac{3}{8}$ before computing.

Problem 3: Use the RDW process to solve a word problem subtracting a fraction from 1.

Maria spent $\frac{4}{7}$ of her money on a book and saved the rest. What fraction of her money did Maria save?

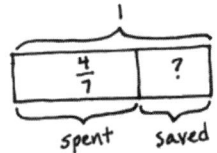

Maria saved $\frac{3}{7}$ of her money.

Solution 1

$1 - \frac{4}{7} = \frac{7}{7} - \frac{4}{7} = \frac{3}{7}$

Solution 2

$\frac{4}{7} + x = 1$

$\frac{4}{7} + \frac{3}{7} = \frac{7}{7}$

$x = \frac{3}{7}$

In this problem, students subtract a fraction from 1. Some may write 1 as $\frac{7}{7}$ and then subtract $\frac{4}{7}$. Alternatively, students may choose to add up to $\frac{7}{7}$.

Problem 4: Use the RDW process to solve a word problem involving the subtraction of fractions.

Mrs. Jones had $1\frac{4}{8}$ pizzas left after a party. After giving some to Gary, she had $\frac{7}{8}$ pizza left. What fraction of a pizza did she give Gary?

She gave $\frac{5}{8}$ pizza to Gary.

Solution 1

$1\frac{4}{8} - \frac{7}{8} = \frac{12}{8} - \frac{7}{8} = \frac{5}{8}$

Solution 2

$1\frac{4}{8} - \frac{7}{8}$

$\frac{8}{8} \quad \frac{4}{8}$

$\frac{8}{8} - \frac{7}{8} = \frac{1}{8}$

$\frac{1}{8} + \frac{4}{8} = \frac{5}{8}$

Lesson 19: Solve word problems involving addition and subtraction of fractions.

A STORY OF UNITS

Lesson 19   4•5

Students can use an adding up method but will likely choose one of the subtracting methods. One way is to rewrite the mixed number as $\frac{12}{8}$ and subtract. The other method subtracts from 1 and adds back the fractional part as practiced in Lesson 17.

Problem Set (10 minutes)

Students should do their personal best to complete the remaining two problems of the Problem Set within the allotted 10 minutes.

Student Debrief (10 minutes)

Lesson Objective: Solve word problems involving addition and subtraction of fractions.

The Student Debrief is intended to invite reflection and active processing of the total lesson experience.

Invite students to review their solutions for the Problem Set. They should check work by comparing answers with a partner before going over answers as a class. Look for misconceptions or misunderstandings that can be addressed in the Debrief. Guide students in a conversation to debrief the Problem Set and process the lesson.

Any combination of the questions below may be used to lead the discussion.

- What strategies did you use to solve the problems in the Problem Set? Did you use the same strategy each time?
- Which problem(s) were the most difficult? How were they difficult? What strategies did you use to persevere?
- Which problem(s) were the least difficult? Why?
- Was it easier to solve Problems 5 and 6 on your own after having completed Problems 1–4 together as a group? Why or why not? Did you use the same strategies that you used in solving Problems 1–4?
- How was Problem 4 different from the other problems?
- What was challenging about Problem 5? About Problem 6?
- How did the Application Problem connect to today's lesson?

NOTES ON MULTIPLE MEANS OF ACTION AND EXPRESSION:

To prepare students working below grade level and others to meaningfully participate in today's work and the closing Student Debrief, quickly review strategies from which students may choose:

- Take apart and redistribute (using a number bond).
- Counting up to subtract.
- Thinking part–whole (using a tape diagram).

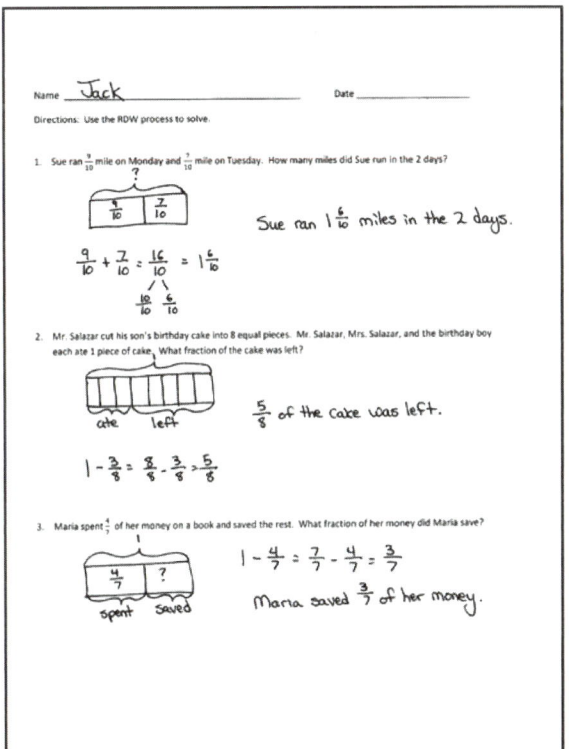

Lesson 19: Solve word problems involving addition and subtraction of fractions.

257

Exit Ticket (3 minutes)

After the Student Debrief, instruct students to complete the Exit Ticket. A review of their work will help with assessing students' understanding of the concepts that were presented in today's lesson and planning more effectively for future lessons. The questions may be read aloud to the students.

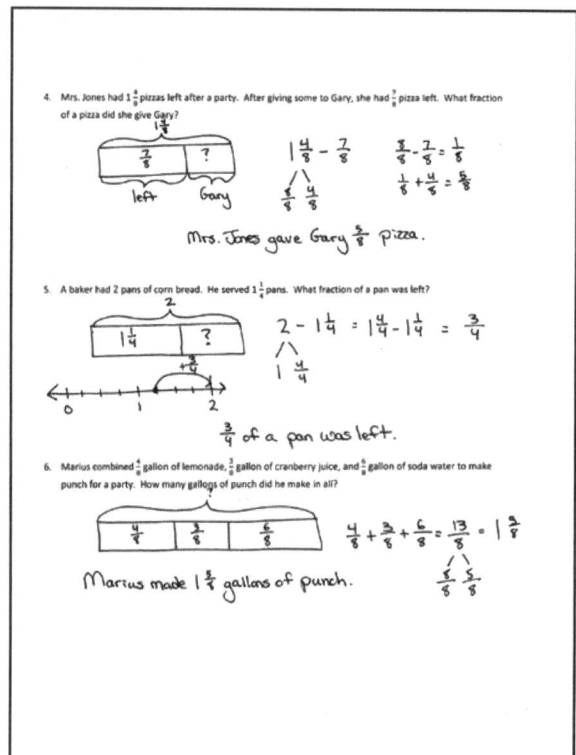

Name _____ Date _____

Use the RDW process to solve.

1. Sue ran $\frac{9}{10}$ mile on Monday and $\frac{7}{10}$ mile on Tuesday. How many miles did Sue run in the 2 days?

2. Mr. Salazar cut his son's birthday cake into 8 equal pieces. Mr. Salazar, Mrs. Salazar, and the birthday boy each ate 1 piece of cake. What fraction of the cake was left?

3. Maria spent $\frac{4}{7}$ of her money on a book and saved the rest. What fraction of her money did Maria save?

4. Mrs. Jones had $1\frac{4}{8}$ pizzas left after a party. After giving some to Gary, she had $\frac{7}{8}$ pizza left. What fraction of a pizza did she give Gary?

5. A baker had 2 pans of corn bread. He served $1\frac{1}{4}$ pans. What fraction of a pan was left?

6. Marius combined $\frac{4}{8}$ gallon of lemonade, $\frac{3}{8}$ gallon of cranberry juice, and $\frac{6}{8}$ gallon of soda water to make punch for a party. How many gallons of punch did he make in all?

Name _____ Date _____

Use the RDW process to solve.

1. Mrs. Smith took her bird to the vet. Tweety weighed $1\frac{3}{10}$ pounds. The vet said that Tweety weighed $\frac{4}{10}$ pound more last year. How much did Tweety weigh last year?

2. Hudson picked $1\frac{1}{4}$ baskets of apples. Suzy picked 2 baskets of apples. How many more baskets of apples did Suzy pick than Hudson?

A STORY OF UNITS

Lesson 19 Homework 4•5

Name _____ Date _____

Use the RDW process to solve.

1. Isla walked $\frac{3}{4}$ mile each way to and from school on Wednesday. How many miles did Isla walk that day?

2. Zach spent $\frac{2}{3}$ hour reading on Friday and $1\frac{1}{3}$ hours reading on Saturday. How much more time did he read on Saturday than on Friday?

3. Mrs. Cashmore bought a large melon. She cut a piece that weighed $1\frac{1}{8}$ pounds and gave it to her neighbor. The remaining piece of melon weighed $\frac{6}{8}$ pound. How much did the whole melon weigh?

Lesson 19: Solve word problems involving addition and subtraction of fractions.

4. Ally's little sister wanted to help her make some oatmeal cookies. First, she put $\frac{5}{8}$ cup of oatmeal in the bowl. Next, she added another $\frac{5}{8}$ cup of oatmeal. Finally, she added another $\frac{5}{8}$ cup of oatmeal. How much oatmeal did she put in the bowl?

5. Marcia baked 2 pans of brownies. Her family ate $1\frac{5}{6}$ pans. What fraction of a pan of brownies was left?

6. Joanie wrote a letter that was $1\frac{1}{4}$ pages long. Katie wrote a letter that was $\frac{3}{4}$ page shorter than Joanie's letter. How long was Katie's letter?

Lesson 19: Solve word problems involving addition and subtraction of fractions.

Lesson 20

Objective: Use visual models to add two fractions with related units using the denominators 2, 3, 4, 5, 6, 8, 10, and 12.

Suggested Lesson Structure

- ■ Fluency Practice   (12 minutes)
- ■ Application Problem   (5 minutes)
- ■ Concept Development   (33 minutes)
- ■ Student Debrief   (10 minutes)
  - **Total Time**   **(60 minutes)**

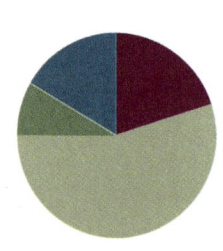

> **A NOTE ON STANDARDS ALIGNMENT:**
>
> In Lessons 20 and 21, students add fractions with related denominators where one denominator is a factor of the other. In Grade 5, students find sums and differences of fractions with unrelated denominators (**5.NF.1**). Because students are able to generate equivalent fractions (**4.NF.1**) from their work in Topics A, B, and C and are very familiar with the idea that units must be the same to be added, this work makes sense and prepares them well for work with decimals in Module 6 where tenths are converted to hundredths and added to hundredths (**4.NBT.5**).

Fluency Practice (12 minutes)

- Count by Equivalent Fractions **4.NF.1**   (6 minutes)
- Add Fractions **4.NF.3**   (3 minutes)
- Subtract Fractions **4.NF.3**   (3 minutes)

Count by Equivalent Fractions (6 minutes)

Note: This activity builds fluency with equivalent fractions. The progression builds in complexity. Work students up to the highest level of complexity in which they can confidently participate.

T: Count by ones to 10 starting at 0.
S: 0, 1, 2, 3, 4, 5, 6, 7, 8, 9, 10.
T: Count by 1 fifths to 10 fifths starting at 0 fifths. (Write as students count.)
S: $\frac{0}{5}, \frac{1}{5}, \frac{2}{5}, \frac{3}{5}, \frac{4}{5}, \frac{5}{5}, \frac{6}{5}, \frac{7}{5}, \frac{8}{5}, \frac{9}{5}, \frac{10}{5}$.

$\frac{0}{5}$	$\frac{1}{5}$	$\frac{2}{5}$	$\frac{3}{5}$	$\frac{4}{5}$	$\frac{5}{5}$	$\frac{6}{5}$	$\frac{7}{5}$	$\frac{8}{5}$	$\frac{9}{5}$	$\frac{10}{5}$
0	$\frac{1}{5}$	$\frac{2}{5}$	$\frac{3}{5}$	$\frac{4}{5}$	1	$\frac{6}{5}$	$\frac{7}{5}$	$\frac{8}{5}$	$\frac{9}{5}$	2
0	$\frac{1}{5}$	$\frac{2}{5}$	$\frac{3}{5}$	$\frac{4}{5}$	1	$1\frac{1}{5}$	$1\frac{2}{5}$	$1\frac{3}{5}$	$1\frac{4}{5}$	2

T: 1 one is the same as how many fifths?
S: 5 fifths.

A STORY OF UNITS

Lesson 20 4•5

T: (Beneath $\frac{5}{5}$, write 1.) 2 ones is the same as how many fifths?

S: 10 fifths.

T: (Beneath $\frac{10}{5}$, write 2.) Count by fifths again from 0 to 2. This time, when you come to the whole number, say the whole number. (Write as students count.)

S: 0, $\frac{1}{5}$, $\frac{2}{5}$, $\frac{3}{5}$, $\frac{4}{5}$, 1, $\frac{6}{5}$, $\frac{7}{5}$, $\frac{8}{5}$, $\frac{9}{5}$, 2.

T: (Point to $\frac{6}{5}$.) Say 6 fifths as a mixed number.

S: $1\frac{1}{5}$.

T: Count by fifths again. This time, convert to whole numbers and mixed numbers. (Write as students count.)

S: 0, $\frac{1}{5}$, $\frac{2}{5}$, $\frac{3}{5}$, $\frac{4}{5}$, 1, $1\frac{1}{5}$, $1\frac{2}{5}$, $1\frac{3}{5}$, $1\frac{4}{5}$, 2.

T: 2 is the same as how many fifths?

S: $\frac{10}{5}$.

T: Let's count backward starting at $\frac{10}{5}$, alternating between fractions and mixed numbers. Try not to look at the board.

S: $\frac{10}{5}$, $1\frac{4}{5}$, $\frac{8}{5}$, $1\frac{2}{5}$, $\frac{6}{5}$, 1, $\frac{4}{5}$, $\frac{3}{5}$, $\frac{2}{5}$, $\frac{1}{5}$, 0.

Add Fractions (3 minutes)

Materials: (S) Personal white board

Note: This fluency activity reviews Lesson 18.

T: (Write $\frac{2}{5} + \frac{1}{5} + \frac{1}{5} = -$.) On your personal white board, write the complete number sentence.

S: (Write $\frac{2}{5} + \frac{1}{5} + \frac{1}{5} = \frac{4}{5}$.)

T: (Write $\frac{5}{8} + \frac{2}{8} + \frac{1}{8} = -$.) Write the complete number sentence.

S: (Write $\frac{5}{8} + \frac{2}{8} + \frac{1}{8} = \frac{8}{8}$.)

T: (Write $\frac{5}{8} + \frac{2}{8} + \frac{1}{8} = \frac{8}{8}$.) Rename 8 eighths as a whole number.

S: $\frac{8}{8} = 1$.

T: (Write $\frac{2}{5} + \frac{2}{5} + \frac{2}{5} = -$.) Write the complete number sentence.

S: (Write $\frac{2}{5} + \frac{2}{5} + \frac{2}{5} = \frac{6}{5}$.)

T: How many fifths are equal to 1?

S: 5 fifths.

T: Write $\frac{6}{5}$ as a mixed number.

Lesson 20: Use visual models to add two fractions with related units using the denominators 2, 3, 4, 5, 6, 8, 10, and 12.

265

S: (Write $\frac{2}{5}+\frac{2}{5}+\frac{2}{5}=\frac{6}{5}=1\frac{1}{5}$.)

Continue the process with $\frac{3}{4}+\frac{2}{4}+\frac{2}{4}$.

Subtract Fractions (3 minutes)

Materials: (S) Personal white board

Note: This fluency activity reviews Lesson 17.

T: (Write $1-\frac{1}{5}=\frac{}{5}$.) How many fifths are in 1?

S: 5 fifths.

T: Write the subtraction sentence. Beneath it, rewrite the subtraction sentence, renaming 1 as fifths.

S: (Write $1-\frac{1}{5}=\frac{}{5}$. Beneath it, write $\frac{5}{5}-\frac{1}{5}=\frac{}{5}$.)

T: Say the subtraction sentence.

S: $1-\frac{1}{5}=\frac{4}{5}$.

Continue with the following possible sequence: $1-\frac{3}{5}$ and $1-\frac{3}{10}$.

T: (Write $1\frac{1}{5}-\frac{4}{5}=\frac{}{5}$.) Write the complete number sentence.

S: (Write $1\frac{1}{5}-\frac{4}{5}=\frac{}{5}$.)

T: Can we take $\frac{4}{5}$ from $\frac{1}{5}$?

S: No.

T: (Break apart $1\frac{1}{5}$, writing $\frac{5}{5}$ as one of the parts.) Take $\frac{4}{5}$ from $\frac{5}{5}$, and solve using an addition sentence.

S: (Break apart $1\frac{1}{5}$ into $\frac{1}{5}$ and $\frac{5}{5}$. Write $\frac{1}{5}+\frac{1}{5}=\frac{2}{5}$.)

Continue with the following possible sequence: $1\frac{3}{8}-\frac{7}{8}$.

Application Problem (5 minutes)

Krista drank $\frac{3}{16}$ of the water in her water bottle in the morning, $\frac{5}{16}$ in the afternoon, and $\frac{3}{16}$ in the evening. What fraction of water was left at the end of the day?

$\frac{3}{16}+\frac{5}{16}+\frac{3}{16}=\frac{11}{16}$

$1-\frac{11}{16}=\frac{16}{16}-\frac{11}{16}=\frac{5}{16}$

$\frac{5}{16}$ of the water in the bottle was left.

Note: This Application Problem builds on Lesson 18 where students added and subtracted two or more addends, as well as Lesson 19 where students solved word problems involving fractions. This problem invites counting on to the whole number as a solution strategy, too.

A STORY OF UNITS   Lesson 20 4•5

Concept Development (33 minutes)

Materials: (S) Personal white board

Problem 1: Add unit fractions with related denominators using tape diagrams.

T: 1 banana + 1 orange = _____?
S: 2 banana-oranges! No, that's not right! We can't add them, because the units are not the same.
T: What do bananas and oranges have in common?
S: They are both fruits.
T: So, what is 1 banana + 1 orange?
S: 2 pieces of fruit.
T: You had to rename, to find a way to name the banana and orange as the same unit.
T: $\frac{1}{3} + \frac{1}{6} =$ _____?
S: The units are different. → The units need to be the same. If the units are different, we cannot add the fractions together.
T: Let's decompose to make like units. Discuss a strategy with your partner.
S: I just know that a third is the same as 2 sixths. → We can draw a tape diagram to represent $\frac{1}{3}$ and another one to represent $\frac{1}{6}$. Then, we can decompose each third into two equal parts. $\frac{1}{3} = \frac{2}{6}$. → I can multiply in my head to rename $\frac{1}{3}$ as $\frac{2}{6}$. → I can use an area model or number line, too.

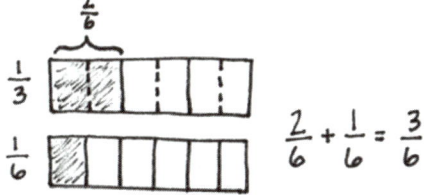

T: Add $\frac{2}{6} + \frac{1}{6}$. How many sixths are there altogether?
S: $\frac{3}{6}$. → $\frac{2}{6} + \frac{1}{6} = \frac{3}{6}$. → And $\frac{3}{6}$ is also $\frac{1}{2}$.
T: (Display $\frac{1}{2} + \frac{1}{8}$.) Draw tape diagrams to represent $\frac{1}{2}$ and $\frac{1}{8}$. Which fraction are we going to decompose?
S: We can decompose the halves into eighths. → You can't decompose eighths into halves, because halves are bigger than eighths. → We don't have enough eighths to compose one half, so we have to convert halves to eighths.
T: How many eighths are in $\frac{1}{2}$?
S: 4 eighths.
T: Add.
S: $\frac{4}{8} + \frac{1}{8} = \frac{5}{8}$.

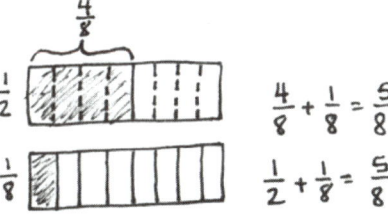

Lesson 20: Use visual models to add two fractions with related units using the denominators 2, 3, 4, 5, 6, 8, 10, and 12.

A STORY OF UNITS   Lesson 20   4•5

Problem 2: Add fractions with related denominators using tape diagrams.

T: (Display $\frac{2}{3} + \frac{3}{12}$.) Draw tape diagrams to show $\frac{2}{3}$ and $\frac{3}{12}$. Is one of the denominators a factor of the other?

S: Yes!

T: Which unit is larger—thirds or twelfths?

S: Thirds.

T: So, which unit do we have to decompose?

S: Thirds.

T: Go ahead and do that.

S: Thirds into twelfths. I can draw dotted vertical lines to show each third decomposed into 4 equal parts since there are 4 times as many twelfths in 1 as there are thirds. → There are $\frac{8}{12}$ shaded.

S: $\frac{8}{12} + \frac{3}{12} = \frac{11}{12}$.

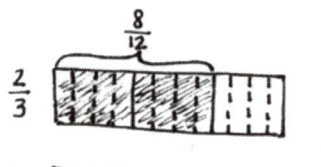

$\frac{8}{12} + \frac{3}{12} = \frac{11}{12}$

A NOTE ON MULTIPLE MEANS OF REPRESENTATION:

Students can also model both fractions on the same tape diagram. Have them model the larger units first and then partition with dotted lines to show the decompositions.

Problem 3: Add fractions with related denominators using a number line.

T: Write $\frac{1}{6} + \frac{3}{12}$. Let's estimate the sum as we draw a number line to model the addition. I'll mark zero. Do I need my number line to go past 1?

S: No. You are adding two small fractions, so it shouldn't go past 1.

T: Yes. Both fractions are less than 1 half. When we add them, the sum will be less than 1.

T: Draw a number line with endpoints 0 and 1. Partition the number line into sixths. Next, partition the number line further into twelfths. Each sixth will be decomposed into how many parts?

S: 2. → There are twice as many twelfths as there are sixths.

T: Use dashed lines to partition each sixth into twelfths.

T: Show the addition of $\frac{1}{6}$ and $\frac{3}{12}$. Start at 0, and hop to $\frac{1}{6}$. Draw another arrow to show the addition of $\frac{3}{12}$. What is the sum?

S: $\frac{5}{12}$.

T: Say the addition sentence with like denominators.

S: $\frac{2}{12} + \frac{3}{12} = \frac{5}{12}$.

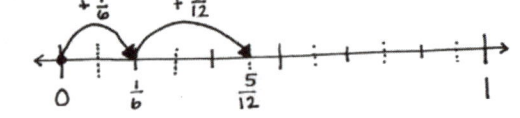

$\frac{1}{6} < \frac{1}{2}$ because $\frac{1}{6} < \frac{3}{6}$

$\frac{3}{12} < \frac{1}{2}$ because $\frac{3}{12} < \frac{6}{12}$

$\frac{1}{6} + \frac{3}{12} = \frac{5}{12}$

$\frac{2}{12} + \frac{3}{12} = \frac{5}{12}$

T: Write $\frac{3}{4} + \frac{5}{8}$. Estimate the sum. Will it be greater than or less than 1?

S: Greater than 1.

Lesson 20: Use visual models to add two fractions with related units using the denominators 2, 3, 4, 5, 6, 8, 10, and 12.

A STORY OF UNITS

Lesson 20 4•5

T: So, our number line has to go past 1. Does it need to go past 2?
S: No. Each fraction is less than 1.
T: Draw a number line. Partition the number line using the larger unit first. Which is the larger unit?
S: Fourths.
T: What's the next step?
S: Make the eighths by putting dashed lines to show each fourth decomposed into 2 eighths. → Just split each fourth into 2 parts.
T: Draw arrows to show the addition. Explain to your partner what you did.

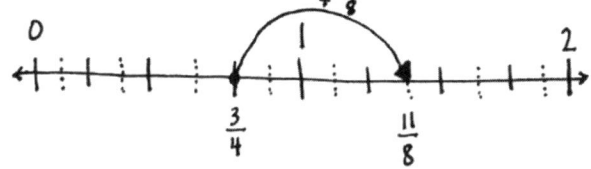

S: I started at 0 and moved to $\frac{3}{4}$. That's equal to $\frac{6}{8}$. Then, I drew an arrow to show the addition of $\frac{5}{8}$ more at $\frac{11}{8}$. → I just started at $\frac{3}{4}$ and added $\frac{5}{8}$.
T: Say the number sentence with like denominators.
S: $\frac{6}{8} + \frac{5}{8} = \frac{11}{8}$.

Problem 4: Add fractions with related denominators without using a model.

T: Today, we learned to add fractions by finding common denominators. We found equivalent fractions using models. Add $\frac{2}{5} + \frac{3}{10}$. Which unit is easiest to decompose?
S: Fifths can be decomposed into tenths.
T: How can we do that without a model? Talk to your partner.
S: We can multiply both the numerator and denominator of $\frac{2}{5}$. → $\frac{2}{5} = \frac{2 \times 2}{5 \times 2} = \frac{4}{10}$.
T: Now, add. Write a complete number sentence.
S: (Write $\frac{4}{10} + \frac{3}{10} = \frac{7}{10}$ or $\frac{2}{5} + \frac{3}{10} = \frac{7}{10}$.)

$\frac{2}{5} + \frac{3}{10} = \frac{4}{10} + \frac{3}{10} = \frac{7}{10}$

$\frac{2}{5} = \frac{2 \times 2}{5 \times 2} = \frac{4}{10}$

Repeat with $\frac{3}{12} + \frac{4}{3}$.

Problem Set (10 minutes)

Students should do their personal best to complete the Problem Set within the allotted 10 minutes. For some classes, it may be appropriate to modify the assignment by specifying which problems they work on first. Some problems do not specify a method for solving. Students should solve these problems using the RDW approach used for Application Problems.

Lesson 20: Use visual models to add two fractions with related units using the denominators 2, 3, 4, 5, 6, 8, 10, and 12.

A STORY OF UNITS   Lesson 20   4•5

Student Debrief (10 minutes)

Lesson Objective: Use visual models to add two fractions with related units using the denominators 2, 3, 4, 5, 6, 8, 10, and 12.

The Student Debrief is intended to invite reflection and active processing of the total lesson experience.

Invite students to review their solutions for the Problem Set. They should check work by comparing answers with a partner before going over answers as a class. Look for misconceptions or misunderstandings that can be addressed in the Debrief. Guide students in a conversation to debrief the Problem Set and process the lesson.

Any combination of the questions below may be used to lead the discussion.

- For Problem 1(a–f), how was drawing tape diagrams helpful?
- In Problem 1(c), did you use sixths as the common denominator? Explain how thirds could be used as the common denominator.
- For Problem 2(a–f), how was drawing a number line helpful?
- For Problem 2(a–f), what strategies did you use to estimate if the sum would be between 0 and 1 or 1 and 2?
- Why is it important to have common denominators when adding fractions? Relate common denominators to adding with mixed units of measurement from Module 2. For example, add 3 meters to 247 centimeters.
- Explain to your partner how to determine the sum of two fractions without drawing a model. What strategies did you use?
- How did the Application Problem connect to today's lesson?

Exit Ticket (3 minutes)

After the Student Debrief, instruct students to complete the Exit Ticket. A review of their work will help with assessing students' understanding of the concepts that were presented in today's lesson and planning more effectively for future lessons. The questions may be read aloud to the students.

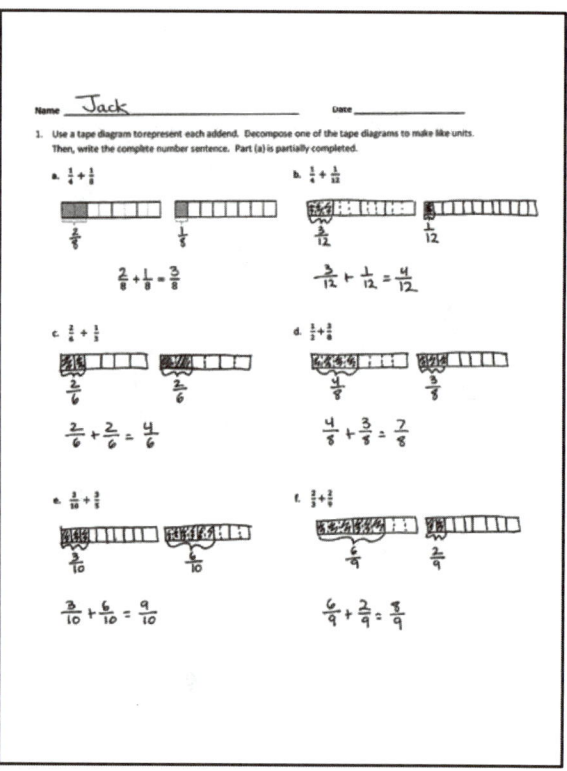

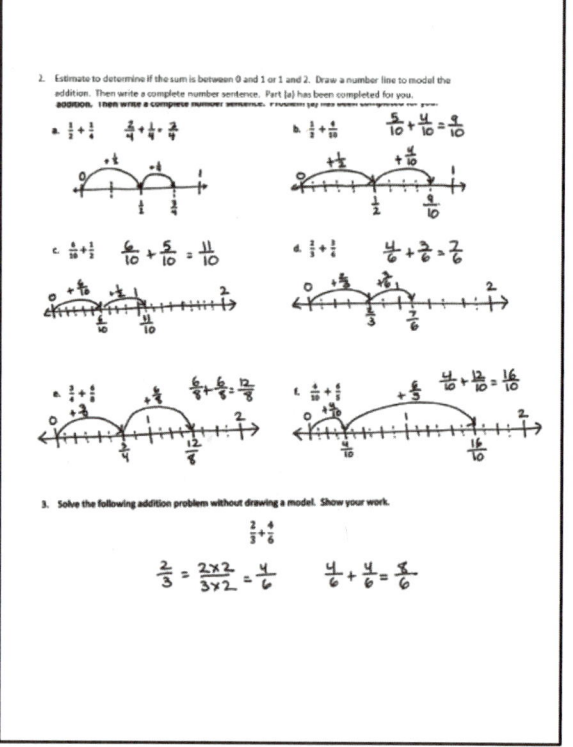

Name _____ Date _____

1. Use a tape diagram to represent each addend. Decompose one of the tape diagrams to make like units. Then, write the complete number sentence. Part (a) is partially completed.

    a. $\frac{1}{4} + \frac{1}{8}$

    b. $\frac{1}{4} + \frac{1}{12}$

    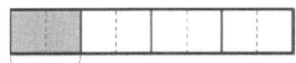 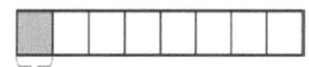

    $\frac{}{8} + \frac{}{8} = \frac{}{8}$

    c. $\frac{2}{6} + \frac{1}{3}$

    d. $\frac{1}{2} + \frac{3}{8}$

    e. $\frac{3}{1\bar{A}} + \frac{3}{5}$

    f. $\frac{2}{3} + \frac{2}{9}$

2. Estimate to determine if the sum is between 0 and 1 or 1 and 2. Draw a number line to model the addition. Then, write a complete number sentence. Part (a) has been completed for you.

   a. $\frac{1}{2} + \frac{1}{4}$ $\frac{2}{4} + \frac{1}{4} = \frac{3}{4}$

   b. $\frac{1}{2} + \frac{4}{10}$

   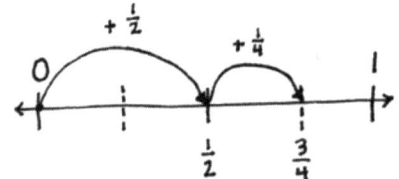

   c. $\frac{6}{10} + \frac{1}{2}$

   d. $\frac{2}{3} + \frac{3}{6}$

   e. $\frac{3}{4} + \frac{6}{8}$

   f. $\frac{4}{10} + \frac{6}{5}$

3. Solve the following addition problem without drawing a model. Show your work.

$$\frac{2}{3} + \frac{4}{6}$$

Lesson 20: Use visual models to add two fractions with related units using the denominators 2, 3, 4, 5, 6, 8, 10, and 12.

Name _____   Date _____

1. Draw a number line to model the addition. Solve, and then write a complete number sentence.

$$\frac{5}{8} + \frac{2}{4}$$

2. Solve without drawing a model.

$$\frac{3}{4} + \frac{1}{2}$$

Lesson 20: Use visual models to add two fractions with related units using the denominators 2, 3, 4, 5, 6, 8, 10, and 12.

Name _____   Date _____

1. Use a tape diagram to represent each addend. Decompose one of the tape diagrams to make like units. Then, write the complete number sentence.

   a. $\frac{1}{3} + \frac{1}{6}$

   b. $\frac{1}{2} + \frac{1}{4}$

   c. $\frac{3}{4} + \frac{1}{8}$

   d. $\frac{1}{4} + \frac{5}{12}$

   e. $\frac{3}{8} + \frac{1}{2}$

   f. $\frac{3}{5} + \frac{3}{10}$

Lesson 20 Homework 4•5

2. Estimate to determine if the sum is between 0 and 1 or 1 and 2. Draw a number line to model the addition. Then, write a complete number sentence. The first one has been completed for you.

   a. $\frac{1}{3} + \frac{1}{6}$ $\frac{2}{6} + \frac{1}{6} = \frac{3}{6}$

   b. $\frac{3}{5} + \frac{7}{10}$

   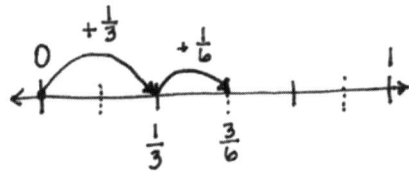

   c. $\frac{5}{12} + \frac{1}{4}$

   d. $\frac{3}{4} + \frac{5}{8}$

   e. $\frac{7}{8} + \frac{3}{4}$

   f. $\frac{1}{6} + \frac{5}{3}$

3. Solve the following addition problem without drawing a model. Show your work.

   $$\frac{5}{6} + \frac{1}{3}$$

Lesson 21

Objective: Use visual models to add two fractions with related units using the denominators 2, 3, 4, 5, 6, 8, 10, and 12.

Suggested Lesson Structure

- ■ Fluency Practice   (12 minutes)
- ■ Application Problem   (5 minutes)
- ■ Concept Development (33 minutes)
- ■ Student Debrief   (10 minutes)
- **Total Time**   **(60 minutes)**

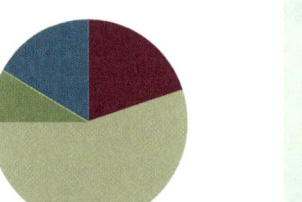

> **A NOTE ON STANDARDS ALIGNMENT:**
>
> In Lesson 21, students add fractions with related denominators where one denominator is a factor of the other. Students are able to generate equivalent fractions (**4.NF.1**) from their work in Topic B. It is a natural progression for students to be exposed to finding sums of fractions with unlike but related denominators where one denominator is a factor of the other. In Grade 5, students find sums and differences of fractions with unlike and unrelated denominators (**5.NF.1**). Lessons 20 and 21 prepare students to work with decimals in Module 6 where students add two fractions with like denominators of 100 (**4.NBT.5**).

Fluency Practice (12 minutes)

- Sprint: Subtract Fractions 4.NF.3   (9 minutes)
- Add Fractions 4.NF.3   (3 minutes)

Sprint: Subtract Fractions (9 minutes)

Materials: (S) Subtract Fractions Sprint

Note: This fluency activity reviews Lesson 17. In the Sprint's final quadrant, starting at Problem 31, there are a few problems which can be simplified (31, 32, 36, 37, 40, 41, and 43). Be accepting of answers in either form.

Add Fractions (3 minutes)

Materials: (S) Personal white board

Note: This fluency activity reviews Lesson 18.

T: (Write $\frac{5}{10} + \frac{3}{10} + \frac{1}{10} = \frac{}{10}$.) On your personal white board, write the complete number sentence.

S: (Write $\frac{5}{10} + \frac{3}{10} + \frac{1}{10} = \frac{9}{10}$.)

T: (Write $\frac{5}{8} + \frac{2}{8} + \frac{1}{8} = \frac{}{8}$.) Write the complete number sentence.

S: (Write $\frac{5}{8} + \frac{2}{8} + \frac{1}{8} = \frac{8}{8}$.)

T: (Write $\frac{5}{8} + \frac{2}{8} + \frac{1}{8} = \frac{8}{8}$.) Rename 8 eighths as a whole number.

S: (Write $\frac{5}{8} + \frac{2}{8} + \frac{1}{8} = \frac{8}{8} = 1$.)

A STORY OF UNITS   Lesson 21   4•5

Continue the process with $\frac{2}{6} + \frac{3}{6} + \frac{1}{6}$.

   T: (Write $\frac{2}{3} + \frac{1}{3} + \frac{2}{3} = \frac{}{3}$.) Complete the equation.

   S: (Write $\frac{2}{3} + \frac{1}{3} + \frac{2}{3} = \frac{5}{3}$.)

   T: How many thirds are in 1?

   S: 3 thirds.

   T: Write $\frac{5}{3}$ as a mixed number.

   S: (Write $\frac{2}{3} + \frac{1}{3} + \frac{2}{3} = \frac{5}{3} = 1\frac{2}{3}$.)

Continue the process with $\frac{5}{8} + \frac{5}{8} + \frac{5}{8}$.

Application Problem (5 minutes)

Two-fifths liter of chemical A was added to $\frac{7}{10}$ liter of chemical B to make chemical C. How many liters of chemical C are there?

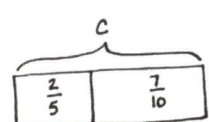 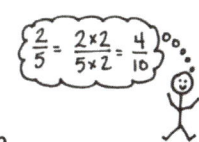

Note: This Application Problem builds on the work of Lesson 20 where students learned to add two fractions with related units. This Application Problem bridges to today's lesson where students again add two fractions with related units, but this time, they use number bonds to write the sums as mixed numbers.

Concept Development (33 minutes)

Materials: (S) Personal white board

Problem 1: Add two fractions with related units modeled with a tape diagram. Use a number bond to rename the sum as a mixed number.

   T: Solve $\frac{3}{8} + \frac{3}{4}$. Work with your partner to draw tape diagrams to represent each fraction. Decompose the larger unit into smaller units as we did in Lesson 20. Solve and write a complete number sentence to show your answer. Explain the process that you used.

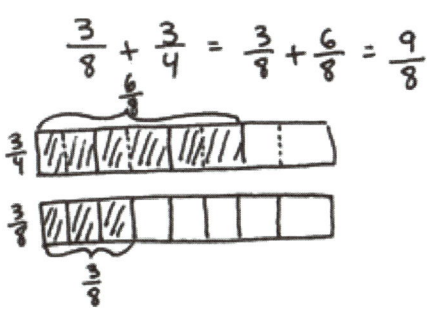

Lesson 21:   Use visual models to add two fractions with related units using the denominators 2, 3, 4, 5, 6, 8, 10, and 12.   277

A STORY OF UNITS   Lesson 21 4•5

S: (Write $\frac{3}{8} + \frac{6}{8} = \frac{9}{8}$.) We drew tape diagrams to show eighths and fourths and then shaded in $\frac{3}{8}$ of one and $\frac{3}{4}$ of the other. We decomposed the larger unit of fourths into eighths and found that $\frac{3}{4} = \frac{6}{8}$.

T: Is $\frac{9}{8}$ greater than 1 or less than 1?

S: It's greater than 1. Since $\frac{8}{8}$ is equal to 1, $\frac{9}{8}$ is greater than 1.

T: Draw a number bond to show $\frac{9}{8}$ as a whole and $\frac{8}{8}$ as a part.

S: (Draw a number bond as shown to the right.)

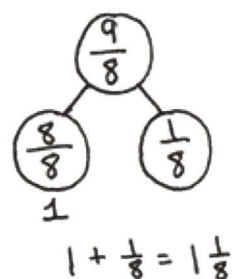

T: In the complete number sentence, show the mixed number equal to $\frac{9}{8}$.

S: (Write $\frac{3}{8} + \frac{6}{8} = \frac{9}{8} = 1\frac{1}{8}$.)

Repeat with $\frac{2}{5} + \frac{7}{10}$ from the Application Problem, drawing the number bond to name the mixed number.

Problem 2: Add two fractions with related units using a number line and number bonds. Use a number bond to rename the sum as a mixed number.

T: Write $\frac{1}{2} + \frac{7}{8}$.

T: Will the sum be greater or less than 1?

S: Greater.

T: Draw a number line, labeling the whole numbers and the larger unit. Decompose the larger unit to show the smaller unit. Show the addition with arrows, and then, write a number sentence. (Allow students time to work.)

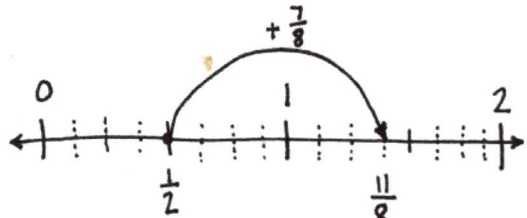

T: With your partner, review the process you used.

S: We estimated that the sum would be greater than 1, since we were adding a half to a fraction greater than 1 half. We drew a number line from 0 to 2 and then labeled the whole numbers. Halves are the larger unit, so we marked and labeled 1 half. Then, we marked the intervals for the eighths. We drew an arrow to show $\frac{1}{2} + \frac{7}{8}$. → $\frac{4}{8} + \frac{7}{8} = \frac{11}{8}$. → $\frac{1}{2} + \frac{7}{8} = \frac{11}{8}$.

T: Draw a number bond to rename $\frac{11}{8}$ as a mixed number. Write the mixed number in the complete number sentence.

S: (Draw a number bond as shown to the right. Write $\frac{11}{8} = \frac{8}{8} + \frac{3}{8} = 1\frac{3}{8}$.)

Repeat with $\frac{5}{6} + \frac{2}{3}$.

> **NOTES ON MULTIPLE MEANS OF REPRESENTATION:**
>
> Ease the task of speaking in English to review the process of adding $\frac{1}{2} + \frac{7}{8}$ for English language learners by providing sentence frames. However, if students are otherwise unable to fully express themselves, allow discussion in their first language, or if writing is easier, have students journal.

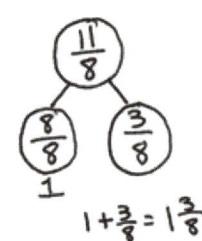

Lesson 21: Use visual models to add two fractions with related units using the denominators 2, 3, 4, 5, 6, 8, 10, and 12.

A STORY OF UNITS — Lesson 21 4•5

Problem 3: Add two fractions with related units without using a model. Express the answer as a mixed number.

T: Write $\frac{3}{4} + \frac{6}{8}$. With a partner, determine the sum of $\frac{3}{4}$ and $\frac{6}{8}$ by converting to equivalent fractions. Explain the process that you used.

S: (Write $\frac{3}{4} = \frac{3 \times 2}{4 \times 2} = \frac{6}{8}$ and $\frac{6}{8} + \frac{6}{8} = \frac{12}{8}$. Explain the process.)

T: Express $\frac{12}{8}$ as a mixed number using a number bond.

S: (Draw a number bond as shown to the right. Write $\frac{12}{8} = \frac{8}{8} + \frac{4}{8} = 1\frac{4}{8}$.)

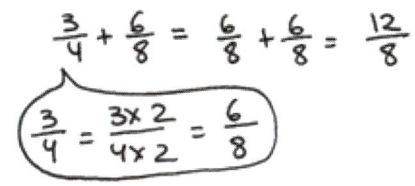

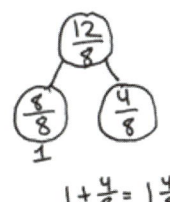

Problem Set (10 minutes)

Students should do their personal best to complete the Problem Set within the allotted 10 minutes. For some classes, it may be appropriate to modify the assignment by specifying which problems they work on first. Some problems do not specify a method for solving. Students should solve these problems using the RDW approach used for Application Problems.

Student Debrief (10 minutes)

Lesson Objective: Use visual models to add two fractions with related units using the denominators 2, 3, 4, 5, 6, 8, 10, and 12.

The Student Debrief is intended to invite reflection and active processing of the total lesson experience.

Invite students to review their solutions for the Problem Set. They should check work by comparing answers with a partner before going over answers as a class. Look for misconceptions or misunderstandings that can be addressed in the Debrief. Guide students in a conversation to debrief the Problem Set and process the lesson.

Any combination of the questions below may be used to lead the discussion.

- What was the complexity of the Problem Set for today's lesson (Lesson 21) as compared to yesterday's Problem Set (Lesson 20)?
- How do number bonds help to show fractions as mixed numbers?

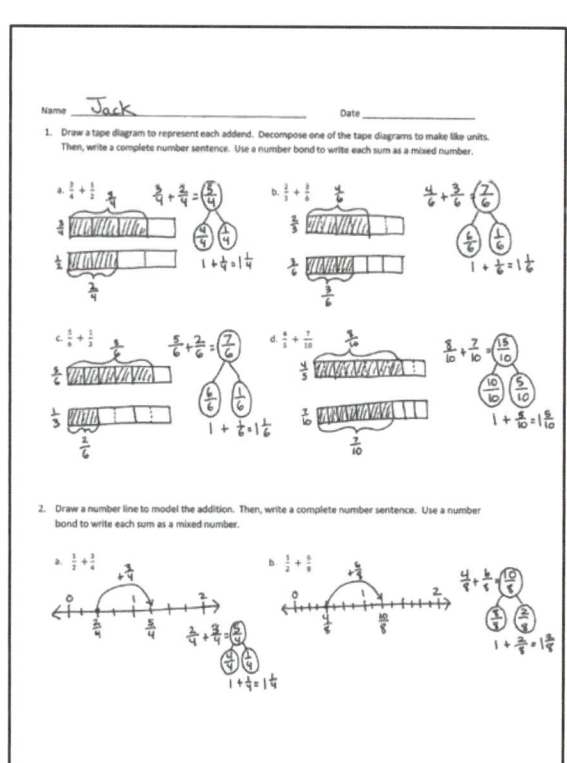

Lesson 21: Use visual models to add two fractions with related units using the denominators 2, 3, 4, 5, 6, 8, 10, and 12.

- What benefit can you see in expressing a fraction as a mixed number or a mixed number as a fraction?
- Compare Problems 1(a) and 2(a). Which strategy worked better for you? Explain.
- How did the Application Problem connect to today's lesson?

Exit Ticket (3 minutes)

After the Student Debrief, instruct students to complete the Exit Ticket. A review of their work will help with assessing students' understanding of the concepts that were presented in today's lesson and planning more effectively for future lessons. The questions may be read aloud to the students.

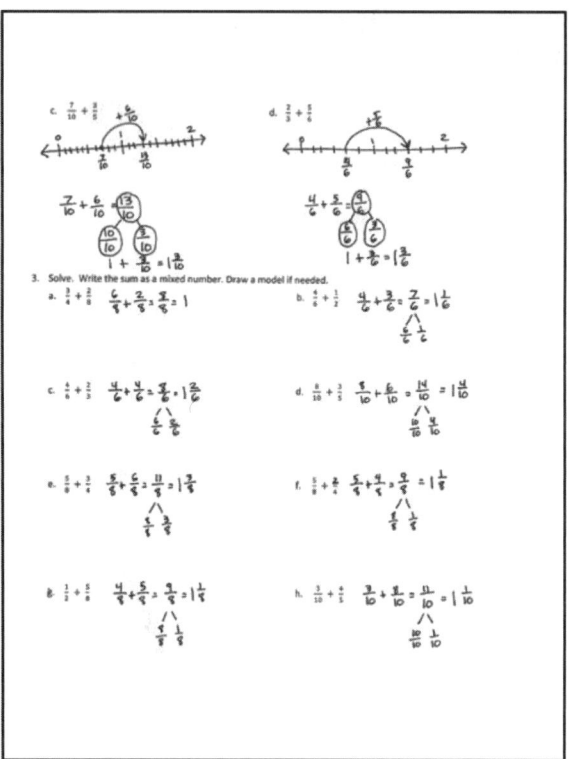

A

Subtract Fractions

Number Correct: _____

1.	$2 - 1 =$		23.	$\frac{4}{3} - \frac{2}{3} =$	
2.	$\frac{2}{2} - \frac{1}{2} =$		24.	$1\frac{1}{3} - \frac{2}{3} =$	
3.	$1 - \frac{1}{2} =$		25.	$1\frac{2}{3} - \frac{1}{3} =$	
4.	$3 - 1 =$		26.	$7 - 4 =$	
5.	$\frac{3}{3} - \frac{1}{3} =$		27.	$\frac{7}{5} - \frac{4}{5} =$	
6.	$1 - \frac{1}{3} =$		28.	$1\frac{2}{5} - \frac{4}{5} =$	
7.	$8 - 1 =$		29.	$1\frac{4}{5} - \frac{2}{5} =$	
8.	$\frac{8}{8} - \frac{1}{8} =$		30.	$5 - 3 =$	
9.	$1 - \frac{1}{8} =$		31.	$\frac{5}{4} - \frac{3}{4} =$	
10.	$5 - 1 =$		32.	$1\frac{1}{4} - \frac{3}{4} =$	
11.	$\frac{5}{5} - \frac{1}{5} =$		33.	$1\frac{3}{4} - \frac{1}{4} =$	
12.	$1 - \frac{1}{5} =$		34.	$1 - \frac{3}{8} =$	
13.	$1 - \frac{2}{5} =$		35.	$1 - \frac{7}{8} =$	
14.	$1 - \frac{4}{5} =$		36.	$1\frac{7}{8} - \frac{3}{8} =$	
15.	$1 - \frac{3}{5} =$		37.	$1\frac{3}{8} - \frac{7}{8} =$	
16.	$1 - \frac{1}{4} =$		38.	$1 - \frac{1}{6} =$	
17.	$1 - \frac{3}{4} =$		39.	$1 - \frac{5}{6} =$	
18.	$1 - \frac{1}{1} =$		40.	$1\frac{5}{6} - \frac{1}{6} =$	
19.	$1 - \frac{9}{1} =$		41.	$1\frac{1}{6} - \frac{5}{6} =$	
20.	$1 - \frac{3}{1} =$		42.	$1 - \frac{5}{12} =$	
21.	$1 - \frac{7}{1} =$		43.	$1\frac{1}{12} - \frac{7}{12} =$	
22.	$4 - 2 =$		44.	$1\frac{4}{15} - \frac{13}{15} =$	

Lesson 21: Use visual models to add two fractions with related units using the denominators 2, 3, 4, 5, 6, 8, 10, and 12.

B

Subtract Fractions

Number Correct: _____
Improvement: _____

1.	$3 - 1 =$	
2.	$\frac{3}{3} - \frac{1}{3} =$	
3.	$1 - \frac{1}{3} =$	
4.	$2 - 1 =$	
5.	$\frac{2}{2} - \frac{1}{2} =$	
6.	$1 - \frac{1}{2} =$	
7.	$6 - 1 =$	
8.	$\frac{6}{6} - \frac{1}{6} =$	
9.	$1 - \frac{1}{6} =$	
10.	$10 - 1 =$	
11.	$\frac{1A}{1A} - \frac{1}{1A} =$	
12.	$1 - \frac{1}{1A} =$	
13.	$1 - \frac{2}{1A} =$	
14.	$1 - \frac{4}{1A} =$	
15.	$1 - \frac{3}{1A} =$	
16.	$1 - \frac{1}{5} =$	
17.	$1 - \frac{4}{5} =$	
18.	$1 - \frac{1}{8} =$	
19.	$1 - \frac{7}{8} =$	
20.	$1 - \frac{3}{8} =$	
21.	$1 - \frac{5}{8} =$	
22.	$5 - 3 =$	

23.	$\frac{5}{4} - \frac{3}{4} =$	
24.	$1\frac{1}{4} - \frac{3}{4} =$	
25.	$1\frac{3}{4} - \frac{1}{4} =$	
26.	$8 - 4 =$	
27.	$\frac{8}{5} - \frac{4}{5} =$	
28.	$1\frac{3}{5} - \frac{4}{5} =$	
29.	$1\frac{4}{5} - \frac{3}{5} =$	
30.	$7 - 5 =$	
31.	$\frac{7}{6} - \frac{5}{6} =$	
32.	$1\frac{1}{6} - \frac{5}{6} =$	
33.	$1\frac{5}{6} - \frac{1}{6} =$	
34.	$1 - \frac{5}{8} =$	
35.	$1 - \frac{7}{8} =$	
36.	$1\frac{7}{8} - \frac{5}{8} =$	
37.	$1\frac{5}{8} - \frac{7}{8} =$	
38.	$1 - \frac{1}{4} =$	
39.	$1 - \frac{3}{4} =$	
40.	$1\frac{3}{4} - \frac{1}{4} =$	
41.	$1\frac{1}{4} - \frac{3}{4} =$	
42.	$1 - \frac{7}{12} =$	
43.	$1\frac{1}{12} - \frac{5}{12} =$	
44.	$1\frac{7}{15} - \frac{11}{15} =$	

Lesson 21: Use visual models to add two fractions with related units using the denominators 2, 3, 4, 5, 6, 8, 10, and 12.

Name _____ Date _____

1. Draw a tape diagram to represent each addend. Decompose one of the tape diagrams to make like units. Then, write a complete number sentence. Use a number bond to write each sum as a mixed number.

   a. $\frac{3}{4} + \frac{1}{2}$

   b. $\frac{2}{3} + \frac{3}{6}$

   c. $\frac{5}{6} + \frac{1}{3}$

   d. $\frac{4}{5} + \frac{7}{10}$

2. Draw a number line to model the addition. Then, write a complete number sentence. Use a number bond to write each sum as a mixed number.

   a. $\frac{1}{2} + \frac{3}{4}$

   b. $\frac{1}{2} + \frac{6}{8}$

Lesson 21: Use visual models to add two fractions with related units using the denominators 2, 3, 4, 5, 6, 8, 10, and 12.

c. $\frac{7}{10} + \frac{3}{5}$

d. $\frac{2}{3} + \frac{5}{6}$

3. Solve. Write the sum as a mixed number. Draw a model if needed.

a. $\frac{3}{4} + \frac{2}{8}$

b. $\frac{4}{6} + \frac{1}{2}$

c. $\frac{4}{6} + \frac{2}{3}$

d. $\frac{8}{10} + \frac{3}{5}$

e. $\frac{5}{8} + \frac{3}{4}$

f. $\frac{5}{8} + \frac{2}{4}$

g. $\frac{1}{2} + \frac{5}{8}$

h. $\frac{3}{10} + \frac{4}{5}$

A STORY OF UNITS

Lesson 21 Exit Ticket 4•5

Name _____ Date _____

Solve. Write a complete number sentence. Use a number bond to write each sum as a mixed number. Use a model if needed.

1. $\frac{1}{4} + \frac{7}{8}$

2. $\frac{2}{3} + \frac{7}{12}$

Lesson 21: Use visual models to add two fractions with related units using the denominators 2, 3, 4, 5, 6, 8, 10, and 12.

A STORY OF UNITS

Lesson 21 Homework 4•5

Name _____ Date _____

1. Draw a tape diagram to represent each addend. Decompose one of the tape diagrams to make like units. Then, write a complete number sentence. Use a number bond to write each sum as a mixed number.

   a. $\frac{7}{8} + \frac{1}{4}$

   b. $\frac{4}{8} + \frac{2}{4}$

   c. $\frac{4}{6} + \frac{1}{2}$

   d. $\frac{3}{5} + \frac{8}{10}$

2. Draw a number line to model the addition. Then, write a complete number sentence. Use a number bond to write each sum as a mixed number.

   a. $\frac{1}{2} + \frac{5}{8}$

   b. $\frac{3}{4} + \frac{3}{8}$

Lesson 21: Use visual models to add two fractions with related units using the denominators 2, 3, 4, 5, 6, 8, 10, and 12.

c. $\frac{4}{10} + \frac{4}{5}$

d. $\frac{1}{3} + \frac{5}{6}$

3. Solve. Write the sum as a mixed number. Draw a model if needed.

a. $\frac{1}{2} + \frac{6}{8}$

b. $\frac{7}{8} + \frac{3}{4}$

c. $\frac{5}{6} + \frac{1}{3}$

d. $\frac{9}{10} + \frac{2}{5}$

e. $\frac{4}{12} + \frac{3}{4}$

f. $\frac{1}{2} + \frac{5}{6}$

g. $\frac{3}{12} + \frac{5}{6}$

h. $\frac{7}{10} + \frac{4}{5}$

A STORY OF UNITS   Mid-Module Assessment Task   4•5

Name _____   Date _____

1. Let each small square represent $\frac{1}{4}$.

   a. Using the same unit, draw and shade the following fractions. Represent each as a sum of unit fractions.

   Example: $\frac{3}{4}$
   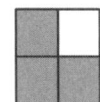
   $\frac{3}{4} = \frac{1}{4} + \frac{1}{4} + \frac{1}{4}$

   i. 1   ii. $\frac{2}{4}$ iii. $\frac{5}{4}$

   b. Record the decompositions of Parts (i) and (iii) using only 2 addends.

   i.

   iii.

   c. Rewrite the equations from Part (a) as the multiplication of a whole number by a unit fraction.

   i.

   ii.

   iii.

288   Module 5:   Fraction Equivalence, Ordering, and Operations

2. a. Using the fractional units shown, identify the fraction of the rectangle that is shaded. Continue this pattern by drawing the next area model in the sequence and identifying the fraction shaded.

   b. Use multiplication to explain why the first two fractions are equivalent.

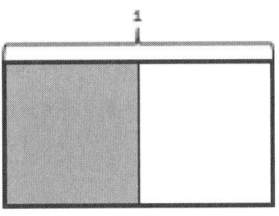

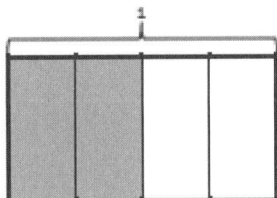

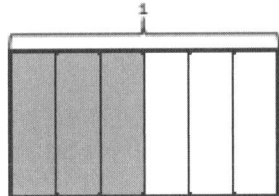

3. Cross out the fraction that is not equivalent to the other three. Show how you know.

   a. $\frac{3}{5} \quad \frac{60}{100} \quad \frac{6}{10} \quad \frac{6}{5}$

   b. $\frac{6}{4} \quad \frac{3}{2} \quad \frac{12}{8} \quad \frac{8}{4}$

   c. $\frac{6}{4} \quad \frac{16}{12} \quad \frac{9}{6} \quad \frac{3}{2}$

4. Fill in the circle with <, =, or > to make a true number sentence. Justify each response by drawing a model (such as an area model or a number line), creating common denominators or numerators, or explaining a comparison to a benchmark fraction.

   a. $\frac{6}{5}$ ◯ $\frac{4}{5}$

   b. $\frac{5}{8}$ ◯ $\frac{5}{10}$

   c. $\frac{5}{5}$ ◯ $\frac{12}{12}$

   d. $\frac{5}{12}$ ◯ $\frac{6}{10}$

   e. $\frac{5}{6}$ ◯ $\frac{3}{4}$

   f. $\frac{8}{3}$ ◯ $\frac{16}{6}$

   g. $\frac{7}{4}$ ◯ $\frac{9}{5}$

   h. $\frac{12}{8}$ ◯ $\frac{11}{6}$

5. Fill in the blanks to make each number sentence true. Draw a number line, a tape diagram, or an area model to represent each problem.

a. _____ = $\frac{5}{12} + \frac{6}{12}$

b. $\frac{53}{100} - \frac{27}{100}$ = _____

c. $\frac{8}{12}$ + _____ = 1

d. $\frac{3}{10} + \frac{6}{10} + \frac{2}{10}$ = _____

e. $1 - \frac{5}{8}$ = _____

f. $\frac{7}{8} - \frac{3}{8}$ = _____

6. Ray, Robin, and Freddy went fishing.

   a. They spent $\frac{1}{6}$ of their money on water, $\frac{4}{6}$ of their money on lunch, and the rest on worms. What fraction of their money was spent on worms? Draw a model, and write an equation to solve.

   b. Robin noticed her water bottle was $\frac{1}{2}$ full and Freddy's was $\frac{3}{4}$ full. Robin said, "My $\frac{1}{2}$ full bottle has more water than your $\frac{3}{4}$ full bottle." Explain how $\frac{1}{2}$ bottle could be more than $\frac{3}{4}$ bottle.

   c. Ray, Robin, and Freddy each had identical containers of worms. Ray used $\frac{3}{8}$ container. Robin used $\frac{6}{8}$ container, and Freddy used $\frac{7}{8}$ container. How many total containers of worms did they use?

   d. Express the number of remaining containers as a product of a whole number and a unit fraction.

   e. Six out of the eight fish they caught were trout. What is another fraction equal to 6 eighths? Write a number sentence, and draw a model to show the two fractions are equal.

Mid-Module Assessment Task
Standards Addressed

Topics A–D

Extend understanding of fraction equivalence and ordering.

4.NF.1 Explain why a fraction a/b is equivalent to a fraction $(n \times a)/(n \times b)$ by using visual fraction models, with attention to how the number and size of the parts differ even though the two fractions themselves are the same size. Use this principle to recognize and generate equivalent fractions.

4.NF.2 Compare two fractions with different numerators and different denominators, e.g., by creating common denominators or numerators, or by comparing to a benchmark fraction such as 1/2. Recognize that comparisons are valid only when the two fractions refer to the same whole. Record the results of comparisons with symbols >, =, or <, and justify the conclusions, e.g., by using a visual fraction model.

Build fractions from unit fractions by applying and extending previous understandings of operations on whole numbers.

4.NF.3 Understand a fraction a/b with $a > 1$ as a sum of fractions $1/b$.

  a. Understand addition and subtraction of fractions as joining and separating parts referring to the same whole.

  b. Decompose a fraction into a sum of fractions with the same denominator in more than one way, recording each decomposition by an equation. Justify decompositions, e.g., by using a visual fraction model. *Examples: 3/8 = 1/8 + 1/8 + 1/8; 3/8 = 1/8 + 2/8; 2 1/8 = 1 + 1 + 1/8 = 8/8 + 8/8 + 1/8.*

  d. Solve word problems involving addition and subtraction of fractions referring to the same whole and having like denominators, e.g., by using visual fraction models and equations to represent the problem.

4.NF.4 Apply and extend previous understandings of multiplication to multiply a fraction by a whole number.

  a. Understand a fraction a/b as a multiple of $1/b$. *For example, use a visual fraction model to represent 5/4 as the product 5 × (1/4), recording the conclusion by the equation 5/4 = 5 × (1/4).*

Evaluating Student Learning Outcomes

A Progression Toward Mastery is provided to describe steps that illuminate the gradually increasing understandings that students develop *on their way to proficiency.* In this chart, this progress is presented from left (Step 1) to right (Step 4). The learning goal for students is to achieve Step 4 mastery. These steps are meant to help teachers and students identify and celebrate what students CAN do now and what they need to work on next.

Module 5: Fraction Equivalence, Ordering, and Operations

Mid-Module Assessment Task 4•5

A Progression Toward Mastery				
Assessment Task Item and Standards Assessed	STEP 1 Little evidence of reasoning without a correct answer. (1 Point)	STEP 2 Evidence of some reasoning without a correct answer. (2 Points)	STEP 3 Evidence of some reasoning with a correct answer or evidence of solid reasoning with an incorrect answer. (3 Points)	STEP 4 Evidence of solid reasoning with a correct answer. (4 Points)
1 4.NF.3ab 4.NF.4a	The student correctly answers fewer than four of the eight parts.	The student correctly answers four or five of the eight parts.	The student correctly answers six or seven of the eight parts.	The student correctly does the following: a. Draws and shades to represent the three given fractions and represents each as a sum of unit fractions: 　i. $1 = \frac{1}{4} + \frac{1}{4} + \frac{1}{4} + \frac{1}{4}$ 　ii. $\frac{2}{4} = \frac{1}{4} + \frac{1}{4}$ 　iii. $\frac{5}{4} = \frac{1}{4} + \frac{1}{4} + \frac{1}{4} + \frac{1}{4} + \frac{1}{4}$ b. Records the decomposition using two addends. (Answers may vary.) 　i. $1 = \frac{3}{4} + \frac{1}{4}$ 　iii. $\frac{5}{4} = \frac{3}{4} + \frac{2}{4}$ c. Rewrites equations as multiplication of a whole number: 　i. $1 = 4 \times \frac{1}{4}$ 　ii. $\frac{2}{4} = 2 \times \frac{1}{4}$ 　iii. $\frac{5}{4} = 5 \times \frac{1}{4}$

A Progression Toward Mastery

2 **4.NF.1**	The student is unable to correctly complete a majority of the problem.	The student is able to correctly identify the fractions naming the three given models but is unable to complete the next model in the sequence and does not correctly explain equivalence using multiplication.	The student is able to correctly identify the fractions naming the three given models and is able to create the next model, as well as identify the appropriate fraction, but offers an incomplete explanation as to why the first two fractions are equivalent.	The student correctly does the following: a. Identifies the shaded fractions as $\frac{1}{2}, \frac{2}{4}, \frac{3}{6}, \frac{4}{8}$ and creates a correct model to represent $\frac{4}{8}$. b. Uses multiplication to explain why $\frac{1}{2}$ and $\frac{2}{4}$ are equivalent: $\frac{1 \times 2}{2 \times 2} = \frac{2}{4}$
3 **4.NF.1**	The student is not able to correctly identify any of the non-equivalent fractions. Explanation or modeling is inaccurate.	The student correctly identifies one of the three non-equivalent fractions. Explanation or modeling is incomplete, or the student does not attempt to show work.	The student correctly identifies two of the three non-equivalent fractions. Explanation or modeling is mostly complete.	The student correctly identifies all three of the non-equivalent fractions and gives complete explanations: a. $\frac{6}{5}$ b. $\frac{8}{4}$ c. $\frac{16}{12}$
4 **4.NF.2**	The student correctly compares three or fewer of the fraction sets with little to no reasoning.	The student correctly compares four or five of the fraction sets with some reasoning.	The student correctly compares six or seven of the fraction sets with solid reasoning. OR The student correctly compares all fraction sets with incomplete reasoning on one or two parts.	The student correctly compares all eight of the fraction sets and justifies all answers using models, common denominators or numerators, or benchmark fractions: a. > b. > c. = d. < e. > f. = g. < h. <

Mid-Module Assessment Task 4•5

A Progression Toward Mastery				
5 **4.NF.3a**	The student correctly completes two or fewer number sentences and does not accurately use models to represent a majority of the problems.	The student correctly completes three number sentences with some accurate modeling to represent the problems.	The student correctly completes four or five number sentences with accurate modeling to represent problems. OR The student correctly completes all number sentences with insufficient models on one or two problems.	The student correctly completes all six number sentences and accurately models each problem using a number line, a tape diagram, or an area model: a. $\frac{11}{12}$ b. $\frac{26}{100}$ c. $\frac{4}{12}$ d. $\frac{11}{10}$ e. $\frac{3}{8}$ f. $\frac{4}{8}$
6 **4.NF.1** **4.NF.2** **4.NF.3abd** **4.NF.4a**	The student correctly completes fewer than three of the five parts with little to no reasoning.	The student correctly completes three of the five parts, providing some reasoning in Part (a), (b), or (c).	The student correctly completes four of the five parts. OR The student correctly completes all five parts but without solid reasoning in Parts (a), (b), or (c).	The student correctly completes all five of the parts: a. Answers $\frac{1}{6}$ and writes an equation and draws a model. b. Accurately explains through words and/or pictures that the two fractions in question refer to two different-size wholes. The water bottle that is half full could be a larger bottle. c. Answers $\frac{16}{8}$ or 2 containers. d. Answers $\frac{8}{8} = 8 \times \frac{1}{8}$. e. Writes a division or multiplication equation and draws a model to represent a fraction equal to $\frac{6}{8}$.

Module 5: Fraction Equivalence, Ordering, and Operations

A STORY OF UNITS — Mid-Module Assessment Task 4•5

Name __Jack_____ Date _____

1. Let each small square represent $\frac{1}{4}$.

    a. Using the same unit, draw and shade the following fractions. Represent each as a sum of unit fractions.

    Example: $\frac{3}{4}$

    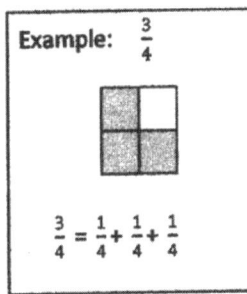

    $\frac{3}{4} = \frac{1}{4} + \frac{1}{4} + \frac{1}{4}$

    i. 1

    $1 = \frac{1}{4} + \frac{1}{4} + \frac{1}{4} + \frac{1}{4}$

    ii. $\frac{2}{4}$

    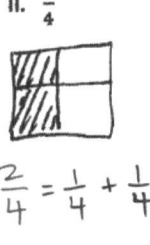

    $\frac{2}{4} = \frac{1}{4} + \frac{1}{4}$

    iii. $\frac{5}{4}$

    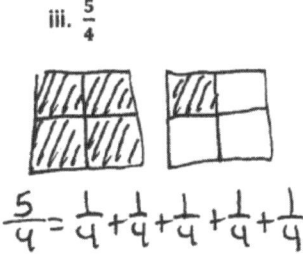

    $\frac{5}{4} = \frac{1}{4} + \frac{1}{4} + \frac{1}{4} + \frac{1}{4} + \frac{1}{4}$

    b. Record the decompositions of Parts (i) and (iii) using only 2 addends.

    i. $1 = \frac{2}{4} + \frac{2}{4}$

    iii. $\frac{5}{4} = \frac{2}{4} + \frac{3}{4}$

    c. Rewrite the equations from Part (a) as the multiplication of a whole number by a unit fraction.

    i. $1 = 4 \times \frac{1}{4}$

    ii. $\frac{2}{4} = 2 \times \frac{1}{4}$

    iii. $\frac{5}{4} = 5 \times \frac{1}{4}$

2. a. Using the fractional units shown, identify the fraction of the rectangle that is shaded. Continue this pattern by drawing the next area model in the sequence and identifying the fraction shaded.

b. Use multiplication to explain why the first two fractions are equivalent.

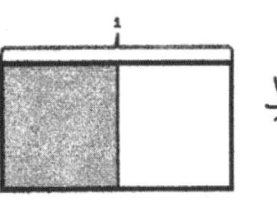 $\frac{1}{2}$

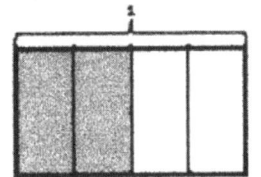 $\frac{2}{4}$

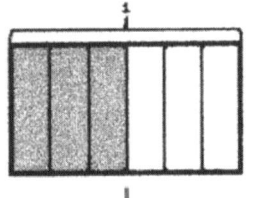 $\frac{3}{6}$

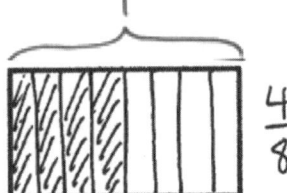 $\frac{4}{8}$

$$\frac{1}{2} = \frac{2}{4}$$

$$\frac{1 \times 2}{2 \times 2} = \frac{2}{4}$$

3. Cross out the fraction that is not equivalent to the other three. Show how you know.

a. $\frac{3}{5}$ $\frac{60}{100}$ $\frac{6}{10}$ ~~$\frac{6}{5}$~~

$$\frac{3 \times 20}{5 \times 20} = \frac{60}{100}$$

$$\frac{6 \times 10}{10 \times 10} = \frac{60}{100}$$

$\frac{3}{5}$ does not equal $\frac{6}{5}$

b. $\frac{6}{4}$ $\frac{3}{2}$ $\frac{12}{8}$ ~~$\frac{8}{4}$~~

$$\frac{3 \times 4}{2 \times 4} = \frac{12}{8}$$

$$\frac{6 \times 2}{4 \times 2} = \frac{12}{8}$$

$\frac{6}{4}$ does not equal $\frac{8}{4}$

c. $\frac{6}{4}$ ~~$\frac{16}{12}$~~ $\frac{9}{6}$ $\frac{3}{2}$

$$\frac{3 \times 3}{2 \times 3} = \frac{9}{6}$$

$$\frac{3 \times 2}{2 \times 2} = \frac{6}{4}$$

$$\frac{3 \times 6}{2 \times 6} = \frac{18}{12}$$

$\frac{18}{12}$ does not equal $\frac{16}{12}$

4. Fill in the circle with <, =, or > to make a true number sentence. Justify each response by drawing a model (such as an area model or number line), creating common denominators or numerators, or explaining a comparison to a benchmark fraction.

a. $\frac{6}{5}$ ⃝> $\frac{4}{5}$

With the same whole, six fifths is more than four fifths.

b. $\frac{5}{8}$ ⃝> $\frac{5}{10}$

With the same size whole, tenths are smaller than eighths. Five tenths are less than five eighths.

c. $\frac{5}{5}$ ⃝= $\frac{12}{12}$

Both fractions are equal to 1 whole.

d. $\frac{5}{12}$ ⃝< $\frac{6}{10}$

$\frac{5}{12}$ is less than $\frac{1}{2}$.
$\frac{6}{10}$ is greater than $\frac{1}{2}$.

e. $\frac{5}{6}$ ⃝> $\frac{3}{4}$

$\frac{5}{6}$ is only $\frac{1}{6}$ from one whole.
$\frac{3}{4}$ is $\frac{1}{4}$ from one whole, so
$\frac{5}{6} > \frac{3}{4}$ since $\frac{5}{6}$ is closer to one whole.

f. $\frac{8}{3}$ ⃝= $\frac{16}{6}$

$\frac{8 \times 2}{3 \times 2} = \frac{16}{6}$

g. $\frac{7}{4}$ ⃝< $\frac{9}{5}$

$\frac{4}{4}$ $\frac{3}{4}$ $\frac{5}{5}$ $\frac{4}{5}$

$\frac{3}{4}$ is $\frac{1}{4}$ from one whole.
$\frac{4}{5}$ is $\frac{1}{5}$ from one whole.
$\frac{3}{4} < \frac{4}{5}$ since $\frac{3}{4}$ is further from one whole.
$\frac{7}{4} < \frac{9}{5}$

h. $\frac{12}{8}$ ⃝< $\frac{11}{6}$

$\frac{12}{8} = 1\frac{4}{8} = 1\frac{1}{2}$

$\frac{8}{8}$ $\frac{4}{8}$

$\frac{11}{6} = 1\frac{5}{6}$

$\frac{6}{6}$ $\frac{5}{6}$ $\frac{5}{6}$ is closer to 1 whole than $\frac{1}{2}$.

5. Fill in the blanks to make each number sentence true. Draw a number line, tape diagram, or area model to represent each problem.

a. $\frac{11}{12} = \frac{5}{12} + \frac{6}{12}$

b. $\frac{53}{100} - \frac{27}{100} = \frac{26}{100}$

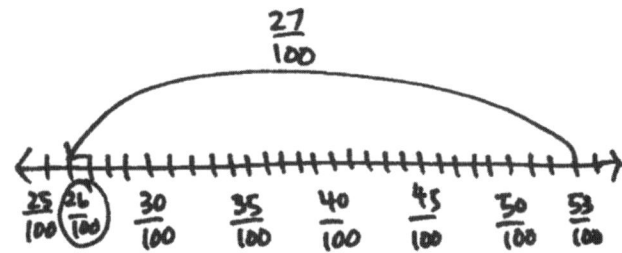

c. $\frac{8}{12} + \frac{4}{12} = 1$

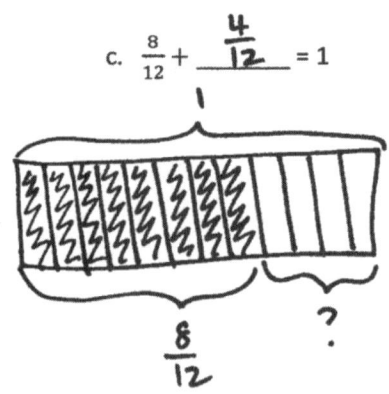

d. $\frac{3}{10} + \frac{6}{10} + \frac{2}{10} = \frac{11}{10}$

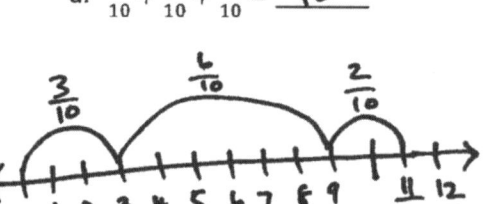

e. $1 - \frac{5}{8} = \frac{3}{8}$ $\frac{8}{8} - \frac{5}{8} = ?$

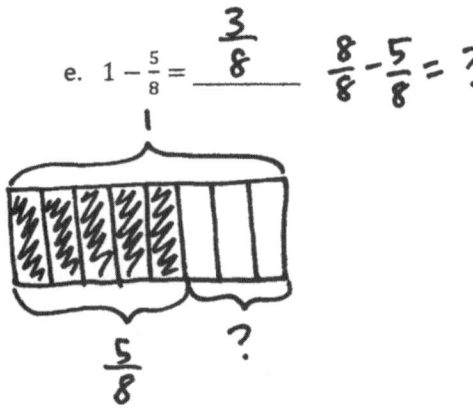

f. $\frac{7}{8} - \frac{3}{8} = \frac{4}{8}$

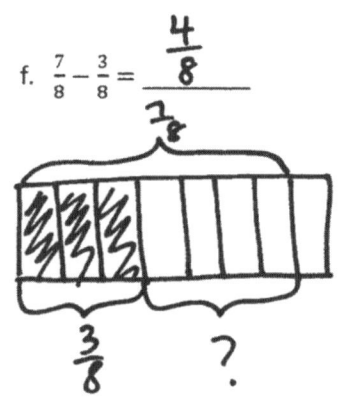

6. Ray, Robin, and Freddy went fishing.

   a. They spent $\frac{1}{6}$ of their money on water, $\frac{4}{6}$ of their money on lunch, and the rest on worms. What fraction of their money was spent on worms? Draw a model and write an equation to solve.

   $\frac{1}{6} + \frac{4}{6} + \frac{1}{6} = \frac{6}{6} = 1$

   They spent $\frac{1}{6}$ of their money on worms.

   b. Robin noticed her water bottle was $\frac{1}{2}$ full and Freddy's was $\frac{3}{4}$ full. Robin said, "My $\frac{1}{2}$ full bottle has more water than your $\frac{3}{4}$ full bottle." Explain how $\frac{1}{2}$ bottle could be more than $\frac{3}{4}$ bottle.

   If Robin's water bottle was bigger than Freddy's, half of her water bottle could be more than $\frac{3}{4}$ of his.

   c. Ray, Robin, and Freddy each had identical containers of worms. Ray used $\frac{3}{8}$ container. Robin used $\frac{6}{8}$ container, and Freddy used $\frac{7}{8}$ container. How many total containers of worms did they use?

   $\frac{3}{8} + \frac{6}{8} + \frac{7}{8} = \frac{16}{8} = 2$

   They used 2 containers of worms.

   d. Express the number of remaining containers as a product of a whole number and a unit fraction.

   $\frac{8}{8} = 8 \times \frac{1}{8}$

   e. Six out of the eight fish they caught were trout. What is another fraction equal to 6 eighths? Write a number sentence and draw a model to show the two fractions are equal.

   $\frac{6}{8} = \frac{3}{4}$

   $\frac{3 \times 2}{4 \times 2} = \frac{6}{8}$

A STORY OF UNITS

Mathematics Curriculum

GRADE 4 • MODULE 5

Topic E
Extending Fraction Equivalence to Fractions Greater Than 1

4.NF.2, 4.NF.3, 4.MD.4, 4.NBT.6, 4.NF.1, 4.NF.4a

Focus Standards:	4.NF.2	Compare two fractions with different numerators and different denominators, e.g., by creating common denominators or numerators, or by comparing to a benchmark fraction such as 1/2. Recognize that comparisons are valid only when the two fractions refer to the same whole. Record the results of comparisons with symbols >, =, or <, and justify the conclusions, e.g., by using a visual fraction model.
	4.NF.3	Understand a fraction a/b with $a > 1$ as a sum of fractions $1/b$.
		a. Understand addition and subtraction of fractions as joining and separating parts referring to the same whole.
		b. Decompose a fraction into a sum of fractions with the same denominator in more than one way, recording each decomposition by an equation. Justify decompositions, e.g., by using a visual fraction model. *Examples: 3/8 = 1/8 + 1/8 + 1/8; 3/8 = 1/8 + 2/8; 2 1/8 = 1 + 1 + 1/8 = 8/8 + 8/8 + 1/8.*
		c. Add and subtract mixed numbers with like denominators, e.g., by replacing each mixed number with an equivalent fraction, and/or by using properties of operations and the relationship between addition and subtraction.
		d. Solve word problems involving addition and subtraction of fractions referring to the same whole and having like denominators, e.g., by using visual fraction models and equations to represent the problem.
	4.MD.4	Make a line plot to display a data set of measurements in fractions of a unit (1/2, 1/4, 1/8). Solve problems involving addition and subtraction of fractions by using information presented in line plots. *For example, from a line plot find and interpret the difference in length between the longest and shortest specimens in an insect collection.*
Instructional Days:	7	
Coherence -Links from:	G3–M5	Fractions as Numbers on the Number Line
-Links to:	G5–M3	Addition and Subtraction of Fractions
	G5–M4	Multiplication and Division of Fractions and Decimal Fractions

A STORY OF UNITS

Topic E

In Topic E, students study equivalence involving both ones and fractional units. In Lesson 22, they use decomposition and visual models to add and subtract fractions less than 1 to and from whole numbers, e.g., $4 + \frac{3}{4} = 4\frac{3}{4}$ and $4 - \frac{3}{4} = (3 + 1) - \frac{3}{4}$, subtracting the fraction from 1 using a number bond and a number line. Lesson 23 has students use addition and multiplication to build fractions greater than 1 and then represent them on the number line. Fractions can be expressed both in mixed units of a whole number and a fraction or simply as a fraction, as pictured below, e.g., $7 \times \frac{1}{3} = \frac{3}{3} + \frac{3}{3} + \frac{1}{3} = 2 \times \frac{3}{3} + \frac{1}{3} = \frac{7}{3} = 2\frac{1}{3}$.

In Lessons 24 and 25, students use decompositions to reason about the various equivalent forms in which a fraction greater than or equal to 1 may be presented, both as fractions and as mixed numbers. In Lesson 24, they decompose, for example, 11 fourths into 8 fourths and 3 fourths, $\frac{11}{4} = \frac{8}{4} + \frac{3}{4}$, or they can think of it as $\frac{11}{4} = \frac{4}{4} + \frac{4}{4} + \frac{3}{4} = 2 \times \frac{4}{4} + \frac{3}{4} = 2\frac{3}{4}$. In Lesson 25, students are then able to decompose the two wholes into 8 fourths so their original number can then be looked at as $\frac{8}{4} + \frac{3}{4}$ or $\frac{11}{4}$. In this way, they see that $2\frac{3}{4} = \frac{11}{4}$. This fact is further reinforced when they plot $\frac{11}{4}$ on the number line and see that it is at the same point as $2\frac{3}{4}$. Unfortunately, the term *improper fraction* carries some baggage. As many have observed, there is nothing *improper* about an improper fraction. Nevertheless, as a mathematical term, it is useful for describing a particular form in which a fraction may be presented (i.e., a fraction is improper if the numerator is greater than or equal to the denominator). Students do need practice in terms of converting between the various forms a fraction may take, but take care not to foster the misconception that every improper fraction *must* be converted to a mixed number.

Students compare fractions greater than 1 in Lessons 26 and 27. They begin by using their understanding of benchmarks to reason about which of two fractions is greater. This activity builds on students' rounding skills, having them identify the whole numbers and the halfway points between them on the number line. The relationship between the numerator and denominator of a fraction is a key concept here as students consider relationships to whole numbers; e.g., a student might reason that $\frac{23}{8}$ is less than

Topic E: Extending Fraction Equivalence to Fractions Greater Than 1

$\frac{29}{10}$ because $\frac{23}{8}$ is 1 eighth less than 3, but $\frac{29}{10}$ is 1 tenth less than 3. They know each fraction is 1 fractional unit away from 3, and since $\frac{1}{8} > \frac{1}{10}$, then $\frac{23}{8} < \frac{29}{10}$. Students progress to finding and using like denominators to compare and order mixed numbers. Once again, students must use reasoning skills as they determine that, when they have two fractions with the same numerator, the larger fraction has a larger unit (or smaller denominator). Conversely, when they have two fractions with the same denominator, the larger one has the larger number of units (or larger numerator).

Lesson 28 concludes the topic with word problems requiring the interpretation of data presented in line plots. Students create line plots to display a given data set that includes fraction and mixed number values. To do this, they apply their skill in comparing mixed numbers, both through reasoning and the use of common numerators or denominators. For example, a data set might contain both $1\frac{5}{9}$ and $\frac{14}{9}$, giving students the opportunity to determine that they must be plotted at the same point. They also use addition and subtraction to solve the problems.

A Teaching Sequence Toward Mastery of Extending Fraction Equivalence to Fractions Greater Than 1
Objective 1: Add a fraction less than 1 to, or subtract a fraction less than 1 from, a whole number using decomposition and visual models.   (Lesson 22)
Objective 2: Add and multiply unit fractions to build fractions greater than 1 using visual models.   (Lesson 23)
Objective 3: Decompose and compose fractions greater than 1 to express them in various forms.   (Lessons 24–25)
Objective 4: Compare fractions greater than 1 by reasoning using benchmark fractions.   (Lesson 26)
Objective 5: Compare fractions greater than 1 by creating common numerators or denominators.   (Lesson 27)
Objective 6: Solve word problems with line plots.   (Lesson 28)

Lesson 22

Objective: Add a fraction less than 1 to, or subtract a fraction less than 1 from, a whole number using decomposition and visual models.

Suggested Lesson Structure

- **Fluency Practice** (12 minutes)
- **Application Problem** (5 minutes)
- **Concept Development** (33 minutes)
- **Student Debrief** (10 minutes)

Total Time **(60 minutes)**

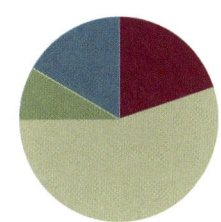

Fluency Practice (12 minutes)

- Sprint: Add Fractions **4.NF.3** (8 minutes)
- Count by Equivalent Fractions **4.NF.1** (4 minutes)

Sprint: Add Fractions (8 minutes)

Materials: (S) Add Fractions Sprint

Note: This fluency activity reviews Lesson 16. This Sprint is designed for students to add fractions and express their answers as fractions greater than one or as mixed numbers. Consider allowing students to not rename fractions and mixed numbers for larger units so that they do not have to perform additional processes while they are focusing on adding fractions.

Count by Equivalent Fractions (4 minutes)

Note: This activity builds fluency with equivalent fractions. The progression builds in complexity. Work students up to the highest level of complexity in which they can confidently participate.

T: Count by twos to 20 starting at 0.
S: 0, 2, 4, 6, 8, 10, 12, 14, 16, 18, 20.
T: Count by 2 tenths to 20 tenths starting at 0 tenths. (Write as students count.)

$\frac{0}{10}$	$\frac{2}{10}$	$\frac{4}{10}$	$\frac{6}{10}$	$\frac{8}{10}$	$\frac{10}{10}$	$\frac{12}{10}$	$\frac{14}{10}$	$\frac{16}{10}$	$\frac{18}{10}$	$\frac{20}{10}$
0	$\frac{2}{10}$	$\frac{4}{10}$	$\frac{6}{10}$	$\frac{8}{10}$	1	$\frac{12}{10}$	$\frac{14}{10}$	$\frac{16}{10}$	$\frac{18}{10}$	2
0	$\frac{2}{10}$	$\frac{4}{10}$	$\frac{6}{10}$	$\frac{8}{10}$	1	$1\frac{2}{10}$	$1\frac{4}{10}$	$1\frac{6}{10}$	$1\frac{8}{10}$	2

S: $\frac{0}{10}, \frac{2}{10}, \frac{4}{10}, \frac{6}{10}, \frac{8}{10}, \frac{10}{10}, \frac{12}{10}, \frac{14}{10}, \frac{16}{10}, \frac{18}{10}, \frac{20}{10}$.

T: 1 is the same as how many tenths?

S: 10 tenths.

T: (Beneath $\frac{10}{10}$, write 1.) 2 is the same as how many tenths?

S: 20 tenths.

T: (Beneath $\frac{20}{10}$, write 2.) Count by 2 tenths again. This time, when you come to the whole number, say the whole number. Start at zero. (Write as students count.)

S: $0, \frac{2}{10}, \frac{4}{10}, \frac{6}{10}, \frac{8}{10}, 1, \frac{12}{10}, \frac{14}{10}, \frac{16}{10}, \frac{18}{10}, 2$.

T: (Point to $\frac{12}{10}$.) Say 12 tenths as a mixed number.

S: $1\frac{2}{10}$.

Continue the process for $\frac{14}{10}, \frac{16}{10}$, and $\frac{18}{10}$.

T: Count by 2 tenths again. This time, convert to whole numbers and mixed numbers. Start at zero. (Write as students count.)

S: $0, \frac{2}{10}, \frac{4}{10}, \frac{6}{10}, \frac{8}{10}, 1, 1\frac{2}{10}, 1\frac{4}{10}, 1\frac{6}{10}, 1\frac{8}{10}, 2$.

T: Let's count by 2 tenths again. After you say 1, alternate between saying the mixed number and the fraction. Start at zero. Try not to look at the board.

S: $0, \frac{2}{10}, \frac{4}{10}, \frac{6}{10}, \frac{8}{10}, 1, 1\frac{2}{10}, \frac{14}{10}, 1\frac{6}{10}, \frac{18}{10}, 2$.

T: 2 is the same as how many tenths?

S: $\frac{20}{10}$.

T: Let's count backward starting at $\frac{20}{10}$, alternating between fractions greater than one and mixed numbers. Try not to look at the board.

S: $\frac{20}{10}, 1\frac{8}{10}, \frac{16}{10}, 1\frac{4}{10}, \frac{12}{10}, 1, \frac{8}{10}, \frac{6}{10}, \frac{4}{10}, \frac{2}{10}, 0$.

> **NOTES ON MULTIPLE MEANS OF ENGAGEMENT:**
>
> Some learners may benefit from counting again and again until they gain fluency. Another way to differentiate the Counting by Equivalent Fractions fluency activity for students working above or below grade level is to grant them more autonomy. Students may enjoy this as a partner activity in which they take turns leading and counting. Students can make individualized choices about when to convert larger units, counting forward and backward, as well as speed.

Application Problem (5 minutes)

Winnie went shopping and spent $\frac{2}{5}$ of the money that was on a gift card. What fraction of the money was left on the card? Draw a number line and a number bond to help show your thinking.

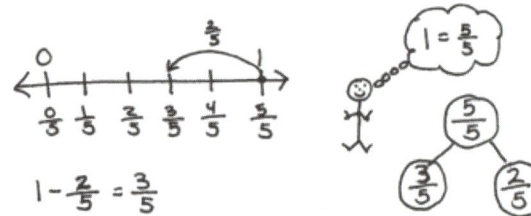

Note: This Application Problem reviews Lesson 17's objective of subtracting a fraction from 1. In this lesson, students subtract from a larger whole number using tape diagrams, number bonds, and a number line to aid in understanding.

A STORY OF UNITS Lesson 22 4•5

Concept Development (33 minutes)

Materials: (S) Personal white board

Problem 1: Add a fraction less than 1 to a whole number using a tape diagram.

T: Answer in mixed units: 2 meters + 5 centimeters is…?
S: 2 meters 5 centimeters.
T: 2 hours + 5 minutes is…?
S: 2 hours 5 minutes.
T: 2 ones + 5 eighths is…?
S: 2 ones and 5 eighths.

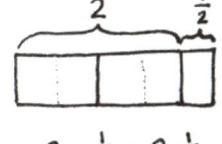

T: (Display $2 + \frac{1}{2}$.) Draw a tape diagram to show 2 ones. To know how large to draw $\frac{1}{2}$, let's partition each whole number into 2 halves.
T: (Demonstrate partitioning the 2 ones with dotted lines.)
T: Partition the ones, and extend your model to add $\frac{1}{2}$. Say a number sentence that adds the whole number to the fraction.
S: $2 + \frac{1}{2} = 2\frac{1}{2}$.
T: In this case, 2 ones plus 1 half gave us a sum that is a mixed number. We have seen mixed numbers often when working with measurement and place value, like when we added hundreds and tens, which are two different units.

Repeat the process with $3 + \frac{2}{3} = 3\frac{2}{3}$.

Problem 2: Subtract a fraction less than 1 from a whole number using a tape diagram.

T: (Display $3 - \frac{1}{4}$.) Draw a tape diagram to represent 3, partitioned as 3 ones. Watch as I subtract $\frac{1}{4}$. (Partition a one into 4 parts. Cross off $\frac{1}{4}$. Trace along the tape diagram with a finger to count the remaining parts.)
T: What is remaining?
S: 2 and 3 fourths. → 2 ones and 3 fourths.
T: Say the complete subtraction sentence.
S: $3 - \frac{1}{4} = 2\frac{3}{4}$.

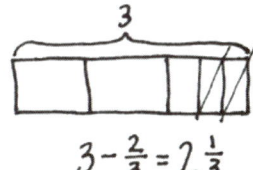

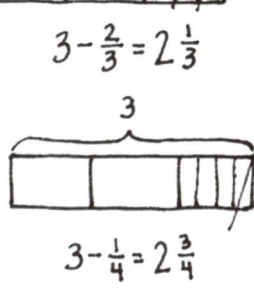

T: Subtract $3 - \frac{2}{3}$. Draw a tape diagram with your partner. Discuss your drawing with your partner.
S: I drew a tape diagram 3 units long. I partitioned the last unit into thirds, and then I crossed off 2 thirds.

Lesson 22: Add a fraction less than 1 to, or subtract a fraction less than 1 from, a whole number using decomposition and visual models.

307

A STORY OF UNITS   Lesson 22   4•5

T: Say the entire number sentence.
S: $3 - \frac{2}{3} = 2\frac{1}{3}$.
T: Discuss what you see happening to the number of ones when you subtract the fraction.
S: It gets smaller. → There are fewer ones. If we started with 3, the answer was 2 and some parts. → Right, so if we had a big number such as $391 - \frac{2}{3}$, we know the whole number would be 1 less, 390, and some parts.
T: What relationship do you see between the fraction being subtracted and the fraction in the answer?
S: They are the same unit. → They are part of one of the whole numbers. → They add together to make 1. That's why the whole number is 1 less in the answer. → Right. In the last problem, we took away $\frac{2}{3}$, and the fraction in the answer was $\frac{1}{3}$. Those add to make 1.

> **NOTES ON MULTIPLE MEANS OF REPRESENTATION:**
>
> Clarify for English language learners multiple meanings for the term *whole*. *Whole* can mean the total or sum as modeled in a number bond. Use *whole number* when referring to a unit in the ones, tens, hundreds, etc.

Problem 3: Given three related numbers, form fact family facts.

T: Write 4, $4\frac{4}{5}$, and $\frac{4}{5}$. These numbers are related. Draw a number bond to show the whole and the parts. Write two addition facts and two subtraction facts that use 4, $4\frac{4}{5}$, and $\frac{4}{5}$. Make a choice as to whether to write your sums and differences to the right or to the left of the equal sign.
S: $4 + \frac{4}{5} = 4\frac{4}{5}$. → $\frac{4}{5} + 4 = 4\frac{4}{5}$. → $4\frac{4}{5} - \frac{4}{5} = 4$. → $4\frac{4}{5} - 4 = \frac{4}{5}$.

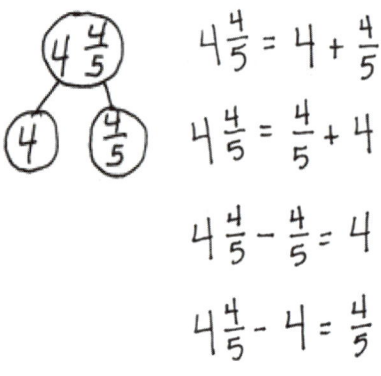

T: We can add and subtract ones and fractions just like we have always done. One number represents the whole, and the other two numbers represent the parts. For each of the following sets of related numbers, write two addition facts and two subtraction facts.

$\frac{3}{4}, 6\frac{3}{4}, 6$ $5, 4\frac{1}{3}, \frac{2}{3}$ $\frac{2}{5}, 4\frac{3}{5}, 5$

Problem 4: Subtract a fraction less than 1 from a whole number using decomposition.

T: Write the expression $5 - \frac{1}{4}$. Discuss a strategy for solving this problem with your partner.
S: We can rename 1 one as 4 fourths, so we have $4\frac{4}{4} - \frac{1}{4}$. → We can make a mixed number so the total is 4 and a fraction. → It's like unbundling a ten to subtract some ones.
T: Draw a number bond for 5 decomposed into two parts, 4 and 4 fourths or 4 and 1. (Allow students time to draw the bond.)

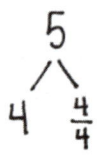

Lesson 22: Add a fraction less than 1 to, or subtract a fraction less than 1 from, a whole number using decomposition and visual models.

A STORY OF UNITS   Lesson 22 4•5

T: Construct a number line to represent $5 - \frac{1}{4}$ with 4 and 5 as endpoints. We are subtracting from $\frac{4}{4}$, so our answer will be more than 4 and less than 5. Draw an arrow to represent $5 - \frac{1}{4}$. Write the number sentence under your number line.

S: (Write $5 - \frac{1}{4} = 4\frac{3}{4}$.)

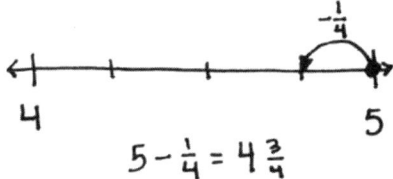

T: Subtract $7 - \frac{3}{5}$. Solve with your partner, drawing a number bond and number line. (Allow students time to solve.)

T: Let's show your thinking using a number sentence. 7 decomposed is...?

S: 6 and $\frac{5}{5}$.

T: (Record the bond under the number sentence.) How many ones remain?

S: 6.

T: (Record 6 in the number sentence.) $\frac{5}{5} - \frac{3}{5}$ is...?

S: $\frac{2}{5}$.

T: So, $\frac{2}{5}$ remains. Add that to 6. The difference is...?

S: $6\frac{2}{5}$.

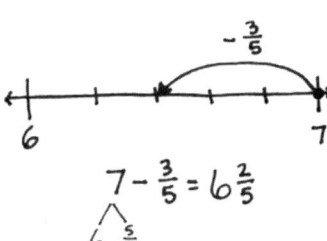

T: Subtract $9 - \frac{5}{12}$. Twelfths are a lot to partition on a number line. Solve this using just a number sentence and a number bond to decompose the total.

S: $9 - \frac{5}{12} = 8\frac{7}{12}$.

Problem Set (10 minutes)

Students should do their personal best to complete the Problem Set within the allotted 10 minutes. For some classes, it may be appropriate to modify the assignment by specifying which problems they work on first. Some problems do not specify a method for solving. Students should solve these problems using the RDW approach used for Application Problems.

Lesson 22: Add a fraction less than 1 to, or subtract a fraction less than 1 from, a whole number using decomposition and visual models.

309

A STORY OF UNITS — Lesson 22 — 4•5

Student Debrief (10 minutes)

Lesson Objective: Add a fraction less than 1 to, or subtract a fraction less than 1 from, a whole number using decomposition and visual models.

The Student Debrief is intended to invite reflection and active processing of the total lesson experience.

Invite students to review their solutions for the Problem Set. They should check work by comparing answers with a partner before going over answers as a class. Look for misconceptions or misunderstandings that can be addressed in the Debrief. Guide students in a conversation to debrief the Problem Set and process the lesson.

Any combination of the questions below may be used to lead the discussion.

- Why is it necessary to decompose the total into ones and a fraction before subtracting? How does that relate to a subtraction problem such as 74 – 28?
- How did knowing how to subtract a fraction from 1 prepare you for this lesson?
- Describe how the whole number is decomposed to subtract a fraction. Use Problem 3(b) to discuss.
- How were number lines and number bonds helpful in representing how to find the difference?
- How did the Application Problem connect to today's lesson?

Exit Ticket (3 minutes)

After the Student Debrief, instruct students to complete the Exit Ticket. A review of their work will help with assessing students' understanding of the concepts that were presented in today's lesson and planning more effectively for future lessons. The questions may be read aloud to the students.

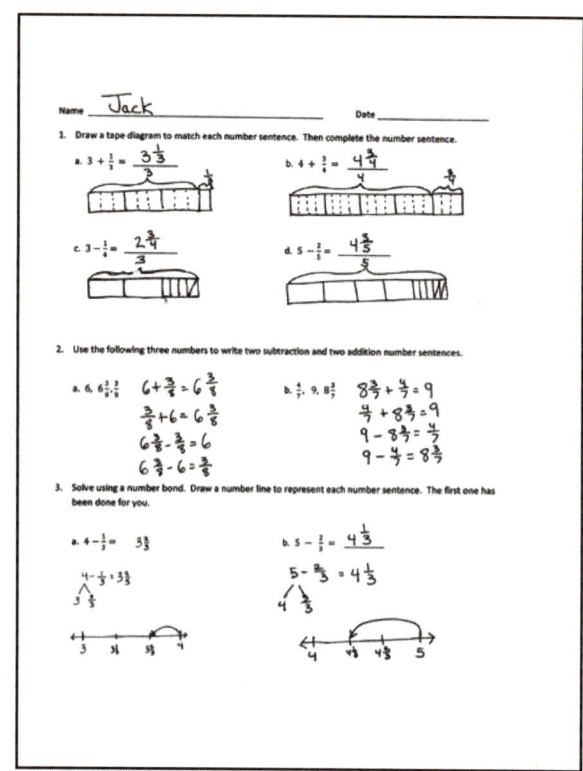

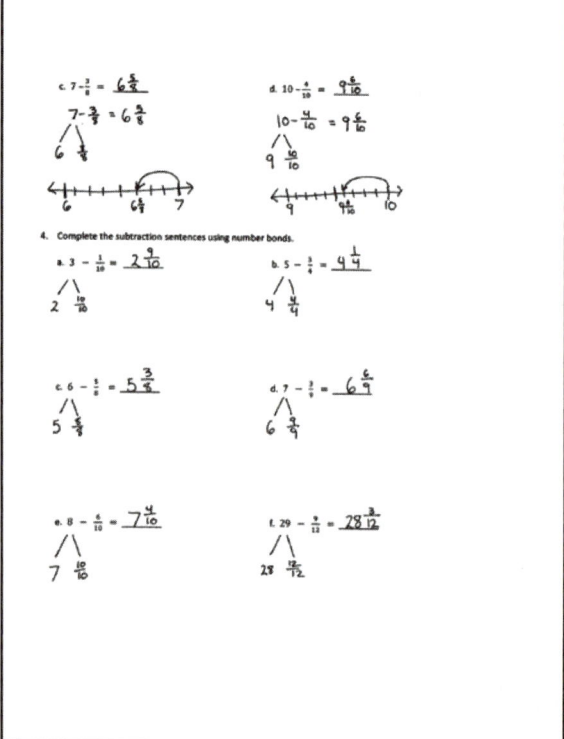

Lesson 22: Add a fraction less than 1 to, or subtract a fraction less than 1 from, a whole number using decomposition and visual models.

A STORY OF UNITS

Lesson 22 Sprint 4•5

A

Add Fractions

Number Correct: _____

1.	$1 + 1 =$	
2.	$\frac{1}{5} + \frac{1}{5} =$	
3.	$2 + 1 =$	
4.	$\frac{2}{5} + \frac{1}{5} =$	
5.	$2 + 2 =$	
6.	$\frac{2}{5} + \frac{2}{5} =$	
7.	$3 + 2 =$	
8.	$\frac{3}{5} + \frac{2}{5} =$	fifths
9.	$\frac{5}{5} =$	
10.	$\frac{3}{5} + \frac{2}{5} =$	
11.	$3 + 2 =$	
12.	$\frac{3}{8} + \frac{2}{8} =$	
13.	$3 + 2 + 2 =$	
14.	$\frac{3}{8} + \frac{2}{8} + \frac{2}{8} =$	
15.	$\frac{3}{8} + \frac{3}{8} + \frac{2}{8} =$	eighths
16.	$\frac{8}{8} =$	
17.	$\frac{3}{8} + \frac{3}{8} + \frac{2}{8} =$	
18.	$2 + 1 + 1 =$	
19.	$\frac{2}{3} + \frac{1}{3} + \frac{1}{3} =$	thirds
20.	$\frac{2}{3} + \frac{1}{3} + \frac{1}{3} =$	$1\frac{}{3}$
21.	$2 + 2 + 2 =$	
22.	$\frac{2}{5} + \frac{2}{5} + \frac{2}{5} =$	fifths
23.	$\frac{2}{5} + \frac{2}{5} + \frac{2}{5} =$	$1\frac{}{5}$
24.	$3 + 3 + 3 =$	
25.	$\frac{3}{8} + \frac{3}{8} + \frac{3}{8} =$	eighths
26.	$\frac{3}{8} + \frac{3}{8} + \frac{3}{8} =$	$1\frac{}{8}$
27.	$\frac{5}{8} + \frac{5}{8} + \frac{5}{8} =$	$1\frac{}{8}$
28.	$1 + 1 + 1 =$	
29.	$\frac{1}{2} + \frac{1}{2} + \frac{1}{2} =$	halves
30.	$\frac{1}{2} + \frac{1}{2} + \frac{1}{2} =$	$1\frac{}{2}$
31.	$4 + 4 + 4 =$	
32.	$\frac{4}{1A} + \frac{4}{1A} + \frac{4}{1A} =$	tenths
33.	$\frac{4}{1A} + \frac{4}{1A} + \frac{4}{1A} =$	$1\frac{}{1A}$
34.	$\frac{6}{1A} + \frac{6}{1A} + \frac{6}{1A} =$	$1\frac{}{1A}$
35.	$2 + 2 + 2 =$	
36.	$\frac{2}{6} + \frac{2}{6} + \frac{2}{6} =$	sixths
37.	$\frac{2}{6} + \frac{2}{6} + \frac{2}{6} =$	
38.	$\frac{3}{6} + \frac{3}{6} + \frac{3}{6} =$	$1\frac{}{6}$
39.	$\frac{5}{12} + \frac{2}{12} + \frac{4}{12} =$	
40.	$\frac{4}{12} + \frac{4}{12} + \frac{4}{12} =$	
41.	$\frac{5}{12} + \frac{5}{12} + \frac{7}{12} =$	$1\frac{}{12}$
42.	$\frac{7}{12} + \frac{9}{12} + \frac{7}{12} =$	$1\frac{}{12}$
43.	$\frac{7}{15} + \frac{8}{15} + \frac{7}{15} =$	$1\frac{}{15}$
44.	$\frac{12}{15} + \frac{8}{15} + \frac{9}{15} =$	$1\frac{}{15}$

Lesson 22: Add a fraction less than 1 to, or subtract a fraction less than 1 from, a whole number using decomposition and visual models.

© 2015 Great Minds. eureka-math.org
G4-M5-TE-B4-1.3.1-01.2016

B

Add Fractions

Number Correct: _____
Improvement: _____

1.	$1 + 1 =$	
2.	$\frac{1}{6} + \frac{1}{6} =$	
3.	$3 + 1 =$	
4.	$\frac{3}{6} + \frac{1}{6} =$	
5.	$3 + 2 =$	
6.	$\frac{3}{6} + \frac{2}{6} =$	
7.	$4 + 2 =$	
8.	$\frac{4}{6} + \frac{2}{6} =$	sixths
9.	$\frac{6}{6} =$	
10.	$\frac{4}{6} + \frac{2}{6} =$	
11.	$5 + 2 =$	
12.	$\frac{5}{8} + \frac{2}{8} =$	
13.	$5 + 1 + 1 =$	
14.	$\frac{5}{8} + \frac{1}{8} + \frac{1}{8} =$	
15.	$\frac{5}{8} + \frac{2}{8} + \frac{1}{8} =$	eighths
16.	$\frac{8}{8} =$	
17.	$\frac{3}{8} + \frac{3}{8} + \frac{2}{8} =$	
18.	$1 + 1 + 2 =$	
19.	$\frac{1}{3} + \frac{1}{3} + \frac{2}{3} =$	thirds
20.	$\frac{1}{3} + \frac{1}{3} + \frac{2}{3} =$	$1\frac{}{3}$
21.	$3 + 3 + 3 =$	
22.	$\frac{3}{8} + \frac{3}{8} + \frac{3}{8} =$	eighths

23.	$\frac{3}{8} + \frac{3}{8} + \frac{3}{8} =$	$1\frac{}{8}$
24.	$1 + 1 + 1 =$	
25.	$\frac{1}{2} + \frac{1}{2} + \frac{1}{2} =$	halves
26.	$\frac{1}{2} + \frac{1}{2} + \frac{1}{2} =$	$1\frac{}{2}$
27.	$2 + 2 + 2 =$	
28.	$\frac{2}{5} + \frac{2}{5} + \frac{2}{5} =$	fifths
29.	$\frac{2}{5} + \frac{2}{5} + \frac{2}{5} =$	$1\frac{}{5}$
30.	$\frac{3}{5} + \frac{3}{5} + \frac{3}{5} =$	$1\frac{}{5}$
31.	$6 + 6 + 6 =$	
32.	$\frac{6}{10} + \frac{6}{10} + \frac{6}{10} =$	tenths
33.	$\frac{6}{10} + \frac{6}{10} + \frac{6}{10} =$	$1\frac{}{10}$
34.	$\frac{5}{10} + \frac{5}{10} + \frac{5}{10} =$	$1\frac{}{10}$
35.	$2 + 2 + 2 =$	
36.	$\frac{2}{6} + \frac{2}{6} + \frac{2}{6} =$	sixths
37.	$\frac{2}{6} + \frac{2}{6} + \frac{2}{6} =$	
38.	$\frac{3}{6} + \frac{3}{6} + \frac{3}{6} =$	$1\frac{}{6}$
39.	$\frac{5}{12} + \frac{3}{12} + \frac{3}{12} =$	
40.	$\frac{5}{12} + \frac{5}{12} + \frac{2}{12} =$	
41.	$\frac{6}{12} + \frac{5}{12} + \frac{6}{12} =$	$1\frac{}{12}$
42.	$\frac{8}{12} + \frac{10}{12} + \frac{5}{12} =$	$1\frac{}{12}$
43.	$\frac{7}{15} + \frac{7}{15} + \frac{8}{15} =$	$1\frac{}{15}$
44.	$\frac{13}{15} + \frac{9}{15} + \frac{7}{15} =$	$1\frac{}{15}$

A STORY OF UNITS

Lesson 22 Problem Set 4•5

Name _____ Date _____

1. Draw a tape diagram to match each number sentence. Then, complete the number sentence.

   a. $3 + \frac{1}{3} =$ _____

   b. $4 + \frac{3}{4} =$ _____

   c. $3 - \frac{1}{4} =$ _____

   d. $5 - \frac{2}{5} =$ _____

2. Use the following three numbers to write two subtraction and two addition number sentences.

   a. $6, 6\frac{3}{8}, \frac{3}{8}$

   b. $\frac{4}{7}, \text{A}, \text{A}\frac{3}{7}$

3. Solve using a number bond. Draw a number line to represent each number sentence. The first one has been done for you.

   a. $4 - \frac{1}{3} =$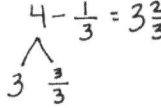

   b. $5 - \frac{2}{3} =$ _____

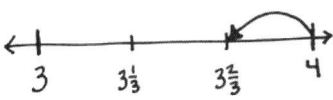

Lesson 22: Add a fraction less than 1 to, or subtract a fraction less than 1 from, a whole number using decomposition and visual models.

313

c. $7 - \frac{3}{8} =$ _____

d. $10 - \frac{4}{10} =$ _____

4. Complete the subtraction sentences using number bonds.

   a. $3 - \frac{1}{10} =$ _____

   b. $5 - \frac{3}{4} =$ _____

   c. $6 - \frac{5}{8} =$ _____

   d. $7 - \frac{3}{9} =$ _____

   e. $8 - \frac{6}{10} =$ _____

   f. $29 - \frac{9}{12} =$ _____

Lesson 22: Add a fraction less than 1 to, or subtract a fraction less than 1 from, a whole number using decomposition and visual models.

A STORY OF UNITS   Lesson 22 Exit Ticket 4•5

Name _____   Date _____

Complete the subtraction sentences using number bonds. Draw a model if needed.

1. $6 - \frac{1}{5} =$ _____

2. $8 - \frac{5}{6} =$ _____

3. $7 - \frac{5}{8} =$ _____

A STORY OF UNITS Lesson 22 Homework 4•5

Name _____ Date _____

1. Draw a tape diagram to match each number sentence. Then, complete the number sentence.

   a. $2 + \frac{1}{4} =$ _____

   b. $3 + \frac{2}{3} =$ _____

   c. $2 - \frac{1}{5} =$ _____

   d. $3 - \frac{3}{4} =$ _____

2. Use the following three numbers to write two subtraction and two addition number sentences.

   a. $4, 4\frac{5}{8}, \frac{5}{8}$

   b. $\frac{2}{7}, 5\frac{5}{7}, 6$

3. Solve using a number bond. Draw a number line to represent each number sentence. The first one has been done for you.

   a. $4 - \frac{1}{3} = 3\frac{2}{3}$

   b. $8 - \frac{5}{6} =$ _____

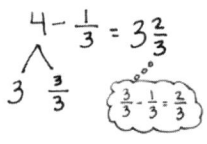

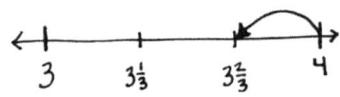

Lesson 22: Add a fraction less than 1 to, or subtract a fraction less than 1 from, a whole number using decomposition and visual models.

A STORY OF UNITS Lesson 22 Homework 4•5

c. $7 - \frac{4}{5} =$ _____

d. $3 - \frac{3}{10} =$ _____

4. Complete the subtraction sentences using number bonds.

   a. $6 - \frac{1}{4} =$ _____

   b. $7 - \frac{2}{10} =$ _____

   c. $5 - \frac{5}{6} =$ _____

   d. $6 - \frac{6}{8} =$ _____

   e. $3 - \frac{7}{8} =$ _____

   f. $26 - \frac{7}{10} =$ _____

Lesson 22: Add a fraction less than 1 to, or subtract a fraction less than 1 from, a whole number using decomposition and visual models.

Lesson 23

Objective: Add and multiply unit fractions to build fractions greater than 1 using visual models.

Suggested Lesson Structure

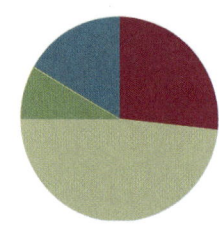

- ■ Fluency Practice   (16 minutes)
- ■ Application Problem (5 minutes)
- ■ Concept Development (29 minutes)
- ■ Student Debrief   (10 minutes)
- **Total Time**   **(60 minutes)**

Fluency Practice (16 minutes)

- Add and Subtract **4.NBT.4**   (4 minutes)
- Count by Equivalent Fractions **4.NF.1**   (4 minutes)
- Add and Subtract Fractions **4.NF.3**   (4 minutes)
- Add to and Subtract from Ones **4.NF.3**   (4 minutes)

Add and Subtract (4 minutes)

Materials: (S) Personal white board

Note: This fluency activity reviews adding and subtracting using the standard algorithm.

- T: (Write 457 thousands 393 ones.) On your personal white board, write this number in standard form.
- S: (Write 457,393.)
- T: (Write 385 thousands 142 ones.) Add this number to 457,393 using the standard algorithm.
- S: (Write 457,393 + 385,142 = 842,535 using the standard algorithm.)

Continue the process for 465,758 + 492,458.

- T: (Write 300 thousands.) On your board, write this number in standard form.
- S: (Write 300,000.)
- T: (Write 137 thousands 623 ones.) Subtract this number from 300,000 using the standard algorithm.
- S: (Write 300,000 − 137,623 = 162,377 using the standard algorithm.)

Continue the process for 534,803 − 235,257.

A STORY OF UNITS   Lesson 23   4•5

Count by Equivalent Fractions (4 minutes)

Note: This activity builds fluency with equivalent fractions. The progression builds in complexity. Work students up to the highest level of complexity in which they can confidently participate.

T: Count by twos to 12, starting at 0.
S: 0, 2, 4, 6, 8, 10, 12.

$\frac{0}{6}$	$\frac{2}{6}$	$\frac{4}{6}$	$\frac{6}{6}$	$\frac{8}{6}$	$\frac{10}{6}$	$\frac{12}{6}$
0	$\frac{2}{6}$	$\frac{4}{6}$	1	$\frac{8}{6}$	$\frac{10}{6}$	2
0	$\frac{2}{6}$	$\frac{4}{6}$	1	$1\frac{2}{6}$	$1\frac{4}{6}$	2

T: Count by 2 sixths to 12 sixths, starting at 0 sixths. (Write as students count.)
S: $\frac{0}{6}, \frac{2}{6}, \frac{4}{6}, \frac{6}{6}, \frac{8}{6}, \frac{10}{6}, \frac{12}{6}$.
T: 1 is the same as how many sixths?
S: 6 sixths.
T: (Beneath $\frac{6}{6}$, write 1.) 2 is the same as how many sixths?
S: 12 sixths.
T: (Beneath $\frac{12}{6}$, write 2.) Count by 2 sixths again. This time, when you come to 1 or 2, say the whole number. Start at zero. (Write as students count.)
S: 0, $\frac{2}{6}, \frac{4}{6}$, 1, $\frac{8}{6}, \frac{10}{6}$, 2
T: (Point at $\frac{8}{6}$.) Say 8 sixths as a mixed number.
S: $1\frac{2}{6}$.

Continue the process for $\frac{10}{6}$.

T: Count by 2 sixths again. This time, convert to whole numbers and mixed numbers. (Write as students count.)
S: 0, $\frac{2}{6}, \frac{4}{6}$, 1, $1\frac{2}{6}, 1\frac{4}{6}$, 2
T: Let's count by 2 sixths again. After you say 1, alternate between saying the mixed number and the fraction. Try not to look at the board.
S: 0, $\frac{2}{6}, \frac{4}{6}$, 1, $1\frac{2}{6}, \frac{10}{6}$, 2
T: 2 is the same as how many sixths?
S: $\frac{12}{6}$.
T: Let's count backward by 2 sixths starting at $\frac{12}{6}$. Alternate between fractions and mixed numbers down to 1, and then continue to count down by 2 sixths to $\frac{0}{6}$. Try not to look at the board.
S: $\frac{12}{6}, 1\frac{4}{6}, \frac{8}{6}$, 1, $\frac{4}{6}, \frac{2}{6}, \frac{0}{6}$.

Lesson 23: Add and multiply unit fractions to build fractions greater than 1 using visual models.

319

A STORY OF UNITS — Lesson 23 4•5

Add and Subtract Fractions (4 minutes)

Materials: (S) Personal white board

Note: This fluency activity reviews Lesson 22.

T: (Draw a number bond with a whole of 2. Write 1 as the known part and $\frac{}{5}$ as the unknown part.) How many fifths are in 1?

S: 5 fifths.

T: (Write $\frac{5}{5}$ as the unknown part. Beneath it, write $2 - \frac{2}{5} = 1 + \frac{}{5}$.) Write the number sentence.

S: (Write $2 - \frac{2}{5} = 1 + \frac{3}{5} = 1\frac{3}{5}$.)

Continue the process for $2 - \frac{1}{3}$, $2 - \frac{3}{4}$ and $2 - \frac{5}{6}$.

T: How much does 3 fifths need to be 1?

S: 2 fifths.

T: (Write $2\frac{3}{5} + \underline{} = 3$.) Write the number sentence, filling in the unknown number.

S: (Write $2\frac{3}{5} + \frac{2}{5} = 3$.)

Continue with the following possible sequence: $1\frac{2}{3} + \underline{} = 2$, $1\frac{1}{4} + \underline{} = 2$, and $1\frac{1}{6} + \underline{} = 2$.

Add to and Subtract from Ones (4 minutes)

Materials: (S) Personal white board

Note: This fluency activity reviews Lesson 22.

T: (Write $1 + \frac{1}{4} = \underline{}$.) Write the complete number sentence.

S: (Write $1 + \frac{1}{4} = 1\frac{1}{4}$.)

Continue the process for $2 + \frac{3}{5}$ and $3 + \frac{3}{10}$.

T: (Write $2 - \frac{1}{3} = \underline{}$.) Draw a number line to match the subtraction problem. Then, beneath it, write the complete number sentence.

S: (Draw a number line as shown to the right. Write $2 - \frac{1}{3} = 1\frac{2}{3}$.)

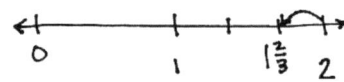

Continue with the following possible sequence: $3 - \frac{3}{4}$ and $3 - \frac{7}{10}$.

Lesson 23: Add and multiply unit fractions to build fractions greater than 1 using visual models.

A STORY OF UNITS   Lesson 23   4•5

Application Problem (5 minutes)

Mrs. Wilcox cut quilt squares and then divided them evenly into 8 piles. She decided to sew together 1 pile each night. After 5 nights, what fraction of the quilt squares was sewn together? Draw a tape diagram or a number line to model your thinking, and then write a number sentence to express your answer.

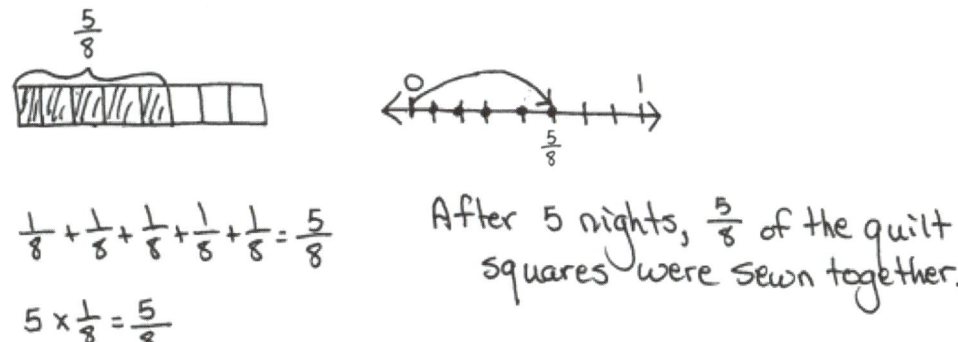

Note: This Application Problem builds on Lesson 3's objective of learning to decompose non-unit fractions and representing them as a whole number times a unit fraction using tape diagrams. Understanding the representation of a non-unit fraction in this way helps students as they learn the content of today's lesson in which they add and multiply unit fractions to compose fractions greater than 1.

Concept Development (29 minutes)

Materials: (S) Personal white board

Problem 1: Multiply a whole number times a unit fraction.

MP.7

T: Write 6 × 2 as an addition sentence showing six groups of 2.
S: (Write 2 + 2 + 2 + 2 + 2 + 2 = 12.)
T: Draw a number line to show 6 twos.
S: (Draw a number line as shown to the right.)
T: Write $6 \times \frac{1}{2}$ as an addition sentence showing six groups of $\frac{1}{2}$.
S: (Write $\frac{1}{2} + \frac{1}{2} + \frac{1}{2} + \frac{1}{2} + \frac{1}{2} + \frac{1}{2} = \frac{6}{2}$.)
T: Draw a number line to show 6 halves.
S: (Draw a number line as shown to the right.)

> **NOTES ON MULTIPLE MEANS OF REPRESENTATION:**
>
> When directing students to draw the number line for 6 × 2, it may be helpful to clarify the directive for English language learners and others by phrasing instructions as "Draw a number line, and label six groups of 2," or, "Model these number sentences on a number line." Students working below grade level and English language learners may also benefit from modeling the first group of two.

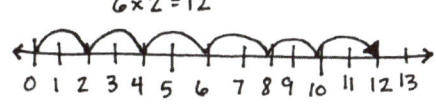

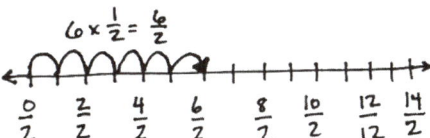

Lesson 23:   Add and multiply unit fractions to build fractions greater than 1 using visual models.

A STORY OF UNITS   Lesson 23   4•5

MP.7

T: Work with your partner to draw parentheses, grouping halves to make ones.

S: We know $\frac{1}{2} + \frac{1}{2} = 1$, so maybe we can make three groups of that. → Yeah, let's draw parentheses around three separate groups of 2 halves.

T: (Place parentheses.) $3 \times \frac{2}{2}$ (point to the number sentence) is equal to...?

S: 3.

T: True or false? $6 \times \frac{1}{2} = 3 \times \frac{2}{2}$. Discuss with your partner.

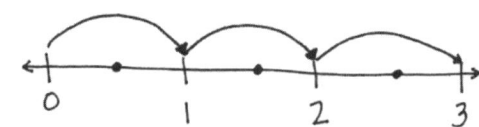

Problem 2: Multiply a whole number times a unit fraction using the associative property.

T: Let's solve $6 \times \frac{1}{2}$ using unit form. $6 \times \frac{1}{2}$ is 6 halves.

T: (Display the number line as pictured.) Do you see three groups of 2 halves?

S: Yes.

T: (Display: 6 halves = (3 × 2) halves = 3 × (2 halves) = $3 \times \left(\frac{2}{2}\right) = 3 \times 1 = 3$.)

T: Discuss this with your partner.

S: It tells us 6 halves equals 3 or $6 \times \frac{1}{2} = 3$. → 3 × (2 halves) and $3 \times \left(\frac{2}{2}\right)$ shows the 3 ones really clearly. → 2 halves make 1, and 3 × 1 = 3.

T: But why did it start with (3 × 2) halves? Why not (2 × 3) halves? Or (1 × 6) halves?

S: Because we want to make ones. 2 halves make 1.

T: How many groups of 2 halves are in 6 halves?

S: 3.

T: So, 6 halves equals 3.

T: (Display $10 \times \frac{1}{5}$.) Solve for 10 fifths using unit form.

S: We want to make groups of 5 fifths to make ones. → 10 fifths is the same as (2 × 5) fifths. 2 × (5 fifths) = $2 \times \left(\frac{5}{5}\right) = 2 \times 1 = 2$.

T: Support your answer with a number line.

S: I can make 10 slides of a fifth. → My arrows show 2 slides of $\frac{5}{5}$. That is equal to 2. → $10 \times \frac{1}{5} = 2 \times \frac{5}{5} = 2$.

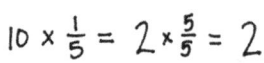

Repeat with $8 \times \frac{1}{4}$.

Problem 3: Express the product of a whole number times a unit fraction as a mixed number.

T: (Display: 9 copies of $\frac{1}{4}$.) 9 fourths. How many fourths make 1?

S: 4 fourths.

A STORY OF UNITS　　　　　　　　　　　　　　　　　　　　　　Lesson 23　4•5

T: To make ones, how many groups of 4 fourths are in 9 fourths?

S: 2.

T: Two groups of 4 fourths makes 8 fourths. There is 1 fourth remaining.

Display: $9 \times \frac{1}{4} = (2 \times \frac{4}{4}) + \frac{1}{4} = 2 + \frac{1}{4} = 2\frac{1}{4}$.

T: Draw a number line with endpoints 0 and 3. Label the ones, and partition fourths. With your partner, show $(2 \times \frac{4}{4}) + \frac{1}{4}$. (Allow students time to draw two slides of $\frac{4}{4}$ and then a slide of $\frac{1}{4}$ as pictured to the right.)

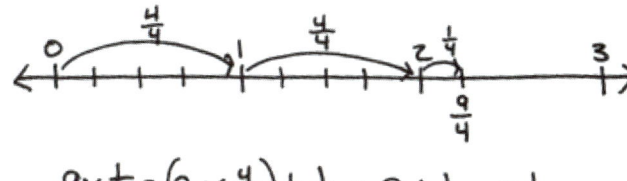

T: With your partner, solve for 8 copies of 1 third.

S: There are two groups of 3 thirds in 8 thirds. That leaves 2 thirds remaining.

→ $8 \times \frac{1}{3} = (2 \times \frac{3}{3}) + \frac{2}{3} = 2\frac{2}{3}$.

Repeat with $7 \times \frac{1}{2}, 13 \times \frac{1}{5}$, and $17 \times \frac{1}{6}$.

Problem Set (10 minutes)

Students should do their personal best to complete the Problem Set within the allotted 10 minutes. For some classes, it may be appropriate to modify the assignment by specifying which problems they work on first. Some problems do not specify a method for solving. Students should solve these problems using the RDW approach used for Application Problems.

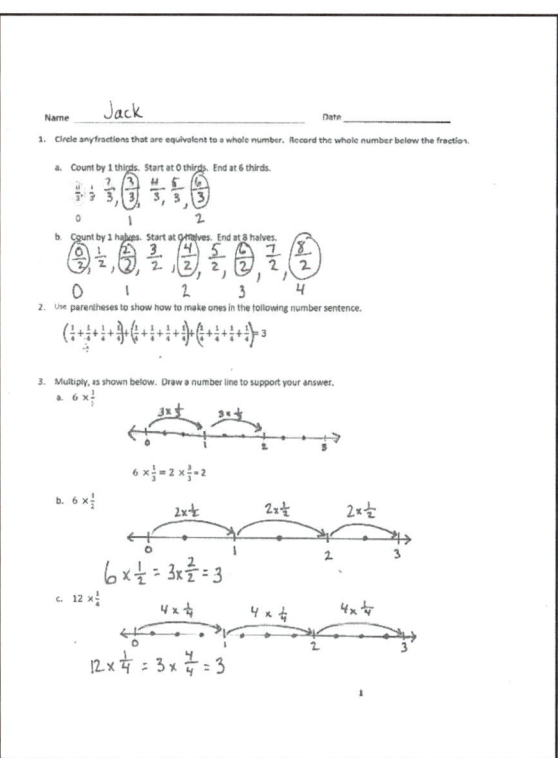

Student Debrief (10 minutes)

Lesson Objective: Add and multiply unit fractions to build fractions greater than 1 using visual models.

The Student Debrief is intended to invite reflection and active processing of the total lesson experience.

Invite students to review their solutions for the Problem Set. They should check work by comparing answers with a partner before going over answers as a class. Look for misconceptions or misunderstandings that can be addressed in the Debrief. Guide students in a conversation to debrief the Problem Set and process the lesson.

Lesson 23: Add and multiply unit fractions to build fractions greater than 1 using visual models.　　323

A STORY OF UNITS — Lesson 23 4•5

Any combination of the questions below may be used to lead the discussion.

- How is your work in Problem 1(a) related to your work in Problem 3(a)? How is adding like-unit fractions related to multiplying unit fractions? Is this true for Problems 1(b) and 3(b)?
- Using Problem 3(a), explain how $6 \times \frac{1}{3}$ is the same as $2 \times \frac{3}{3}$.
- Explain why Problems 3(b) and 3(c) equal the same whole number.
- Which is greater, $6 \times \frac{1}{3}$ or $6 \times \frac{1}{2}$?
- How are parentheses helpful as you solve Problem 2?
- Look at Problem 2 and Problem 3. Is there a way to tell when the product will be a whole number before multiplying? Explain your thinking.
- How did the Application Problem connect to today's lesson?

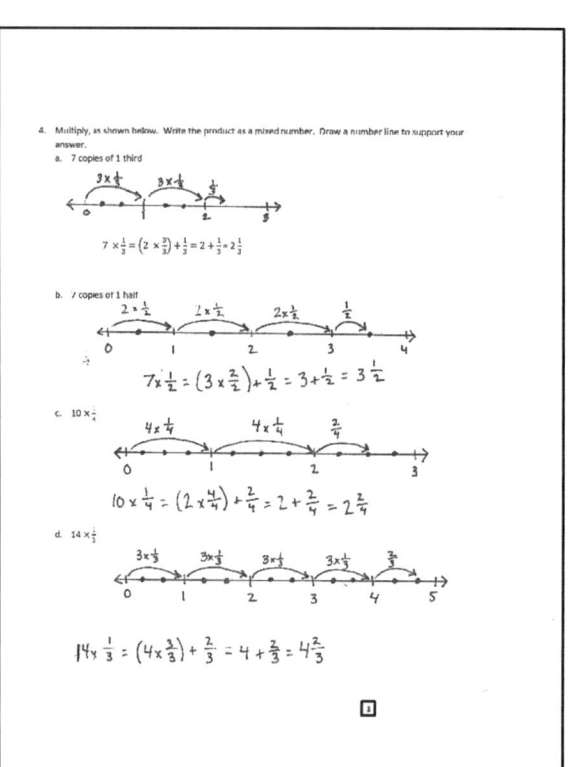

Exit Ticket (3 minutes)

After the Student Debrief, instruct students to complete the Exit Ticket. A review of their work will help with assessing students' understanding of the concepts that were presented in today's lesson and planning more effectively for future lessons. The questions may be read aloud to the students.

Lesson 23: Add and multiply unit fractions to build fractions greater than 1 using visual models.

A STORY OF UNITS   Lesson 23 Problem Set   4•5

Name _____ Date _____

1. Circle any fractions that are equivalent to a whole number. Record the whole number below the fraction.

   a. Count by 1 thirds. Start at 0 thirds. End at 6 thirds.

   $\left(\dfrac{0}{3}\right), \dfrac{1}{3},$

   0

   b. Count by 1 halves. Start at 0 halves. End at 8 halves.

2. Use parentheses to show how to make ones in the following number sentence.

   $\dfrac{1}{4}+\dfrac{1}{4}+\dfrac{1}{4}+\dfrac{1}{4}+\dfrac{1}{4}+\dfrac{1}{4}+\dfrac{1}{4}+\dfrac{1}{4}+\dfrac{1}{4}+\dfrac{1}{4}+\dfrac{1}{4}+\dfrac{1}{4}=3$

3. Multiply, as shown below. Draw a number line to support your answer.

   a. $6 \times \dfrac{1}{3}$

   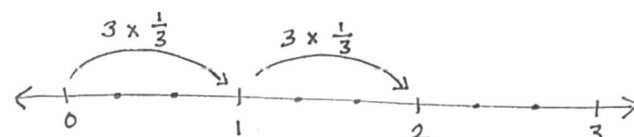

   $6 \times \dfrac{1}{3} = 2 \times \dfrac{3}{3} = 2$

   b. $6 \times \dfrac{1}{2}$

   c. $12 \times \dfrac{1}{4}$

Lesson 23: Add and multiply unit fractions to build fractions greater than 1 using visual models.

4. Multiply, as shown below. Write the product as a mixed number. Draw a number line to support your answer.

   a. 7 copies of 1 third

   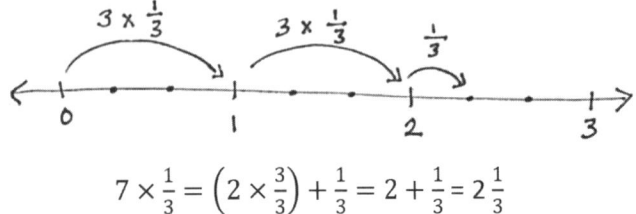

   $$7 \times \frac{1}{3} = \left(2 \times \frac{3}{3}\right) + \frac{1}{3} = 2 + \frac{1}{3} = 2\frac{1}{3}$$

   b. 7 copies of 1 half

   c. $10 \times \frac{1}{4}$

   d. $14 \times \frac{1}{3}$

Name _____   Date _____

Multiply and write the product as a mixed number. Draw a number line to support your answer.

1. $8 \times \frac{1}{2}$

2. 7 copies of 1 fourth

3. $13 \times \frac{1}{3}$

A STORY OF UNITS   Lesson 23 Homework 4•5

Name _____   Date _____

1. Circle any fractions that are equivalent to a whole number. Record the whole number below the fraction.

   a. Count by 1 fourths. Start at 0 fourths. Stop at 6 fourths.

      $\left(\dfrac{0}{4}\right)$, $\dfrac{1}{4}$,

      0

   b. Count by 1 sixths. Start at 0 sixths. Stop at 14 sixths.

2. Use parentheses to show how to make ones in the following number sentence.

   $$\dfrac{1}{3}+\dfrac{1}{3}+\dfrac{1}{3}+\dfrac{1}{3}+\dfrac{1}{3}+\dfrac{1}{3}+\dfrac{1}{3}+\dfrac{1}{3}+\dfrac{1}{3}+\dfrac{1}{3}+\dfrac{1}{3}+\dfrac{1}{3}=4$$

3. Multiply, as shown below. Draw a number line to support your answer.

   a.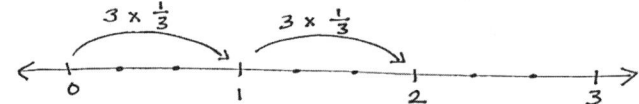

      $6 \times \dfrac{1}{3} = 2 \times \dfrac{3}{3} = 2$

   b. $10 \times \dfrac{1}{2}$

   c. $8 \times \dfrac{1}{4}$

A STORY OF UNITS   Lesson 23 Homework   4•5

4. Multiply, as shown below. Write the product as a mixed number. Draw a number line to support your answer.

   a. 7 copies of 1 third

   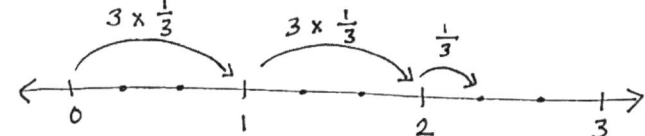

   $$7 \times \frac{1}{3} = \left(2 \times \frac{3}{3}\right) + \frac{1}{3} = 2 + \frac{1}{3} = 2\frac{1}{3}$$

   b. 7 copies of 1 fourth

   c. 11 groups of 1 fifth

   d. $7 \times \frac{1}{2}$

   e. $9 \times \frac{1}{5}$

Lesson 23: Add and multiply unit fractions to build fractions greater than 1 using visual models.

A STORY OF UNITS   Lesson 24   4•5

Lesson 24

Objective: Decompose and compose fractions greater than 1 to express them in various forms.

Suggested Lesson Structure

- ■ Fluency Practice   (12 minutes)
- ■ Application Problem   (6 minutes)
- ■ Concept Development   (32 minutes)
- ■ Student Debrief   (10 minutes)

Total Time   **(60 minutes)**

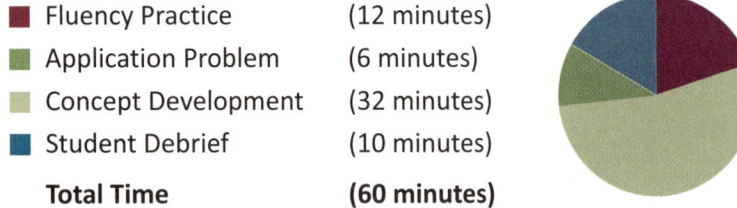

Fluency Practice (12 minutes)

- Add and Subtract **4.NBT.4**   (3 minutes)
- Count by Equivalent Fractions **4.NF.1**   (3 minutes)
- Add and Subtract Fractions **4.NF.3**   (3 minutes)
- Multiply Fractions on a Number Line **4.NF.4**   (3 minutes)

Add and Subtract (3 minutes)

Materials: (S) Personal white board

Note: This fluency activity reviews adding and subtracting using the standard algorithm.

- T: (Write 547 thousands 936 ones.) On your personal white board, write this number in standard form.
- S: (Write 547,936.)
- T: (Write 270 thousands 654 ones.) Add this number to 547,936 using the standard algorithm.
- S: (Write 547,936 + 270,654 = 818,590 using the standard algorithm.)

Continue the process for 547,239 + 381,798.

- T: (Write 500 thousands.) On your board, write this number in standard form.
- S: (Write 500,000.)
- T: (Write 213 thousands 724 ones.) Subtract this number from 500,000 using the standard algorithm.
- S: (Write 500,000 − 213,724 = 286,276 using the standard algorithm.)

Continue the process for 635,704 − 395,615.

330   Lesson 24:   Decompose and compose fractions greater than 1 to express them in various forms.

A STORY OF UNITS   Lesson 24 4•5

Count by Equivalent Fractions (3 minutes)

Note: This activity reviews Lesson 23.

T: Count by ones to 10, starting at 0.
S: 0, 1, 2, 3, 4, 5, 6, 7, 8, 9, 10.
T: Count by halves to 10 halves, starting at 0 halves. (Write as students count.)

$\frac{0}{2}$	$\frac{1}{2}$	$\frac{2}{2}$	$\frac{3}{2}$	$\frac{4}{2}$	$\frac{5}{2}$	$\frac{6}{2}$	$\frac{7}{2}$	$\frac{8}{2}$	$\frac{9}{2}$	$\frac{10}{2}$
0	$\frac{1}{2}$	1	$\frac{3}{2}$	2	$\frac{5}{2}$	3	$\frac{7}{2}$	4	$\frac{9}{2}$	5
0	$\frac{1}{2}$	1	$1\frac{1}{2}$	2	$2\frac{1}{2}$	3	$3\frac{1}{2}$	4	$4\frac{1}{2}$	5

S: $\frac{0}{2}, \frac{1}{2}, \frac{2}{2}, \frac{3}{2}, \frac{4}{2}, \frac{5}{2}, \frac{6}{2}, \frac{7}{2}, \frac{8}{2}, \frac{9}{2}, \frac{10}{2}$.
T: 1 is the same as how many halves?
S: 2 halves.
T: (Beneath $\frac{2}{2}$, write 1.) 2 is the same as how many halves?
S: 4 halves.
T: (Beneath $\frac{4}{2}$, write 2.)

Continue the process for 3, 4, and 5.

T: Count by halves again. This time, when you come to the whole numbers, say them. Start at zero. (Write as students count.)
S: $0, \frac{1}{2}, 1, \frac{3}{2}, 2, \frac{5}{2}, 3, \frac{7}{2}, 4, \frac{9}{2}, 5$.
T: Count by halves once more. This time, convert to whole numbers and mixed numbers. Start at zero. (Write as students count.)
S: $0, \frac{1}{2}, 1, 1\frac{1}{2}, 2, 2\frac{1}{2}, 3, 3\frac{1}{2}, 4, 4\frac{1}{2}, 5$.

Add and Subtract Fractions (3 minutes)

Materials: (S) Personal white board

Note: This fluency activity reviews Lesson 22.

T: (Draw a number bond with a whole of 5. Write 4 as the known part and $\frac{}{3}$ as the unknown part.) How many thirds are in 1?
S: 3 thirds.

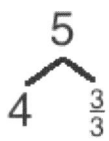

$5 - \frac{2}{3} = 4\frac{1}{3}$

$4\frac{1}{3} + \underline{} = 5$

T: (Write $\frac{3}{3}$ as the unknown part. Beneath it, write $5 - \frac{2}{3} = 4 + \frac{1}{3}$.) Write the number sentence.
S: (Write $5 - \frac{2}{3} = 4 + \frac{1}{3} = 4\frac{1}{3}$.)

Lesson 24:   Decompose and compose fractions greater than 1 to express them in various forms.

Continue with the following possible sequence: $4-\frac{3}{5}$ and $5-\frac{3}{4}$.

T: How much more does 1 third need to equal 1?

S: 2 thirds.

T: (Write $4\frac{1}{3}+$ ___ $= 5$.) Write the number sentence, filling in the unknown part.

S: (Write $4\frac{1}{3}+\frac{2}{3}=5$.)

Continue with the following possible sequence: $3\frac{2}{5}+$ ___ $= 4$ and $4\frac{1}{4}+$ ___ $= 5$.

Multiply Fractions on a Number Line (3 minutes)

Materials: (S) Personal white board

Note: This fluency activity reviews Lesson 23.

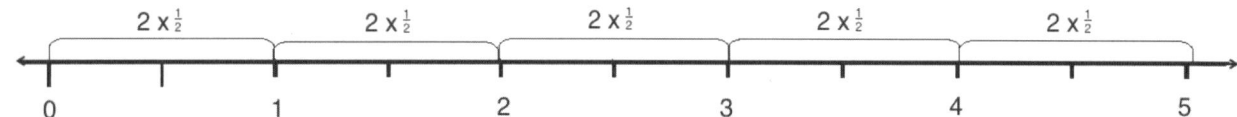

T: Draw a number line on your personal white board.

S: (Draw a number line.)

T: (Write $10 \times \frac{1}{2}$.) How many halves will you record on your number line?

S: 10 halves.

T: Starting at 0, draw tick marks on your number line to represent 10 halves.

S: (Draw 11 tick marks equally spaced on the number line.)

T: (Write ___ $\times \frac{1}{2} = 1$.) How many halves are in 1?

S: 2 halves.

T: (Write $2 \times \frac{1}{2} = 1$.) Label as many ones as possible, and record each with multiplication.

S: (Label 5 ones, and record each slide on the number line as $2 \times \frac{1}{2}$.)

T: How many times did you write $2 \times \frac{1}{2}$?

S: 5 times.

T: (Write $10 \times \frac{1}{2} = 5 \times \frac{1}{2} =$ ___.) Fill in the unknown numerator and unknown whole number.

S: (Write $10 \times \frac{1}{2} = 5 \times \frac{2}{2} = 5$.)

Continue with the following possible sequence: $12 \times \frac{1}{3}$.

A STORY OF UNITS   Lesson 24 4•5

Application Problem (6 minutes)

Shelly read her book for $\frac{1}{2}$ hour each afternoon for 9 days. How many hours did Shelly spend reading in all 9 days?

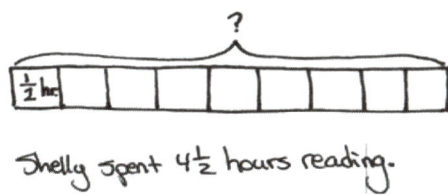

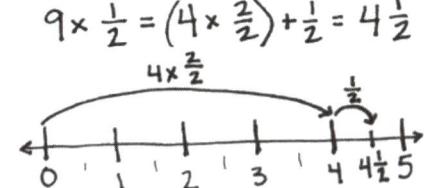

Note: This Application Problem relates back to Lesson 23 where students multiplied unit fractions to build fractions greater than 1. Ask students who are struggling to explain their number sentence to draw a number line.

Concept Development (32 minutes)

Materials: (S) Personal white board

Problem 1: Rename fractions as mixed numbers using decomposition.

> **NOTES ON MULTIPLE MEANS OF REPRESENTATION:**
> Some learners may need explicit instructions for counting by 3 thirds and later by 5 fifths. It might be helpful to scaffold the count by directing students to first count by threes. Then, have them count by 3 thirds. If needed, do the same for counting by 5 fifths.

T: (Display $\frac{7}{3}$.) How many thirds make 1?

S: 3 thirds.

T: Count by 3 thirds.

S: 3 thirds, 6 thirds, 9 thirds.

T: Stop. We only have 7 thirds. Decompose $\frac{7}{3}$ using a bond to show $\frac{6}{3}$ and the remaining fraction.

S: (Draw a bond.)

T: Use the bond to write an addition sentence for $\frac{7}{3}$.

S: $\frac{7}{3} = \frac{6}{3} + \frac{1}{3}$.

T: Rename using whole numbers.

S: $\frac{6}{3} + \frac{1}{3} = 2 + \frac{1}{3}$.

T: (Write $\frac{7}{3} = 2\frac{1}{3}$.) Let's use a number line to model that. Draw a number line with endpoints 0 and 3. Decompose each whole number into thirds, and plot $\frac{7}{3}$. Start at zero. Slide $\frac{6}{3}$. Slide $\frac{1}{3}$. $\frac{7}{3}$ is equal to...?

S: $2\frac{1}{3}$.

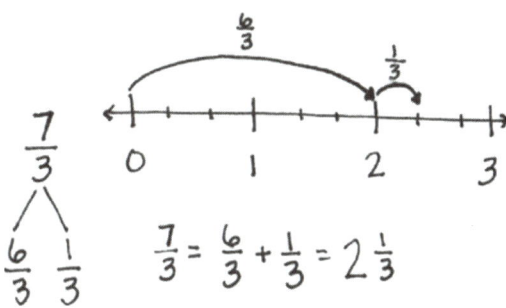

Repeat the process with $\frac{13}{5}$.

Lesson 24: Decompose and compose fractions greater than 1 to express them in various forms.   333

Problem 2: Convert a fraction into a mixed number using multiplication.

T: (Display $\frac{7}{3}$.) How many groups of 3 thirds are in 7 thirds?

S: 2.

T: We write two groups of 3 thirds as $2 \times \frac{3}{3}$. (Record as shown to the right.) How many thirds remain?

S: $\frac{1}{3}$.

T: What is $2 + \frac{1}{3}$?

S: $2\frac{1}{3}$.

T: True or false: $2\frac{1}{3} = \frac{7}{3}$?

S: True.

T: (Display $\frac{10}{4}$.) How many groups of $\frac{4}{4}$ are in $\frac{10}{4}$?

S: 2.

T: Record a number sentence to show $\frac{10}{4}$ as a mixed number.

S: $\frac{10}{4} = (2 \times \frac{4}{4}) + \frac{2}{4} = 2\frac{2}{4}$.

T: Watch as I write that number sentence a new way: $\frac{10}{4} = \frac{2 \times 4}{4} + \frac{2}{4}$. Discuss this number sentence with your partner.

S: The 2 is in the numerator! → But if I multiply 2 times 4, I get 8. So, $\frac{8}{4}$ is the same as $\frac{2 \times 4}{4}$. → $2 \times \frac{4}{4} = \frac{2 \times 4}{4}$.

T: With your partner, write the following fractions as mixed numbers using multiplication: $\frac{12}{5}, \frac{20}{6}, \frac{35}{10}$ and $\frac{26}{12}$.

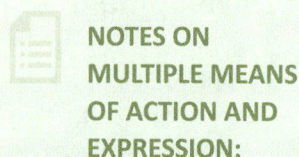

$$\frac{7}{3} = \left(2 \times \frac{3}{3}\right) + \frac{1}{3}$$
$$= 2 + \frac{1}{3}$$
$$= 2\frac{1}{3}$$

NOTES ON MULTIPLE MEANS OF ACTION AND EXPRESSION:

Guide students to identify and strengthen personally efficient strategies for converting a fraction to a mixed number. After practice with number bonds and multiplication, facilitate student reflection and self-assessment. Students might ask themselves, "Which method is easier for me? Which is fastest?"

Problem Set (10 minutes)

Students should do their personal best to complete the Problem Set within the allotted 10 minutes. For some classes, it may be appropriate to modify the assignment by specifying which problems they work on first. Some problems do not specify a method for solving. Students should solve these problems using the RDW approach used for Application Problems.

Student Debrief (10 minutes)

Lesson Objective: Decompose and compose fractions greater than 1 to express them in various forms.

The Student Debrief is intended to invite reflection and active processing of the total lesson experience.

Invite students to review their solutions for the Problem Set. They should check work by comparing answers with a partner before going over answers as a class. Look for misconceptions or misunderstandings that can be addressed in the Student Debrief. Guide students in a conversation to debrief the Problem Set and process the lesson.

Any combination of the questions below may be used to lead the discussion.

- How can drawing a number line help you when converting a fraction to a mixed number?
- How can decomposing a fraction into two parts help you rename each fraction?
- In Problem 1, how did you decide what your two parts should be? Use a specific example to explain.
- Compare the strategies you used in Problem 1 with the strategies you used in Problem 2. In the example in Problem 1(a) and Problem 2(a), how is using a number bond of $\frac{9}{3}$ and $\frac{2}{3}$ related to $\frac{3 \times 3}{3} + \frac{2}{3}$?
- In Problem 3, which fractions were the easiest for you to convert? Which were the most challenging? Why?
- How did the Application Problem connect to today's lesson?

Exit Ticket (3 minutes)

After the Student Debrief, instruct students to complete the Exit Ticket. A review of their work will help with assessing students' understanding of the concepts that were presented in today's lesson and planning more effectively for future lessons. The questions may be read aloud to the students.

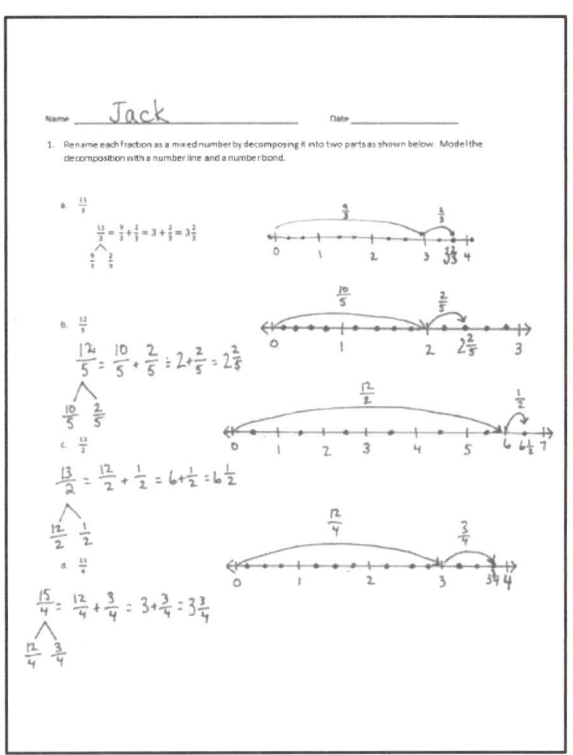

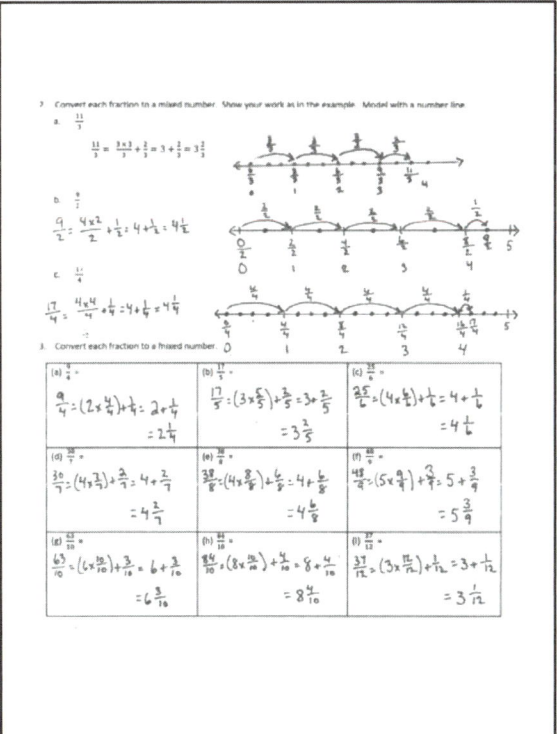

Name _____   Date _____

1. Rename each fraction as a mixed number by decomposing it into two parts as shown below. Model the decomposition with a number line and a number bond.

   a.

   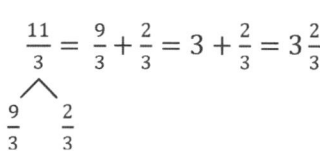   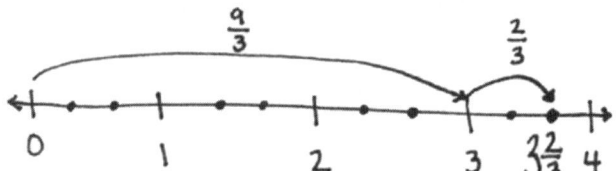

   b. $\frac{12}{5}$

   c. $\frac{13}{2}$

   d. $\frac{15}{4}$

A STORY OF UNITS Lesson 24 Problem Set 4•5

2. Convert each fraction to a mixed number. Show your work as in the example. Model with a number line.

   a. $\frac{11}{3}$

   $\frac{11}{3} = \frac{3 \times 3}{3} + \frac{2}{3} = 3 + \frac{2}{3} = 3\frac{2}{3}$

   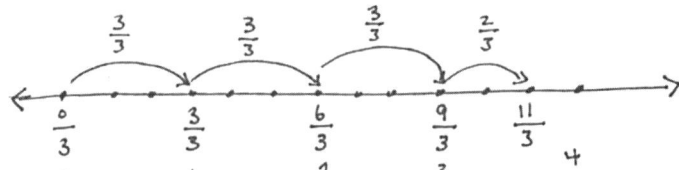

   b. $\frac{9}{2}$

   c. $\frac{14}{4}$

3. Convert each fraction to a mixed number.

a. $\frac{9}{4} =$	b. $\frac{14}{5} =$	c. $\frac{25}{6} =$
d. $\frac{37}{7} =$	e. $\frac{38}{8} =$	f. $\frac{49}{9} =$
g. $\frac{63}{14} =$	h. $\frac{14}{14} =$	i. $\frac{37}{12} =$

Lesson 24: Decompose and compose fractions greater than 1 to express them in various forms.

337

A STORY OF UNITS   Lesson 24 Exit Ticket   4•5

Name _____   Date _____

1. Rename the fraction as a mixed number by decomposing it into two parts. Model the decomposition with a number line and a number bond.

$$\frac{17}{5}$$

2. Convert the fraction to a mixed number. Model with a number line.

$$\frac{13}{3}$$

3. Convert the fraction to a mixed number.

$$\frac{11}{4}$$

A STORY OF UNITS

Lesson 24 Homework 4•5

Name _____ Date _____

1. Rename each fraction as a mixed number by decomposing it into two parts as shown below. Model the decomposition with a number line and a number bond.

   a. $\frac{11}{3}$

   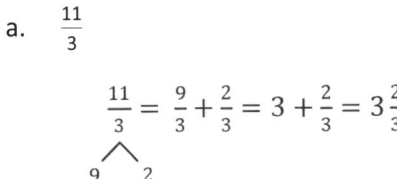

   $\frac{11}{3} = \frac{9}{3} + \frac{2}{3} = 3 + \frac{2}{3} = 3\frac{2}{3}$

   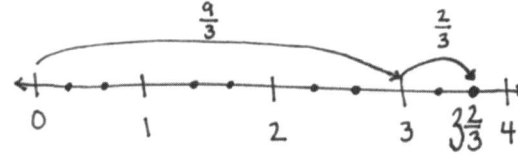

   b. $\frac{13}{4}$

   c. $\frac{16}{5}$

   d. $\frac{15}{2}$

   e. $\frac{17}{3}$

Lesson 24: Decompose and compose fractions greater than 1 to express them in various forms.

2. Convert each fraction to a mixed number. Show your work as in the example. Model with a number line.

   a. $\frac{11}{3}$

   $\frac{11}{3} = \frac{3 \times 3}{3} + \frac{2}{3} = 3 + \frac{2}{3} = 3\frac{2}{3}$

   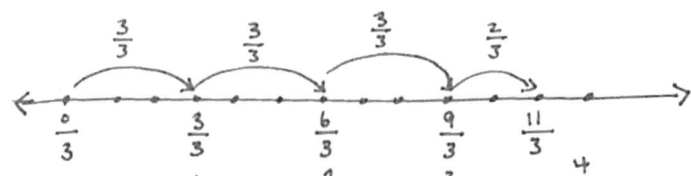

   b. $\frac{13}{2}$

   c. $\frac{17}{4}$

3. Convert each fraction to a mixed number.

a. $\frac{14}{3} =$	b. $\frac{17}{4} =$	c. $\frac{21}{5} =$
d. $\frac{26}{6} =$	e. $\frac{23}{7} =$	f. $\frac{31}{8} =$
g. $\frac{51}{9} =$	h. $\frac{14}{11} =$	i. $\frac{45}{12} =$

A STORY OF UNITS

Lesson 25 4•5

Lesson 25

Objective: Decompose and compose fractions greater than 1 to express them in various forms.

Suggested Lesson Structure

- **Fluency Practice** (12 minutes)
- **Application Problem** (6 minutes)
- **Concept Development** (32 minutes)
- **Student Debrief** (10 minutes)

Total Time **(60 minutes)**

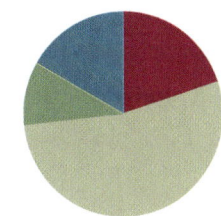

Fluency Practice (12 minutes)

- How Many Ones? **4.NF.1**   (3 minutes)
- Add and Subtract Fractions **4.NF.3**   (4 minutes)
- Change Fractions to Mixed Numbers **4.NF.4**   (5 minutes)

How Many Ones? (3 minutes)

Materials: (S) Personal white board

Note: This fluency activity prepares students for Lesson 25.

T: I'll say a fraction; you will say the number of ones it is equal to. 2 halves.
S: 1 one.
T: 4 halves.
S: 2 ones.
T: 6 halves.
S: 3 ones.
T: (Write $\frac{10}{2} = $ ___.) On your personal white board, write the answer.
S: (Write $\frac{10}{2} = 5$.)

Continue with the following possible sequence: $\frac{10}{10}, \frac{20}{10}, \frac{30}{10}, \frac{60}{10}, \frac{3}{3}, \frac{6}{3}, \frac{9}{3}, \frac{15}{3}, \frac{4}{4}, \frac{8}{4}, \frac{16}{4}, \frac{12}{12}, \frac{24}{12}, \frac{36}{12}$, and $\frac{120}{12}$.

Lesson 25: Decompose and compose fractions greater than 1 to express them in various forms.

341

A STORY OF UNITS   Lesson 25   4•5

Add and Subtract Fractions (4 minutes)

Materials: (S) Personal white board

Note: This fluency activity reviews Lesson 22.

    T: (Draw a number bond with a whole of 3. Write 2 as the known part and $\frac{}{4}$ as the unknown part.) How many fourths are in 1?

    S: 4 fourths.

    T: (Write $\frac{4}{4}$ as the unknown part. Beneath it, write $3 - \frac{1}{4} = 2 + \frac{}{4}$.) Write the number sentence.

    S: (Write $3 - \frac{1}{4} = 2 + \frac{3}{4} = 2\frac{3}{4}$.)

Continue with the following possible sequence: $5 - \frac{5}{6}, 7 - \frac{2}{5}$, and $5 - \frac{3}{8}$.

    T: How much does 3 fourths need in order to equal 1?

    S: 1 fourth.

    T: (Write $2\frac{3}{4} + \_\_ = 3$.) Write the number sentence, filling in the unknown number. $2\frac{3}{4} + \_\_ = 3$

    S: (Write $2\frac{3}{4} + \frac{1}{4} = 3$.)

Continue with the following possible sequence: $4\frac{1}{6} + \_\_ = 5, 6\frac{3}{5} + \_\_ = 7$, and $4\frac{5}{8} + \_\_ = 5$.

Change Fractions to Mixed Numbers (5 minutes)

Materials: (S) Personal white board

Note: This fluency activity reviews Lesson 24.

    T: Draw a number line with endpoints 0 and 3.

    T: (Write $\frac{9}{4}$.) Say the fraction.

    S: 9 fourths.

    T: Decompose each whole number into fourths by marking each fourth with a dot. Label $\frac{9}{4}$.

    T: How many fourths are in 1?

    S: 4 fourths.

    T: 2?

    S: 8 fourths.

    T: 3?

    S: 12 fourths.

    T: Label each whole number both as a fraction and whole number.

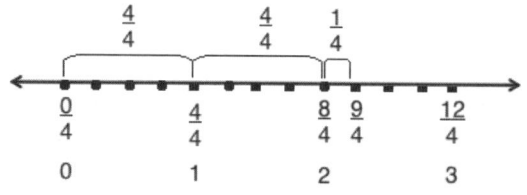

Lesson 25: Decompose and compose fractions greater than 1 to express them in various forms.

A STORY OF UNITS   Lesson 25   4•5

T: How many groups of 4 fourths are in 9 fourths?
S: Two groups.
T: (Write $\frac{9}{4} = 2 \times \frac{4}{4} + \frac{1}{4} = 2 + \frac{1}{4}$.) Fill in the unknown numerator, and write $\frac{9}{4}$ as a mixed number.
S: (Write $\frac{9}{4} = \frac{8}{4} + \frac{1}{4} = 2 + \frac{1}{4} = 2\frac{1}{4}$.)
T: (Write $\frac{9}{4} = \frac{\_ \times 4}{4} + \frac{1}{4} = 2 + \frac{1}{4} = 2\frac{1}{4}$.) Fill in the numerator's unknown factor to make the number sentence true.
S: (Write $\frac{9}{4} = \frac{2 \times 4}{4} + \frac{1}{4} = 2 + \frac{1}{4} = 2\frac{1}{4}$.)

Continue the process for the following possible sequence: $\frac{16}{5}$ and $\frac{15}{4}$.

Application Problem (6 minutes)

Mrs. Fowler knew that the perimeter of the soccer field was $\frac{1}{6}$ mile. Her goal was to walk two miles while watching her daughter's game. If she walked around the field 13 times, did she meet her goal? Explain your thinking.

$\frac{13}{6} = \frac{12}{6} + \frac{1}{6} = 2\frac{1}{6}$

$\frac{13}{6} = (2 \times \frac{6}{6}) + \frac{1}{6} = 2\frac{1}{6}$

Mrs. Fowler met her goal. 2 miles would be 12 laps, and she walked 13. $2\frac{1}{6}$ miles is greater than 2 miles.

Note: This Application Problem builds on Lesson 24 where students learned to convert a fraction to a mixed number. Knowing how to make this conversion leads to today's lesson in which students use what they know about mixed numbers to convert to a fraction greater than 1.

Concept Development (32 minutes)

Materials: (S) Personal white board

Problem 1: Model with a number line to convert a mixed number into a fraction greater than 1.

T: (Display $2\frac{1}{6}$.) Use a number bond to decompose $2\frac{1}{6}$ into ones and sixths. How many sixths are in 2 ones?
S: 12 sixths.

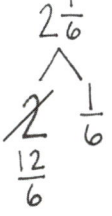

Lesson 25: Decompose and compose fractions greater than 1 to express them in various forms.   343

T: $\frac{12}{6} + \frac{1}{6}$ equals...?

S: $\frac{13}{6}$.

T: To check our work, let's draw a number line with 0 and 3 as endpoints. Use dots to decompose each whole number into sixths. Locate $2\frac{1}{6}$.

T: Point to zero. Slide your finger to 1. How many sixths are there from 0 to 1?

S: 6 sixths. (Record $\frac{6}{6}$ above the arrow from 0 to 1.)

T: Slide your finger from 1 to 2. How many sixths are there from 1 to 2?

S: 6 sixths. (Record $\frac{6}{6}$ above the arrow from 1 to 2.)

T: Slide your finger to $2\frac{1}{6}$. Say an addition sentence representing our movements. (Slide finger as students say the sentence.)

S: $\frac{6}{6} + \frac{6}{6} + \frac{1}{6} = \frac{13}{6}$.

T: $2\frac{1}{6}$ is equal to...?

S: $\frac{13}{6}$.

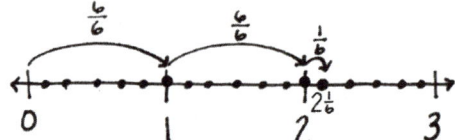

T: Notice, in the Application Problem, you converted a fraction greater than 1, $\frac{13}{6}$, to a mixed number, $2\frac{1}{6}$. Now you've converted a mixed number to a fraction greater than 1.

T: With your partner, convert $3\frac{1}{3}$ to a fraction greater than 1. Draw a number line to model your work.

Problem 2: Use multiplication to convert a mixed number to a fraction.

T: (Display $4\frac{1}{4}$.) Draw a number bond for $4\frac{1}{4}$, separating the ones and the fourths as two parts.

T: 1 one equals 4 fourths, so 2 ones equals 2 × (4 fourths). What is 4 ones equal to? Write your answer in unit form.

S: 4 ones = 4 × (4 fourths).

T: Write that number sentence numerically, and add the remaining 1 fourth. What is the total number of fourths?

S: $4\frac{1}{4} = 4 + \frac{1}{4} = (4 \times \frac{4}{4}) + \frac{1}{4} = \frac{16}{4} + \frac{1}{4} = \frac{17}{4}$.

NOTES ON MULTIPLE MEANS OF ACTION AND EXPRESSION:

When explaining multiples of $\frac{4}{4}$ to English language learners, check for understanding. Explain in students' first language, if needed, and couple the language with visual aids, such as the following:

$1 = \frac{4}{4}$

$2 = 2 \times \frac{4}{4}$

$3 = 3 \times \frac{4}{4}$

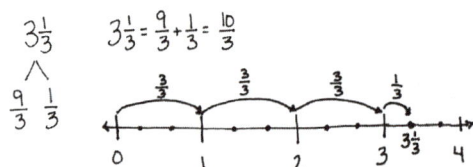

Lesson 25: Decompose and compose fractions greater than 1 to express them in various forms.

A STORY OF UNITS   Lesson 25   4•5

T: With your partner, convert $2\frac{2}{3}$ into a fraction greater than 1 using multiplication.

S: $2\frac{2}{3} = \left(2 \times \frac{3}{3}\right) + \frac{2}{3} = \frac{6}{3} + \frac{2}{3} = \frac{8}{3}$

T: Compare your number sentence with mine. $2\frac{2}{3} = \left(\frac{2 \times 3}{3}\right) + \frac{2}{3} = \frac{6}{3} + \frac{2}{3} = \frac{8}{3}$.

S: Instead of showing $\left(2 \times \frac{3}{3}\right)$, it's written as $\left(\frac{2 \times 3}{3}\right)$. Both mean the same. They both equal $\frac{6}{3}$, and both are read as "two times 3 thirds."

Repeat the process with $5\frac{3}{4}$.

Problem 3: Use mental math to convert a mixed number into a fraction greater than 1.

T: Write $3\frac{4}{5}$. Create a picture in your head. How many ones, and how many fifths?

S: 3 ones and 4 fifths.

T: How many fifths are there in 3 ones?

S: $1 = \frac{5}{5}, 2 = \frac{10}{5}$, and $3 = \frac{15}{5}$. → $\frac{5}{5}, \frac{10}{5}, \frac{15}{5}$.

T: Plus $\frac{4}{5}$ is...?

S: $\frac{19}{5}$.

Repeat the process with $4\frac{2}{3}$.

Problem Set (10 minutes)

Students should do their personal best to complete the Problem Set within the allotted 10 minutes. For some classes, it may be appropriate to modify the assignment by specifying which problems they work on first. Some problems do not specify a method for solving. Students should solve these problems using the RDW approach used for Application Problems.

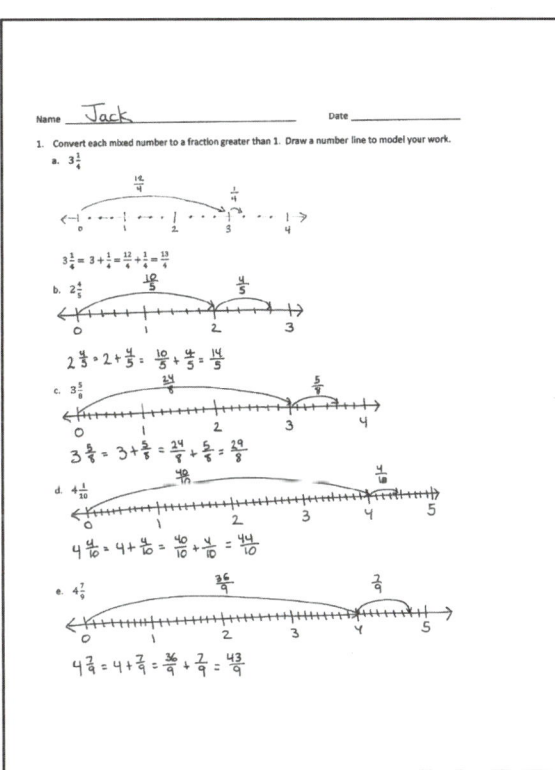

Student Debrief (10 minutes)

Lesson Objective: Decompose and compose fractions greater than 1 to express them in various forms.

The Student Debrief is intended to invite reflection and active processing of the total lesson experience.

Invite students to review their solutions for the Problem Set. They should check work by comparing answers with a partner before going over answers as a class. Look for misconceptions or misunderstandings that can be addressed in the Student Debrief. Guide students in a conversation to debrief the Problem Set and process the lesson.

Lesson 25:   Decompose and compose fractions greater than 1 to express them in various forms.

Any combination of the questions below may be used to lead the discussion.

- Explain to your partner how you solved Problems 1(b), 2(b), and 3(b). Did you use the same strategies to solve or different strategies?
- How was the work from previous lessons helpful in converting from a mixed number to a fraction greater than 1?
- How does the number line help to show the conversion from a mixed number to a fraction greater than 1?
- How did the Application Problem connect to today's lesson?

Exit Ticket (3 minutes)

After the Student Debrief, instruct students to complete the Exit Ticket. A review of their work will help with assessing students' understanding of the concepts that were presented in today's lesson and planning more effectively for future lessons. The questions may be read aloud to the students.

A STORY OF UNITS　　　　　　　　　　　　　　　　　　　　Lesson 25 Problem Set 4•5

Name _____ Date _____

1. Convert each mixed number to a fraction greater than 1. Draw a number line to model your work.

   a. $3\frac{1}{4}$

   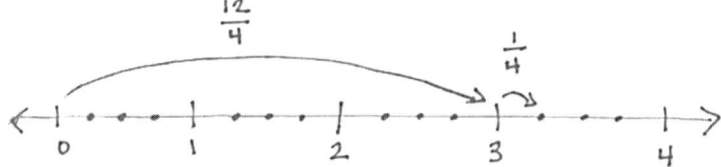

   $3\frac{1}{4} = 3 + \frac{1}{4} = \frac{12}{4} + \frac{1}{4} = \frac{13}{4}$

   b. $2\frac{4}{5}$

   c. $3\frac{5}{8}$

   d. $4\frac{4}{10}$

   e. $4\frac{7}{9}$

Lesson 25: Decompose and compose fractions greater than 1 to express them in various forms.

347

2. Convert each mixed number to a fraction greater than 1. Show your work as in the example. (Note: $3 \times \frac{4}{4} = \frac{3 \times 4}{4}$.)

   a. $3\frac{3}{4}$

   $$3\frac{3}{4} = 3 + \frac{3}{4} = \left(3 \times \frac{4}{4}\right) + \frac{3}{4} = \frac{12}{4} + \frac{3}{4} = \frac{15}{4}$$

   b. $4\frac{1}{3}$

   c. $4\frac{3}{5}$

   d. $4\frac{6}{8}$

3. Convert each mixed number to a fraction greater than 1.

a. $2\frac{3}{4}$	b. $2\frac{2}{5}$	c. $3\frac{3}{6}$
d. $3\frac{3}{8}$	e. $3\frac{1}{10}$	f. $4\frac{3}{8}$
g. $5\frac{2}{3}$	h. $6\frac{1}{2}$	i. $7\frac{2}{10}$

Name _____ Date _____

Convert each mixed number to a fraction greater than 1.

1. $3\frac{1}{5}$

2. $2\frac{3}{5}$

3. $4\frac{2}{9}$

Name _____ Date _____

1. Convert each mixed number to a fraction greater than 1. Draw a number line to model your work.

   a. $3\frac{1}{4}$

   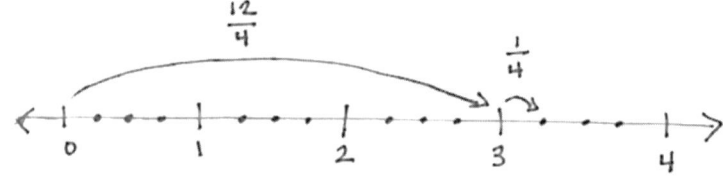

   $3\frac{1}{4} = 3 + \frac{1}{4} = \frac{12}{4} + \frac{1}{4} = \frac{13}{4}$

   b. $4\frac{2}{5}$

   c. $5\frac{3}{8}$

   d. $3\frac{7}{10}$

   e. $6\frac{2}{9}$

A STORY OF UNITS   Lesson 25 Homework   4•5

2. Convert each mixed number to a fraction greater than 1. Show your work as in the example.

   (Note: $3 \times \frac{4}{4} = \frac{3 \times 4}{4}$.)

   a. $3\frac{3}{4}$

   $3\frac{3}{4} = 3 + \frac{3}{4} = (3 \times \frac{4}{4}) + \frac{3}{4} = \frac{12}{4} + \frac{3}{4} = \frac{15}{4}$

   b. $5\frac{2}{3}$

   c. $4\frac{1}{5}$

   d. $3\frac{7}{8}$

3. Convert each mixed number to a fraction greater than 1.

a. $2\frac{1}{3}$	b. $2\frac{3}{4}$	c. $3\frac{2}{5}$
d. $3\frac{1}{6}$	e. $4\frac{5}{12}$	f. $4\frac{2}{5}$
g. $4\frac{1}{10}$	h. $5\frac{1}{5}$	i. $5\frac{5}{6}$
j. $6\frac{1}{4}$	k. $7\frac{1}{2}$	l. $7\frac{11}{12}$

Lesson 25: Decompose and compose fractions greater than 1 to express them in various forms.

A STORY OF UNITS　　　　　　　　　　　　　　　　　　　　　　Lesson 26　4•5

Lesson 26

Objective: Compare fractions greater than 1 by reasoning using benchmark fractions.

Suggested Lesson Structure

- Fluency Practice　　(10 minutes)
- Application Problem　(5 minutes)
- Concept Development　(35 minutes)
- Student Debrief　　(10 minutes)

Total Time　　(60 minutes)

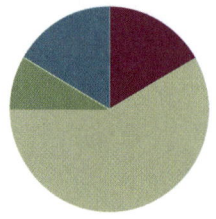

Fluency Practice (10 minutes)

- Change Fractions to Mixed Numbers　**4.NF.4**　　(4 minutes)
- Change Mixed Numbers to Fractions　**4.NF.4**　　(6 minutes)

Change Fractions to Mixed Numbers (4 minutes)

Materials: (S) Personal white board

Note: This fluency activity reviews Lesson 24.

T: (Write $\frac{4}{3}$.) Say the fraction.
S: 4 thirds.
T: (Draw a number bond with $\frac{4}{3}$ as the whole.) How many thirds are in 1?
S: 3 thirds.
T: (Write $\frac{3}{3}$ as a part. Write $\frac{-}{3}$ as the other part.) Write the remaining part, filling in the unknown numerator.
S: (Write $\frac{1}{3}$ as the unknown part.)
T: (Cross out $\frac{3}{3}$, and write 1 beneath it. Write $\frac{4}{3}$ = ___.) Write $\frac{4}{3}$ as a mixed number.
S: (Write $\frac{4}{3} = 1\frac{1}{3}$.)

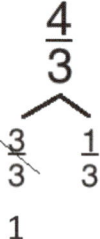

Continue with the following possible sequence: $\frac{9}{4}$, $\frac{14}{5}$, and $\frac{11}{3}$.

A STORY OF UNITS   Lesson 26   4•5

Change Mixed Numbers to Fractions (6 minutes)

Materials: (S) Personal white board

Note: This fluency activity reviews Lesson 25.

T: (Write $4\frac{2}{3}$) $4\frac{2}{3}$ is between which two whole numbers?
S: 4 and 5.
T: Draw a number line, and label 0, 1, 2, 3, 4, and 5.
S: (Draw a number line. Label 0, 1, 2, 3, 4, and 5.)
T: Decompose each whole number into thirds.
T: How many thirds are in 1?
S: 3 thirds.
T: 2?
S: 6 thirds.
T: 4?
S: 12 thirds.
T: Label 12 thirds on your number line.
S: (Draw an arrow from 0 to 4. Above the arrow, write $\frac{12}{3}$.)
T: (Write $4\frac{2}{3} = \frac{\_}{3} + \frac{2}{3}$.) Fill in the unknown numerator in the number sentence.
S: (Write $4\frac{2}{3} = \frac{12}{3} + \frac{2}{3}$.)
T: (Write $4\frac{2}{3} = \frac{12}{3} + \frac{2}{3} = \frac{\_}{3}$.) Label the slide from 4 to $4\frac{2}{3}$ on your number line. Then, complete the number sentence.
S: (Draw and label an arrow from 4 to $\frac{2}{3}$ more than 4. Write $4\frac{2}{3} = \frac{12}{3} + \frac{2}{3} = \frac{14}{3}$.)

Continue with the following possible sequence: $2\frac{3}{4}$ and $2\frac{3}{5}$.

> **NOTES ON MULTIPLE MEANS OF REPRESENTATION:**
>
> English language learners may need explicit instruction in reading and speaking mixed numbers. Teach students to read the whole number, say "and," and then read the fraction.

Lesson 26: Compare fractions greater than 1 by reasoning using benchmark fractions.

A STORY OF UNITS

Lesson 26 4•5

Application Problem (5 minutes)

Barbara needed $3\frac{1}{4}$ cups of flour for her recipe. If she measured $\frac{1}{4}$ cup at a time, how many times did she have to fill the measuring cup?

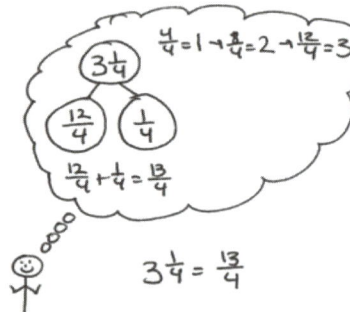

Barbara will have to fill the measuring cup 13 times.

Note: This Application Problem builds on the learning from Lesson 25. Students convert $3\frac{1}{4}$ to $\frac{13}{4}$ to determine that Barbara would have to fill the measuring cup 13 times. In Lesson 26, students compare fractions greater than 1. This Application Problem extends to the Concept Development of today's lesson as students compare $\frac{13}{4}$, $\frac{9}{2}$, and $3\frac{6}{8}$. It also hints at division by a unit fraction, a Grade 5 standard.

Concept Development (35 minutes)

Materials: (S) Personal white board

Problem 1: Compare mixed numbers and fractions on a number line using benchmark fractions.

T: Barbara needed $\frac{13}{4}$ cups of flour, her friend Jeanette needed $\frac{9}{2}$ cups, and her friend Robert needed $3\frac{6}{8}$ cups. Let's compare the amounts using a number line.

T: Draw a number line with the endpoints of 3 and 5. In the Application Problem, we found that $\frac{13}{4}$ equals $3\frac{1}{4}$. Find 3 on the number line. Imagine the fourths. Mark $\frac{1}{4}$ past 3. That shows where $3+\frac{1}{4}$ is located. Label $\frac{13}{4}$.

T: Plot $\frac{9}{2}$ on the number line. Work with a partner. How many ones are in $\frac{9}{2}$? How many remaining halves?

S: There are four groups of 2 halves in 9. → There are 4 ones and $\frac{1}{2}$ more. → We can find 4 on the number line and then mark $\frac{1}{2}$ past the 4.

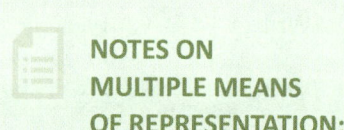

NOTES ON MULTIPLE MEANS OF REPRESENTATION:

Some learners may benefit from scaffolded questioning to convert $\frac{9}{2}$ to a mixed number. Ask, "How many halves make 1?" Then, say, "Count by 2 halves. 2 halves, 4 halves, 6 halves, 8 halves. Stop. We only have $\frac{9}{2}$. Decompose $\frac{9}{2}$ using a bond with $\frac{8}{2}$ and the remaining fraction."

354   Lesson 26:   Compare fractions greater than 1 by reasoning using benchmark fractions.

A STORY OF UNITS   Lesson 26   4•5

T: Label $\frac{9}{2}$. Is 9 halves greater than or less than 13 fourths?

S: Greater than, of course. There are 4 ones in $\frac{9}{2}$. There are only 3 ones in $\frac{13}{4}$.

T: Plot and label $3\frac{6}{8}$. Explain to a partner how this is done.

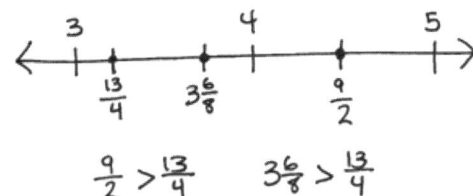

S: We can find the ones, 3, and then picture in our minds where $\frac{6}{8}$ more would be. $\frac{6}{8}$ is $\frac{2}{8}$ greater than $\frac{1}{2}$ since $\frac{4}{8} = \frac{1}{2}$. → $3\frac{6}{8}$ is between $3\frac{1}{2}$ and 4. → $\frac{6}{8} = \frac{3}{4}$. → $3\frac{6}{8} = 3\frac{3}{4}$.

T: Compare $3\frac{6}{8}$ and $\frac{13}{4}$.

S: $3\frac{6}{8}$ is greater than $3\frac{1}{2}$. $\frac{13}{4}$ is less than $3\frac{1}{2}$. → $3\frac{6}{8}$ is greater than $\frac{13}{4}$.

Repeat with $\frac{58}{8}$, $7\frac{5}{8}$, and $\frac{30}{4}$.

Problem 2: Compare two mixed numbers or two fractions greater than 1.

T: (Display $\frac{29}{7}$ and $\frac{31}{8}$.) Can we compare these fractions easily?

S: No. The denominators are different. → They are not mixed numbers. Mixed numbers would be easier to compare.

T: To compare them, let's rewrite $\frac{29}{7}$ and $\frac{31}{8}$ as mixed numbers.

S: 4 copies of 7 sevenths is $\frac{28}{7}$, so 29 sevenths must be $4\frac{1}{7}$.
→ $\frac{28}{7} + \frac{1}{7} = 4\frac{1}{7}$.

S: 3 copies of 8 eighths is $\frac{24}{8}$, so 31 eighths must be $3\frac{7}{8}$.
→ $\frac{24}{8} + \frac{7}{8} = 3\frac{7}{8}$.

$\frac{29}{7} = \frac{28}{7} + \frac{1}{7} = 4\frac{1}{7}$

$\frac{31}{8} = \frac{24}{8} + \frac{7}{8} = 3\frac{7}{8}$

T: Compare $4\frac{1}{7}$ and $3\frac{7}{8}$ using the words *a little bit more* and *a little bit less*.

S: $4\frac{1}{7}$ is a little bit more than 4. $3\frac{7}{8}$ is a little bit less than 4.

T: Write a comparison statement for $\frac{29}{7}$ and $\frac{31}{8}$.

S: $\frac{29}{7} > \frac{31}{8}$.

T: Write $5\frac{7}{8}$ and $5\frac{9}{10}$. Name the whole numbers these are between.

S: 5 and 6.

Lesson 26: Compare fractions greater than 1 by reasoning using benchmark fractions.

A STORY OF UNITS Lesson 26 4•5

MP.7

T: They both have 5 ones. Since the ones are the same, we look to the fractional units to compare. Compare $\frac{7}{8}$ and $\frac{9}{10}$.

S: $\frac{7}{8}$ is 1 eighth away from 6. → $\frac{9}{10}$ is 1 tenth away from 6. → $\frac{1}{10}$ is less than $\frac{1}{8}$, which means that $5\frac{9}{10}$ will be closer to 6 than $5\frac{7}{8}$. → $5\frac{7}{8} < 5\frac{9}{10}$.

T: Compare $\frac{43}{8}$ and $\frac{35}{6}$.

S: $\frac{43}{8} = 5\frac{3}{8}$. $\frac{35}{6} = 5\frac{5}{6}$. Now we can compare $\frac{3}{8}$ and $\frac{5}{6}$ because both mixed numbers have the same number of ones. $\frac{3}{8} < \frac{1}{2}$ and $\frac{5}{6} > \frac{1}{2}$. So, $5\frac{3}{8} < 5\frac{5}{6}$, and that means $\frac{43}{8} < \frac{35}{6}$.

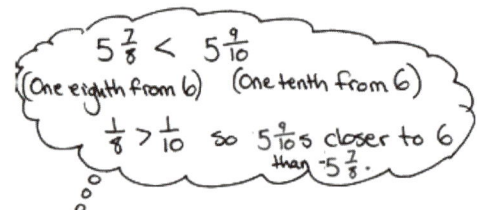

Problem Set (10 minutes)

Students should do their personal best to complete the Problem Set within the allotted 10 minutes. For some classes, it may be appropriate to modify the assignment by specifying which problems they work on first. Some problems do not specify a method for solving. Students should solve these problems using the RDW approach used for Application Problems.

Student Debrief (10 minutes)

Lesson Objective: Compare fractions greater than 1 by reasoning using benchmark fractions.

The Student Debrief is intended to invite reflection and active processing of the total lesson experience.

Invite students to review their solutions for the Problem Set. They should check work by comparing answers with a partner before going over answers as a class. Look for misconceptions or misunderstandings that can be addressed in the Student Debrief. Guide students in a conversation to debrief the Problem Set and process the lesson.

Any combination of the questions below may be used to lead the discussion.

- When comparing the mixed numbers and fractions on the Problem Set, which strategies did you use? Were some strategies easier than others? Was it helpful to think about benchmark fractions?
- Why is it often easier to compare mixed numbers than to compare fractions greater than 1?
- How does this lesson relate to earlier lessons? How did earlier lessons help you to understand this lesson?

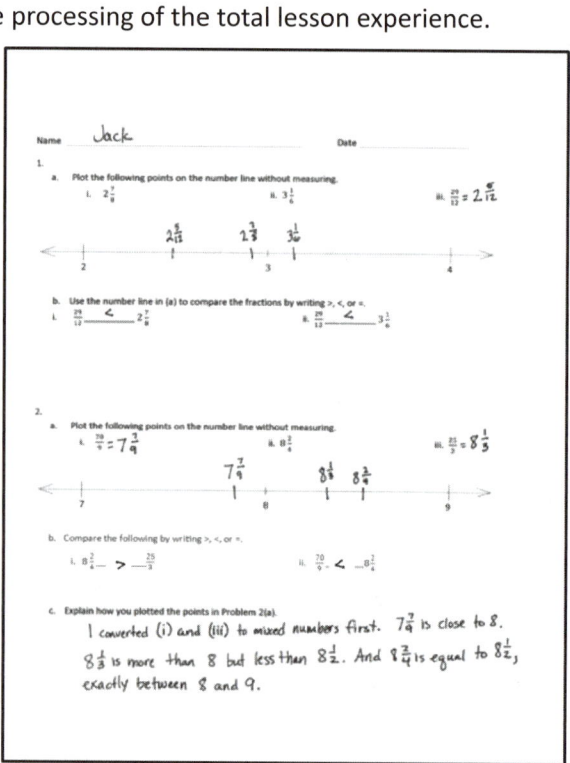

356 | Lesson 26: Compare fractions greater than 1 by reasoning using benchmark fractions.

- In what way is Problem 3(a) easier than 3(b)?
- At first glance, Problem 3(j) looks really difficult. What makes it easier to solve?
- How did the Application Problem connect to today's lesson?

Exit Ticket (3 minutes)

After the Student Debrief, instruct students to complete the Exit Ticket. A review of their work will help with assessing students' understanding of the concepts that were presented in today's lesson and planning more effectively for future lessons. The questions may be read aloud to the students.

3. Compare the fractions given below by writing >, <, or =. Give a brief explanation for each answer, referring to benchmark fractions.

a. $5\frac{1}{3} > 4\frac{2}{5}$
5 is greater than 4 so $5\frac{1}{3}$ is greater than $4\frac{2}{5}$.

b. $\frac{12}{6} < \frac{25}{12}$
$\frac{12}{6}$ is the same as 2, and $\frac{25}{12}$ is more than 2.

c. $\frac{18}{7} < \frac{17}{5}$
$\frac{17}{5}$ is more than 3, but you would need $\frac{21}{7}$ to make 3 so $\frac{18}{7}$ has to be less.

d. $5\frac{1}{3} < 5\frac{5}{8}$
I knew $5\frac{5}{8}$ is less than $5\frac{1}{2}$ but $5\frac{5}{8}$ is more than $5\frac{1}{2}$.

e. $6\frac{2}{3} > 6\frac{3}{5}$
$6\frac{2}{3}$ is less than $6\frac{1}{2}$, but $6\frac{2}{3}$ is more.

f. $\frac{31}{8} > \frac{32}{9}$
$\frac{31}{8}$ is the same as 4. It would only take $\frac{28}{7}$ to make 4, so $\frac{31}{7}$ must be greater.

g. $\frac{31}{10} < \frac{25}{8}$
$\frac{31}{10}$ is $3\frac{1}{10}$, and $\frac{25}{8}$ is $3\frac{1}{8}$. $\frac{1}{8}$ is bigger than $\frac{1}{10}$!

h. $\frac{39}{12} > \frac{19}{6}$
$\frac{39}{12}$ is $3\frac{3}{12}$ and $\frac{19}{6}$ is $3\frac{1}{6}$. They are both close to 3, but $3\frac{3}{12}$ is just like $3\frac{1}{4}$ so it is greater.

i. $\frac{49}{50} < 3\frac{90}{100}$
$\frac{49}{50}$ is not even a whole, but $3\frac{90}{100} = 3 + \frac{90}{100}$ so it is greater.

j. $5\frac{5}{12} < 5\frac{51}{100}$
They both have mixed numbers with a 5, but $\frac{5}{12}$ is less than $\frac{1}{2}$ and $\frac{51}{100}$ is greater than $\frac{1}{2}$, so $5\frac{5}{12} < 5\frac{51}{100}$.

Lesson 26: Compare fractions greater than 1 by reasoning using benchmark fractions.

A STORY OF UNITS   Lesson 26 Problem Set 4•5

Name _____ Date _____

1. a. Plot the following points on the number line without measuring.

   i. $2\frac{7}{8}$ ii. $3\frac{1}{6}$ iii. $\frac{2\bar{A}}{12}$

   <----+----------------------+----------------------+---->
        2   3   4

   b. Use the number line in Problem 1(a) to compare the fractions by writing >, <, or =.

   i. $\frac{2\bar{A}}{12}$ _____ $2\frac{7}{8}$ ii. $\frac{2\bar{A}}{12}$ _____ $3\frac{1}{6}$

2. a. Plot the following points on the number line without measuring.

   i. $\frac{70}{9}$ ii. $8\frac{2}{4}$ iii. $\frac{25}{3}$

   <----+----------------------+----------------------+---->
        7   8   9

   b. Compare the following by writing >, <, or =.

   i. $8\frac{2}{4}$ _____ $\frac{25}{3}$ ii. $\frac{70}{9}$ _____ $8\frac{2}{4}$

   c. Explain how you plotted the points in Problem 2(a).

358   Lesson 26:   Compare fractions greater than 1 by reasoning using benchmark fractions.

3. Compare the fractions given below by writing >, <, or =. Give a brief explanation for each answer, referring to benchmark fractions.

a. $5\frac{1}{3}$ _____ $4\frac{3}{4}$

b. $\frac{12}{6}$ _____ $\frac{25}{12}$

c. $\frac{1\bar{A}}{7}$ _____ $\frac{1\bar{A}}{5}$

d. $5\frac{2}{5}$ _____ $5\frac{5}{8}$

e. $6\frac{2}{3}$ _____ $6\frac{3}{7}$

f. $\frac{31}{7}$ _____ $\frac{32}{8}$

g. $\frac{31}{1\bar{A}}$ _____ $\frac{25}{8}$

h. $\frac{3\bar{A}}{12}$ _____ $\frac{1\bar{A}}{6}$

i. $\frac{4\bar{A}}{5\bar{A}}$ _____ $3\frac{90}{1\bar{A}\bar{A}}$

j. $5\frac{5}{12}$ _____ $5\frac{51}{1\bar{A}\bar{A}}$

A STORY OF UNITS   Lesson 26 Exit Ticket 4•5

Name _____   Date _____

Compare the fractions given below by writing >, <, or =.

Give a brief explanation for each answer, referring to benchmark fractions.

1. $3\frac{2}{3}$ _____ $3\frac{4}{6}$

2. $\frac{12}{3}$ _____ $\frac{21}{7}$

3. $\frac{11}{6}$ _____ $\frac{5}{4}$

4. $3\frac{2}{5}$ _____ $3\frac{3}{11}$

A STORY OF UNITS   Lesson 26 Homework   4•5

Name _____   Date _____

1. a. Plot the following points on the number line without measuring.

   i. $2\frac{1}{6}$ ii. $3\frac{3}{4}$ iii. $\frac{33}{9}$

   <----+----------------+----------------+---->
        2   3   4

   b. Use the number line in Problem 1(a) to compare the fractions by writing >, <, or =.

   i. $\frac{33}{9}$ _____ $2\frac{1}{6}$ ii. $\frac{33}{9}$ _____ $3\frac{3}{4}$

2. a. Plot the following points on the number line without measuring.

   i. $\frac{65}{8}$ ii. $8\frac{5}{6}$ iii. $\frac{2\overline{A}}{4}$

   <----+----------------+----------------+---->
        7   8   9

   b. Compare the following by writing >, <, or =.

   i. $8\frac{5}{6}$ _____ $\frac{65}{8}$ ii. $\frac{2\overline{A}}{4}$ _____ $\frac{65}{8}$

   c. Explain how you plotted the points in Problem 2(a).

Lesson 26: Compare fractions greater than 1 by reasoning using benchmark fractions.   361

3. Compare the fractions given below by writing >, <, or =. Give a brief explanation for each answer, referring to benchmark fractions.

   a. $5\frac{1}{3}$ _____ $5\frac{3}{4}$

   b. $\frac{12}{4}$ _____ $\frac{25}{8}$

   c. $\frac{1\bar{A}}{6}$ _____ $\frac{1\bar{A}}{4}$

   d. $5\frac{3}{5}$ _____ $5\frac{5}{1\bar{A}}$

   e. $6\frac{3}{4}$ _____ $6\frac{3}{5}$

   f. $\frac{33}{6}$ _____ $\frac{34}{7}$

   g. $\frac{23}{1\bar{A}}$ _____ $\frac{2\bar{A}}{8}$

   h. $\frac{2\bar{A}}{12}$ _____ $\frac{15}{6}$

   i. $2\frac{4\bar{A}}{5\bar{A}}$ _____ $2\frac{99}{1\bar{A}\bar{A}}$

   j. $6\frac{5}{9}$ _____ $6\frac{4\bar{A}}{1\bar{A}\bar{A}}$

Lesson 27

Objective: Compare fractions greater than 1 by creating common numerators or denominators.

Suggested Lesson Structure

- ■ Fluency Practice (12 minutes)
- ■ Application Problem (6 minutes)
- ■ Concept Development (32 minutes)
- ■ Student Debrief (10 minutes)
- **Total Time** **(60 minutes)**

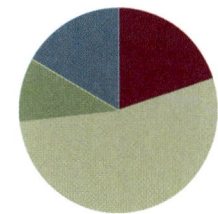

Fluency Practice (12 minutes)

- Add and Subtract Fractions **4.NF.3** (3 minutes)
- Change Fractions to Mixed Numbers **4.NF.4** (4 minutes)
- Change Mixed Numbers to Fractions **4.NF.4** (5 minutes)

Add and Subtract Fractions (3 minutes)

Materials: (S) Personal white board

Note: This fluency activity reviews Lesson 22.

T: (Draw a number bond with a whole of 5. Write 4 as the known part and $\frac{}{10}$ as the unknown part.) How many tenths are in 1?

S: 10 tenths.

T: (Write $\frac{10}{10}$ as the unknown part. Beneath it, write $5 - \frac{3}{10} = 4 + \frac{}{10}$.) Write the number sentence.

S: (Write $5 - \frac{3}{10} = 4 + \frac{7}{10} = 4\frac{7}{10}$.)

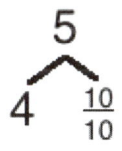

Continue with the following possible sequence: $6 - \frac{5}{8}$, $5 - \frac{3}{4}$, and $4 - \frac{7}{12}$.

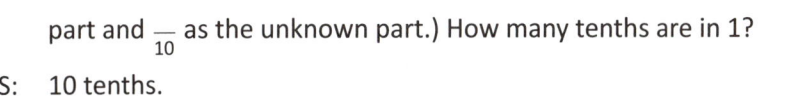

T: How much does 7 tenths need to be 1?

S: 3 tenths.

A STORY OF UNITS   Lesson 27 4•5

T: (Write $4\frac{7}{10} + \_\_ = 5$.) Write the number sentence, filling in the unknown number.

S: (Write $4\frac{7}{10} + \frac{3}{10} = 5$.)

Continue with the following possible sequence: $5\frac{3}{8} + \_\_ = 6$, $4\frac{1}{4} + \_\_ = 5$, and $3\frac{5}{12} + \_\_ = 4$.

Change Fractions to Mixed Numbers (4 minutes)

Materials: (S) Personal white board

Note: This fluency activity reviews Lesson 24.

T: (Write $\frac{7}{5}$.) Say the fraction.

S: 7 fifths.

T: (Draw a number bond with a whole of $\frac{7}{5}$.) How many fifths are in 1?

S: 5 fifths.

T: (Write $\frac{5}{5}$ as the known part.) Write the unknown part.

S: (Write $\frac{2}{5}$ as the unknown part.)

T: (Cross out $\frac{5}{5}$, and write 1 beneath it.) Write $\frac{7}{5}$ as a mixed number.

S: (Write $\frac{7}{5} = 1\frac{2}{5}$.)

Continue with the following possible sequence: $\frac{9}{4}$ and $\frac{11}{3}$.

Change Mixed Numbers to Fractions (5 minutes)

Materials: (S) Personal white board

Note: This fluency activity reviews Lesson 25.

T: (Write $1\frac{1}{2}$.) Say the mixed number.

S: 1 and 1 half.

T: (Draw a number bond with a whole of $1\frac{1}{2}$. Write $\frac{1}{2}$ as the known part.) Write the unknown part.

S: (Write $\frac{2}{2}$ as the unknown part.)

T: (Write $\frac{2}{2}$ as the unknown part. Write $1\frac{1}{2} = \frac{1}{2}$.) Write $1\frac{1}{2}$ as a fraction.

S: (Write $1\frac{1}{2} = \frac{3}{2}$.)

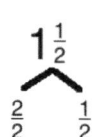

Continue with the following possible sequence: $1\frac{2}{3}$ and $2\frac{1}{5}$.

Lesson 27: Compare fractions greater than 1 by creating common numerators or denominators.

A STORY OF UNITS   Lesson 27   4•5

Application Problem (6 minutes)

Jeremy ran 27 laps on a track that was $\frac{1}{8}$ mile long. Jimmy ran 15 laps on a track that was $\frac{1}{4}$ mile long. Who ran farther?

$\frac{27}{8} = \frac{3 \times 8}{8} + \frac{3}{8} = 3\frac{3}{8}$

$\frac{15}{4} = \frac{3 \times 4}{4} + \frac{3}{4} = 3\frac{3}{4}$

$3\frac{3}{8} < 3\frac{3}{4}$ so $\frac{27}{8} < \frac{15}{4}$

Jimmy ran farther.

Note: This Application Problem builds from Lesson 26 where students compared fractions greater than 1 using benchmark fractions. In today's lesson, students compare fractions by creating common numerators or denominators. The fractions $\frac{27}{8}$ and $\frac{15}{4}$, used in the Application Problem, lend themselves to comparison using either benchmark fractions or common numerators or related denominators.

NOTES ON MULTIPLE MEANS OF REPRESENTATION:

Because precision in modeling is critical when comparing, it may be helpful to provide aligned parallel tape diagram templates of equal length that students can partition.

Concept Development (32 minutes)

Materials: (S) Personal white board

Problem 1: Model, using a tape diagram, the comparison of two mixed numbers having related denominators.

T: (Display $3\frac{3}{8}$ and $3\frac{3}{4}$.) Look at the mixed numbers from the Application Problem. You compared fractions by thinking about the size of units. Can you remember another way to compare fractions?

S: We can use common denominators.

Lesson 27: Compare fractions greater than 1 by creating common numerators or denominators.

365

A STORY OF UNITS   Lesson 27 4•5

T: Yes! Four is a factor of 8. We can convert fourths to eighths by decomposing each fourth to make eighths. Draw a tape diagram to model the comparison of $\frac{3}{4}$ and $\frac{3}{8}$.

S: (Draw as shown to the right.) $\frac{3}{4} = \frac{6}{8}$. $\frac{6}{8} > \frac{3}{8}$. So, $3\frac{3}{8} < 3\frac{3}{4}$.

T: With your partner, draw a tape diagram to compare $2\frac{2}{6}$ and $2\frac{3}{12}$.

S: $\frac{2}{6} = \frac{4}{12}$. So, $2\frac{2}{6} > 2\frac{3}{12}$.

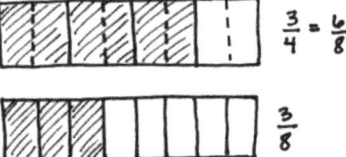

Repeat, using a number line to make like units, to compare $4\frac{1}{3}$ and $4\frac{2}{9}$ and then $5\frac{1}{4}$ and $5\frac{3}{8}$.

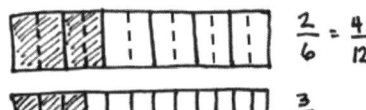

Problem 2: Compare two fractions with unrelated denominators.

T: Discuss a strategy to use to compare $4\frac{3}{4}$ and $\frac{23}{5}$.

S: We need to convert $\frac{23}{5}$ to a mixed number. $\frac{23}{5} = 4\frac{3}{5}$. Both mixed numbers have the same number of ones, so I can use the denominators to compare because the numerators are the same. Fourths are bigger than fifths, so 3 fourths would be greater than 3 fifths. $4\frac{3}{4} > \frac{23}{5}$.

T: Yes. That is the same strategy we used in the Application Problem. This time, use the area model to show $\frac{3}{4}$ is greater than $\frac{3}{5}$. Draw two same-sized rectangles representing 1 one. Partition one area into fourths using vertical lines. Partition one area into fifths using horizontal lines. Make like denominators.

MP.4

S: I'll draw fifths horizontally on the fourths, and fourths vertically on the fifths. We made twentieths!

T: Compare the twentieths to prove $4\frac{3}{4} > 4\frac{3}{5}$.

S: $\frac{3}{4} = \frac{15}{20}$ and $\frac{3}{5} = \frac{12}{20}$. $\frac{15}{20} > \frac{12}{20}$. So, $4\frac{3}{4} > 4\frac{3}{5}$.

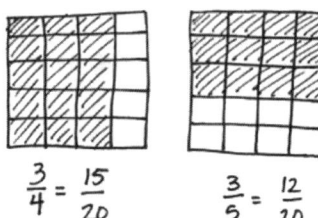

Repeat, using the area model to make like units, to compare $2\frac{2}{3}$ and $2\frac{3}{5}$.

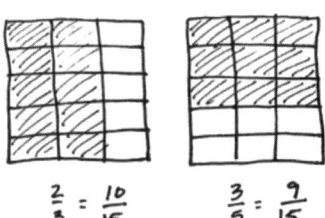

Problem 3: Compare two fractions.

T: (Display $3\frac{7}{10}$ and $\frac{18}{5}$.) Compare by finding like denominators. What is different about this comparison?

S: One number is written as a mixed number and the other as a fraction.

T: Work together with your partner. One of you is Partner A, and the other is Partner B.

T: Partner A, convert $3\frac{7}{10}$ to a fraction, and compare it to 18 fifths. Partner B, convert $\frac{18}{5}$ to a mixed number, and compare it to 3 and 7 tenths.

366 Lesson 27: Compare fractions greater than 1 by creating common numerators or denominators.

T: Consider using multiplication to solve. Take turns discussing how you solved. Did you both get the same answer?

S: Partner A: $\frac{30}{10} = 3$. $\frac{30}{10} + \frac{7}{10} = \frac{37}{10}$. We have $\frac{37}{10}$ and $\frac{18}{5}$. Fifths and tenths are related denominators. Convert $\frac{18}{5}$ to tenths. $\frac{18 \times 2}{5 \times 2} = \frac{36}{10}$. $\frac{37}{10} > \frac{36}{10}$. So, $3\frac{7}{10} > \frac{18}{5}$.

S: Partner B: There are 3 copies of 5 fifths in $\frac{18}{5}$. $\frac{15}{5} + \frac{3}{5} = \frac{18}{5}$. $\frac{18}{5} = 3\frac{3}{5}$. Convert the fifths to tenths. $\frac{3 \times 2}{3 \times 2} = \frac{6}{10}$. $3\frac{3}{5} = 3\frac{6}{10}$. $3\frac{7}{10} > 3\frac{6}{10}$. So, $3\frac{7}{10} > \frac{18}{5}$.

S: We both found that $3\frac{7}{10} > \frac{18}{5}$. → Our answers were the same because we both were converting to equivalent amounts. We just expressed it in a different way.

T: Compare $7\frac{3}{5}$ and $7\frac{4}{6}$.

T: We can make like denominators using multiplication. 30 is a multiple of both 5 and 6. We know that because 5 times 6 is …?

S: 30.

T: Let's rename each fraction using multiplication to have 30 as the new denominator. (Demonstrate as shown to the right.)

S: $7\frac{18}{30} < 7\frac{20}{30}$; then, $7\frac{3}{5} < 7\frac{4}{6}$.

T: Show how we can also make like numerators of 12. (Demonstrate as shown to the right.)

S: $7\frac{12}{20} < 7\frac{12}{18}$ because eighteenths are larger units than twentieths, so $7\frac{3}{5} < 7\frac{4}{6}$.

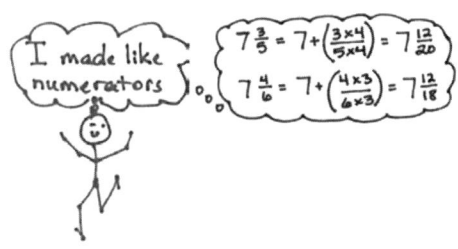

Note: Do not overemphasize the use of like numerators since like denominators is a much bigger idea mathematically that is used when like units are compared or added and subtracted.

Other possible examples could include $4\frac{2}{3}$, $4\frac{7}{12}$; $4\frac{3}{4}$, $\frac{29}{6}$; $\frac{25}{9}$, $\frac{14}{5}$.

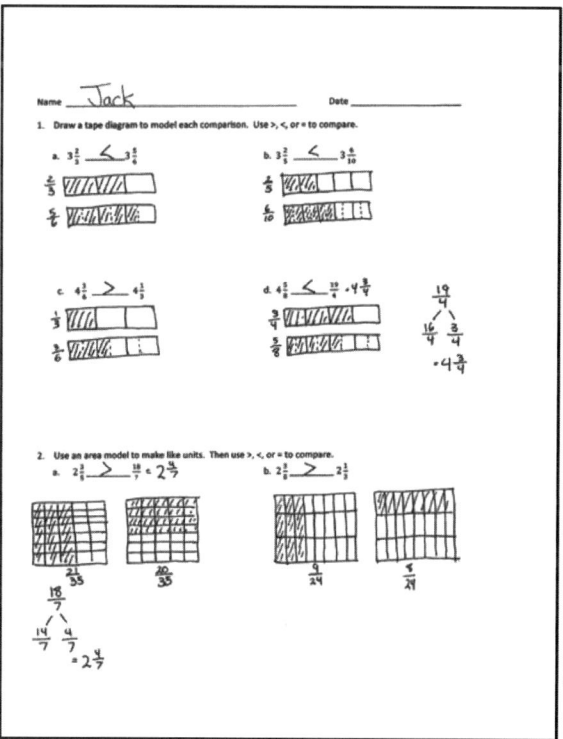

Lesson 27: Compare fractions greater than 1 by creating common numerators or denominators.

A STORY OF UNITS　　　　　　　　　　　　　　　　　　　　　　　　Lesson 27　4•5

Problem Set (10 minutes)

Students should do their personal best to complete the Problem Set within the allotted 10 minutes. For some classes, it may be appropriate to modify the assignment by specifying which problems they work on first. Some problems do not specify a method for solving. Students should solve these problems using the RDW approach used for Application Problems.

Student Debrief (10 minutes)

Lesson Objective: Compare fractions greater than 1 by creating common numerators or denominators.

The Student Debrief is intended to invite reflection and active processing of the total lesson experience.

Invite students to review their solutions for the Problem Set. They should check work by comparing answers with a partner before going over answers as a class. Look for misconceptions or misunderstandings that can be addressed in the Student Debrief. Guide students in a conversation to debrief the Problem Set and process the lesson.

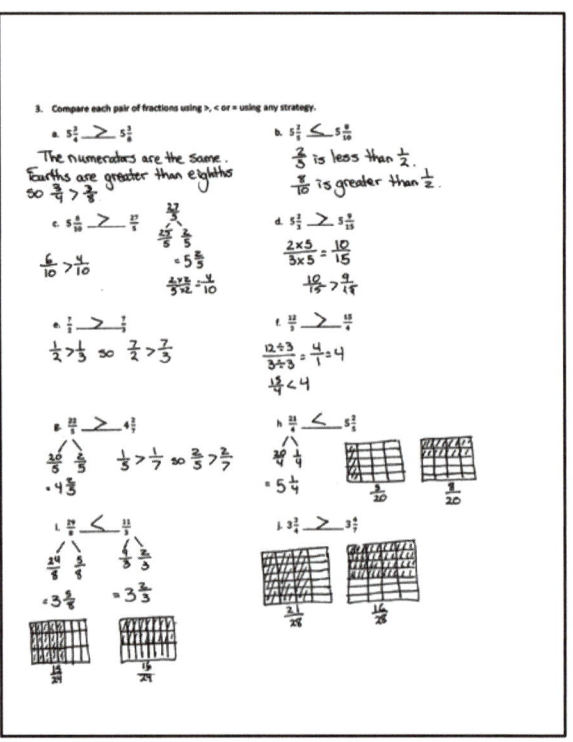

Any combination of the questions below may be used to lead the discussion.

- How did the tape diagram help to solve Problem 1 (a) and (b)? Why is it important to make sure the whole for both tape diagrams is the same size?
- Who converted to a mixed number or a fraction greater than 1 before finding like units for Problem 3(c)? Was it easier to compare mixed numbers or fractions greater than 1 for this particular problem? (Note: Finding mixed numbers first, one could use a benchmark fraction of 1 half to compare $\frac{6}{1A}$ to $\frac{2}{5}$ without needing to find like units.) Is it more efficient to compare fractions greater than 1 or mixed numbers?
- In Problem 3(e), the added complexity was that the denominators were not related, as in the previous problems. What strategy did you use to solve? Did you solve by finding like numerators or by drawing an area model to find like denominators?

NOTES ON MULTIPLE MEANS OF ACTION AND EXPRESSION:

The number line may also be used to compare fractions with related denominators. Decompose the unit just as practiced in Topic C. The number line comparison allows students to consider the entire number they are comparing and not just the fractional parts. Show how the number line can be partitioned just like the tape diagram by aligning them on top of each other.

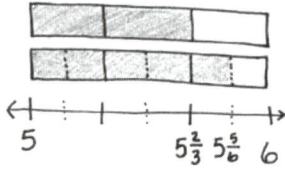

368　　　Lesson 27:　　Compare fractions greater than 1 by creating common numerators or denominators.

- Were there any problems in Problem 3 that you could compare without renaming or without drawing a model? How were you able to mentally compare them?
- How did having to compare a mixed number to a fraction add to the complexity of Problem 2(a)?
- How did the Application Problem connect to today's lesson?

Exit Ticket (3 minutes)

After the Student Debrief, instruct students to complete the Exit Ticket. A review of their work will help with assessing students' understanding of the concepts that were presented in today's lesson and planning more effectively for future lessons. The questions may be read aloud to the students.

Name _____   Date _____

1. Draw a tape diagram to model each comparison. Use >, <, or = to compare.

    a. $3\frac{2}{3}$ _____ $3\frac{5}{6}$

    b. $3\frac{2}{5}$ _____ $3\frac{6}{10}$

    c. $4\frac{3}{6}$ _____ $4\frac{1}{3}$

    d. $4\frac{5}{8}$ _____ $\frac{10}{4}$

2. Use an area model to make like units. Then, use >, <, or = to compare.

    a. $2\frac{3}{5}$ _____ $\frac{10}{7}$

    b. $2\frac{3}{8}$ _____ $2\frac{1}{3}$

3. Compare each pair of fractions using >, <, or = using any strategy.

a. $5\frac{3}{4}$ _____ $5\frac{3}{8}$

b. $5\frac{2}{5}$ _____ $5\frac{8}{10}$

c. $5\frac{6}{10}$ _____ $\frac{20}{5}$

d. $5\frac{2}{3}$ _____ $5\frac{9}{15}$

e. $\frac{7}{2}$ _____ $\frac{7}{3}$

f. $\frac{12}{3}$ _____ $\frac{15}{4}$

g. $\frac{22}{5}$ _____ $4\frac{2}{7}$

h. $\frac{21}{4}$ _____ $5\frac{2}{5}$

i. $\frac{20}{8}$ _____ $\frac{11}{3}$

j. $3\frac{3}{4}$ _____ $3\frac{4}{7}$

Lesson 27: Compare fractions greater than 1 by creating common numerators or denominators.

A STORY OF UNITS Lesson 27 Exit Ticket 4•5

Name _____ Date _____

Compare each pair of fractions using >, <, or = using any strategy.

1. $4\dfrac{3}{8}$ _____ $4\dfrac{1}{4}$

2. $3\dfrac{4}{5}$ _____ $3\dfrac{9}{10}$

3. $2\dfrac{1}{3}$ _____ $2\dfrac{2}{5}$

4. $10\dfrac{2}{5}$ _____ $10\dfrac{3}{4}$

A STORY OF UNITS

Lesson 27 Homework 4•5

Name _____ Date _____

1. Draw a tape diagram to model each comparison. Use >, <, or = to compare.

   a. $2\frac{3}{4}$ _____ $2\frac{7}{8}$

   b. $10\frac{2}{6}$ _____ $10\frac{1}{3}$

   c. $5\frac{3}{8}$ _____ $5\frac{1}{4}$

   d. $2\frac{5}{9}$ _____ $\frac{21}{3}$

2. Use an area model to make like units. Then, use >, <, or = to compare.

   a. $2\frac{4}{5}$ _____ $\frac{11}{4}$

   b. $2\frac{3}{5}$ _____ $2\frac{2}{3}$

Lesson 27: Compare fractions greater than 1 by creating common numerators or denominators.

373

Lesson 27 Homework 4•5

3. Compare each pair of fractions using >, <, or = using any strategy.

   a. $6\frac{1}{2}$ _____ $6\frac{3}{8}$

   b. $7\frac{5}{6}$ _____ $7\frac{11}{12}$

   c. $3\frac{6}{10}$ _____ $3\frac{2}{5}$

   d. $2\frac{2}{5}$ _____ $2\frac{8}{15}$

   e. $\frac{10}{3}$ _____ $\frac{10}{4}$

   f. $\frac{12}{4}$ _____ $\frac{10}{3}$

   g. $\frac{30}{9}$ _____ $4\frac{2}{12}$

   h. $\frac{23}{4}$ _____ $5\frac{2}{3}$

   i. $\frac{30}{8}$ _____ $3\frac{7}{12}$

   j. $10\frac{3}{4}$ _____ $10\frac{4}{6}$

A STORY OF UNITS — Lesson 28 — 4•5

Lesson 28

Objective: Solve word problems with line plots.

Suggested Lesson Structure

- ■ Fluency Practice (12 minutes)
- ■ Concept Development (38 minutes)
- ■ Student Debrief (10 minutes)
- **Total Time** **(60 minutes)**

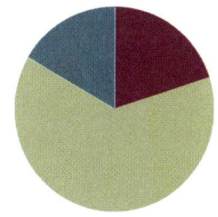

Fluency Practice (12 minutes)

- Change Mixed Numbers to Fractions **4.NF.4** (6 minutes)
- Compare Fractions **4.NF.2** (6 minutes)

Change Mixed Numbers to Fractions (6 minutes)

Materials: (S) Personal white board

Note: This fluency activity reviews Lesson 25.

- T: (Write $1\frac{3}{4}$.) Say the mixed number.
- S: 1 and 3 fourths.
- T: (Draw a number bond for $1\frac{3}{4}$. Write $\frac{3}{4}$ as a part.) Complete the bond.
- S: (Write $\frac{4}{4}$ as the unknown part.)
- T: (Write $1\frac{3}{4} = \frac{}{4}$.) Complete the number sentence.
- S: (Write $1\frac{3}{4} = \frac{7}{4}$.)

Continue the process for the following possible sequence: $1\frac{4}{5}$, $2\frac{1}{4}$, and $4\frac{5}{6}$.

Compare Fractions (6 minutes)

Materials: (S) Personal white board

Note: This fluency activity reviews Lessons 26 and 27.

NOTES ON MULTIPLE MEANS OF REPRESENTATION:

The Change Mixed Numbers to Fractions fluency activity may be a good opportunity for English language learners to practice speaking mixed numbers, particularly if they omit *and* as they say the whole number and the fraction.

Lesson 28: Solve word problems with line plots. 375

T: (Project the number line with endpoints 1 and 2 and $1\frac{1}{2}$ as the midpoint.) Copy the number line.

S: (Copy the number line with endpoints 1 and 2 and $1\frac{1}{2}$ as the midpoint.)

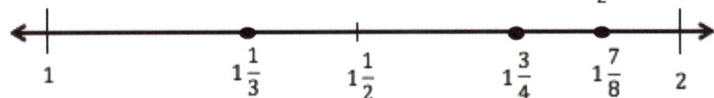

T: (Write $1\frac{1}{3}$ and $1\frac{3}{4}$.) Plot $1\frac{1}{3}$ and $1\frac{3}{4}$ on the number line.

S: (Plot and label $1\frac{1}{3}$ between 1 and $1\frac{1}{2}$, and $1\frac{3}{4}$ between $1\frac{1}{2}$ and 2.)

T: (Write $1\frac{1}{3}$ ___ $1\frac{3}{4}$.) Write a greater than or less than sign to make the number sentence true.

S: (Write $1\frac{1}{3} < 1\frac{3}{4}$.)

T: $1\frac{3}{4}$ is the same as 1 and how many eighths?

S: $1\frac{6}{8}$.

T: Plot $1\frac{7}{8}$ on your number line.

S: (Write $1\frac{7}{8}$ between $1\frac{3}{4}$ and 2.)

T: (Write $1\frac{7}{8}$ ___ $1\frac{3}{4}$.) Write a greater than or less than sign to make the number sentence true.

S: (Write $1\frac{7}{8} > 1\frac{3}{4}$.)

Continue with the following possible sequence using the same number line: $1\frac{5}{12}$, $1\frac{5}{6}$, and $1\frac{2}{3}$.

Continue with the following possible sequence using the following number line: $2\frac{1}{4}$, $2\frac{5}{8}$, and $2\frac{5}{6}$.

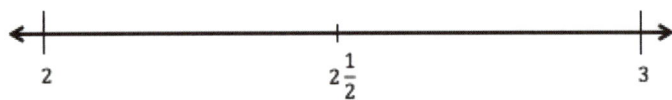

Concept Development (38 minutes)

Materials: (S) Personal white board, Problem Set

Note: Today's Problem Set is used throughout the Concept Development. The teacher guides the construction and interpretation of a line plot. As students complete each problem, the teacher might debrief with students about their solutions. Students have had prior exposure to creating and interpreting line plots in Grades 2 and 3.

Problem 1

Display the table from the Problem Set.

Student	Distance (in miles)
Joe	$2\frac{1}{2}$
Arianna	$1\frac{3}{4}$
Bobbi	$2\frac{1}{8}$
Morgan	$1\frac{5}{8}$
Jack	$2\frac{5}{8}$
Saisha	$2\frac{1}{4}$
Tyler	$2\frac{2}{4}$
Jenny	$\frac{5}{8}$
Anson	$2\frac{2}{8}$
Chandra	$2\frac{4}{8}$

T: This table shows the distance that Ms. Smith's fourth graders were able to run before stopping for a rest. Tell your partner what you notice about the data.

S: It has the names of students and the distances they ran as a mixed number. → Some of the fractions have different denominators. → I can see fractions that are equivalent. → The distance is measured in miles.

T: Let's create a **line plot** to show the information. Discuss with your partner what you might remember about line plots from Grade 3: How does a line plot represent data?

S: It's like a number line. → We don't put points on the line, but we make marks above the line. → Yeah. The X's go above the line because sometimes there are a lot of X's at one number. → It's like a bar graph because the tallest column shows the most.

T: Discuss with your partner what the endpoints will be for the number line.

S: The largest fraction is $2\frac{5}{8}$, and the smallest is $\frac{5}{8}$, so we could use 0 and 3.

T: To create a number line using a ruler, we need to decide what measurement on the ruler we can use to mark off the distances students ran. What is the smallest unit of measurement in the chart?

S: 1 eighth mile.

T: Let's see. If I mark off eighth miles from 0 to 3 using an eighth of an inch on a ruler, the increments are very small! Discuss with your partner another length unit that we could use to mark the eighth miles.

S: Let's use inches. Those are nice and big! → There are 24 eighths between 0 and 3. Our paper isn't 24 inches wide. → What if we double the eighth inch to fourth inch marks?

T: Draw a line, and make hash marks at every $\frac{1}{4}$ inch to represent each eighth mile. Then, label the whole numbers. (Allow students time to work.)

T: Below the line, write "Distance (in miles)" to tell what unit our line plot shows. (Allow students time to work.)

T: The line plot needs a title to tell what it shows. Tell your neighbor a title for this line plot and record it above the line, leaving some space for the data.

S: We can title it "Distance Ms. Smith's Fourth Graders Ran". (Record the title.)

T: Data on a line plot is marked with an X. We need to tell what each X will represent by providing a key. Below the line plot, record "X = 1 student". (Allow students time to work.)

Lesson 28: Solve word problems with line plots.

A STORY OF UNITS   Lesson 28   4•5

T: Now, mark each student's distance using an X above a point on the number line that shows the distance they ran in miles. Label that point on the number line with the unit *eighths*. Tell your partner what you notice.

S: One student ran almost 3 miles! → Some students ran the same distance. → Some distances were measured using different fractional units. I converted fourths and halves to eighths. → Most students ran between 2 and 3 miles.

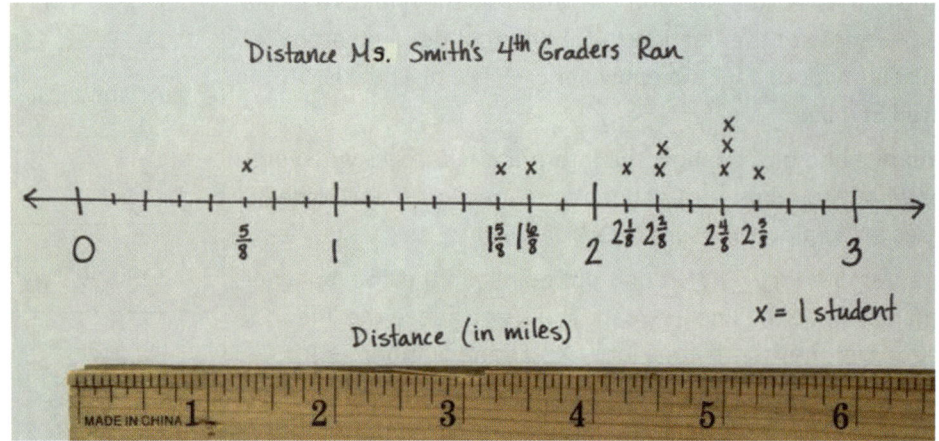

Problem 2

Circulate as students work. When the class is ready, stop students and debrief Problem 2. If preferred, ask questions such as the following:

T: For Problems 2(a) and 2(b), did you refer to the table or the line plot?

T: For Problem 2(b), make a comparison statement for the distance Jack ran compared to Jenny.

T: What strategy did you use for Problem 2(c)? Did you count on the number line or use renaming a fraction to solve?

T: What previous knowledge about subtracting fractions or subtracting mixed units helped you to solve Problem 2(d)?

T: The line plot works just like a number line. I can tell that Arianna ran farther than Morgan. For Problem 2(e), how can you confirm that?

T: For Problem 2(g), comparing eighths and tenths requires a large denominator, like fortieths or eightieths. Using what you know about equivalent fractions to eighths, how could renaming Ms. Smith's distance to fourths make the comparison to Mr. Reynolds's distance simpler?

NOTES ON MULTIPLE MEANS OF ACTION AND EXPRESSION:

Scaffold the word problems on the Problem Set for students working below grade level with questioning. For example, for Problem 2(d) ask, "What was the longest distance run? The shortest? What is the difference, in miles, between the longest and shortest distance run?"

Additionally, students may benefit from organizing data in a table before solving, for example, Problem 2(b).

378   Lesson 28:   Solve word problems with line plots.

A STORY OF UNITS

Lesson 28 4•5

Problem Set (10 minutes)

Students should do their personal best to complete Problem 3 of the Problem Set within the allotted 10 minutes. For some classes, it may be appropriate to modify the assignment by specifying which problems they work on first. Some problems do not specify a method for solving. Students should solve these problems using the RDW approach used for Application Problems.

Student Debrief (10 minutes)

Lesson Objective: Solve word problems with line plots.

The Student Debrief is intended to invite reflection and active processing of the total lesson experience.

Invite students to review their solutions for the Problem Set. They should check work by comparing answers with a partner before going over answers as a class. Look for misconceptions or misunderstandings that can be addressed in the Student Debrief. Guide students in a conversation to debrief the Problem Set and process the lesson.

Any combination of the questions below may be used to lead the discussion.

- For Problem 2(g), which strategy did you use to compare the two distances? Would you be able to determine the correct answer if you answered Problem 2(f) incorrectly? Why or why not?
- Let's share some of the questions that you wrote for Problem 3. Were there similarities in the questions that you and your partner wrote? Were there differences? Explain.
- How is a **line plot** useful in showing data? By simply looking at the line plot, what can you tell about the distances that students ran?
- What might be some reasons to use a line plot to display data rather than using a chart or table?

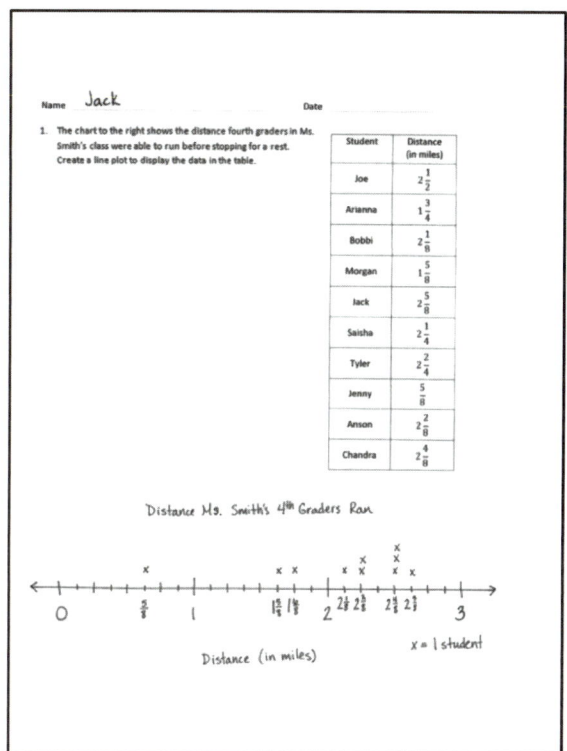

Exit Ticket (3 minutes)

After the Student Debrief, instruct students to complete the Exit Ticket. A review of their work will help with assessing students' understanding of the concepts that were presented in today's lesson and planning more effectively for future lessons. The questions may be read aloud to the students.

Lesson 28: Solve word problems with line plots.

Name _____   Date _____

1. The chart to the right shows the distance fourth graders in Ms. Smith's class were able to run before stopping for a rest. Create a line plot to display the data in the table.

Student	Distance (in miles)
Joe	$2\frac{1}{2}$
Arianna	$1\frac{3}{4}$
Bobbi	$2\frac{1}{8}$
Morgan	$1\frac{5}{8}$
Jack	$2\frac{5}{8}$
Saisha	$2\frac{1}{4}$
Tyler	$2\frac{2}{4}$
Jenny	$\frac{5}{8}$
Anson	$2\frac{2}{8}$
Chandra	$2\frac{4}{8}$

2. Solve each problem.
   a. Who ran a mile farther than Jenny?

   b. Who ran a mile less than Jack?

   c. Two students ran exactly $2\frac{1}{4}$ miles. Identify the students. How many quarter miles did each student run?

   d. What is the difference, in miles, between the longest and shortest distance run?

   e. Compare the distances run by Arianna and Morgan using >, <, or =.

   f. Ms. Smith ran twice as far as Jenny. How far did Ms. Smith run? Write her distance as a mixed number.

   g. Mr. Reynolds ran $1\frac{3}{10}$ miles. Use >, <, or = to compare the distance Mr. Reynolds ran to the distance that Ms. Smith ran. Who ran farther?

3. Using the information in the table and on the line plot, develop and write a question similar to those above. Solve, and then ask your partner to solve. Did you solve in the same way? Did you get the same answer?

A STORY OF UNITS   Lesson 28 Exit Ticket 4•5

Name _____   Date _____

Mr. O'Neil asked his students to record the length of time they read over the weekend. The times are listed in the table.

1. At the bottom of the page, make a line plot of the data.

2. One of the students read $\frac{3}{4}$ hour on Friday, $\frac{3}{4}$ hour on Saturday, and $\frac{3}{4}$ hour on Sunday. How many hours did that student read over the weekend? Name that student.

Student	Length of time (in hours)
Robin	$\frac{1}{2}$
Bill	1
Katrina	$\frac{3}{4}$
Kelly	$1\frac{3}{4}$
Mary	$1\frac{1}{2}$
Gail	$2\frac{1}{4}$
Scott	$1\frac{3}{4}$
Ben	$2\frac{2}{4}$

Lesson 28: Solve word problems with line plots.

A STORY OF UNITS

Lesson 28 Homework 4•5

Name _____ Date _____

1. A group of students measured the lengths of their shoes. The measurements are shown in the table. Make a line plot to display the data.

Student	Length of time (in hours)
Collin	$8\frac{1}{2}$
Dickon	$7\frac{3}{4}$
Ben	$7\frac{1}{2}$
Martha	$7\frac{3}{4}$
Lilias	8
Susan	$8\frac{1}{2}$
Frances	$7\frac{3}{4}$
Mary	$8\frac{3}{4}$

2. Solve each problem.

   a. Who has a shoe length 1 inch longer than Dickon's?

   b. Who has a shoe length 1 inch shorter than Susan's?

Lesson 28: Solve word problems with line plots.

383

c. How many quarter inches long is Martha's shoe length?

d. What is the difference, in inches, between Lilias's and Martha's shoe lengths?

e. Compare the shoe length of Ben and Frances using >, <, or =.

f. How many students had shoes that measured less than 8 inches?

g. How many students measured the length of their shoes?

h. Mr. Jones's shoe length was $\frac{25}{2}$ inches. Use >, <, or = to compare the length of Mr. Jones's shoe to the length of the longest student shoe length. Who had the longer shoe?

3. Using the information in the table and on the line plot, write a question you could solve by using the line plot. Solve.

A STORY OF UNITS

Mathematics Curriculum

GRADE 4 • MODULE 5

Topic F
Addition and Subtraction of Fractions by Decomposition

4.NF.3c, 4.MD.2

Focus Standard:	4.NF.3c	Understand a fraction a/b with $a > 1$ as a sum of fractions $1/b$. c. Add and subtract mixed numbers with like denominators, e.g., by replacing each mixed number with an equivalent fraction, and/or by using properties of operations and the relationship between addition and subtraction.
Instructional Days:	6	
Coherence -Links from:	G3–M5	Fractions as Numbers on the Number Line
-Links to:	G5–M3	Addition and Subtraction of Fractions

Topic F provides students with the opportunity to use their understandings of fraction addition and subtraction as they explore mixed number addition and subtraction by decomposition.

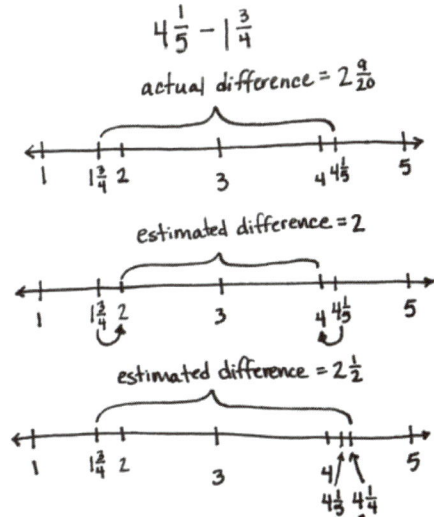

Lesson 29 focuses on the process of using benchmark numbers to estimate sums and differences of mixed numbers. Students once again call on their understanding of benchmark fractions as they determine, prior to performing the actual operation, what a reasonable outcome will be. One student might use benchmark whole numbers and reason, for example, that the difference between $4\frac{1}{5}$ and $1\frac{3}{4}$ is close to 2 because $4\frac{1}{5}$ is closer to 4 than 5, $1\frac{3}{4}$ is closer to 2 than 1, and the difference between 4 and 2 is 2. Another student might use familiar benchmark fractions and reason that the answer will be closer to $2\frac{1}{2}$ since $4\frac{1}{5}$ is about $\frac{1}{4}$ more than 4 and $1\frac{3}{4}$ is about $\frac{1}{4}$ less than 2, making the difference about a half more than 2, or $2\frac{1}{2}$.

In Lesson 30, students begin adding a mixed number to a fraction using unit form. They add like units, applying their Grades 1 and 2 understanding of completing a unit to add when the sum of the fractional units exceeds 1. Students ask, "How many more do we need to make one?" rather than "How many more do we need to make ten?" as was the case in Grade 1. A number bond decomposes the fraction to make one and can be modeled on the number line or using the arrow way, as shown to the right. Alternatively, a number bond can be used after adding like units, when the sum results in a mixed number with a fraction greater than 1, to decompose the fraction greater than 1 into ones and fractional units.

Directly applying what was learned in Lesson 30, Lesson 31 starts with adding like units, e.g., ones with ones and fourths with fourths, to add two mixed numbers. Students can, again, choose to make one before finding the sum or to decompose the sum to result in a proper mixed number.

Lessons 32 and 33 follow the same sequence for subtraction. In Lesson 32, students simply subtract a fraction from a mixed number, using three main strategies both when there are and there are not enough fractional units. They count back or up, subtract from 1, or take one out to subtract from 1. In Lesson 33, students apply these strategies after subtracting the ones first. They model subtraction of mixed numbers using a number line or the arrow way.

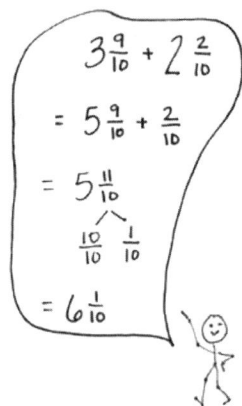

In Lesson 34, students learn another strategy for subtraction by decomposing the total into a whole number and a fraction greater than one to either subtract a fraction or a mixed number.

A STORY OF UNITS   Topic F 4•5

A Teaching Sequence Toward Mastery of Addition and Subtraction of Fractions by Decomposition

Objective 1: Estimate sums and differences using benchmark numbers.
(Lesson 29)

Objective 2: Add a mixed number and a fraction.
(Lesson 30)

Objective 3: Add mixed numbers.
(Lesson 31)

Objective 4: Subtract a fraction from a mixed number.
(Lesson 32)

Objective 5: Subtract a mixed number from a mixed number.
(Lesson 33)

Objective 6: Subtract mixed numbers.
(Lesson 34)

Lesson 29

Objective: Estimate sums and differences using benchmark numbers.

Suggested Lesson Structure

- ■ Fluency Practice   (12 minutes)
- ■ Application Problem   (3 minutes)
- ■ Concept Development   (35 minutes)
- ■ Student Debrief   (10 minutes)
- **Total Time**   **(60 minutes)**

Fluency Practice (12 minutes)

- Count by Equivalent Fractions **4.NF.1**   (6 minutes)
- Change Fractions to Mixed Numbers **4.NF.4**   (6 minutes)

Count by Equivalent Fractions (6 minutes)

Note: This activity reviews Lesson 24. The progression builds in complexity. Build students to the highest level of complexity in which they can confidently participate.

T: Count by twos to 16, starting at 0.
S: 0, 2, 4, 6, 8, 10, 12, 14, 16.
T: Count by 2 fourths to 16 fourths, starting at 0 fourths. (Write as students count.)

$\frac{0}{4}$	$\frac{2}{4}$	$\frac{4}{4}$	$\frac{6}{4}$	$\frac{8}{4}$	$\frac{10}{4}$	$\frac{12}{4}$	$\frac{14}{4}$	$\frac{16}{4}$
0	$\frac{2}{4}$	1	$\frac{6}{4}$	2	$\frac{10}{4}$	3	$\frac{14}{4}$	4
0	$\frac{2}{4}$	1	$1\frac{2}{4}$	2	$2\frac{2}{4}$	3	$3\frac{2}{4}$	4

S: $\frac{0}{4}, \frac{2}{4}, \frac{4}{4}, \frac{6}{4}, \frac{8}{4}, \frac{10}{4}, \frac{12}{4}, \frac{14}{4}, \frac{16}{4}$.
T: 1 is the same as how many fourths?
S: 4 fourths.
T: (Beneath $\frac{4}{4}$, write 1.)

A STORY OF UNITS   Lesson 29 4•5

Continue the process for 2, 3, and 4.

   T: Count by 2 fourths again. This time, when you come to the whole numbers, say the ones. Start at zero. (Write as students count.)
   S: $0, \frac{2}{4}, 1, \frac{6}{4}, 2, \frac{10}{4}, 3, \frac{14}{4}, 4$.
   T: (Point to $\frac{6}{4}$.) Say $\frac{6}{4}$ as a mixed number.
   S: $1\frac{2}{4}$.

Continue the process for $\frac{10}{4}$ and $\frac{14}{4}$.

   T: Count by 2 fourths again. This time, convert to whole numbers and mixed numbers. Start at zero. (Write as students count.)
   S: $0, \frac{2}{4}, 1, 1\frac{2}{4}, 2, 2\frac{2}{4}, 3, 3\frac{2}{4}, 4$.

Change Fractions to Mixed Numbers (6 minutes)

Materials: (S) Personal white board

Note: This fluency activity reviews Lesson 24.

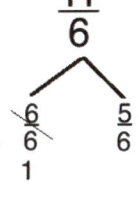

   T: (Write $\frac{11}{6}$.) Say the fraction.
   S: 11 sixths.
   T: (Draw a number bond with $\frac{11}{6}$ as the whole.) How many sixths are in 1?
   S: 6 sixths.
   T: (Write $\frac{6}{6}$ as a part. Write $\frac{\_}{6}$ as the other part.) Write the unknown part.
   S: (Write $\frac{5}{6}$ as the unknown part.)
   T: (Cross out $\frac{6}{6}$, and write 1 beneath it. Write $\frac{11}{6}$ = _____.) Write $\frac{11}{6}$ as a mixed number.
   S: (Write $\frac{11}{6} = 1\frac{5}{6}$.)

Continue with the following possible sequence: $\frac{17}{6}, \frac{15}{4}$, and $\frac{29}{8}$.

Application Problem (3 minutes)

Both Allison and Jennifer jogged on Sunday. When asked about their distances, Allison said, "I ran $2\frac{7}{8}$ miles this morning and $3\frac{3}{8}$ miles this afternoon. So, I ran a total of about 6 miles," and Jennifer said, "I ran $3\frac{1}{10}$ miles this morning and $3\frac{3}{10}$ miles this evening. I ran a total of $6\frac{4}{10}$ miles."

How do their answers differ? Discuss with your partner.

Lesson 29:   Estimate sums and differences using benchmark numbers.   389

A STORY OF UNITS   Lesson 29   4•5

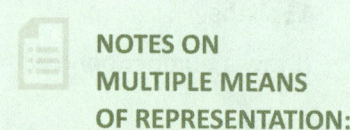

A [2⅞ | 3⅜] ≈ 6

J [3 1/10 | 3 3/10] 6 4/10

Their answers differ because Allison estimated how far she ran. She said "about 6." Jennifer gave an exact answer because she added.

Note: This Application Problem prepares students for today's Concept Development by prompting them to think about and discuss exact answers and estimates. Therefore, student conversations should include exact and approximate reflections.

Concept Development (35 minutes)

Materials: (T) 4-inch piece of string (S) Personal white board

Problem 1: Estimate the sum or difference of two mixed numbers by rounding each fraction.

T: What does it mean to estimate?

S: We don't find the exact answer. → We find numbers at about the same value that are easier to work with. → We find an answer that is close but not exact. → If we estimate, it doesn't have to be exact.

T: Write $3\frac{1}{5} + 4\frac{8}{9}$. Let's estimate the sum.

T: Round $3\frac{1}{5}$. Think about benchmark numbers.

S: $3\frac{1}{5}$ is close to 3. → It's a little bit more than 3. → It's $\frac{1}{5}$ more than 3. → I round down to 3.

T: Round $4\frac{8}{9}$.

S: $4\frac{8}{9}$ is close to 5. → It's a little less than 5. → It's $\frac{1}{9}$ less than 5. → I round up to 5.

T: Quickly show $3\frac{1}{5}$ and $4\frac{8}{9}$ on a number line with endpoints at 3 and 5, only marking whole numbers and the two addends.

S: (Construct and label a number line.)

T: Notice how close the mixed numbers are to the rounded numbers. What is the estimated sum?

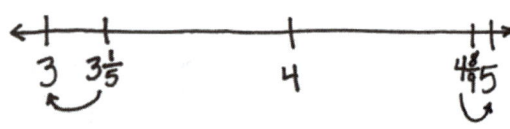

S: 3 + 5 = 8. Eight is our estimate.

T: What if we were to estimate the difference?

Lesson 29: Estimate sums and differences using benchmark numbers.

S: We would still round to 3 and 5 and subtract 3 from 5. The difference of $4\frac{8}{9}$ and $3\frac{1}{5}$ is about 2.

T: Talk to your partner: Will the actual difference be a little more than 2 or a little less than 2?

S: A little less because you can see from the number line that the difference is greater when we rounded. → A little less because the number line shows that the distance between $3\frac{1}{5}$ and $4\frac{8}{9}$ is less than 2.

Problem 2: Round two mixed numbers to the nearest half or whole number, and then find the sum.

T: Write $8\frac{9}{10} + 2\frac{4}{8}$. What's $8\frac{9}{10}$ rounded to the nearest one?

S: 9.

T: How about $2\frac{4}{8}$? Do we need to round $2\frac{4}{8}$?

S: No. $2\frac{4}{8}$ is the same as $2\frac{1}{2}$. Can I keep it as $2\frac{1}{2}$?

T: Yes. $9 + 2\frac{1}{2}$ is …?

S: It's just 11 and then another half—$11\frac{1}{2}$. → Well, I can think of 9 on a number line, and then I can picture adding two and a half more. Two more makes 11. → $11 + \frac{1}{2} = 11\frac{1}{2}$.

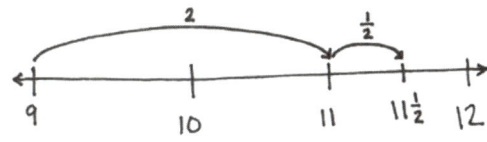

T: Why is your estimate greater than the actual sum? Talk to your partner.

S: It's greater because we rounded up $8\frac{9}{10}$. We made it bigger. → Our estimate is greater than the actual amount because we rounded 9 tenths up to 1. → We didn't round $2\frac{4}{8}$ at all, but we did round up $8\frac{9}{10}$ by $\frac{1}{10}$, so our actual answer will be $\frac{1}{10}$ less than our estmate.

Problem 3: Estimate the difference of two fractions greater than 1.

T: Write $\frac{15}{4}$ and $\frac{22}{7}$. What do you notice about these fractions?

S: They have different units. → They are more than 1.

T: Go ahead and convert each to a mixed number.

S: $\frac{15}{4} = \left(3 \times \frac{4}{4}\right) + \frac{3}{4} = 3\frac{3}{4}$, and $\frac{22}{7} = \left(3 \times \frac{7}{7}\right) + \frac{1}{7} = 3\frac{1}{7}$.

T: Round $3\frac{1}{7}$ to the nearest one. Round $3\frac{3}{4}$ to the nearest one.

S: 3. 4.

T: 4 – 3?

S: 1.

T: How else could you round to be more precise?

S: I could round $3\frac{3}{4}$ to $3\frac{1}{2}$ and $3\frac{1}{7}$ to 3. The estimated difference would be $\frac{1}{2}$.

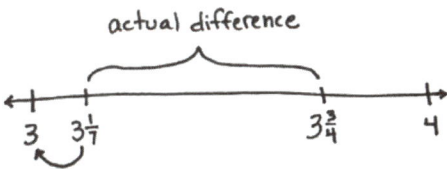

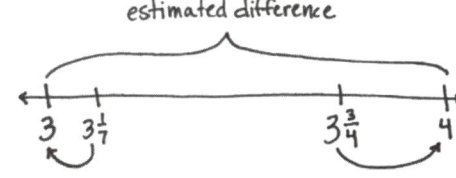

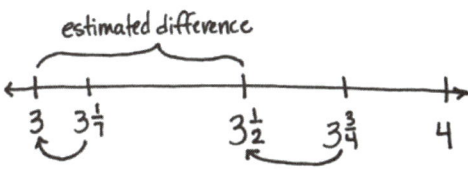

Lesson 29: Estimate sums and differences using benchmark numbers.

A STORY OF UNITS

Lesson 29 4•5

MP.4

T: Discuss with your partner. Which estimate is closer?

S: One-half is closer. I know that because I took a little away from $3\frac{3}{4}$ to get $3\frac{1}{2}$ and a little away from $3\frac{1}{7}$ to get 3. Taking away a little from each means the difference is almost the same. I can see that on a number line.

To verify that final statement (or to make it), take a string, and stretch it from $3\frac{1}{7}$ to $3\frac{3}{4}$ on the number line. Then, without adjusting its length at all, move it to the left to now match 3 and 3 and a half. The difference is about the same.

Problem 4: Use benchmark numbers or mental math to estimate the sum and difference of two mixed numbers.

T: (Write $18\frac{7}{12}$ and $17\frac{3}{8}$.) Estimate the sum using benchmark numbers or mental math. Discuss your strategy with a partner.

S: $18\frac{7}{12}$ is close to $18\frac{1}{2}$, and $17\frac{3}{8}$ is close to $17\frac{1}{2}$. I can add the whole numbers first to get 35. 2 halves make 1. 35 and 1 is 36. → $18\frac{1}{2} + 17\frac{1}{2} = 35 + \left(\frac{1}{2}+\frac{1}{2}\right) = 36$. The sum is about 36.

T: Now, estimate the difference of the same two numbers.

S: I can round to 19 and 17. → But that's rounding up and down, which makes the estimated difference bigger. Remember the example in the last problem? → I can just count up from $17\frac{1}{2}$ to $18\frac{1}{2}$, one. → There are two halves between them. Two halves make 1

Problem Set (10 minutes)

Students should do their personal best to complete the Problem Set within the allotted 10 minutes. For some classes, it may be appropriate to modify the assignment by specifying which problems they work on first. Some problems do not specify a method for solving. Students should solve these problems using the RDW approach used for Application Problems.

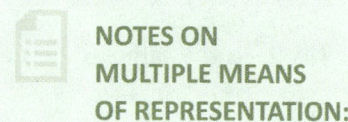

NOTES ON
MULTIPLE MEANS
OF REPRESENTATION:

Scaffold finding the sum and difference of $18\frac{7}{12}$ and $17\frac{3}{8}$ for students working below grade level by chunking. First, isolate the fractions. Guide students to find the benchmark closest to $\frac{7}{12}$. Then, reintroduce the whole numbers.

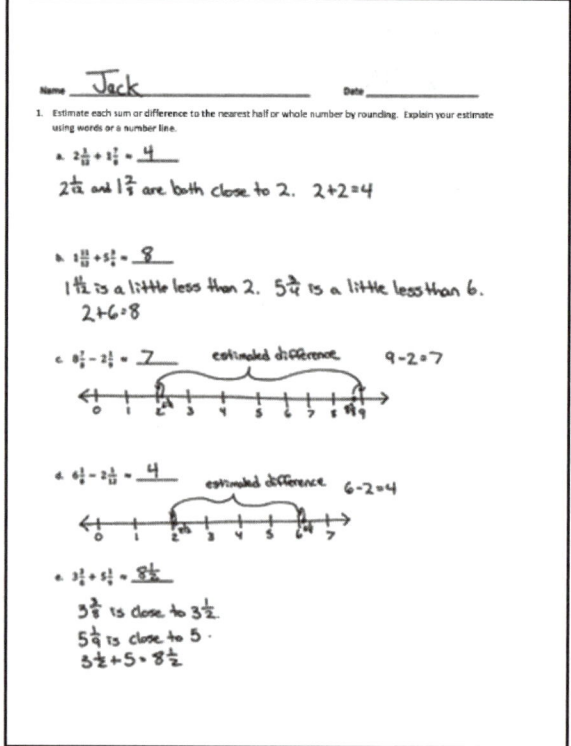

Lesson 29: Estimate sums and differences using benchmark numbers.

Student Debrief (10 minutes)

Lesson Objective: Estimate sums and differences using benchmark numbers.

The Student Debrief is intended to invite reflection and active processing of the total lesson experience.

Invite students to review their solutions for the Problem Set. They should check work by comparing answers with a partner before going over answers as a class. Look for misconceptions or misunderstandings that can be addressed in the Student Debrief. Guide students in a conversation to debrief the Problem Set and process the lesson.

Any combination of the questions below may be used to lead the discussion.

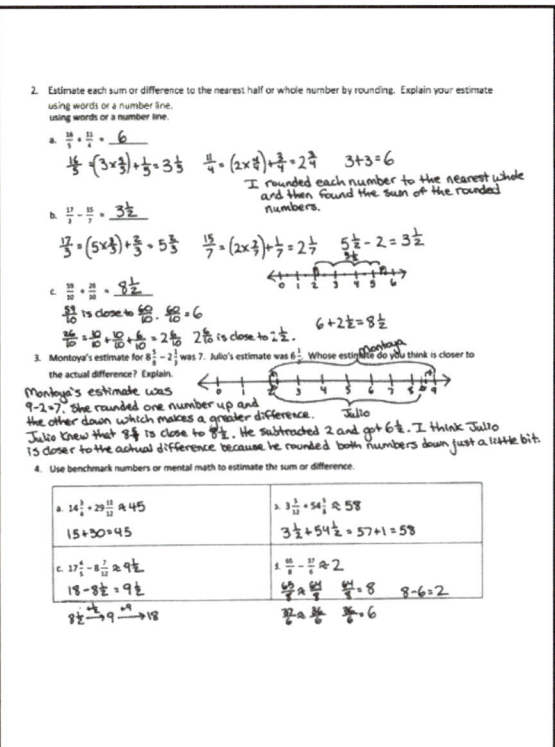

- In Problems 1(a) and 1(b), all fractions could be rounded up or down by one unit fraction. Which of the two estimates is closer to the actual amount?
- If one of the two fractions in Problem 1(a) was rounded down to half, the estimate would be more accurate than rounding both to the nearest one. How do you decide which fraction rounds up and which one rounds down?
- Did your partner have the same estimates as you in Problem 2? Why or why not? Whose estimates are closer to the actual answers?
- Think about Problem 3. When would estimates need to be very close to the actual answer? When might estimates be acceptable if the numbers were rounded to the closest whole number?
- Some students estimated 45 or $44\frac{3}{4}$ for Problem 4(a). Some students estimated 9 or $9\frac{1}{2}$ for Problem 4(c). Which answer for each problem is more reasonable? How does someone determine how accurate the answer is?
- What prior knowledge about fractions did you use as you completed the problems in the Problem Set?
- What tools did you use to help you estimate?

Exit Ticket (3 minutes)

After the Student Debrief, instruct students to complete the Exit Ticket. A review of their work will help with assessing students' understanding of the concepts that were presented in today's lesson and planning more effectively for future lessons. The questions may be read aloud to the students.

Name _____   Date _____

1. Estimate each sum or difference to the nearest half or whole number by rounding. Explain your estimate using words or a number line.

   a. $2\frac{1}{12} + 1\frac{7}{8} \approx$ _____

   b. $1\frac{11}{12} + 5\frac{3}{4} \approx$ _____

   c. $8\frac{7}{8} - 2\frac{1}{9} \approx$ _____

   d. $6\frac{1}{8} - 2\frac{1}{12} \approx$ _____

   e. $3\frac{3}{8} + 5\frac{1}{9} \approx$ _____

A STORY OF UNITS — Lesson 29 Problem Set 4•5

2. Estimate each sum or difference to the nearest half or whole number by rounding. Explain your estimate using words or a number line.

   a. $\dfrac{16}{5} + \dfrac{11}{4} \approx$ _____

   b. $\dfrac{17}{3} - \dfrac{15}{7} \approx$ _____

   c. $\dfrac{59}{10} + \dfrac{26}{10} \approx$ _____

3. Montoya's estimate for $8\dfrac{5}{8} - 2\dfrac{1}{3}$ was 7. Julio's estimate was $6\dfrac{1}{2}$. Whose estimate do you think is closer to the actual difference? Explain.

4. Use benchmark numbers or mental math to estimate the sum or difference.

a. $14\dfrac{3}{4} + 29\dfrac{11}{12}$	b. $3\dfrac{5}{12} + 54\dfrac{5}{8}$
c. $17\dfrac{4}{5} - 8\dfrac{7}{12}$	d. $\dfrac{65}{8} - \dfrac{37}{6}$

Lesson 29: Estimate sums and differences using benchmark numbers.

A STORY OF UNITS

Lesson 29 Exit Ticket 4•5

Name _____ Date _____

Estimate each sum or difference to the nearest half or whole number by rounding. Explain your estimate using words or a number line.

1. $2\frac{9}{10} + 2\frac{1}{4} \approx$ _____

2. $11\frac{8}{9} - 3\frac{3}{8} \approx$ _____

Lesson 29: Estimate sums and differences using benchmark numbers.

Name _____ Date _____

1. Estimate each sum or difference to the nearest half or whole number by rounding. Explain your estimate using words or a number line.

   a. $3\frac{1}{10} + 1\frac{3}{4} \approx$ _____

   b. $2\frac{9}{10} + 4\frac{4}{5} \approx$ _____

   c. $9\frac{9}{10} - 5\frac{1}{5} \approx$ _____

   d. $4\frac{1}{9} - 1\frac{1}{10} \approx$ _____

   e. $6\frac{3}{12} + 5\frac{1}{9} \approx$ _____

Lesson 29: Estimate sums and differences using benchmark numbers.

A STORY OF UNITS

Lesson 29 Homework 4•5

2. Estimate each sum or difference to the nearest half or whole number by rounding. Explain your estimate using words or a number line.

   a. $\frac{16}{3} + \frac{17}{8} \approx$ _____

   b. $\frac{17}{3} - \frac{15}{4} \approx$ _____

   c. $\frac{57}{8} + \frac{26}{8} \approx$ _____

3. Gina's estimate for $7\frac{5}{8} - 2\frac{1}{2}$ was 5. Dominick's estimate was $5\frac{1}{2}$. Whose estimate do you think is closer to the actual difference? Explain.

4. Use benchmark numbers or mental math to estimate the sum or difference.

a. $10\frac{3}{4} + 12\frac{11}{12}$	b. $2\frac{7}{10} + 23\frac{3}{8}$
c. $15\frac{9}{12} - 8\frac{11}{12}$	d. $\frac{56}{7} - \frac{31}{8}$

398   Lesson 29:   Estimate sums and differences using benchmark numbers.

Lesson 30

Objective: Add a mixed number and a fraction.

Suggested Lesson Structure

- ■ Fluency Practice   (12 minutes)
- ■ Application Problem (5 minutes)
- ■ Concept Development (33 minutes)
- ■ Student Debrief   (10 minutes)
- **Total Time**   **(60 minutes)**

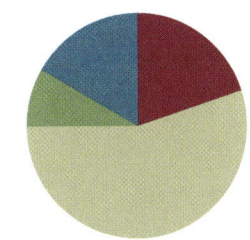

Fluency Practice (12 minutes)

- Sprint: Change Fractions to Mixed Numbers 4.NF.4   (8 minutes)
- Compare Fractions 4.NF.2   (4 minutes)

Sprint: Change Fractions to Mixed Numbers (8 minutes)

Materials: Change Fractions to Mixed Numbers Sprint

Note: This fluency activity reviews Lesson 24.

Compare Fractions (4 minutes)

Materials: (S) Personal white board

Note: This fluency activity reviews Lesson 26.

- T: (Write $\frac{19}{5}$.) How many ones are in 19 fifths?
- S: 3 ones.
- T: Between what two whole numbers is 19 fifths?
- S: 3 and 4.
- T: (Write $\frac{19}{5}$ ___ $\frac{12}{3}$.) Write a greater than or less than sign to compare the fractions.
- S: (Write $\frac{19}{6} < \frac{12}{3}$.)
- T: How do you know?
- S: $\frac{12}{3}$ equals 4. $\frac{19}{5}$ is between 3 and 4.

Continue with the following possible sequence: $\frac{25}{4}$ ___ $\frac{20}{5}$, $\frac{25}{4}$ ___ $\frac{26}{5}$, $\frac{26}{3}$ ___ $\frac{32}{4}$, and $\frac{26}{3}$ ___ $\frac{19}{2}$.

> **NOTES ON MULTIPLE MEANS OF ENGAGEMENT:**
>
> Consider preceding the Compare Fractions fluency activity with a counting by fifths, thirds, and fourths activity to increase student confidence and participation.

Lesson 30: Add a mixed number and a fraction.   399

A STORY OF UNITS

Lesson 30   4•5

Application Problem (5 minutes)

One board measures 2 meters 70 centimeters. Another measures 87 centimeters. What is the total length of the two boards expressed in meters and centimeters?

Solution A
2 m 70 cm + 87 cm = 2 m 100 cm + 57 cm
                         /\
                       30 cm 57 cm   = 3 m 57 cm

[Diagram: bar labeled L, divided into 2m 70cm and 87cm; L = 3m 57cm]

The total length of the two boards is 3m 57cm.

Solution B
2 m 70 cm + 87 cm = 2 m + 157 cm = 3 m 57 cm
                       /\
                   100 cm 57 cm

Note: This Application Problem anticipates the addition of a fraction and a mixed number using a measurement context. Solution A shows a solution whereby students decomposed 87 centimeters to complete the unit of one meter and added on the remaining centimeters. Solution B shows a solution whereby students added all of the centimeters and decomposed the sum.

Concept Development (33 minutes)

Materials: (S) Personal white board

Problem 1: Use unit form and the number line to add a mixed number and a fraction having sums of fractional units less than or equal to 1.

- T: Write $2\frac{3}{8} + \frac{3}{8}$.
- T: Say the expression using unit form.
- S: 2 ones 3 eighths + 3 eighths.
- T: What are the units involved in this problem?
- S: Ones and eighths.
- T: When we add numbers, we add like units. (Point to the mixed numbers and demonstrate.) How many ones are there in all?

2 ones 3 eighths + 3 eighths = 2 ones 6 eighths

- S: 2 ones.
- T: How many eighths are there in all?
- S: 6 eighths.
- T: 2 ones + 6 eighths is…?
- S:  $2\frac{6}{8}$.

> **NOTES ON MULTIPLE MEANS OF ENGAGEMENT:**
>
> English language learners and others may benefit from explicit instruction and additional practice speaking mixed numbers in unit language. If time is a consideration, prepare students beforehand to increase confidence and participation.

400   Lesson 30:   Add a mixed number and a fraction.

EUREKA MATH

A STORY OF UNITS   Lesson 30   4•5

T: Show the addition using a number line. Start at $2\frac{3}{8}$, and then add $\frac{3}{8}$ more. Notice how the ones stay the same and the fractional units are simply added together since their sum is less than 1.

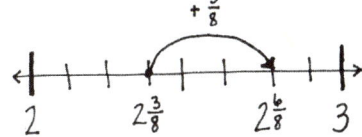

T: Write $2\frac{3}{8} + \frac{5}{8}$. Add like units. How many ones? How many eighths?

S: 2 ones and 8 eighths.

T: Show the addition using a number line. Start at $2\frac{3}{8}$. Add $\frac{5}{8}$ more.

S: Hey! When I add $\frac{5}{8}$ more, it equals 3.

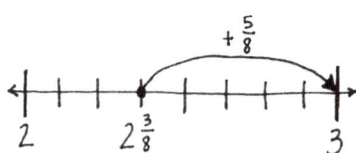

T: The fractional units have a sum of 1. $\frac{3}{8} + \frac{5}{8} = \frac{8}{8} = 1$.

Problem 2: Complete a unit of one to add a mixed number and a fraction.

T: To add fractional units, sometimes we complete a unit of 1. We look for fractions that have a sum of 1. If a fraction is equal to 1, what do we know about the numerator and denominator?

S: They are the same number.

T: (Write $\frac{1}{4}$.) How much more to make one?

S: $\frac{3}{4}$.

T: Explain.

S: To make a whole number with fourths, four parts are needed. 1 fourth + 3 fourths = 4 fourths.

T: Write $\frac{3}{8}$. What fraction can be added to make one or a unit of 1?

S: $\frac{5}{8}$.

T: Explain.

S: I think about 3 + ? = 8. The answer is 5. Since our units are eighths, the answer is 5 eighths.

T: Write $3\frac{1}{8}$. How many more eighths make one?

S: $\frac{7}{8}$.

T: How do you know?

S: $\frac{1}{8} + \frac{7}{8} = \frac{8}{8}$. $3 + \frac{8}{8} = 4$.

> **NOTES ON MULTIPLE MEANS OF REPRESENTATION:**
>
> To support English language learners and students working below grade level, couple the request of "How much more to make one?" with a tape diagram such as the following:
>
>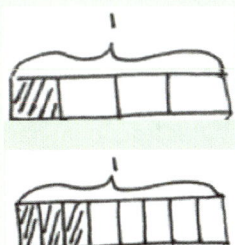
>
>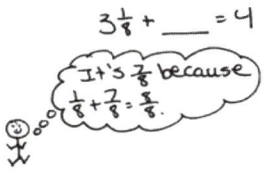

Lesson 30: Add a mixed number and a fraction.   401

A STORY OF UNITS   Lesson 30   4•5

T: Show this on a number line. Start at $3\frac{1}{8}$, and then add $\frac{7}{8}$ more.

Let students practice with the following: $4\frac{4}{5} +$ _____ $= 5$ and $6 = 5\frac{1}{8} +$ _____. Encourage them to solve mentally.

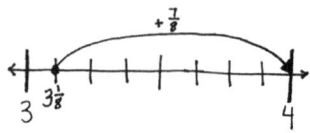

Problem 3: Decompose a sum of a mixed number and a fraction with sums of fractional units greater than 1.

T: (Write $5\frac{2}{4} + \frac{3}{4}$.) Right away, we see that the sum of the fourths is greater than 1.

T: The sum of the ones is...?

S: 5.

T: The sum of the fourths is...?

S: 5 fourths.

T: Decompose 5 fourths to make one. Use a number bond.

S: $\frac{5}{4} = \frac{4}{4} + \frac{1}{4}$.

T: (Write the following.)
$5\frac{2}{4} + \frac{3}{4} = 5 + \frac{5}{4} = 5 + \frac{4}{4} + \frac{1}{4} = 6\frac{1}{4}$.

$$5\frac{2}{4} + \frac{3}{4} = 5 + \frac{5}{4} = 6\frac{1}{4}$$
$$\diagup \diagdown$$
$$\frac{4}{4} \quad \frac{1}{4}$$

T: Explain to your partner how we got a sum of $6\frac{1}{4}$.

S: We added like units. We added ones to ones and fourths to fourths. We changed 5 fourths to make 1 and 1 fourth and added $5 + 1\frac{1}{4}$. The sum is $6\frac{1}{4}$.

Let students practice adding like units to find the sum using the following: $7\frac{2}{5} + \frac{4}{5}$ and $3\frac{5}{12} + 1\frac{11}{12}$.

Problem 4: Decompose a fractional addend to make one before finding the sum.

T: (Write $5\frac{2}{4} + \frac{3}{4}$.) We can also decompose to make one in the same way that we did earlier in the lesson.

T: What fractional part added to $5\frac{2}{4}$ makes the next whole number?

S: $\frac{2}{4}$.

T: Decompose $\frac{3}{4}$ into parts so that $\frac{2}{4}$ is one of the parts.

S: $\frac{3}{4} = \frac{2}{4} + \frac{1}{4}$.

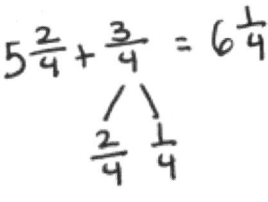

T: Write a number bond to show this. $5\frac{2}{4} + \frac{2}{4} = 6$. We add $\frac{1}{4}$ to 6 to get $6\frac{1}{4}$.

T: We can use the arrow way to show this clearly. Instead of drawing a number line, we can draw arrows to show the sum. $5\frac{2}{4} + \frac{2}{4} = 6$ and $6 + \frac{1}{4} = 6\frac{1}{4}$. Notice how we added each part of the number bond to find the total.

$$5\frac{2}{4} \xrightarrow{+\frac{2}{4}} 6 \xrightarrow{+\frac{1}{4}} 6\frac{1}{4}$$

Let students practice with the following: $3\frac{7}{8} + \frac{3}{8}$ and $9\frac{11}{12} + \frac{5}{12}$.

Lesson 30: Add a mixed number and a fraction.

A STORY OF UNITS   Lesson 30   4•5

Problem Set (10 minutes)

Students should do their personal best to complete the Problem Set within the allotted 10 minutes. For some classes, it may be appropriate to modify the assignment by specifying which problems they work on first. Some problems do not specify a method for solving. Students should solve these problems using the RDW approach used for Application Problems.

Student Debrief (10 minutes)

Lesson Objective: Add a mixed number and a fraction.

The Student Debrief is intended to invite reflection and active processing of the total lesson experience.

Invite students to review their solutions for the Problem Set. They should check work by comparing answers with a partner before going over answers as a class. Look for misconceptions or misunderstandings that can be addressed in the Student Debrief. Guide students in a conversation to debrief the Problem Set and process the lesson.

Any combination of the questions below may be used to lead the discussion.

- Explain how decomposing mixed numbers helps you to find their sum.
- Explain how you solved Problem 1(d).
- Explain the challenge in solving Problem 4(d). What strategy did you use?
- If you were unsure of any answer on this Problem Set, what could you do to see if your answer is reasonable? Would drawing a picture or estimating the sum or difference be helpful?
- How does Problem 4(g) relate to the Application Problem?

Exit Ticket (3 minutes)

After the Student Debrief, instruct students to complete the Exit Ticket. A review of their work will help with assessing students' understanding of the concepts that were presented in today's lesson and planning more effectively for future lessons. The questions may be read aloud to the students.

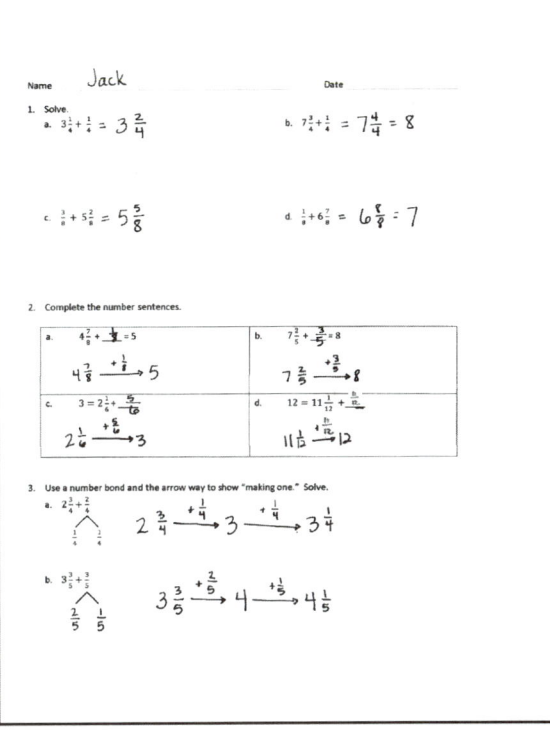

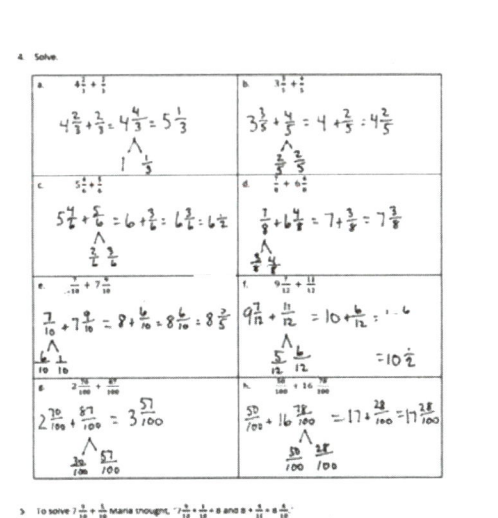

Lesson 30: Add a mixed number and a fraction.   403

A

Number Correct: _____

Change Fractions to Mixed Numbers

1.	$3 = 2 + \_\_\_$	
2.	$\frac{3}{2} = \frac{2}{2} + \frac{}{2}$	
3.	$\frac{3}{2} = 1 + \frac{}{2}$	
4.	$\frac{3}{2} = 1\frac{}{2}$	
5.	$5 = 4 + \_\_\_$	
6.	$\frac{5}{4} = \frac{4}{4} + \frac{}{4}$	
7.	$\frac{5}{4} = 1 + \frac{}{4}$	
8.	$\frac{5}{4} = 1\frac{}{4}$	
9.	$4 = \_\_\_ + 1$	
10.	$\frac{4}{3} = \frac{}{3} + \frac{1}{3}$	
11.	$\frac{4}{3} = 1 + \frac{}{3}$	
12.	$\frac{4}{3} = \frac{}{}\frac{1}{3}$	
13.	$7 = \_\_\_ + 2$	
14.	$\frac{7}{5} = \frac{}{5} + \frac{2}{5}$	
15.	$\frac{7}{5} = 1 + \frac{}{5}$	
16.	$\frac{7}{5} = 1\frac{}{5}$	
17.	$\frac{8}{5} = 1\frac{}{5}$	
18.	$\frac{9}{5} = 1\frac{}{5}$	
19.	$\frac{6}{5} = 1\frac{}{5}$	
20.	$\frac{10}{5} =$	
21.	$\frac{}{5} = \frac{10}{5} + \frac{1}{5}$	
22.	$\frac{}{5} = 2 + \frac{1}{5}$	

23.	$\frac{6}{3} =$	
24.	$\frac{}{3} = \frac{6}{3} + \frac{2}{3}$	
25.	$\frac{8}{3} = \frac{6}{3} + \frac{}{3}$	
26.	$\frac{8}{3} = 2 + \frac{}{3}$	
27.	$\frac{8}{3} = 2\frac{}{3}$	
28.	$\frac{}{4} = \frac{8}{4} + \frac{1}{4}$	
29.	$\frac{}{4} = 2 + \frac{1}{4}$	
30.	$\frac{9}{4} = \frac{}{}\frac{1}{4}$	
31.	$\frac{11}{4} = \frac{}{}\frac{3}{4}$	
32.	$\frac{8}{3} = \frac{}{3} + \frac{2}{3}$	
33.	$\frac{8}{3} = \frac{6}{3} + \frac{}{3}$	
34.	$\frac{8}{3} = \_\_\_ + \frac{2}{3}$	
35.	$\frac{8}{3} = \frac{}{}\frac{2}{3}$	
36.	$\frac{14}{5} = \frac{10}{5} + \frac{}{5}$	
37.	$\frac{14}{5} = \_\_\_ + \frac{4}{5}$	
38.	$\frac{14}{5} = 2\frac{}{5}$	
39.	$\frac{13}{5} = 2\frac{}{5}$	
40.	$\frac{9}{8} = 1 + \frac{}{8}$	
41.	$\frac{15}{8} = 1 + \frac{}{8}$	
42.	$\frac{17}{12} = \frac{}{12} + \frac{5}{12}$	
43.	$\frac{11}{8} = 1 + \frac{}{8}$	
44.	$\frac{17}{12} = 1 + \frac{}{12}$	

B

Change Fractions to Mixed Numbers

1.	6 = 5 + ___	
2.	$\frac{6}{5} = \frac{5}{5} + \frac{}{5}$	
3.	$\frac{6}{5} = 1 + \frac{}{5}$	
4.	$\frac{6}{5} = 1\frac{}{5}$	
5.	4 = 3 + ___	
6.	$\frac{4}{3} = \frac{3}{3} + \frac{}{3}$	
7.	$\frac{4}{3} = 1 + \frac{}{3}$	
8.	$\frac{4}{3} = 1\frac{}{3}$	
9.	5 = ___ + 1	
10.	$\frac{5}{4} = \frac{}{4} + \frac{1}{4}$	
11.	$\frac{5}{4} = 1 + \frac{}{4}$	
12.	$\frac{5}{4} = \frac{}{4}\frac{1}{4}$	
13.	8 = ___ + 3	
14.	$\frac{8}{5} = \frac{}{5} + \frac{3}{5}$	
15.	$\frac{8}{5} = 1 + \frac{}{5}$	
16.	$\frac{8}{5} = 1\frac{}{5}$	
17.	$\frac{9}{5} = 1\frac{}{5}$	
18.	$\frac{6}{5} = 1\frac{}{5}$	
19.	$\frac{7}{5} = 1\frac{}{5}$	
20.	$\frac{6}{3} =$	
21.	$\frac{}{3} = \frac{6}{3} + \frac{1}{3}$	
22.	$\frac{}{3} = 2 + \frac{1}{3}$	
23.	$\frac{4}{2} =$	
24.	$\frac{}{2} = \frac{4}{2} + \frac{1}{2}$	
25.	$\frac{5}{2} = \frac{4}{2} + \frac{}{2}$	
26.	$\frac{5}{2} = 2 + \frac{}{2}$	
27.	$\frac{5}{2} = 2\frac{}{2}$	
28.	$\frac{}{5} = \frac{10}{5} + \frac{1}{5}$	
29.	$\frac{}{5} = 2 + \frac{1}{5}$	
30.	$\frac{11}{5} = \underline{}\frac{1}{5}$	
31.	$\frac{13}{5} = \underline{}\frac{3}{5}$	
32.	$\frac{5}{3} = \frac{}{3} + \frac{1}{3}$	
33.	$\frac{5}{2} = \frac{4}{2} + \frac{}{2}$	
34.	$\frac{5}{2} = \underline{} + \frac{1}{2}$	
35.	$\frac{5}{2} = \underline{}\frac{1}{2}$	
36.	$\frac{12}{5} = \frac{10}{5} + \frac{}{5}$	
37.	$\frac{12}{5} = \underline{} + \frac{2}{5}$	
38.	$\frac{12}{5} = 2\frac{}{5}$	
39.	$\frac{14}{5} = 2\frac{}{5}$	
40.	$\frac{9}{8} = 1 + \frac{}{8}$	
41.	$\frac{11}{8} = 1 + \frac{}{8}$	
42.	$\frac{19}{12} = \frac{}{12} + \frac{7}{12}$	
43.	$\frac{15}{8} = 1 + \frac{}{8}$	
44.	$\frac{19}{12} = 1 + \frac{}{12}$	

Name _____   Date _____

1. Solve.

    a. $3\frac{1}{4} + \frac{1}{4}$

    b. $7\frac{3}{4} + \frac{1}{4}$

    c. $\frac{3}{8} + 5\frac{2}{8}$

    d. $\frac{1}{8} + 6\frac{7}{8}$

2. Complete the number sentences.

a.	$4\frac{7}{8} + \_\_\_ = 5$		b.	$7\frac{2}{5} + \_\_\_ = 8$
c.	$3 = 2\frac{1}{6} + \_\_\_$		d.	$12 = 11\frac{1}{12} + \_\_\_$

3. Use a number bond and the arrow way to show how to make one. Solve.

    a. $2\frac{3}{4} + \frac{2}{4}$

    $\frac{1}{4}\quad \frac{1}{4}$

    b. $3\frac{3}{5} + \frac{3}{5}$

Lesson 30: Add a mixed number and a fraction.

Lesson 30 Problem Set 4•5

4. Solve.

a.	$4\frac{2}{3} + \frac{2}{3}$	b.	$3\frac{3}{5} + \frac{4}{5}$
c.	$5\frac{4}{6} + \frac{5}{6}$	d.	$\frac{7}{8} + 6\frac{4}{8}$
e.	$\frac{7}{10} + 7\frac{9}{10}$	f.	$9\frac{7}{12} + \frac{11}{12}$
g.	$2\frac{70}{100} + \frac{87}{100}$	h.	$\frac{50}{100} + 16\frac{78}{100}$

Lesson 30: Add a mixed number and a fraction.

5. To solve $7\frac{9}{10} + \frac{5}{10}$, Maria thought, "$7\frac{9}{10} + \frac{1}{10} = 8$ and $8 + \frac{4}{10} = 8\frac{4}{10}$."

   Paul thought, "$7\frac{9}{10} + \frac{5}{10} = 7\frac{14}{10} = 7 + \frac{10}{10} + \frac{4}{10} = 8\frac{4}{10}$." Explain why Maria and Paul are both right.

Name _____ Date _____

Solve.

1. $3\frac{2}{5} +$ _____ $= 4$

2. $2\frac{3}{8} + \frac{7}{8}$

A STORY OF UNITS　　　　　　　　　　　　　　　　Lesson 30 Homework　4•5

Name _____　Date _____

1. Solve.

   a. $4\frac{1}{3} + \frac{1}{3}$

   b. $5\frac{1}{4} + \frac{2}{4}$

   c. $\frac{2}{6} + 3\frac{4}{6}$

   d. $\frac{5}{8} + 7\frac{3}{8}$

2. Complete the number sentences.

a. $3\frac{5}{6} +$ _____ $= 4$	b. $5\frac{3}{7} +$ _____ $= 6$
c. $5 = 4\frac{1}{8} +$ _____	d. $15 = 14\frac{4}{12} +$ _____

3. Draw a number bond and the arrow way to show how to make one. Solve.

   a. $2\frac{4}{5} + \frac{2}{5}$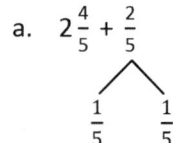

   b. $3\frac{2}{3} + \frac{2}{3}$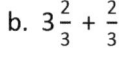

   c. $4\frac{4}{6} + \frac{5}{6}$

   $2\frac{4}{5} \xrightarrow{+\frac{1}{5}} 3 \xrightarrow{+\frac{1}{5}} 3\frac{1}{5}$

410　　Lesson 30:　Add a mixed number and a fraction.

4. Solve.

a.	$2\frac{3}{5} + \frac{3}{5}$	b.	$3\frac{6}{8} + \frac{4}{8}$
c.	$5\frac{4}{6} + \frac{3}{6}$	d.	$\frac{7}{10} + 6\frac{6}{10}$
e.	$\frac{5}{10} + 8\frac{9}{10}$	f.	$7\frac{8}{12} + \frac{11}{12}$
g.	$3\frac{90}{100} + \frac{58}{100}$	h.	$\frac{60}{100} + 14\frac{79}{100}$

Lesson 30: Add a mixed number and a fraction.

5. To solve $4\frac{8}{10} + \frac{3}{10}$, Carmen thought, "$4\frac{8}{10} + \frac{2}{10} = 5$, and $5 + \frac{1}{10} = 5\frac{1}{10}$."
Benny thought, "$4\frac{8}{10} + \frac{3}{10} = 4\frac{11}{10} = 4 + \frac{10}{10} + \frac{1}{10} = 5\frac{1}{10}$." Explain why Carmen and Benny are both right.

Lesson 31

Objective: Add mixed numbers.

Suggested Lesson Structure

- ■ Fluency Practice (12 minutes)
- ■ Application Problem (5 minutes)
- ■ Concept Development (33 minutes)
- ■ Student Debrief (10 minutes)

Total Time **(60 minutes)**

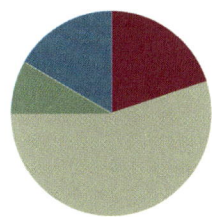

Fluency Practice (12 minutes)

- Sprint: Change Fractions to Mixed Numbers 4.NF.4 (8 minutes)
- Compare Fractions 4.NF.2 (4 minutes)

Sprint: Change Fractions to Mixed Numbers (8 minutes)

Materials: (S) Change Fractions to Mixed Numbers Sprint

Note: This fluency activity reviews Lesson 24.

Compare Fractions (4 minutes)

Materials: (S) Personal white board

Note: This fluency activity reviews Lesson 26.

- T: (Write $\frac{26}{6}$.) How many ones are in 26 sixths?
- S: 4 ones.
- T: Between what two whole numbers is 26 sixths?
- S: 4 and 5.
- T: (Write $\frac{26}{6}$ __ $\frac{20}{5}$.) Write a greater than or less than sign to compare the fractions.
- S: (Write $\frac{26}{6} > \frac{20}{5}$.)
- T: How do you know?
- S: $\frac{20}{5}$ equals 4. $\frac{26}{6}$ is between 4 and 5.

> **NOTES ON MULTIPLE MEANS OF ENGAGEMENT:**
>
> Consider preceding the Compare Fractions fluency activity with a counting by fifths, thirds, and fourths activity to increase student confidence and participation.

Continue with the following possible sequence: $\frac{31}{5}$ __ $\frac{24}{6}, \frac{31}{5}$ __ $\frac{28}{4}, \frac{65}{8}$ __ $\frac{48}{6}$, and $\frac{57}{8}$ __ $\frac{23}{3}$.

A STORY OF UNITS — Lesson 31 4•5

Application Problem (5 minutes)

Marta has 2 meters 80 centimeters of cotton cloth and 3 meters 87 centimeters of linen cloth. What is the total length of both pieces of cloth?

Solution A
2m 80cm + 3m 87cm = 3m + 3m 67cm
       /\
   20cm 67cm   = 6m 67cm

[tape diagram labeled L with 2m 80cm | 3m 87cm]

Solution B
2m 80cm + 3m 87cm = 5m + 167cm
                        /\
                   100cm 67cm
                              = 6m 67cm

L = 6m 67cm

The total length of both pieces of cloth is 6m 67cm.

Note: This Application Problem anticipates the adding of two mixed numbers using a measurement context. Solution A shows a solution whereby students decomposed 87 centimeters to complete the unit of one meter and added on the remaining centimeters. Solution B shows a solution whereby students added all of the centimeters and decomposed the sum.

Concept Development (33 minutes)

Materials: (S) Personal white board

Problem 1: Add mixed numbers combining like units.

MP.2

T: Write $2\frac{1}{8} + 1\frac{5}{8}$. Let's find the sum.
T: Say the expression using unit form.
S: 2 ones 1 eighth plus 1 one 5 eighths.
T: What are the units involved in this problem?
S: Ones and eighths.
T: When we add numbers, we add like units. (Point to the mixed numbers, and demonstrate.) How many ones are there in all?
S: 3 ones.
T: How many eighths are there in all?
S: 6 eighths.
T: 3 ones + 6 eighths is…?
S: $3\frac{6}{8}$.
T: (Write $2\frac{3}{4} + 3\frac{1}{4} = 2 + \frac{3}{4} + 3 + \frac{1}{4}$. Pause to allow students to analyze.) From our previous work, we know $2\frac{3}{4} + 3\frac{1}{4} = 2 + \frac{3}{4} + 3 + \frac{1}{4}$. True?

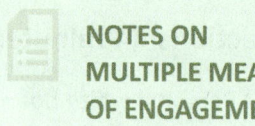

NOTES ON MULTIPLE MEANS OF ENGAGEMENT:

English language learners and others may benefit from explicit instruction and additional practice speaking mixed numbers in unit language. If time is a consideration, prepare students beforehand to increase confidence and participation.

S: Yes!

T: So, we do *not* have to write all that down.

T: The sum of the ones is...?

S: 5 ones.

T: The sum of the fourths is...?

S: 4 fourths.

T: (Write.)
$$2\frac{3}{4} + 3\frac{1}{4} = 5 + \frac{4}{4}$$
$$= 6.$$

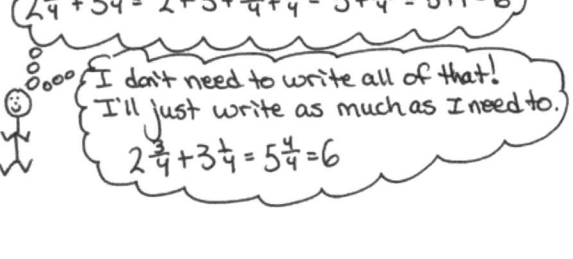

T: Explain to your partner how we got a sum of 6.

S: Easy. 2 ones and 3 ones is 5. Three fourths and 1 fourth is one. Five and 1 makes 6. → We just put the like units together, ones with ones and fourths with fourths. → Add the ones. Add the fractions.

Let students practice with the following: $5\frac{1}{3} + 6\frac{2}{3}$ and $21\frac{2}{5} + 10\frac{2}{5}$. Encourage them to write the solution using the shorter recording method, e.g., $5\frac{1}{3} + 6\frac{2}{3} = 11 + \frac{3}{3} = 12$ and $21\frac{2}{5} + 10\frac{2}{5} = 31 + \frac{4}{5} = 31\frac{4}{5}$, but if they must decompose each addend as a sum, let them. Encourage them to think in terms of what fractional part gets them to the next whole number.

Problem 2: Add mixed numbers when the sum of the fractional units is greater than 1 by combining like units.

T: (Write $2\frac{5}{8} + 3\frac{5}{8}$.) Right away, we see that the sum of the eighths is greater than 1.

T: The sum of the ones is...?

S: 5.

T: The sum of the eighths is...?

S: 10 eighths.

T: Take out 8 eighths to make one.

S: $1\frac{2}{8}$. → $\frac{8}{8}$ and $\frac{2}{8}$. (Record with a number bond.)

T: (Write the following.)
$$2\frac{5}{8} + 3\frac{5}{8} = 5 + \frac{10}{8}$$
$$= 5 + \frac{8}{8} + \frac{3}{8}$$
$$= 6\frac{2}{8}$$

T: Explain to your partner how we got a sum of $6\frac{2}{8}$.

S: We added like units. We added ones to ones and eighths to eighths. Then, we changed 10 eighths to make 1 and 2 eighths and added $5 + 1\frac{2}{8} = 6\frac{2}{8}$.

T: Use a number line to model the addition of like units.

A STORY OF UNITS　　　　　　　　　　　　　　　　　　　　　　　　　　　　Lesson 31　4•5

Students may show slides on the number line in different ways depending on their fluency with the addition of like units. Accept representations that are logical and follow the path of the number sentence.

Two samples are shown.

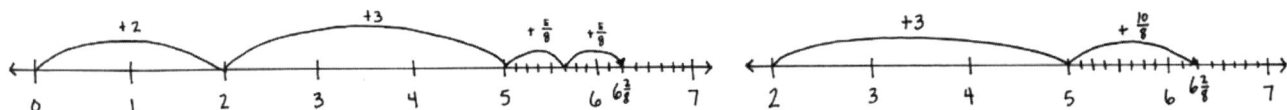

Let students practice with the following: $2\frac{2}{5} + 2\frac{4}{5}$ and $3\frac{5}{12} + 1\frac{11}{12}$. Allow students to work mentally to solve, if they can, without recording the breakdown of steps.

Problem 3: Add mixed numbers when the sum of the fractional units is greater than 1 by making one.

T: (Write $5\frac{5}{8} + 6\frac{5}{8}$.) We can also add the ones first and decompose to make one in the same way we learned to make ten in the first and second grades.

T: 5 and 6 is…?

S: 11.

T: (Write $11\frac{5}{8} + \frac{5}{8}$.) How much does 5 eighths need to make one?

S: 3 eighths. (Decompose $\frac{5}{8}$ as $\frac{3}{8}$ and $\frac{2}{8}$ as shown to the right.)

T: We can use the arrow way to show this clearly. Instead of drawing a number line, we can draw arrows to show the sum. $11\frac{5}{8} + \frac{3}{8}$ is…? (Model the arrow way while speaking.)

S: 12. (Record 12, and draw the next arrow.)

T: $12 + \frac{2}{8}$ is…? (Record as modeled to the right.)

S: $12\frac{2}{8}$.

T: $5\frac{5}{8} + 6\frac{5}{8} = 11\frac{5}{8} + \frac{5}{8}$

$= 11\frac{5}{8} + \frac{3}{8} + \frac{2}{8}$

$= 12\frac{2}{8}$

Let students practice with $3\frac{7}{8} + 4\frac{3}{8}$ and $9\frac{11}{12} + 10\frac{5}{12}$. Again, students may want to add more steps in the recording, e.g., $5\frac{5}{8} + 6\frac{5}{8} = 11\frac{5}{8} + \frac{5}{8} = 11\frac{8}{8} + \frac{2}{8} = 12\frac{2}{8}$. Gently encourage them to stop recording the steps they are able to easily complete mentally.

T: (Write $4\frac{2}{3} + 3\frac{1}{3} + 5\frac{2}{3}$.) The sum of the ones is…?

S: 12.

T: The sum of the thirds is…?

S: 5 thirds.

T: Record your work.

S: $4\frac{2}{3} + 3\frac{1}{3} + 5\frac{2}{3} = 12 + \frac{5}{3}$

$= 13\frac{2}{3}.$

Please note that this is not the only way to record this sum. Students might break the problem down into more or fewer steps, use a number bond, or do mental math.

Problem 4: Record the addition of mixed numbers.

T: How much you write down of your calculation is up to you. Some of you may write down each step in a detailed way, while others may do a lot of the work mentally. Write down what you need to so that you can keep track of the problem. At times, I write down more than at other times depending on the problem and even on my mood.

T: (Write $4\frac{7}{12} + 16\frac{9}{12}$.) Solve this problem. The goal is to write down only as much as necessary.

Below are some different recordings that might be seen. Students vary in their ability to do mental math. Be mindful that some students may think they are doing more math by writing as much as possible. Work to bring thoughtfulness to each student's experience. Be sure to check periodically to make sure that students can explain their thinking.

$4\frac{7}{12} + 16\frac{9}{12} = 21\frac{4}{12}$

$4\frac{7}{12} + 16\frac{9}{12} = 20 + \frac{16}{12}$
$= 20 + 1 + \frac{4}{12}$
$= 21\frac{4}{12}$

$4\frac{7}{12} + 16\frac{9}{12} = 20\frac{7}{12} + \frac{9}{12}$

$\frac{5}{12} \quad \frac{4}{12}$

$= 21\frac{4}{12}$

T: The ssum is…?

S: $21\frac{4}{12}$.

T: Share your way of recording with a partner. If you did your work mentally, explain to your partner how you did it. Did you use the same strategies or different strategies? Remember to use the strategy that makes the most sense to you.

Problem Set (10 minutes)

Students should do their personal best to complete the Problem Set within the allotted 10 minutes. For some classes, it may be appropriate to modify the assignment by specifying which problems they work on first. Some problems do not specify a method for solving. Students should solve these problems using the RDW approach used for Application Problems.

A STORY OF UNITS Lesson 31 4•5

Student Debrief (10 minutes)

Lesson Objective: Add mixed numbers.

The Student Debrief is intended to invite reflection and active processing of the total lesson experience.

Invite students to review their solutions for the Problem Set. They should check work by comparing answers with a partner before going over answers as a class. Look for misconceptions or misunderstandings that can be addressed in the Student Debrief. Guide students in a conversation to debrief the Problem Set and process the lesson.

Any combination of the questions below may be used to lead the discussion.

- Explain how decomposing mixed numbers helps you find their sum.
- Explain how you solved Problem 1(c).
- Explain the methods you chose for solving Problems 4(a), 4(b), and 4(c). Did you use the same methods as your partner?
- How is adding 4 tens 7 ones and 6 tens 9 ones like adding 4 ones 7 twelfths and 6 ones 9 twelfths? How is it different?
- If you were unsure of any answer on this Problem Set, what could you do to see if your answer is reasonable? Would drawing a picture or estimating the sum or difference be helpful?
- How did the Application Problem connect to today's lesson?

Exit Ticket (3 minutes)

After the Student Debrief, instruct students to complete the Exit Ticket. A review of their work will help with assessing students' understanding of the concepts that were presented in today's lesson and planning more effectively for future lessons. The questions may be read aloud to the students.

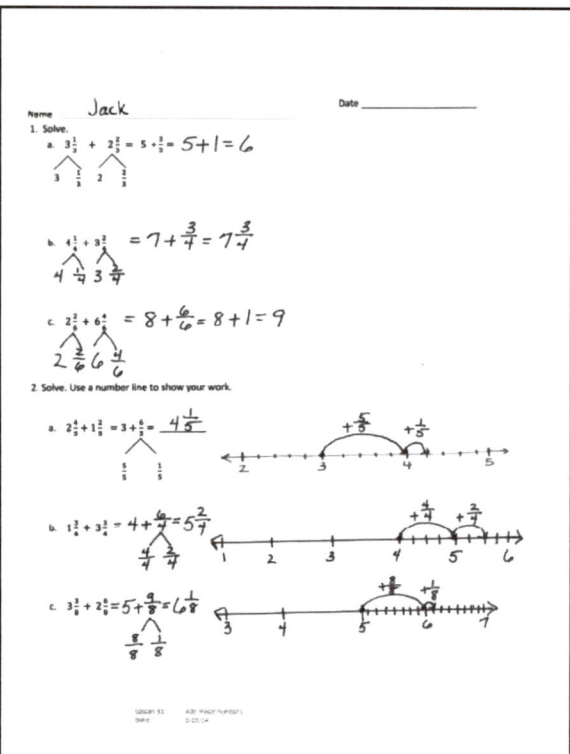

418   Lesson 31:   Add mixed numbers.

A

Number Correct: _____

Change Fractions to Mixed Numbers

1.	$3 + 1 =$		23.	$1\frac{3}{8} = \frac{}{8}$	
2.	$\frac{3}{3} + \frac{1}{3} = \frac{}{3}$		24.	$2 + \frac{1}{3} = 2\frac{}{3}$	
3.	$1 + \frac{1}{3} = \frac{}{3}$		25.	$\frac{6}{3} + \frac{1}{3} = \frac{}{3}$	
4.	$1\frac{1}{3} = \frac{}{3}$		26.	$2 + \frac{1}{3} = \frac{}{3}$	
5.	$5 + 1 =$		27.	$2\frac{1}{3} = \frac{}{3}$	
6.	$\frac{5}{5} + \frac{1}{5} = \frac{}{5}$		28.	$2 + \frac{1}{5} = 2\frac{}{5}$	
7.	$1 + \frac{1}{5} = \frac{}{5}$		29.	$\frac{10}{5} + \frac{1}{5} = \frac{}{5}$	
8.	$1\frac{1}{5} = \frac{}{5}$		30.	$2 + \frac{1}{5} = \frac{}{5}$	
9.	$2 + 1 =$		31.	$2\frac{1}{5} = \frac{}{5}$	
10.	$\frac{2}{2} + \frac{1}{2} = \frac{}{2}$		32.	$\frac{8}{4} + \frac{3}{4} = \frac{}{4}$	
11.	$1 + \frac{1}{2} = \frac{}{2}$		33.	$2 + \frac{3}{4} = \frac{}{4}$	
12.	$1\frac{1}{2} = \frac{}{2}$		34.	$2\frac{3}{4} = \frac{}{4}$	
13.	$\frac{4}{4} + \frac{1}{4} = \frac{}{4}$		35.	$\frac{12}{3} + \frac{2}{3} = \frac{}{3}$	
14.	$1 + \frac{1}{4} = \frac{}{4}$		36.	$4 + \frac{2}{3} = \frac{}{3}$	
15.	$1\frac{1}{4} = \frac{}{4}$		37.	$4\frac{2}{3} = \frac{}{3}$	
16.	$1\frac{3}{4} = \frac{}{4}$		38.	$3 + \frac{3}{5} = \frac{}{5}$	
17.	$\frac{5}{5} + \frac{1}{5} = \frac{}{5}$		39.	$3 + \frac{1}{2} = \frac{}{2}$	
18.	$1 + \frac{1}{5} = \frac{}{5}$		40.	$4 + \frac{3}{4} = \frac{}{4}$	
19.	$1\frac{1}{5} = \frac{}{5}$		41.	$2 + \frac{1}{6} = \frac{}{6}$	
20.	$1\frac{3}{5} = \frac{}{5}$		42.	$2 + \frac{5}{8} = \frac{}{8}$	
21.	$\frac{8}{8} + \frac{3}{8} = \frac{}{8}$		43.	$2\frac{4}{5} = \frac{}{5}$	
22.	$1 + \frac{3}{8} = \frac{}{8}$		44.	$3\frac{7}{8} = \frac{}{8}$	

B

Change Fractions to Mixed Numbers

Number Correct: _____

Improvement: _____

1.	$4 + 1 =$		23.	$1\frac{5}{8} = \frac{}{8}$	
2.	$\frac{4}{4} + \frac{1}{4} = \frac{}{4}$		24.	$2 + \frac{1}{2} = 2\frac{}{2}$	
3.	$1 + \frac{1}{4} = \frac{}{4}$		25.	$\frac{4}{2} + \frac{1}{2} = \frac{}{2}$	
4.	$1\frac{1}{4} = \frac{}{4}$		26.	$2 + \frac{1}{2} = \frac{}{2}$	
5.	$2 + 1 =$		27.	$2\frac{1}{2} = \frac{}{2}$	
6.	$\frac{2}{2} + \frac{1}{2} = \frac{}{2}$		28.	$2 + \frac{1}{4} = 2\frac{}{4}$	
7.	$1 + \frac{1}{2} = \frac{}{2}$		29.	$\frac{8}{4} + \frac{1}{4} = \frac{}{4}$	
8.	$1\frac{1}{2} = \frac{}{2}$		30.	$2 + \frac{1}{4} = \frac{}{4}$	
9.	$5 + 1 =$		31.	$2\frac{1}{4} = \frac{}{4}$	
10.	$\frac{5}{5} + \frac{1}{5} = \frac{}{5}$		32.	$\frac{6}{3} + \frac{2}{3} = \frac{}{3}$	
11.	$1 + \frac{1}{5} = \frac{}{5}$		33.	$2 + \frac{2}{3} = \frac{}{3}$	
12.	$1\frac{1}{5} = \frac{}{5}$		34.	$2\frac{2}{3} = \frac{}{3}$	
13.	$\frac{3}{3} + \frac{1}{3} = \frac{}{3}$		35.	$\frac{12}{4} + \frac{3}{4} = \frac{}{4}$	
14.	$1 + \frac{1}{3} = \frac{}{3}$		36.	$3 + \frac{3}{4} = \frac{}{4}$	
15.	$1\frac{1}{3} = \frac{}{3}$		37.	$3\frac{3}{4} = \frac{}{4}$	
16.	$1\frac{2}{3} = \frac{}{3}$		38.	$3 + \frac{4}{5} = \frac{}{5}$	
17.	$\frac{10}{10} + \frac{1}{10} = \frac{}{10}$		39.	$4 + \frac{1}{2} = \frac{}{2}$	
18.	$1 + \frac{1}{10} = \frac{}{10}$		40.	$4 + \frac{2}{3} = \frac{}{3}$	
19.	$1\frac{1}{10} = \frac{}{10}$		41.	$3 + \frac{1}{6} = \frac{}{6}$	
20.	$1\frac{7}{10} = \frac{}{10}$		42.	$2 + \frac{7}{8} = \frac{}{8}$	
21.	$\frac{8}{8} + \frac{5}{8} = \frac{}{8}$		43.	$2\frac{3}{5} = \frac{}{5}$	
22.	$1 + \frac{5}{8} = \frac{}{8}$		44.	$2\frac{7}{8} = \frac{}{8}$	

A STORY OF UNITS Lesson 31 Problem Set 4•5

Name _____ Date _____

1. Solve.

   a. $3\frac{1}{3} + 2\frac{2}{3} = 5 + \frac{3}{3} =$

   3 — $\frac{1}{3}$ 2 — $\frac{2}{3}$

   b. $4\frac{1}{4} + 3\frac{2}{4}$

   c. $2\frac{2}{6} + 6\frac{4}{6}$

2. Solve. Use a number line to show your work.

   a. $2\frac{4}{5} + 1\frac{2}{5} = 3 + \frac{6}{5} =$ _____

   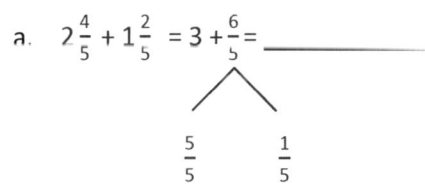

   b. $1\frac{3}{4} + 3\frac{3}{4}$

   c. $3\frac{3}{8} + 2\frac{6}{8}$

EUREKA MATH Lesson 31: Add mixed numbers.

3. Solve. Use the arrow way to show how to make one.

   a. $2\frac{4}{6} + 1\frac{5}{6} = 3\frac{4}{6} + \frac{5}{6} =$
   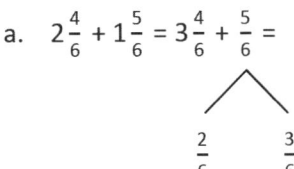

   b. $1\frac{3}{4} + 3\frac{3}{4}$

   c. $3\frac{3}{8} + 2\frac{6}{8}$

4. Solve. Use whichever method you prefer.

   a. $1\frac{3}{5} + 3\frac{4}{5}$

   b. $2\frac{6}{8} + 3\frac{7}{8}$

   c. $3\frac{8}{12} + 2\frac{7}{12}$

A STORY OF UNITS   **Lesson 31 Exit Ticket** 4•5

Name _____ Date _____

Solve.

1. $2\frac{3}{8} + 1\frac{5}{8}$

2. $3\frac{4}{5} + 2\frac{3}{5}$

Lesson 31: Add mixed numbers.

A STORY OF UNITS   **Lesson 31 Homework**   **4•5**

Name _____ Date _____

1. Solve.

   a. $2\frac{1}{3} + 1\frac{2}{3} = 3 + \frac{3}{3} = $

   (number bond: $2\frac{1}{3}$ breaks into 2 and $\frac{1}{3}$; $1\frac{2}{3}$ breaks into 1 and $\frac{2}{3}$)

   b. $2\frac{2}{5} + 2\frac{2}{5}$

   c. $3\frac{3}{8} + 1\frac{5}{8}$

2. Solve. Use a number line to show your work.

   a. $2\frac{2}{4} + 1\frac{3}{4} = 3 + \frac{5}{4} = $ _____

   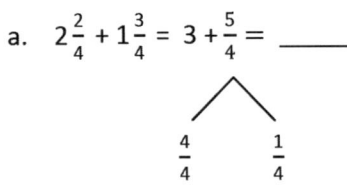
   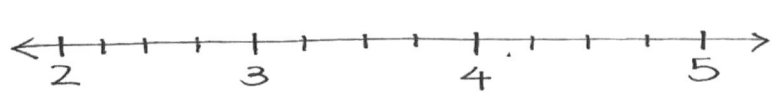

   b. $3\frac{4}{6} + 2\frac{5}{6}$

   c. $1\frac{9}{12} + 1\frac{7}{12}$

424   Lesson 31:   Add mixed numbers.

3. Solve. Use the arrow way to show how to make one.

   a. $2\frac{3}{4} + 1\frac{3}{4} = 3\frac{3}{4} + \frac{3}{4} =$

   (bond: $\frac{3}{4}$ splits into $\frac{1}{4}$ and $\frac{2}{4}$)

   $3\frac{3}{4} \xrightarrow{+\frac{1}{4}} 4 \longrightarrow$

   b. $2\frac{7}{8} + 3\frac{4}{8}$

   c. $1\frac{7}{9} + 4\frac{5}{9}$

4. Solve. Use whichever method you prefer.

   a. $1\frac{4}{5} + 1\frac{3}{5}$

   b. $3\frac{8}{10} + 1\frac{5}{10}$

   c. $2\frac{5}{7} + 3\frac{6}{7}$

Lesson 31: Add mixed numbers.

Lesson 32

Objective: Subtract a fraction from a mixed number.

Suggested Lesson Structure

- ■ Fluency Practice   (12 minutes)
- ■ Application Problem   (3 minutes)
- ■ Concept Development   (35 minutes)
- ■ Student Debrief   (10 minutes)
- **Total Time**   **(60 minutes)**

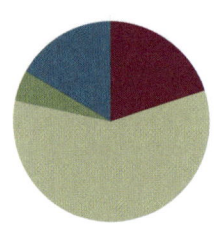

Fluency Practice (12 minutes)

- Count by Equivalent Fractions 4.NF.1   (5 minutes)
- Change Mixed Numbers to Fractions 4.NF.4   (4 minutes)
- Add Mixed Numbers 4.NF.3   (3 minutes)

Count by Equivalent Fractions (5 minutes)

Note: This activity reviews Lessons 24 and 25. The progression builds in complexity. Work students up to the highest level of complexity in which they can confidently participate.

T: Count by twos to 18, starting at 0.
S: 0, 2, 4, 6, 8, 10, 12, 14, 16, 18.
T: Count by 2 sixths to 18 sixths, starting at 0 sixths. (Write as students count.)
S: $\frac{0}{6}, \frac{2}{6}, \frac{4}{6}, \frac{6}{6}, \frac{8}{6}, \frac{10}{6}, \frac{12}{6}, \frac{14}{6}, \frac{16}{6}, \frac{18}{6}$

$\frac{0}{6}$	$\frac{2}{6}$	$\frac{4}{6}$	$\frac{6}{6}$	$\frac{8}{6}$	$\frac{10}{6}$	$\frac{12}{6}$	$\frac{14}{6}$	$\frac{16}{6}$	$\frac{18}{6}$
0	$\frac{2}{6}$	$\frac{4}{6}$	1	$\frac{8}{6}$	$\frac{10}{6}$	2	$\frac{14}{6}$	$\frac{16}{6}$	3
0	$\frac{2}{6}$	$\frac{4}{6}$	1	$1\frac{2}{6}$	$1\frac{4}{6}$	2	$2\frac{2}{6}$	$2\frac{4}{6}$	3
0	$\frac{1}{3}$	$\frac{2}{3}$	1	$1\frac{1}{3}$	$1\frac{2}{3}$	2	$2\frac{1}{3}$	$2\frac{2}{3}$	3

T: Zero is the same as how many sixths?
S: 0 sixths.

A STORY OF UNITS   Lesson 32 4•5

T: (Beneath $\frac{0}{6}$, write 0.) 1 is the same as how many sixths?

S: 6 sixths.

T: (Beneath $\frac{6}{6}$, write 1.)

Continue this process for 2 and 3.

T: Count by 2 sixths again. This time, when you come to the whole number, say the whole number. Start at zero. (Write as students count.)

S: 0, $\frac{2}{6}$, $\frac{4}{6}$, 1, $\frac{8}{6}$, $\frac{10}{6}$, 2, $\frac{14}{6}$, $\frac{16}{6}$, 3.

T: (Point to $\frac{8}{6}$.) Say $\frac{8}{6}$ as a mixed number.

S: $1\frac{2}{6}$.

Continue this process for $\frac{10}{6}$, $\frac{14}{6}$, and $\frac{16}{6}$.

T: Count by 2 sixths again. This time, convert to whole numbers and mixed numbers. Start at zero. (Write as students count.)

S: 0, $\frac{2}{6}$, $\frac{4}{6}$, 1, $1\frac{2}{6}$, $1\frac{4}{6}$, 2, $2\frac{2}{6}$, $2\frac{4}{6}$, 3.

Possibly extend, having students rename sixths as thirds.

Change Mixed Numbers to Fractions (4 minutes)

Materials: (S) Personal white board

Note: This fluency activity reviews Lesson 25.

T: (Write $1\frac{4}{5}$.) Say the mixed number.

S: 1 and 4 fifths.

T: (Draw a number bond with $1\frac{4}{5}$ as the whole. Write $\frac{4}{5}$ as the known part. Write $\frac{-}{5}$ as the other part.) Write the unknown part, filling in the numerator.

S: (Write $\frac{5}{5}$ as the unknown part.)

T: (Write $\frac{5}{5}$ as the unknown part. Write $1\frac{4}{5} = \frac{-}{5}$.) Fill in the numerator.

S: (Write $1\frac{4}{5} = \frac{9}{5}$.)

Continue with the following possible sequence: $2\frac{1}{4}$ and $3\frac{5}{6}$.

Lesson 32: Subtract a fraction from a mixed number.   427

A STORY OF UNITS   Lesson 32   4•5

Add Mixed Numbers (3 minutes)

Note: This fluency activity reviews Lesson 31.

T: (Write $5\frac{1}{3} + 2\frac{1}{3}$.) On your personal white board, add like units to solve.

S: (Write $5 + \frac{1}{3} + 2 + \frac{1}{3} = 7\frac{2}{3}$.)

Continue with the following possible sequence: $4\frac{3}{5} + 2\frac{1}{5}$ and $6\frac{3}{5} + 2\frac{3}{5}$.

Application Problem (3 minutes)

Meredith had 2 m 65 cm of ribbon. She used 87 cm of the ribbon. How much ribbon did she have left?

Solution A
2 m 65 cm − 87 cm = 2 m − 22 cm = 1 m 78 cm.
     /\
  65cm 22cm

Solution B
2 m 65 cm − 87 cm = 265 cm − 87 cm = 178 cm = 1 m 78 cm.
              /\   /\
         200cm 65cm   1m 78cm

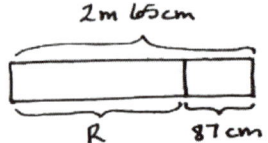

R = 1 m 78 cm
Meridith has 1 m 78 cm ribbon left.

Solution C
2 m 65 cm − 87 cm = 1 m 65 cm + 13 cm = 1 m 78 cm.
      /\
  1m 65cm 1m

Note: This Application Problem anticipates the subtraction of a fraction from a mixed number using a measurement context. In Solution A, 87 cm is decomposed as 65 cm and 22 cm to count back to 2 and then to subtract the remaining centimeters. In Solution B, the total is decomposed into smaller units before subtracting. In Solution C, the one is taken out of 2 m 65 cm, and 87 centimeters is subtracted from 1. The remaining 13 centimeters is then added to 1 m 65 cm.

Concept Development (35 minutes)

Materials: (S) Personal white board

Problem 1: Subtract a fraction from a mixed number by counting back.

T: 3 oranges 2 apples − 1 apple is…?
S: 3 oranges 1 apple.
T: 3 dogs 2 puppies − 1 puppy is…?
S: 3 dogs 1 puppy.
T: 3 ones 2 fifths − 1 fifth is…?
S: 3 ones 1 fifth.

Lesson 32: Subtract a fraction from a mixed number.

A STORY OF UNITS   Lesson 32 4•5

T: (Write $3\frac{4}{5} - \frac{3}{5}$ including the number bond as shown.)

T: Do we have enough fifths to subtract 3 fifths?

S: Yes!

T: Solve the problem.

S: $3\frac{4}{5} - \frac{3}{5} = 3\frac{1}{5}$.

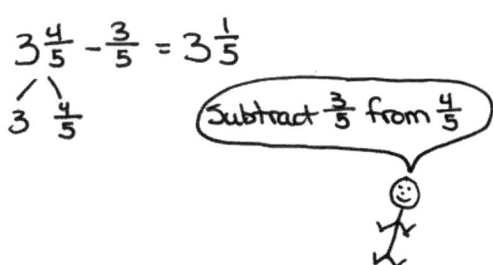

T: Draw a number line to model the subtraction. What will the endpoints of the number line be? How will you partition the whole number?

S: The endpoints will be 3 and 4. We will partition the whole number into fifths.

T: Start at $3\frac{4}{5}$. Subtract $\frac{3}{5}$. Say the number sentence again.

S: $3\frac{4}{5} - \frac{3}{5} = 3\frac{1}{5}$.

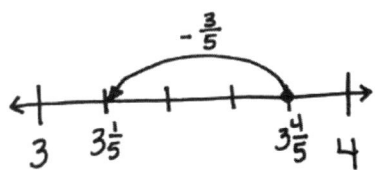

T: Try $4\frac{9}{10} - \frac{3}{10}$. We can count back by a tenth 3 times from $4\frac{9}{10}$ to find the answer. Draw a number line, and use it to explain the difference to your partner.

S: $4\frac{9}{10} - \frac{3}{10} = 4\frac{6}{10}$. There are 4 ones. 9 tenths − 3 tenths = 6 tenths.

T: Try $4\frac{1}{5} - \frac{2}{5}$. Model with a number line, and try using the arrow way.

S: $4\frac{1}{5} - \frac{2}{5} = 3\frac{4}{5}$. Counting back 1 fifth, we get 4 ones. Counting back 1 more fifth, and we get $3\frac{4}{5}$.

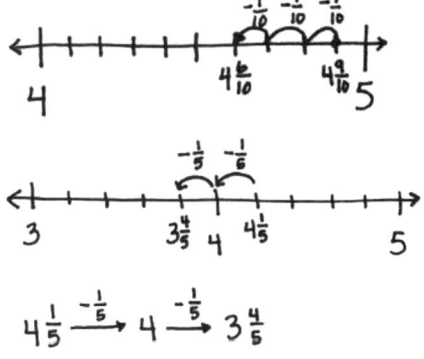

Let students quickly practice with the following: $4\frac{11}{12} - \frac{3}{12}$ and $3\frac{2}{6} - \frac{3}{6}$.

Problem 2: Subtract a fraction less than 1 from a whole number by decomposing the subtrahend.

T: (Write $4\frac{1}{5} - \frac{3}{5}$.) Do we have enough fifths to subtract 3 fifths?

S: No!

T: (Show $\frac{3}{5}$ decomposed as $\frac{1}{5}$ and $\frac{2}{5}$ as pictured to the right.)

MP.4

T: Does $\frac{1}{5} + \frac{2}{5}$ have the same value as $\frac{3}{5}$? (Point to the parts of the bond.)

S: Yes!

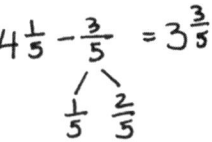

T: Now do we have enough fifths?

S: No. It's still $\frac{3}{5}$. We can't take that from $\frac{1}{5}$.

Lesson 32: Subtract a fraction from a mixed number.

A STORY OF UNITS   Lesson 32 4•5

T: Look at the parts. Let's take away one part at a time. Draw a number line to model the subtraction.

T: Solve $4\frac{1}{5} - \frac{1}{5}$. Count back 1 fifth on the number line.

S: That's 4.

T: Now, subtract $\frac{2}{5}$ from 4. Talk to your partner.

S: We already know how to do that, $3\frac{5}{5} - \frac{2}{5} = 3\frac{3}{5}$. → $1 - \frac{2}{5}$ is $\frac{3}{5}$, so $4 - \frac{2}{5}$ is $3\frac{3}{5}$.

T: We can also use the arrow way. Start with $4\frac{1}{5}$, count back $\frac{1}{5}$ to get to 4, and then count back $\frac{2}{5}$ more to get $3\frac{3}{5}$. (Shown above to the right.)

T: Write $3\frac{3}{5} - \frac{4}{5}$. First, decompose $\frac{4}{5}$ into two parts, count back to 3, and then subtract the other part.

S: I see. We take away one part of $\frac{4}{5}$ at a time. $\frac{4}{5} = \frac{3}{5} + \frac{1}{5}$. $3\frac{3}{5} - \frac{3}{5} = 3$. $3 - \frac{1}{5} = 2\frac{4}{5}$.

T: Model on a number line, and then model using arrows.

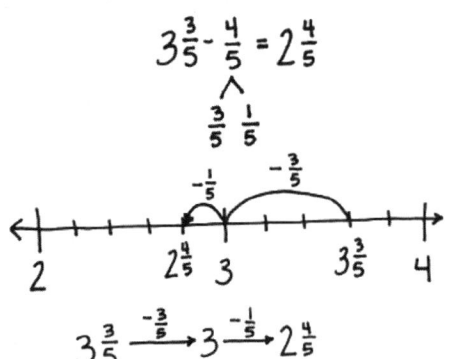

Let students practice with the following: $4\frac{5}{10} - \frac{7}{10}$, $2\frac{2}{12} - \frac{7}{12}$, $3\frac{7}{10} - \frac{9}{10}$, $2\frac{1}{4} - \frac{3}{4}$.

Problem 3: Decompose the total to take out 1 when subtracting a fraction from a mixed number when there are not enough fractional units.

T: (Write $3\frac{1}{5} - \frac{3}{5}$, including the number bond as shown to the right.)

T: Do you have enough fifths to subtract $\frac{3}{5}$?

S: No. → This is the same problem as before.

T: Let's try a different strategy to solve. Talk to your partner. Where can we get more fifths?

S: From $3\frac{1}{5}$.

T: Decompose $3\frac{1}{5}$ by taking out one. We have $2\frac{1}{5}$ and 1. (Record using a number bond.)

T: Take $\frac{3}{5}$ from 1. How many are left?

S: $\frac{2}{5}$.

T: We have $\frac{2}{5}$ left plus $2\frac{1}{5}$, which is equal to $2\frac{3}{5}$. Let's show this using the arrow way.

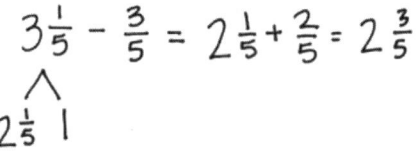

Let students practice with the following: $12\frac{1}{4} - \frac{3}{4}$ and $7\frac{3}{10} - \frac{9}{10}$.

Lesson 32: Subtract a fraction from a mixed number.

Problem Set (10 minutes)

Students should do their personal best to complete the Problem Set within the allotted 10 minutes. For some classes, it may be appropriate to modify the assignment by specifying which problems they work on first. Some problems do not specify a method for solving. Students should solve these problems using the RDW approach used for Application Problems.

Student Debrief (10 minutes)

Lesson Objective: Subtract a fraction from a mixed number.

The Student Debrief is intended to invite reflection and active processing of the total lesson experience.

Invite students to review their solutions for the Problem Set. They should check work by comparing answers with a partner before going over answers as a class. Look for misconceptions or misunderstandings that can be addressed in the Student Debrief. Guide students in a conversation to debrief the Problem Set and process the lesson.

> **NOTE ON MULTIPLE MEANS OF REPRESENTATION:**
>
> There are other strategies for subtracting a fraction from a mixed number. Gauge students. Those who quickly show mastery of one strategy can be encouraged to understand and try others.
>
> Those who struggle to master a method might be better off working with the decomposition modeled in Lesson 34 since it most closely resembles regrouping with whole number subtraction.
>
> This connection may well strengthen their understanding of and skill with whole number subtraction, which may also be weak.

Any combination of the questions below may be used to lead the discussion.

- Use Problems 2(a) and 3(c) to compare the different methods to subtract when there are not enough fractional units.
- How is 7 tens 3 ones – 9 ones like 7 ones 3 tenths – 9 tenths? How is it different?
- Tell your partner the process of subtracting a fraction from a mixed number when regrouping is necessary.
- Here is another way to solve $3\frac{1}{5} - \frac{3}{5}$. A student wrote this (write $3\frac{1}{5} - \frac{3}{5} = 3\frac{3}{5} - 1 = 2\frac{3}{5}$). What was he thinking? (See the illustration of student's thinking below. Compare this method to whole number compensation such as 153 – 98 = 155 – 100.)

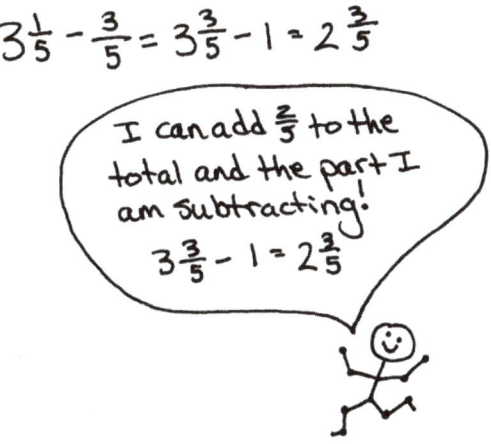

Lesson 32

Exit Ticket (3 minutes)

After the Student Debrief, instruct students to complete the Exit Ticket. A review of their work will help with assessing students' understanding of the concepts that were presented in today's lesson and planning more effectively for future lessons. The questions may be read aloud to the students.

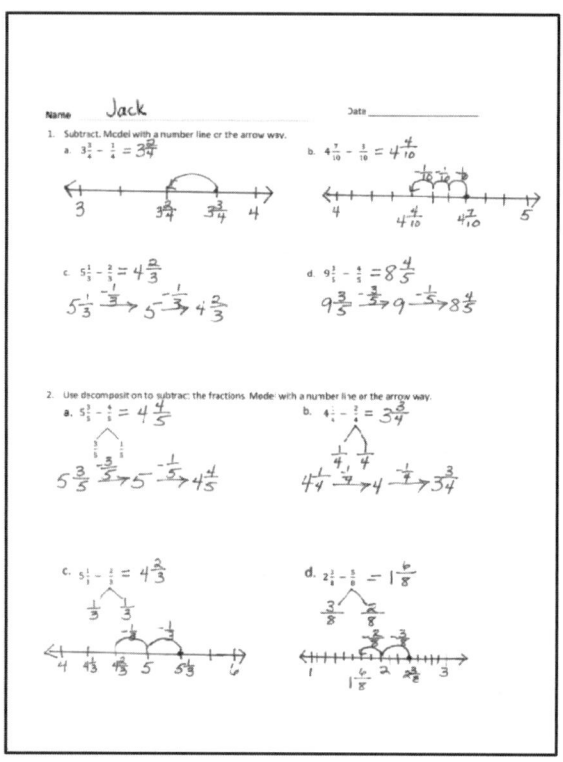

A STORY OF UNITS   Lesson 32 Problem Set   4•5

Name _____   Date _____

1. Subtract. Model with a number line or the arrow way.

   a. $3\frac{3}{4} - \frac{1}{4}$

   b. $4\frac{7}{10} - \frac{3}{10}$

   c. $5\frac{1}{3} - \frac{2}{3}$

   d. $9\frac{3}{5} - \frac{4}{5}$

2. Use decomposition to subtract the fractions. Model with a number line or the arrow way.

   a. $5\frac{3}{5} - \frac{4}{5}$

   $\frac{3}{5} \quad \frac{1}{5}$

   b. $4\frac{1}{4} - \frac{2}{4}$

   c. $5\frac{1}{3} - \frac{2}{3}$

   d. $2\frac{3}{8} - \frac{5}{8}$

Lesson 32: Subtract a fraction from a mixed number.

3. Decompose the total to subtract the fractions.

   a. $3\frac{1}{8} - \frac{3}{8} = 2\frac{1}{8} + \frac{5}{8} = 2\frac{6}{8}$

   $2\frac{1}{8} \quad 1$

   b. $5\frac{1}{8} - \frac{7}{8}$

   c. $5\frac{3}{5} - \frac{4}{5}$

   d. $5\frac{4}{6} - \frac{5}{6}$

   e. $6\frac{4}{12} - \frac{7}{12}$

   f. $9\frac{1}{8} - \frac{5}{8}$

   g. $7\frac{1}{6} - \frac{5}{6}$

   h. $8\frac{3}{10} - \frac{4}{10}$

   i. $12\frac{3}{5} - \frac{4}{5}$

   j. $11\frac{2}{6} - \frac{5}{6}$

A STORY OF UNITS — **Lesson 32 Exit Ticket** 4•5

Name _____ Date _____

Solve.

1. $10\frac{5}{6} - \frac{4}{6}$

2. $8\frac{3}{8} - \frac{6}{8}$

Lesson 32: Subtract a fraction from a mixed number.

Name _____   Date _____

1. Subtract. Model with a number line or the arrow way.

   a. $6\frac{3}{5} - \frac{1}{5}$

   b. $4\frac{9}{12} - \frac{7}{12}$

   c. $7\frac{1}{4} - \frac{3}{4}$

   d. $8\frac{3}{8} - \frac{5}{8}$

2. Use decomposition to subtract the fractions. Model with a number line or the arrow way.

   a. $2\frac{2}{5} - \frac{4}{5}$

   $\frac{2}{5} \quad \frac{2}{5}$

   b. $2\frac{1}{3} - \frac{2}{3}$

   c. $4\frac{1}{6} - \frac{4}{6}$

   d. $3\frac{3}{6} - \frac{5}{6}$

Lesson 32: Subtract a fraction from a mixed number.

e. $9\frac{3}{8} - \frac{7}{8}$

f. $7\frac{1}{12} - \frac{6}{12}$

g. $10\frac{1}{8} - \frac{5}{8}$

h. $9\frac{4}{12} - \frac{7}{12}$

i. $11\frac{3}{5} - \frac{4}{5}$

j. $17\frac{1}{9} - \frac{5}{9}$

3. Decompose the total to subtract the fractions.

a. $4\frac{1}{8} - \frac{3}{8} = 3\frac{1}{8} + \frac{5}{8} = 3\frac{6}{8}$

   $3\frac{1}{8} \quad 1$

b. $5\frac{2}{5} - \frac{3}{5}$

c. $7\frac{1}{8} - \frac{3}{8}$

d. $3\frac{3}{9} - \frac{4}{9}$

e. $6\frac{3}{12} - \frac{7}{12}$

f. $2\frac{5}{9} - \frac{8}{9}$

Lesson 33

Objective: Subtract a mixed number from a mixed number.

Suggested Lesson Structure

- ■ Fluency Practice (12 minutes)
- ■ Application Problem (5 minutes)
- ■ Concept Development (33 minutes)
- ■ Student Debrief (10 minutes)
- **Total Time** **(60 minutes)**

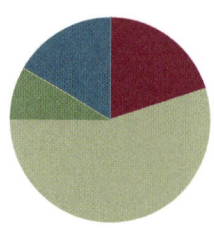

Fluency Practice (12 minutes)

- Sprint: Change Mixed Numbers to Fractions **4.NF.4**   (9 minutes)
- Subtract Fractions from Whole Numbers **4.NF.3**   (3 minutes)

Sprint: Change Mixed Numbers to Fractions (9 minutes)

Materials: (S) Change Mixed Numbers to Fractions Sprint

Note: This fluency activity reviews Lesson 25.

Subtract Fractions from Whole Numbers (3 minutes)

Materials: (S) Personal white board

Note: This fluency activity reviews Lesson 17.

T: (Write $3 - \frac{2}{5}$.) Break apart the whole number and solve.

S: (Write $3 - \frac{2}{5} = 2\frac{3}{5}$.)

Continue with the following possible sequence: $5 - \frac{3}{4}$ and $9 - \frac{7}{10}$.

A STORY OF UNITS   Lesson 33   4•5

Application Problem (5 minutes)

Jeannie's pumpkin had a weight of 3 kg 250 g in August and 4 kg 125 g in October. What was the difference in weight from August to October?

Note: This Application Problem anticipates the subtraction of a mixed number from a mixed number using a measurement context. Solution A shows counting up using the arrow way. Solution B shows subtracting 3 kilograms from 4 kilograms first, and then subtracting 250 grams from the total remaining 1,125 grams.

Concept Development (33 minutes)

Materials: (S) Personal white board

Problem 1: Subtract a mixed number from a mixed number by counting up.

T: (Write $4\frac{3}{8} - 2\frac{5}{8}$.) Let's count up to solve.

T: Draw a number line with endpoints 2 and 5. Label $2\frac{5}{8}$. What fractional part can we add to get to the next one? $2\frac{5}{8}$ plus what is 3?

S: $\frac{3}{8}$.

T: Show a slide from $2\frac{5}{8}$ to 3. Next, count up from 3 to the whole number in $4\frac{3}{8}$.

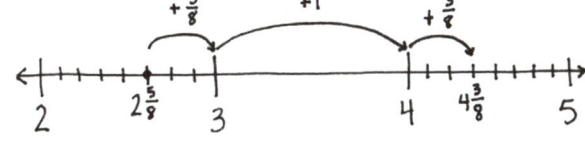

> **NOTES ON MULTIPLE MEANS OF REPRESENTATION:**
>
> In Grade 1, students relate subtraction to addition by counting up (**1.OA.6**). For example, 12 – 8 is easily solved by counting up from 8 to 12: 8, 9, 10, 11, 12.
>
> As students become more adept at using a unit of 10, they see they can get to the next ten, rather than counting by ones. For example, 12 – 8 can be solved by thinking, "8 and 2 is 10. 2 more is 12. The unknown part is 4."
>
> In Grade 2, students apply this strategy to subtract larger numbers (**2.NBT.7**).
>
> For example, 120 – 80 can be solved by thinking, "80 plus 20 is 100, and 20 more is 120. The unknown part is 40." Their use of the ten in Grade 1 has evolved into a place value strategy in Grade 2. Here in Grade 4, it evolves yet again as students use fractional units rather than place value units.

Lesson 33: Subtract a mixed number from a mixed number.

A STORY OF UNITS   Lesson 33 4•5

S: (Draw an arrow from 3 to 4.) We added 1.
T: Count up to $4\frac{3}{8}$.
S: (Draw an arrow from 4 to $4\frac{3}{8}$.) We added $\frac{3}{8}$ more.
T: What is $\frac{3}{8} + 1 + \frac{3}{8}$?
S: $\frac{3}{8} + 1 + \frac{3}{8} = 1\frac{6}{8}$.
T: Use the arrow way to track our recording.
S: $2\frac{5}{8} + \frac{3}{8} = 3, 3 + 1 = 4$, and $4 + \frac{3}{8} = 4\frac{3}{8}$. We counted up $\frac{3}{8}$, 1, and $\frac{3}{8}$. That's $1\frac{6}{8}$.

$2\frac{5}{8} \xrightarrow{+\frac{3}{8}} 3 \xrightarrow{+1} 4 \xrightarrow{+\frac{3}{8}} 4\frac{3}{8}$

Let students practice with the following: $2\frac{5}{12} - 1\frac{8}{12}$ and $9\frac{2}{6} - 3\frac{5}{6}$.

Problem 2: Subtract a mixed number from a mixed number when there are not enough fractional units by first subtracting the whole numbers and then decomposing the subtrahend.

T: (Write $11\frac{1}{5} + 2\frac{3}{5}$.) When we add mixed numbers, we add the like units. We could add the ones first and then the fifths.
T: (Write $11\frac{1}{5} + 2\frac{3}{5}$.) When we subtract mixed numbers, we can subtract the ones first. What subtraction expression remains?
S: $9\frac{1}{5} - \frac{3}{5}$.
T: Just like yesterday, decompose 3 fifths as $\frac{1}{5}$ and $\frac{2}{5}$ (as pictured to the right).
T: $9\frac{1}{5} - \frac{1}{5}$ is...? (Record using the arrow way, as seen to the right.)
S: 9.

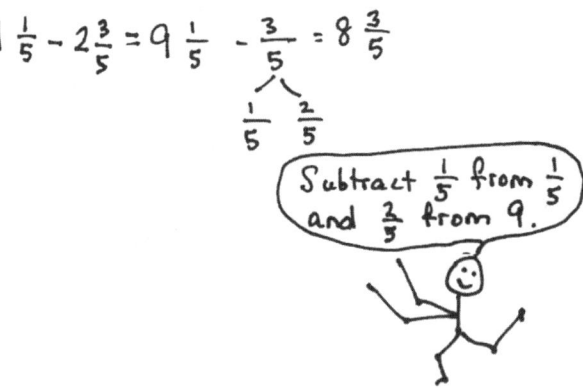

T: Count back $\frac{2}{5}$ from 9. $9 - \frac{2}{5}$ is...? (Record with the second arrow.)
S: $8\frac{3}{5}$.
T: (Write $9\frac{1}{5} - \frac{3}{5} = 9 - \frac{2}{5} = 8\frac{3}{5}$.)

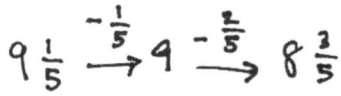

T: Explain to your partner why this is true.
S: It's like counting back! → We subtract a fifth from $9\frac{1}{5}$, and then we subtract $\frac{2}{5}$ from 9. → First, we renamed $\frac{3}{5}$ as $\frac{1}{5}$ and $\frac{2}{5}$. Then, we subtracted in two steps. → It looks like we subtracted $\frac{1}{5}$ from both numbers and got $9 - \frac{2}{5}$, which is just easier.
T: Use a number line to model the steps of counting backward from $11\frac{1}{5}$ to subtract $2\frac{3}{5}$.

A STORY OF UNITS   Lesson 33   4•5

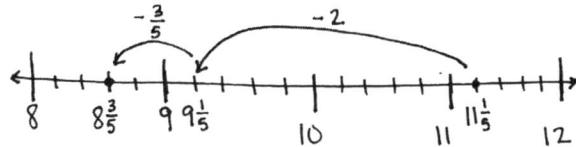

S: (Draw as shown to the right, or draw to match the arrow way recording.)

Let students practice with the following:

$4\frac{1}{8} - 1\frac{7}{8}$ and $7\frac{5}{12} - 3\frac{9}{12}$. Those who struggle with subtracting from a whole number with automaticity can break apart the whole number using Lesson 32's strategy until gaining mastery, e.g., $4\frac{1}{8} - 2\frac{7}{8} = 2\frac{1}{8} - \frac{7}{8} = 1\frac{9}{8} - \frac{7}{8} = 1\frac{1}{8}$ Have them share their work with a partner, explaining their solution.

Problem 3: Subtract a mixed number from a mixed number when there are not enough fractional units by decomposing a whole number into fractional parts.

T: (Write $11\frac{1}{5} - 2\frac{3}{5}$.) Let's solve using a different strategy.

T: Subtract the whole numbers.

S: $11\frac{1}{5} - 2\frac{3}{5} = 9\frac{1}{5} - \frac{3}{5}$.

T: Decompose $9\frac{1}{5}$ by taking out one.

S: (Draw a number bond to show $8\frac{1}{5}$ and 1.)

T: $1 - \frac{3}{5}$ is...?

S: $\frac{2}{5}$.

T: $8\frac{1}{5} + \frac{2}{5}$ is...?

S: $8\frac{1}{5} + \frac{2}{5} = 8\frac{3}{5}$. That's the same answer as before. We just found it in a different way.

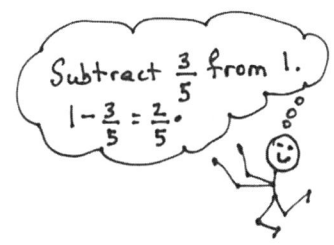

Let students practice with the following: $4\frac{1}{8} - 1\frac{7}{8}$ and $7\frac{5}{12} - 3\frac{9}{12}$. Encourage students to practice this strategy of subtracting from 1, but don't belabor its use with students. Allow them to use any strategy that makes sense to them and enables them to correctly solve the problem, explaining the steps to their partner. Ask those who finish early to solve using an alternative strategy to strengthen their number sense.

Problem Set (10 minutes)

Students should do their personal best to complete the Problem Set within the allotted 10 minutes. For some classes, it may be appropriate to modify the assignment by specifying which problems they work on first. Some problems do not specify a method for solving. Students should solve these problems using the RDW approach used for Application Problems.

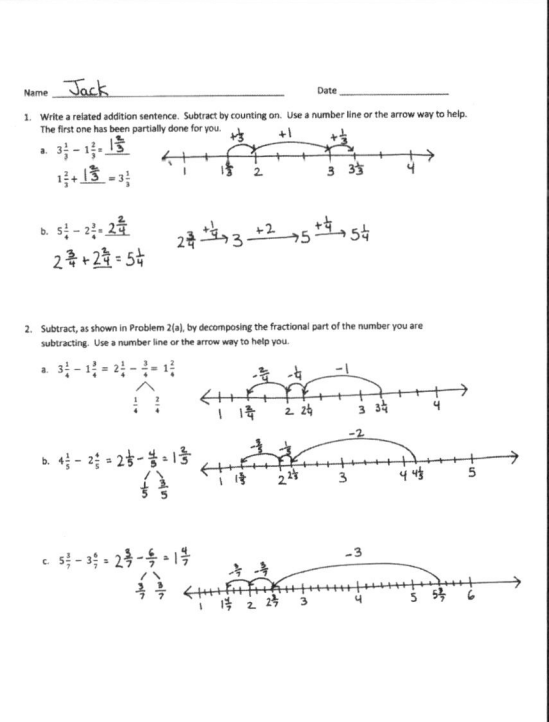

Lesson 33:   Subtract a mixed number from a mixed number.   441

Student Debrief (10 minutes)

Lesson Objective: Subtract a mixed number from a mixed number.

The Student Debrief is intended to invite reflection and active processing of the total lesson experience.

Invite students to review their solutions for the Problem Set. They should check work by comparing answers with a partner before going over answers as a class. Look for misconceptions or misunderstandings that can be addressed in the Student Debrief. Guide students in a conversation to debrief the Problem Set and process the lesson.

Any combination of the questions below may be used to lead the discussion.

- Can you accurately subtract mixed numbers by subtracting the fraction first, or must you always subtract the whole numbers first? Give an example to explain.
- When subtracting mixed numbers, what is the advantage of subtracting the whole numbers first?
- Which strategy do you prefer to use, decomposing the number we are subtracting as we did in Problem 2 of the Concept Development or taking from 1, as we did in Problem 3? Discuss the advantages of the strategy as you explain your preference.
- Which strategies did you choose to solve Problem 4(a–d) of the Problem Set? Explain how you decided which strategy to use.
- What learning from Lesson 32 was used in this lesson? How can subtracting a mixed number from a mixed number be similar to subtracting a fraction from a mixed number?
- How did our Application Problem relate to today's lesson?

Exit Ticket (3 minutes)

After the Student Debrief, instruct students to complete the Exit Ticket. A review of their work will help with assessing students' understanding of the concepts that were presented in today's lesson and planning more effectively for future lessons. The questions may be read aloud to the students.

A

Change Fractions to Mixed Numbers

Number Correct: _____

1.	$2 + 1 =$		23.	$1\frac{5}{6} = \frac{}{6}$	
2.	$\frac{2}{2} + \frac{1}{2} = \frac{}{2}$		24.	$2 + \frac{1}{2} = 2\frac{}{2}$	
3.	$1 + \frac{1}{2} = \frac{}{2}$		25.	$\frac{4}{2} + \frac{1}{2} = \frac{}{2}$	
4.	$1\frac{1}{2} = \frac{}{2}$		26.	$2 + \frac{1}{2} = \frac{}{2}$	
5.	$4 + 1 =$		27.	$2\frac{1}{2} = \frac{}{2}$	
6.	$\frac{4}{4} + \frac{1}{4} = \frac{}{4}$		28.	$2 + \frac{1}{4} = 2\frac{}{4}$	
7.	$1 + \frac{1}{4} = \frac{}{4}$		29.	$\frac{8}{4} + \frac{1}{4} = \frac{}{4}$	
8.	$1\frac{1}{4} = \frac{}{4}$		30.	$2 + \frac{1}{4} = \frac{}{4}$	
9.	$3 + 1 =$		31.	$2\frac{1}{4} = \frac{}{4}$	
10.	$\frac{3}{3} + \frac{1}{3} = \frac{}{3}$		32.	$\frac{9}{3} + \frac{2}{3} = \frac{}{3}$	
11.	$1 + \frac{1}{3} = \frac{}{3}$		33.	$3 + \frac{2}{3} = \frac{}{3}$	
12.	$1\frac{1}{3} = \frac{}{3}$		34.	$3\frac{2}{3} = \frac{}{3}$	
13.	$\frac{5}{5} + \frac{1}{5} = \frac{}{5}$		35.	$\frac{16}{4} + \frac{3}{4} = \frac{}{4}$	
14.	$1 + \frac{1}{5} = \frac{}{5}$		36.	$4 + \frac{3}{4} = \frac{}{4}$	
15.	$1\frac{1}{5} = \frac{}{5}$		37.	$4\frac{3}{4} = \frac{}{4}$	
16.	$1\frac{2}{5} = \frac{}{5}$		38.	$3 + \frac{2}{5} = \frac{}{5}$	
17.	$1\frac{4}{5} = \frac{}{5}$		39.	$4 + \frac{1}{2} = \frac{}{2}$	
18.	$1\frac{3}{5} = \frac{}{5}$		40.	$3 + \frac{3}{4} = \frac{}{4}$	
19.	$\frac{4}{4} + \frac{3}{4} = \frac{}{4}$		41.	$3 + \frac{1}{6} = \frac{}{6}$	
20.	$1 + \frac{3}{4} = \frac{}{4}$		42.	$3 + \frac{5}{8} = \frac{}{8}$	
21.	$\frac{6}{6} + \frac{5}{6} = \frac{}{6}$		43.	$3\frac{4}{5} = \frac{}{5}$	
22.	$1 + \frac{5}{6} = \frac{}{6}$		44.	$4\frac{7}{8} = \frac{}{8}$	

Lesson 33: Subtract a mixed number from a mixed number.

B

Change Mixed Numbers to Fractions

Number Correct: _____
Improvement: _____

1.	$5 + 1 =$	
2.	$\frac{5}{5} + \frac{1}{5} = \frac{}{5}$	
3.	$1 + \frac{1}{5} = \frac{}{5}$	
4.	$1\frac{1}{5} = \frac{}{5}$	
5.	$3 + 1 =$	
6.	$\frac{3}{3} + \frac{1}{3} = \frac{}{3}$	
7.	$1 + \frac{1}{3} = \frac{}{3}$	
8.	$1\frac{1}{3} = \frac{}{3}$	
9.	$4 + 1 =$	
10.	$\frac{4}{4} + \frac{1}{4} = \frac{}{4}$	
11.	$1 + \frac{1}{4} = \frac{}{4}$	
12.	$1\frac{1}{4} = \frac{}{4}$	
13.	$\frac{10}{10} + \frac{1}{10} = \frac{}{10}$	
14.	$1 + \frac{1}{10} = \frac{}{10}$	
15.	$1\frac{1}{10} = \frac{}{10}$	
16.	$1\frac{2}{10} = \frac{}{10}$	
17.	$1\frac{4}{10} = \frac{}{10}$	
18.	$1\frac{3}{10} = \frac{}{10}$	
19.	$\frac{3}{3} + \frac{2}{3} = \frac{}{3}$	
20.	$1 + \frac{2}{3} = \frac{}{3}$	
21.	$\frac{8}{8} + \frac{7}{8} = \frac{}{8}$	
22.	$1 + \frac{7}{8} = \frac{}{8}$	

23.	$1\frac{7}{8} = \frac{}{8}$	
24.	$2 + \frac{1}{2} = 2\frac{}{2}$	
25.	$\frac{4}{2} + \frac{1}{2} = \frac{}{2}$	
26.	$2 + \frac{1}{2} = \frac{}{2}$	
27.	$2\frac{1}{2} = \frac{}{2}$	
28.	$2 + \frac{1}{3} = 2\frac{}{3}$	
29.	$\frac{6}{3} + \frac{1}{3} = \frac{}{3}$	
30.	$2 + \frac{1}{3} = \frac{}{3}$	
31.	$2\frac{1}{3} = \frac{}{3}$	
32.	$\frac{12}{4} + \frac{3}{4} = \frac{}{4}$	
33.	$3 + \frac{3}{4} = \frac{}{4}$	
34.	$3\frac{3}{4} = \frac{}{4}$	
35.	$\frac{12}{3} + \frac{2}{3} = \frac{}{3}$	
36.	$4 + \frac{2}{3} = \frac{}{3}$	
37.	$4\frac{2}{3} = \frac{}{3}$	
38.	$3 + \frac{3}{5} = \frac{}{5}$	
39.	$5 + \frac{1}{2} = \frac{}{2}$	
40.	$3 + \frac{2}{3} = \frac{}{3}$	
41.	$3 + \frac{1}{8} = \frac{}{8}$	
42.	$3 + \frac{1}{6} = \frac{}{6}$	
43.	$3\frac{2}{5} = \frac{}{5}$	
44.	$4\frac{5}{6} = \frac{}{6}$	

Lesson 33 Problem Set 4•5

Name _____ Date _____

1. Write a related addition sentence. Subtract by counting on. Use a number line or the arrow way to help. The first one has been partially done for you.

   a. $3\frac{1}{3} - 1\frac{2}{3} =$ _____

   $1\frac{2}{3} +$ _____ $= 3\frac{1}{3}$

   b. $5\frac{1}{4} - 2\frac{3}{4} =$ _____

2. Subtract, as shown in Problem 2(a), by decomposing the fractional part of the number you are subtracting. Use a number line or the arrow way to help you.

   a. $3\frac{1}{4} - 1\frac{3}{4} = 2\frac{1}{4} - \frac{3}{4} = 1\frac{2}{4}$

   b. $4\frac{1}{5} - 2\frac{4}{5}$

   c. $5\frac{3}{7} - 3\frac{6}{7}$

Lesson 33: Subtract a mixed number from a mixed number.

3. Subtract, as shown in Problem 3(a), by decomposing to take one out.

   a. $5\frac{3}{5} - 2\frac{4}{5} = 3\frac{3}{5} - \frac{4}{5}$

   $2\frac{3}{5} \quad 1$

   b. $4\frac{3}{6} - 3\frac{5}{6}$

   c. $8\frac{3}{10} - 2\frac{7}{10}$

4. Solve using any method.

   a. $6\frac{1}{4} - 3\frac{3}{4}$

   b. $5\frac{1}{8} - 2\frac{7}{8}$

   c. $8\frac{3}{12} - 3\frac{8}{12}$

   d. $5\frac{1}{100} - 2\frac{97}{100}$

A STORY OF UNITS

Lesson 33 Exit Ticket 4•5

Name _____   Date _____

Solve using any strategy.

1. $4\frac{2}{3} - 2\frac{1}{3}$

2. $12\frac{5}{8} - 8\frac{7}{8}$

A STORY OF UNITS

Lesson 33 Homework 4•5

Name _____ Date _____

1. Write a related addition sentence. Subtract by counting on. Use a number line or the arrow way to help. The first one has been partially done for you.

   a. $3\frac{2}{5} - 1\frac{4}{5} = $ _____

   $1\frac{4}{5} + $ _____ $= 3\frac{2}{5}$

   b. $5\frac{3}{8} - 2\frac{5}{8}$

2. Subtract, as shown in Problem 2(a) below, by decomposing the fractional part of the number you are subtracting. Use a number line or the arrow way to help you.

   a. $4\frac{1}{5} - 1\frac{3}{5} = 3\frac{1}{5} - \frac{3}{5} = 2\frac{3}{5}$

   $\frac{1}{5} \quad \frac{2}{5}$

   b. $4\frac{1}{7} - 2\frac{4}{7}$

   c. $5\frac{5}{12} - 3\frac{8}{12}$

Lesson 33: Subtract a mixed number from a mixed number.

A STORY OF UNITS   Lesson 33 Homework 4•5

3. Subtract, as shown in 3(a) below, by decomposing to take one out.

   a. $5\frac{5}{8} - 2\frac{7}{8} = 3\frac{5}{8} - \frac{7}{8} =$

   $2\frac{5}{8} \quad 1$

   b. $4\frac{3}{12} - 3\frac{8}{12}$

   c. $9\frac{1}{10} - 6\frac{9}{10}$

4. Solve using any strategy.

   a. $6\frac{1}{9} - 4\frac{3}{9}$

   b. $5\frac{3}{10} - 3\frac{6}{10}$

   c. $8\frac{7}{12} - 5\frac{9}{12}$

   d. $7\frac{4}{100} - 2\frac{42}{100}$

Lesson 33: Subtract a mixed number from a mixed number.

A STORY OF UNITS   Lesson 34   4•5

Lesson 34

Objective: Subtract mixed numbers.

Suggested Lesson Structure

- Fluency Practice   (12 minutes)
- Application Problem   (5 minutes)
- Concept Development   (33 minutes)
- Student Debrief   (10 minutes)

Total Time   **(60 minutes)**

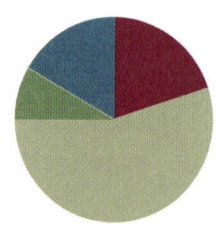

Fluency Practice (12 minutes)

- Sprint: Change Mixed Numbers to Fractions **4.NF.4**   (9 minutes)
- Subtract Fractions from Whole Numbers **4.NF.3**   (3 minutes)

Sprint: Change Mixed Numbers to Fractions (9 minutes)

Materials: (S) Change Mixed Numbers to Fractions Sprint

Note: This fluency activity reviews Lesson 25.

Subtract Fractions from Whole Numbers (3 minutes)

Materials: (S) Personal white board

Note: This fluency activity reviews Lesson 17.

    T: (Write $6 - \frac{4}{5}$.) Break apart the whole number and solve.

    S: (Write $6 - \frac{4}{5} = 5\frac{1}{5}$.)

Continue with the following possible sequence: $7 - \frac{5}{9}$ and $10 - \frac{5}{12}$.

450   Lesson 34:   Subtract mixed numbers.

A STORY OF UNITS

Lesson 34 4•5

Application Problem (5 minutes)

There were $4\frac{1}{8}$ pizzas. Benny took $\frac{2}{8}$ of a pizza. How many pizzas are left?

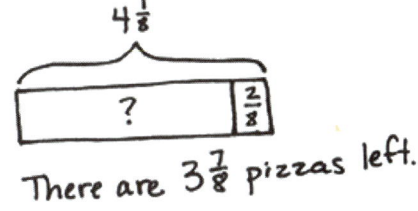

Solution A

$4\frac{1}{8} \xrightarrow{-\frac{1}{8}} 4 \xrightarrow{-\frac{1}{8}} 3\frac{7}{8}$

Solution B

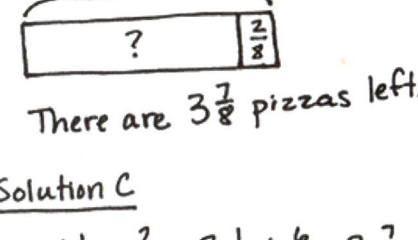

There are $3\frac{7}{8}$ pizzas left.

Solution C

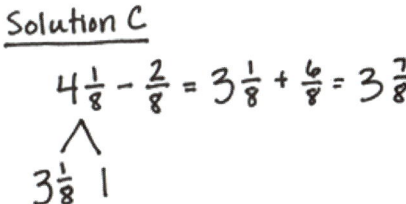

Note: This Application Problem reviews decomposition to subtract from a mixed number. This bridges to subtracting mixed numbers from mixed numbers.

Concept Development (33 minutes)

Materials: (S) Personal white board

Problem 1: Subtract a fraction from a mixed number by taking out 1 when there are not enough fractional units.

- T: (Write $8\frac{1}{10} - \frac{8}{10}$.) Do we have enough tenths to subtract 8 tenths?
- S: No!
- T: Let's decompose 8 ones 1 tenth by taking out 10 tenths from 8. How many ones and tenths make up the two parts of my number bond?
- S: 7 ones 11 tenths.
- T: (Record a number bond for $8\frac{1}{10}$.) Subtract.
- S: $7\frac{11}{10} - \frac{8}{10} = 7\frac{3}{10}$.
- T: Model the subtraction on a number line. Rename $8\frac{1}{10}$ and make one slide of $\frac{8}{10}$.
- S: (Draw a number line, and subtract as shown to the right.)

> **NOTES ON MULTIPLE MEANS OF REPRESENTATION:**
>
> The strategy presented here involves the decomposition of a higher value unit, the same process used in the standard algorithm when 8 tens 1 one would be renamed as 7 tens 11 ones to subtract 2 tens 8 ones.
>
> This connection is made in the Debrief. Students who struggle with this strategy may benefit from calling out the connection sooner if their understanding of renaming with whole number subtraction has a conceptual foundation.

$8\frac{1}{10} - \frac{8}{10} = 7\frac{11}{10} - \frac{8}{10} = 7\frac{3}{10}$

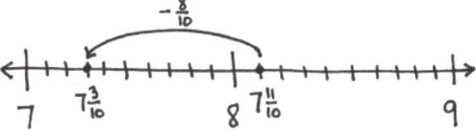

Let students practice with the following:

$6\frac{2}{8} - \frac{7}{8}$, $5\frac{1}{4} - \frac{3}{4}$, and $4\frac{2}{6} - \frac{5}{6}$.

Lesson 34: Subtract mixed numbers.

451

Problem 2: Subtract a mixed number from a mixed number by taking out 1 when there are not enough fractional units.

T: (Write $11\frac{1}{5} - 2\frac{3}{5}$.)

T: Subtract the whole numbers. What new subtraction expression remains?

S: $9\frac{1}{5} - \frac{3}{5}$.

T: (Write $9\frac{1}{5} - \frac{3}{5}$.)

T: Think back to the last problem you solved. What strategy did you use?

S: We renamed the first mixed number, or the total, from which we were subtracting.

T: Decompose 9 ones 1 fifth by taking out 5 fifths to make 6 fifths. How many ones and fifths are in the total?

S: (Record a number bond for $9\frac{1}{5}$.) 8 ones 6 fifths.

T: (Record a number bond for $9\frac{1}{5}$.) Subtract $8\frac{6}{5} - \frac{3}{5}$.

S: $8\frac{6}{5} - \frac{3}{5} = 8\frac{3}{5}$.

T: Explain to your partner why this is true. Draw a number line to explain your thinking.

S: It's like regrouping, so we have enough fifths to subtract. → We subtract 2 ones first. We can rename $9\frac{1}{5}$ as $8\frac{6}{5}$ and easily subtract $\frac{3}{5}$.

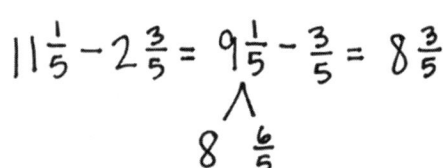

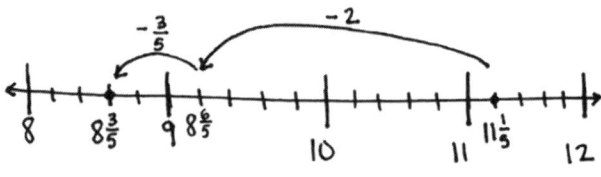

MP.3

Let students practice with the following: $4\frac{1}{8} - 1\frac{7}{8}$ and $7\frac{5}{12} - 3\frac{9}{12}$. Those who are struggling may need to record core steps to keep track of their thinking, e.g., $4\frac{1}{8} - 2\frac{7}{8} = 2\frac{1}{8} - \frac{7}{8} = 1\frac{9}{8} - \frac{7}{8} = 1\frac{2}{8}$, or to continue modeling with a number line. Have them share their work with a partner, explaining their solution.

Problem 3: Rename the total to subtract a mixed number from a mixed number when there are not enough actional units using the arrow way.

T: Solve $4\frac{1}{5} - 2\frac{4}{5}$. Tell your partner the first step.

S: Subtract the ones.

T: (Record subtracting 2 using the arrow way.) Say the number sentence.

S: $4\frac{1}{5} - 2\frac{4}{5} = 2\frac{1}{5} - \frac{4}{5}$.

T: Tell your partner the next step.

S: Rename $2\frac{1}{5}$ as $1\frac{6}{5}$, and subtract $\frac{4}{5}$.

T: (Record subtracting $\frac{4}{5}$ using the arrow way.) What is the difference?

S: $1\frac{2}{5}$.

T: Discuss with your partner what you have learned about mixed number subtraction that can help you solve without recording the number bond.

A STORY OF UNITS   Lesson 34 4•5

S: The arrow way lets me keep track of the steps in subtracting. → I can use counting backward. Subtracting $\frac{1}{5}$ gets me to 2, and then I just count back 3 more fifths. → I could rename the mixed number as a fraction greater than 1. So, $2\frac{1}{5}$ is the same as $\frac{11}{5}$. $\frac{11}{5} - \frac{4}{5}$ is easy to think of in my head.

Let students practice with the following: $9\frac{3}{8} - 7\frac{5}{8}$, $6\frac{2}{7} - 3\frac{6}{7}$, and $7\frac{3}{10} - 2\frac{4}{10}$. Encourage students to solve mentally, recording only as much as they need to keep track of the problem. Have students share their work with their partner to explain their solution.

Problem Set (10 minutes)

Students should do their personal best to complete the Problem Set within the allotted 10 minutes. For some classes, it may be appropriate to modify the assignment by specifying which problems they work on first. Some problems do not specify a method for solving. Students should solve these problems using the RDW approach used for Application Problems.

Student Debrief (10 minutes)

Lesson Objective: Subtract mixed numbers.

The Student Debrief is intended to invite reflection and active processing of the total lesson experience.

Invite students to review their solutions for the Problem Set. They should check work by comparing answers with a partner before going over answers as a class. Look for misconceptions or misunderstandings that can be addressed in the Student Debrief. Guide students in a conversation to debrief the Problem Set and process the lesson.

Any combination of the questions below may be used to lead the discussion.

- With your partner, compare and contrast the methods you used for solving Problem 3. Did you find that your partner used a method more efficient than your method? How can you be sure your methods are efficient and effective?
- Solve Problem 2(b) again; this time, do not subtract the ones first. What is more challenging about this method? What could be advantageous about this method?
- How can estimation be used when checking your work for this Problem Set?

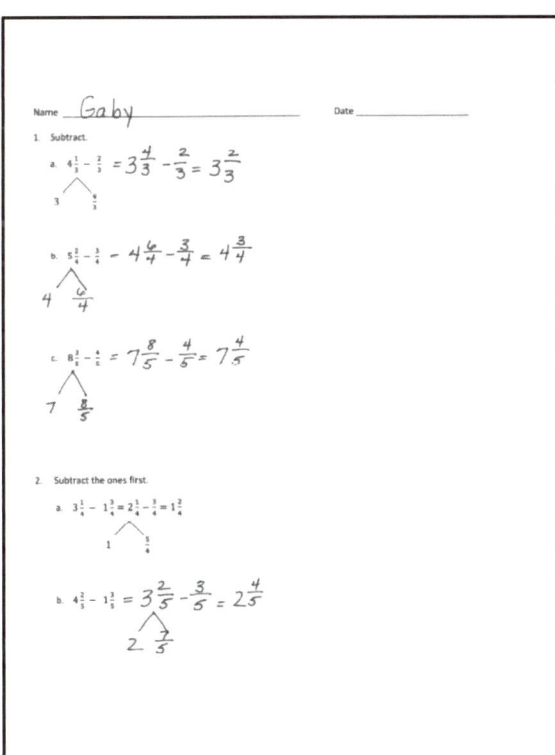

Lesson 34: Subtract mixed numbers.

- How is renaming to subtract 3 tens 8 ones from 6 tens 4 ones similar to how you solved for Problem 3(b)? Explain.

- We know $11 - 8 = 13 - 10 = 3$. What was added to the total and the part being subtracted? Think about this solution to Problem 3(c). How did this person solve Problem 3(c)?

$$8\tfrac{3}{12} - 3\tfrac{8}{12} = 8\tfrac{7}{12} - 4 = 4\tfrac{7}{12}$$

Exit Ticket (3 minutes)

After the Student Debrief, instruct students to complete the Exit Ticket. A review of their work will help with assessing students' understanding of the concepts that were presented in today's lesson and planning more effectively for future lessons. The questions may be read aloud to the students.

A STORY OF UNITS Lesson 34 Sprint 4•5

Number Correct: _____

A
Change Mixed Numbers to Fractions

1.	4 = 3 + ___		23.	$\frac{8}{4} =$	
2.	$\frac{4}{3} = \frac{3}{3} + \frac{}{3}$		24.	$\frac{}{4} = \frac{8}{4} + \frac{3}{4}$	
3.	$\frac{4}{3} = 1 + \frac{}{3}$		25.	$\frac{11}{4} = \frac{8}{4} + \frac{}{4}$	
4.	$\frac{4}{3} = 1\frac{}{3}$		26.	$\frac{11}{4} = 2 + \frac{}{4}$	
5.	6 = 5 + ___		27.	$\frac{11}{4} = 2\frac{}{4}$	
6.	$\frac{6}{5} = \frac{5}{5} + \frac{}{5}$		28.	$\frac{}{3} = \frac{6}{3} + \frac{1}{3}$	
7.	$\frac{6}{5} = 1 + \frac{}{5}$		29.	$\frac{}{3} = 2 + \frac{1}{3}$	
8.	$\frac{6}{5} = 1\frac{}{5}$		30.	$\frac{7}{3} = \underline{\quad}\frac{1}{3}$	
9.	5 = ___ + 1		31.	$\frac{8}{3} = \underline{\quad}\frac{2}{3}$	
10.	$\frac{5}{4} = \frac{}{4} + \frac{1}{4}$		32.	$\frac{17}{5} = \frac{}{5} + \frac{2}{5}$	
11.	$\frac{5}{4} = 1 + \frac{}{4}$		33.	$\frac{17}{5} = \frac{15}{5} + \frac{}{5}$	
12.	$\frac{5}{4} = \underline{\quad}\frac{1}{4}$		34.	$\frac{17}{5} = \underline{\quad} + \frac{2}{5}$	
13.	8 = ___ + 3		35.	$\frac{17}{5} = \underline{\quad}\frac{2}{5}$	
14.	$\frac{8}{5} = \frac{}{5} + \frac{3}{5}$		36.	$\frac{13}{6} = \frac{12}{6} + \frac{}{6}$	
15.	$\frac{8}{5} = 1 + \frac{}{5}$		37.	$\frac{13}{6} = \underline{\quad} + \frac{1}{6}$	
16.	$\frac{8}{5} = 1\frac{}{5}$		38.	$\frac{13}{6} = 2\frac{}{6}$	
17.	$\frac{7}{5} = 1\frac{}{5}$		39.	$\frac{17}{6} = 2\frac{}{6}$	
18.	$\frac{6}{5} = 1\frac{}{5}$		40.	$\frac{9}{8} = 1 + \frac{}{8}$	
19.	$\frac{9}{5} = 1\frac{}{5}$		41.	$\frac{13}{8} = 1 + \frac{}{8}$	
20.	$\frac{10}{5} =$		42.	$\frac{19}{10} = 1 + \frac{}{10}$	
21.	$\frac{}{5} = \frac{10}{5} + \frac{4}{5}$		43.	$\frac{19}{12} = \frac{}{12} + \frac{7}{12}$	
22.	$\frac{}{5} = 2 + \frac{4}{5}$		44.	$\frac{11}{6} = 1 + \frac{}{6}$	

Lesson 34: Subtract mixed numbers. 455

B

Change Mixed Numbers to Fractions

Number Correct: _____

Improvement: _____

1.	$5 = 4 + \_\_\_$		23.	$\frac{6}{3} =$	
2.	$\frac{5}{4} = \frac{4}{4} + \frac{}{4}$		24.	$\frac{}{3} = \frac{6}{3} + \frac{2}{3}$	
3.	$\frac{5}{4} = 1 + \frac{}{4}$		25.	$\frac{8}{3} = \frac{6}{3} + \frac{}{3}$	
4.	$\frac{5}{4} = 1\frac{}{4}$		26.	$\frac{8}{3} = 2 + \frac{}{3}$	
5.	$3 = 2 + \_\_\_$		27.	$\frac{8}{3} = 2\frac{}{3}$	
6.	$\frac{3}{2} = \frac{2}{2} + \frac{}{2}$		28.	$\frac{}{10} = \frac{20}{10} + \frac{1}{10}$	
7.	$\frac{3}{2} = 1 + \frac{}{2}$		29.	$\frac{}{10} = 2 + \frac{1}{10}$	
8.	$\frac{3}{2} = 1\frac{}{2}$		30.	$\frac{21}{10} = \frac{}{}\frac{1}{10}$	
9.	$9 = \_\_\_ + 1$		31.	$\frac{27}{10} = \frac{}{}\frac{7}{10}$	
10.	$\frac{9}{8} = \frac{}{8} + \frac{1}{8}$		32.	$\frac{13}{6} = \frac{}{6} + \frac{1}{6}$	
11.	$\frac{9}{8} = 1 + \frac{}{8}$		33.	$\frac{13}{6} = \frac{12}{6} + \frac{}{6}$	
12.	$\frac{9}{8} = \frac{}{}\frac{1}{8}$		34.	$\frac{13}{6} = \_\_\_ + \frac{1}{6}$	
13.	$9 = \_\_\_ + 4$		35.	$\frac{13}{6} = \frac{}{}\frac{1}{6}$	
14.	$\frac{9}{5} = \frac{}{5} + \frac{4}{5}$		36.	$\frac{17}{8} = \frac{16}{8} + \frac{}{8}$	
15.	$\frac{9}{5} = 1 + \frac{}{5}$		37.	$\frac{17}{8} = \frac{}{8} + \frac{1}{8}$	
16.	$\frac{9}{5} = 1\frac{}{5}$		38.	$\frac{17}{8} = 2\frac{}{8}$	
17.	$\frac{8}{5} = 1\frac{}{5}$		39.	$\frac{21}{8} = 2\frac{}{8}$	
18.	$\frac{7}{5} = 1\frac{}{5}$		40.	$\frac{7}{6} = 1 + \frac{}{6}$	
19.	$\frac{6}{5} = 1\frac{}{5}$		41.	$\frac{11}{6} = 1 + \frac{}{6}$	
20.	$\frac{8}{4} =$		42.	$\frac{13}{5} = 2 + \frac{}{5}$	
21.	$\frac{}{4} = \frac{8}{4} + \frac{1}{4}$		43.	$\frac{17}{12} = \frac{}{12} + \frac{5}{12}$	
22.	$\frac{}{4} = 2 + \frac{1}{4}$		44.	$\frac{13}{8} = 1 + \frac{}{8}$	

Lesson 34: Subtract mixed numbers.

A STORY OF UNITS

Lesson 34 Problem Set 4•5

Name _____   Date _____

1. Subtract.

    a. $4\frac{1}{3} - \frac{2}{3}$

    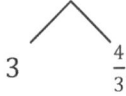

    b. $5\frac{2}{4} - \frac{3}{4}$

    c. $8\frac{3}{5} - \frac{4}{5}$

2. Subtract the ones first.

    a. $3\frac{1}{4} - 1\frac{3}{4} = 2\frac{1}{4} - \frac{3}{4} = 1\frac{2}{4}$

    b. $4\frac{2}{5} - 1\frac{3}{5}$

Lesson 34: Subtract mixed numbers.

c. $5\frac{2}{6} - 3\frac{5}{6}$

d. $9\frac{3}{5} - 2\frac{4}{5}$

3. Solve using any strategy.

    a. $7\frac{3}{8} - 2\frac{5}{8}$

    b. $6\frac{4}{10} - 3\frac{8}{10}$

    c. $8\frac{3}{12} - 3\frac{8}{12}$

    d. $14\frac{2}{50} - 6\frac{43}{50}$

A STORY OF UNITS — **Lesson 34 Exit Ticket** 4•5

Name _____ Date _____

Solve.

1. $7\frac{1}{6} - 2\frac{4}{6}$

2. $12\frac{5}{8} - 3\frac{7}{8}$

Lesson 34: Subtract mixed numbers.

Name _____ Date _____

1. Subtract

   a. $5\frac{1}{4} - \frac{3}{4}$

   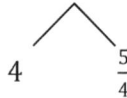
   4      $\frac{5}{4}$

   b. $6\frac{3}{8} - \frac{6}{8}$

   c. $7\frac{4}{6} - \frac{5}{6}$

2. Subtract the ones first.

   a. $4\frac{1}{5} - 1\frac{3}{5} = 3\frac{1}{5} - \frac{3}{5} = 2\frac{3}{5}$

   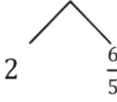
   2      $\frac{6}{5}$

   b. $4\frac{3}{6} - 2\frac{5}{6}$

c. $8\frac{3}{8} - 2\frac{5}{8}$

d. $13\frac{3}{12} - 8\frac{7}{12}$

3. Solve using any strategy.

   a. $7\frac{3}{12} - 4\frac{9}{12}$

   b. $9\frac{6}{12} - 5\frac{8}{12}$

   c. $17\frac{2}{16} - 9\frac{7}{16}$

   d. $12\frac{5}{100} - 8\frac{14}{100}$

Lesson 34: Subtract mixed numbers.

A STORY OF UNITS

Mathematics Curriculum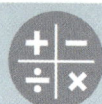

GRADE 4 • MODULE 5

Topic G
Repeated Addition of Fractions as Multiplication

4.NF.4, 4.OA.2, 4.MD.2, 4.MD.4

Focus Standard:	4.NF.4	Apply and extend previous understandings of multiplication to multiply a fraction by a whole number.
		a. Understand a fraction a/b as a multiple of $1/b$. For example, use a visual fraction model to represent 5/4 as the product 5 × (1/4), recording the conclusion by the equation 5/4 = 5 × (1/4).
		b. Understand a multiple of a/b as a multiple of $1/b$, and use this understanding to multiply a fraction by a whole number. For example, use a visual fraction model to express 3 × (2/5) as 6 × (1/5), recognizing this product as 6/5. (In general, n × (a/b) = (n × a)/b.)
		c. Solve word problems involving multiplication of a fraction by a whole number, e.g., by using visual fraction models and equations to represent the problem. For example, if each person at a party will eat 3/8 of a pound of roast beef, and there will be 5 people at the party, how many pounds of roast beef will be needed? Between what two whole numbers does your answer lie?
Instructional Days:	6	
Coherence -Links from:	G3–M5	Fractions as Numbers on the Number Line
-Links to:	G5–M3	Addition and Subtraction of Fractions
	G5–M4	Multiplication and Division of Fractions and Decimal Fractions

A STORY OF UNITS   Topic G   4•5

Topic G extends the concept of representing repeated addition as multiplication, applying this familiar concept to work with fractions.

Multiplying a whole number times a fraction was introduced in Topic A as students learned to decompose fractions, e.g., $\frac{3}{5} = \frac{1}{5} + \frac{1}{5} + \frac{1}{5} = 3 \times \frac{1}{5}$. In Lessons 35 and 36, students use the associative property, as exemplified below, to multiply a whole number times a mixed number.

3 bananas + 3 bananas + 3 bananas + 3 bananas
= 4 × 3 bananas
= 4 × (3 × 1 banana) = (4 × 3) × 1 banana = 12 bananas
3 fifths + 3 fifths + 3 fifths + 3 fifths
= 4 × 3 fifths
= 4 × (3 fifths) = (4 × 3) fifths = 12 fifths

$4 \times \frac{3}{5}$

$4 \times \left(3 \times \frac{1}{5}\right) = (4 \times 3) \times \frac{1}{5} = \frac{4 \times 3}{5} = \frac{12}{5}$

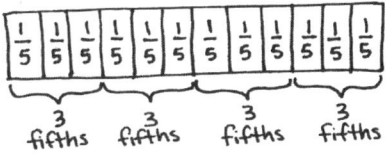

$4 \times (3 \text{ fifths}) = (4 \times 3) \text{ fifths}$
$= 12 \text{ fifths}$

$4 \times 3 \text{ fifths} = 12 \text{ fifths}$
$4 \times \frac{3}{5} = \frac{12}{5}$

Students may have never before considered that 3 bananas = 3 × 1 banana, but it is an understanding that connects place value, whole number work, measurement conversions, and fractions, e.g., 3 hundreds = 3 × 1 hundred or 3 feet = 3 × (1 foot); 1 foot = 12 inches; therefore, 3 feet = 3 × (12 inches) = (3 × 12) inches = 36 inches.

Students explore the use of the distributive property in Lessons 37 and 38 to multiply a whole number by a mixed number. They see the multiplication of each part of a mixed number by the whole number and use the appropriate strategies to do so. As students progress through each lesson, they are encouraged to record only as much as they need to keep track of the math. As shown below, there are multiple steps when using the distributive property, and students can become lost in those steps. Efficiency in solving is encouraged.

$2 \times 3\frac{1}{5} = (2 \times 3) + (2 \times \frac{1}{5})$
$= 6 + \frac{2}{5} = 6\frac{2}{5}$

$4 \times 9\frac{3}{4} = 36 + \frac{12}{4}$
$= 36 + 3$
$= 39$

$5 \times 3\frac{3}{4} = 5 \times (3 + \frac{3}{4}) = (5 \times 3) + (5 \times \frac{3}{4}) = 15 + \frac{5 \times 3}{4} = 15 + \frac{15}{4} = 15 + 3\frac{3}{4} = 18\frac{3}{4}$

Topic G: Repeated Addition of Fractions as Multiplication

A STORY OF UNITS · Topic G · 4•5

In Lesson 39, students build their problem-solving skills by solving multiplicative comparison word problems involving mixed numbers, e.g., "Jennifer bought 3 times as much meat on Saturday as she did on Monday. If she bought $1\frac{1}{2}$ pounds on Monday, what is the total amount of meat bought for the two days?" They create and use tape diagrams to represent these problems before using various strategies to solve them numerically.

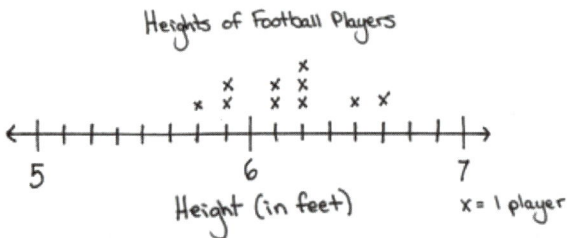

In Lesson 40, students solve word problems involving multiplication of a fraction by a whole number. Additionally, students work with data presented in line plots.

A Teaching Sequence Toward Mastery of Repeated Addition of Fractions as Multiplication
Objective 1: Represent the multiplication of *n* times *a/b* as (*n* × *a*)/*b* using the associative property and visual models. (Lessons 35–36)
Objective 2: Find the product of a whole number and a mixed number using the distributive property. (Lessons 37–38)
Objective 3: Solve multiplicative comparison word problems involving fractions. (Lesson 39)
Objective 4: Solve word problems involving the multiplication of a whole number and a fraction including those involving line plots. (Lesson 40)

Lesson 35

Objective: Represent the multiplication of *n* times *a/b* as (*n* × *a*)/*b* using the associative property and visual models.

Suggested Lesson Structure

- ■ Fluency Practice   (12 minutes)
- ■ Application Problem   (5 minutes)
- ■ Concept Development   (33 minutes)
- ■ Student Debrief   (10 minutes)
- **Total Time**   **(60 minutes)**

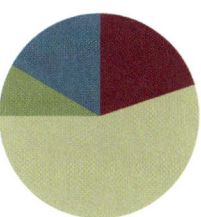

Fluency Practice (12 minutes)

- Add and Subtract **4.NBT.4**   (4 minutes)
- Count by Equivalent Fractions **4.NF.1**   (4 minutes)
- Add and Subtract Mixed Numbers **4.NF.3**   (4 minutes)

Add and Subtract (4 minutes)

Materials: (S) Personal white board

Note: This fluency activity reviews adding and subtracting using the standard algorithm.

- T: (Write 676 thousands 696 ones.) On your personal white boards, write this number in standard form.
- S: (Write 676,696.)
- T: (Write 153 thousands 884 ones.) Add this number to 676,696 using the standard algorithm.
- S: (Write 676,696 + 153,884 = 830,580 using the standard algorithm.)

Continue the process for 678,717 + 274,867.

- T: (Write 300 thousands.) On your boards, write this number in standard form.
- S: (Write 300,000.)
- T: (Write 134 thousands 759 ones.) Subtract this number from 300,000 using the standard algorithm.
- S: (Write 300,000 − 134,759 = 165,241 using the standard algorithm.)

Continue the process for 734,902 − 477,479.

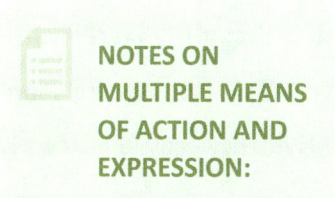

NOTES ON MULTIPLE MEANS OF ACTION AND EXPRESSION:

Some learners may benefit from using grid paper or a place value chart to organize numbers up to 1 million as they add and subtract.

Lesson 35: Represent the multiplication of *n* times *a/b* as (*n* × *a*)/*b* using the associative property and visual models.

A STORY OF UNITS Lesson 35 4•5

Count by Equivalent Fractions (4 minutes)

Note: This activity reviews Lesson 24. The progression builds in complexity. Work students up to the highest level of complexity in which they can confidently participate.

T: Count by threes to 15, starting at 0.
S: 0, 3, 6, 9, 12, 15.
T: Count by 3 fifths to 15 fifths, starting at 0 fifths. (Write as students count.)
S: $\frac{0}{5}, \frac{3}{5}, \frac{6}{5}, \frac{9}{5}, \frac{12}{5}, \frac{15}{5}$.
T: 1 one is the same as how many fifths?
S: 5 fifths.
T: 2 ones?
S: 10 fifths.
T: 3 ones?
S: 15 fifths.
T: (Beneath $\frac{15}{5}$, write 3.) Count by 3 fifths again. This time, when you come to the whole number, say the whole number. Start at zero. (Write as students count.)
S: $0, \frac{3}{5}, \frac{6}{5}, \frac{9}{5}, \frac{12}{5}, 3$.
T: (Point to $\frac{6}{5}$.) Say $\frac{6}{5}$ as a mixed number.
S: $1\frac{1}{5}$.

Continue the process for $\frac{9}{5}$ and $\frac{12}{5}$.

T: Count by 3 fifths again. This time, convert to whole numbers and mixed numbers. Start at zero. (Write as students count.)
S: $0, \frac{3}{5}, 1\frac{1}{5}, 1\frac{4}{5}, 2\frac{2}{3}, 3$.

Add and Subtract Mixed Numbers (4 minutes)

Materials: (S) Personal white board

Note: This fluency activity reviews Lesson 31 and Lesson 33. Allow students to solve using any strategy.

T: (Write $5\frac{5}{10} + 3\frac{2}{10} = $ ___.) Decompose the mixed numbers and solve.
S: (Write $5\frac{5}{10} + 3\frac{2}{10} = 8\frac{7}{10}$.)

Continue with the following possible sequence: $2\frac{3}{5} + 2\frac{2}{5}$, $10\frac{3}{5} + 5\frac{4}{5}$, and $7\frac{2}{3} + 3\frac{2}{3}$

T: (Write $10\frac{7}{10} - 5\frac{4}{10} = $ ___.) Decompose the mixed numbers and solve.
S: (Write $10\frac{7}{10} - 5\frac{4}{10} = 5\frac{3}{10}$.)

Continue with the following possible sequence: $6\frac{2}{3} - 3\frac{2}{3}$, $6\frac{1}{3} - 4\frac{2}{3}$, $10\frac{1}{5} - 4\frac{3}{5}$, and $6\frac{3}{8} - 2\frac{7}{8}$.

466   Lesson 35:   Represent the multiplication of n times a/b as (n × a)/b using the associative property and visual models.

A STORY OF UNITS — Lesson 35 4•5

Application Problem (5 minutes)

Mary Beth is knitting scarves that are 1 meter long. If she knits 54 centimeters of a scarf each night for 3 nights, how many scarves will she complete? How much more does she need to knit to complete another scarf?

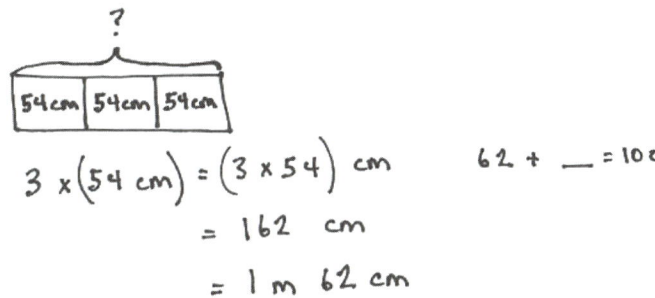

NOTES ON MULTIPLE MEANS OF REPRESENTATION:

Modeling the change in the association of the 54 centimeters with the factor of 3 prepares students to interpret fraction multiplication using the associative property, (e.g., 3 × 4 fifths = $3 \times \frac{4}{5} = (3 \times 4)$ fifths = $\frac{3 \times 4}{5}$.

Students might also benefit from understanding that 3 × 54 cm = 3 × (54 × 1 cm), just as $3 \times \frac{4}{5} = 3 \times \left(4 \times \frac{1}{5}\right)$.

Since notation can become a barrier for students, be prepared to adjust it when necessary. The Concept Development aims to keep it very simple.

Note: This Application Problem prepares students to think about how a fractional unit behaves like any other unit in a multiplication sentence, e.g., 3 × 4 wheels = 12 wheels, 3 × 54 centimeters = 162 centimeters, and 3 × 4 fifths = 12 fifths or $\frac{3 \times 4}{5}$.

Concept Development (33 minutes)

Materials: (S) Personal white board

Problem 1: Use the associative property to solve $n \times \frac{a}{b}$ in unit form.

T: Write a multiplication number sentence to show four copies of 3 centimeters.

S: (Write 4 × 3 centimeters = 12 centimeters.)

T: (Write 4 × (3 centimeters).) I put parentheses around 3 centimeters to show that 3 is telling the number of centimeters in one group, but to solve, we moved the parentheses. Show me where you moved them to.

S: (Write (4 × 3) centimeters = 12 centimeters.)

T: Yes, you used the associative property by associating the 3 with the number of groups rather than the unit of centimeters.

Lesson 35: Represent the multiplication of n times a/b as (n × a)/b using the associative property and visual models.

467

A STORY OF UNITS   Lesson 35   4•5

T: Write a multiplication number sentence to show four copies of 3 fifths in unit form.

S: (Write 4 × 3 fifths = 12 fifths.)

T: (Write 4 × (3 fifths) = (4 × 3) fifths.) Is this true?

S: Yes, that's the associative property.

T: Draw a tape diagram to show four copies of 3 fifths.

S: (Draw a tape diagram.)

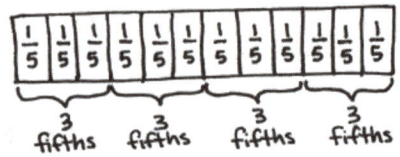

$4 \times (3 \text{ fifths}) = (4 \times 3) \text{ fifths}$
$= 12 \text{ fifths}$

Repeat with three copies of 5 sixths and four copies of 3 eighths, associating the factors and drawing a matching tape diagram.

Problem 2: Use the associative property to solve $n \times \frac{a}{b}$ numerically.

T: (Display $4 \times \frac{3}{5}$.) Say this expression.

S: Four times 3 fifths.

T: Write it in unit form.

S: (Write 4 × 3 fifths.) We just did this problem!

T: (Write 4 × 3 fifths = 12 fifths and $4 \times \frac{3}{5} = \frac{12}{5}$, as shown to the right.) Compare these number sentences. Are these true? Discuss with your partner.

$4 \times 3 \text{ fifths} = 12 \text{ fifths}$

$4 \times \frac{3}{5} = \frac{12}{5}$

S: Yes, the top was solved in unit form, and the bottom used numbers.

T: (Write $4 \times (3 \times \frac{1}{5}) = 4 \times 3$ fifths.) We can say $4 \times (3 \times \frac{1}{5}) = 4 \times 3$ fifths. On your personal board, move the parentheses to associate the factors of 4 and 3.

S: (Write $(4 \times 3) \times \frac{1}{5}$.)

T: And the value is…?

S: $\frac{12}{5}$.

T: (Write $4 \times (3 \times \frac{1}{5}) = (4 \times 3) \times \frac{1}{5} = \frac{12}{5}$.) Is 4 groups of 3 fifths the same as 12 fifths?

S: Yes.

T: (Display $5 \times \frac{3}{4}$.) Say this expression.

S: Five times 3 fourths.

T: Keep the unit form in mind as you solve numerically. Record only as much as you need.

S: $5 \times \frac{3}{4} = \frac{15}{4}$.

> **NOTES ON MULTIPLE MEANS OF REPRESENTATION:**
>
> When using the associative property to solve $4 \times \frac{3}{5}$, some students may proficiently solve mentally, while others may need visual support to solve, including step-by-step guidance. For example, before asking for the value of $(4 \times 3) \times \frac{1}{5}$, it might be helpful to ask, "What is $12 \times \frac{1}{5}$?"

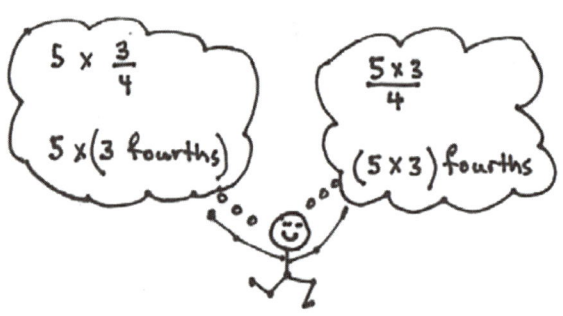

Lesson 35: Represent the multiplication of n times a/b as (n × a)/b using the associative property and visual models.

T: Yes, and as I thought of this as 5 times 3 fourths, I wrote down $5 \times \frac{3}{4} = \frac{5 \times 3}{4} = \frac{15}{4}$. Why is my number sentence true?

S: When you associated the factors, fourths became the unit, and we write the unit fourths as the denominator.

T: Yes. I think of 5 × (3 fourths) as $5 \times \frac{3}{4}$ and (5 × 3) fourths as $\frac{5 \times 3}{4}$. Both have the same value—12 fourths.

Repeat with $8 \times \frac{2}{3}$ and $12 \times \frac{3}{10}$.

Problem Set (10 minutes)

Students should do their personal best to complete the Problem Set within the allotted 10 minutes. For some classes, it may be appropriate to modify the assignment by specifying which problems they work on first. Some problems do not specify a method for solving. Students should solve these problems using the RDW approach used for Application Problems.

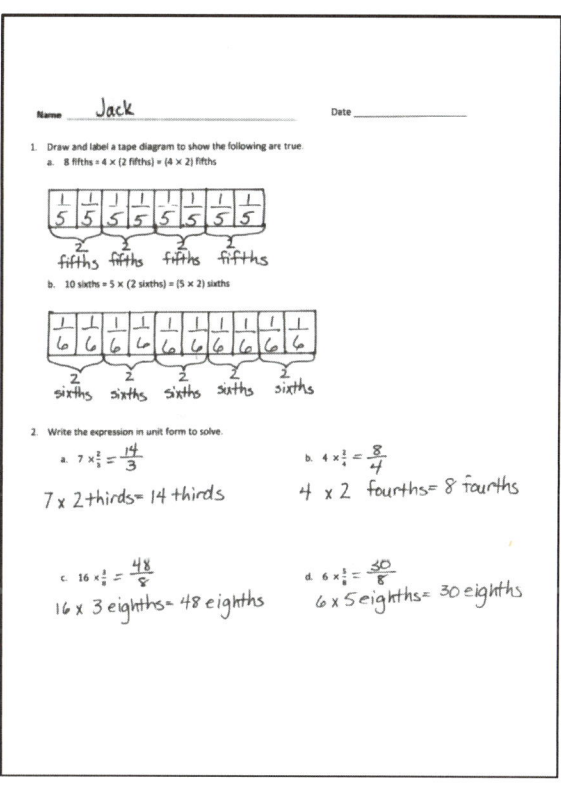

Student Debrief (10 minutes)

Lesson Objective: Represent the multiplication of n times a/b as (n × a)/b using the associative property and visual models.

The Student Debrief is intended to invite reflection and active processing of the total lesson experience.

Invite students to review their solutions for the Problem Set. They should check work by comparing answers with a partner before going over answers as a class. Look for misconceptions or misunderstandings that can be addressed in the Student Debrief. Guide students in a conversation to debrief the Problem Set and process the lesson.

Any combination of the questions below may be used to lead the discussion.

- How do the tape diagrams that you drew in Problems 1(a) and 1(b) help with the understanding that there are different ways to express fractions?
- How did you record your solutions to Problem 3(a–f)?
- Look at your answers for Problem 3(c) and 3(d). Convert each answer to a mixed number. What do you notice? How are the expressions in Problem 3(c) and 3(d) similar?
- How does moving the parentheses change the meaning of the expression? Use the tape diagrams in Problem 1 to help you explain.

A STORY OF UNITS   Lesson 35   4•5

- Explain to a partner how you solved Problem 4.
- What significant math vocabulary did we use today to communicate precisely?
- How does the Application Problem relate to today's Concept Development?

Exit Ticket (3 minutes)

After the Student Debrief, instruct students to complete the Exit Ticket. A review of their work will help with assessing students' understanding of the concepts that were presented in today's lesson and planning more effectively for future lessons. The questions may be read aloud to the students.

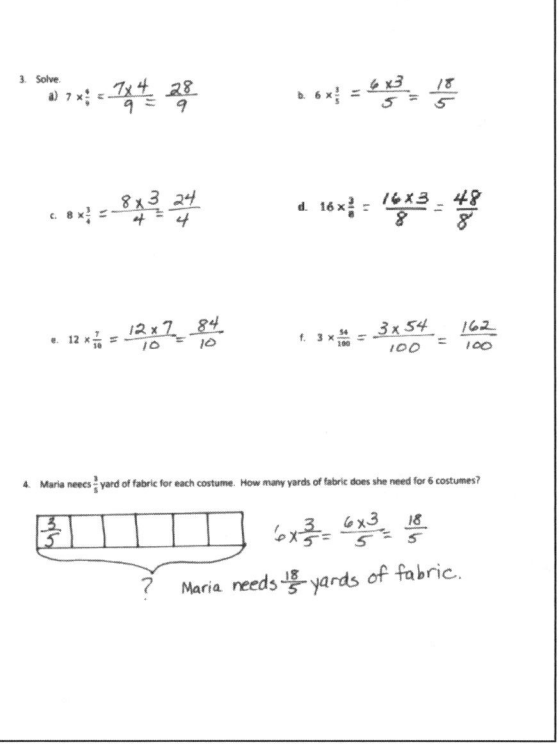

Lesson 35: Represent the multiplication of *n* times *a*/*b* as (*n* × *a*)/*b* using the associative property and visual models.

A STORY OF UNITS

Lesson 35 Problem Set 4•5

Name _____ Date _____

1. Draw and label a tape diagram to show the following are true.

    a. 8 fifths = 4 × (2 fifths) = (4 × 2) fifths

    b. 10 sixths = 5 × (2 sixths) = (5 × 2) sixths

2. Write the expression in unit form to solve.

    a. $7 \times \frac{2}{3}$

    b. $4 \times \frac{2}{4}$

    c. $16 \times \frac{3}{8}$

    d. $6 \times \frac{5}{8}$

Lesson 35: Represent the multiplication of n times a/b as (n × a)/b using the associative property and visual models.

3. Solve.

    a. $7 \times \frac{4}{9}$

    b. $6 \times \frac{3}{5}$

    c. $8 \times \frac{3}{4}$

    d. $16 \times \frac{3}{8}$

    e. $12 \times \frac{7}{10}$

    f. $3 \times \frac{54}{100}$

4. Maria needs $\frac{3}{5}$ yard of fabric for each costume. How many yards of fabric does she need for 6 costumes?

Name _____ Date _____

1. Solve using unit form.

   $5 \times \frac{2}{3}$

2. Solve.

   $11 \times \frac{5}{6}$

Lesson 35: Represent the multiplication of n times a/b as (n × a)/b using the associative property and visual models.

A STORY OF UNITS

Lesson 35 Homework 4•5

Name _____ Date _____

1. Draw and label a tape diagram to show the following are true.

   a. 8 thirds = 4 × (2 thirds) = (4 × 2) thirds

   b. 15 eighths = 3 × (5 eighths) = (3 × 5) eighths

2. Write the expression in unit form to solve.

   a. $10 \times \frac{2}{5}$

   b. $3 \times \frac{5}{6}$

   c. $9 \times \frac{4}{9}$

   d. $7 \times \frac{3}{4}$

474   Lesson 35:   Represent the multiplication of n times a/b as (n × a)/b using the associative property and visual models.

3. Solve.

   a. $6 \times \frac{3}{4}$

   b. $7 \times \frac{5}{8}$

   c. $13 \times \frac{2}{3}$

   d. $18 \times \frac{2}{3}$

   e. $14 \times \frac{7}{10}$

   f. $7 \times \frac{14}{100}$

4. Mrs. Smith bought some orange juice. Each member of her family drank $\frac{2}{3}$ cup for breakfast. There are five people in her family. How many cups of orange juice did they drink?

Lesson 36

Objective: Represent the multiplication of *n* times *a/b* as (*n* × *a*)/*b* using the associative property and visual models.

Suggested Lesson Structure

- ■ Fluency Practice   (10 minutes)
- ■ Application Problem   (5 minutes)
- ■ Concept Development   (35 minutes)
- ■ Student Debrief   (10 minutes)
- **Total Time**   **(60 minutes)**

Fluency Practice (10 minutes)

- Count by Equivalent Fractions **4.NF.1**   (5 minutes)
- Multiply Fractions **4.NF.4**   (5 minutes)

Count by Equivalent Fractions (5 minutes)

Note: This activity reviews Lessons 24 and 25. The progression builds in complexity. Work students up to the highest level of complexity in which they can confidently participate.

T: Count by threes to 30, starting at 0.
S: 0, 3, 6, 9, 12, 15, 18, 21, 24, 27, 30.
T: Count by 3 tenths to 30 tenths, starting at 0 tenths. (Write as students count.)

$\frac{0}{10}$	$\frac{3}{10}$	$\frac{6}{10}$	$\frac{9}{10}$	$\frac{12}{10}$	$\frac{15}{10}$	$\frac{18}{10}$	$\frac{21}{10}$	$\frac{24}{10}$	$\frac{27}{10}$	$\frac{30}{10}$
0	$\frac{3}{10}$	$\frac{6}{10}$	$\frac{9}{10}$	$\frac{12}{10}$	$\frac{15}{10}$	$\frac{18}{10}$	$\frac{21}{10}$	$\frac{24}{10}$	$\frac{27}{10}$	3
0	$\frac{3}{10}$	$\frac{6}{10}$	$\frac{9}{10}$	$1\frac{2}{10}$	$1\frac{5}{10}$	$1\frac{8}{10}$	$2\frac{1}{10}$	$2\frac{4}{10}$	$2\frac{7}{10}$	3

S: $\frac{0}{10}, \frac{3}{10}, \frac{6}{10}, \frac{9}{10}, \frac{12}{10}, \frac{15}{10}, \frac{18}{10}, \frac{21}{10}, \frac{24}{10}, \frac{27}{10}, \frac{30}{10}$.
T: Name the fraction that's equal to a whole number.
S: 30 tenths.
T: (Point to $\frac{30}{10}$.) 30 tenths is how many ones?
S: 3 ones.

A STORY OF UNITS   Lesson 36   4•5

T: (Beneath $\frac{30}{10}$, write 3 ones.) Count by 3 tenths again. This time, when you come to the whole number, say the whole number. Start at zero. (Write as students count.)

S: $0, \frac{3}{10}, \frac{6}{10}, \frac{9}{10}, \frac{12}{10}, \frac{15}{10}, \frac{18}{10}, \frac{21}{10}, \frac{24}{10}, \frac{27}{10}, 3$.

T: (Point to $\frac{12}{10}$.) Say $\frac{12}{10}$ as a mixed number.

S: $1\frac{2}{10}$.

Continue the process for $1\frac{5}{10}, 1\frac{8}{10}, 2\frac{1}{10}, 2\frac{4}{10}$, and $2\frac{7}{10}$.

T: Count by 3 tenths again. This time, convert to whole numbers and mixed numbers. Start at zero. (Write as students count.)

S: $0, \frac{3}{10}, \frac{6}{10}, \frac{9}{10}, 1\frac{2}{10}, 1\frac{5}{10}, 1\frac{8}{10}, 2\frac{1}{10}, 2\frac{4}{10}, 2\frac{7}{10}, 3$.

Multiply Fractions (5 minutes)

Materials: (S) Personal white board

Note: This fluency activity reviews Lesson 35.

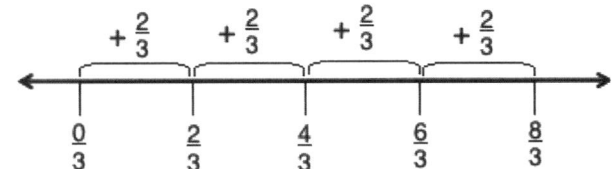

T: Draw a number line.

S: (Draw a number line.)

T: (Write $4 \times \frac{2}{3}$.) Starting with zero, mark four intervals of $\frac{2}{3}$ on the number line.

S: (Mark $\frac{2}{3}, \frac{4}{3}, \frac{6}{3}$, and $\frac{8}{3}$ on the number line.)

T: What's $4 \times \frac{2}{3}$?

S: $\frac{8}{3}$.

T: (Write $3 \times \frac{2}{3} = $ ___.) Complete the number sentence.

S: (Write $3 \times \frac{2}{3} = \frac{6}{3}$.)

T: (Write $2 \times \frac{2}{3} = $ ___.) Complete the number sentence.

S: (Write $2 \times \frac{2}{3} = \frac{4}{3}$.)

Continue with the following possible sequence: $4 \times \frac{2}{5}$ and $5 \times \frac{3}{4}$.

Lesson 36:   Represent the multiplication of n times a/b as (n × a)/b using the associative property and visual models.   477

A STORY OF UNITS   Lesson 36   4•5

Application Problem (5 minutes)

Rhonda exercised for $\frac{5}{6}$ hour every day for 5 days. How many total hours did Rhonda exercise?

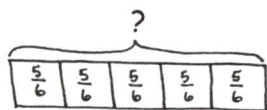

$5 \times \frac{5}{6} = \frac{25}{6} = 4\frac{1}{6}$

$\frac{24}{6} \quad \frac{1}{6}$

Rhonda exercised for $4\frac{1}{6}$ hours.

> **NOTES ON MULTIPLE MEANS OF ENGAGEMENT:**
>
> Adjust the Application Problem to challenge students working above grade level. For example, ask, "How many total hours and minutes did Rhonda exercise?"

Note: This Application Problem builds on the learning from the previous lesson where students multiplied a whole number by a fraction.

Concept Development (35 minutes)

Materials: (S) Personal white board

Problem 1: Rewrite a repeated addition problem as *n* times *a/b*.

T: Look back to the tape diagram we drew for the Application Problem. Say an addition sentence that represents this model.

S: $\frac{5}{6}+\frac{5}{6}+\frac{5}{6}+\frac{5}{6}+\frac{5}{6} = \frac{25}{6}$.

T: Write it as a multiplication sentence.

S: $5 \times \frac{5}{6} = \frac{25}{6}$.

T: Which is more efficient? $\frac{5}{6}+\frac{5}{6}+\frac{5}{6}+\frac{5}{6}+\frac{5}{6}$ or $5 \times \frac{5}{6}$? Discuss with your partner.

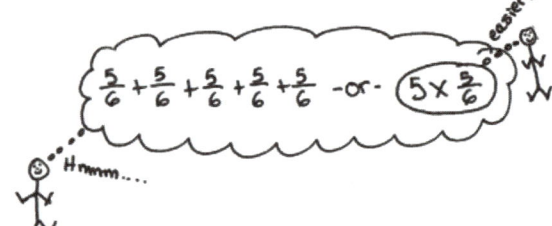

S: $5 \times \frac{5}{6}$. It doesn't take as long to write. → Multiplication is usually more efficient because making groups is easier than counting by fives, especially if there are a lot of copies.

T: How do we solve $5 \times \frac{5}{6}$?

S: We know $5 \times \frac{5}{6}$ can be solved like this: $\frac{5 \times 5}{6} = \frac{25}{6}$. → It's 5 × 5 sixths, so that is 25 sixths.

Repeat with $\frac{3}{5}+\frac{3}{5}+\frac{3}{5}+\frac{3}{5}$, drawing a tape diagram and solving using multiplication.

MP.2

478   Lesson 36: Represent the multiplication of *n* times *a/b* as (*n* × *a*)/*b* using the associative property and visual models.

A STORY OF UNITS   Lesson 36   4•5

Problem 2: Solve n times a/b as (n × a)/b.

T: (Project $6 \times \frac{3}{8}$.) Say this expression in unit form.

S: 6 × 3 eighths.

T: (6 × 3) eighths = $\frac{6 \times 3}{8}$, yes?

S: Yes!

T: Use this way of recording this time.

S: $6 \times \frac{3}{8} = \frac{6 \times 3}{8} = \frac{18}{8}$.

T: Rename as a mixed number.

S: $\frac{18}{8} = \frac{16}{8} + \frac{2}{8} = 2\frac{2}{8}$.

Repeat with $\frac{3}{8} \times 5$ and $9 \times \frac{4}{5}$.

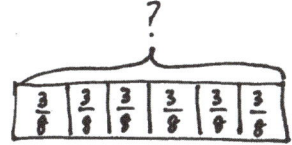

Problem 3: Solve a word problem involving the multiplication of fractions.

T: The serving size for cereal is $\frac{2}{3}$ cup. Each of 27 students in health class measured out one serving to eat for breakfast. If a box of cereal contained 16 cups, how many boxes of cereal were needed?

T: Draw what you know, and write a number sentence to solve.

S: $27 \times \frac{2}{3} = \frac{27 \times 2}{3} = \frac{54}{3}$.

T: As a mixed number?

S: Hmm. Those numbers are bigger than I am used to converting.

T: We want to know how many groups of $\frac{3}{3}$ there are in $\frac{54}{3}$. Three times what number is close to or equal to 54? To find that out, I can divide. 54 ÷ 3 = 18. The answer is 18 cups of cereal, so how many boxes are needed?

S: Two boxes because 1 box serves 16 cups, but the class needs 18 cups.

> **NOTES ON MULTIPLE MEANS OF ENGAGEMENT:**
>
> Empower students working below grade level to solve on-level word problems by using strategies such as the distributive property or decomposition. For example, students who are challenged by 27 × 2 can multiply (20 + 7) × 2 = (20 × 2) + (7 × 2) = 40 + 14, or 9 × (3 × 2).

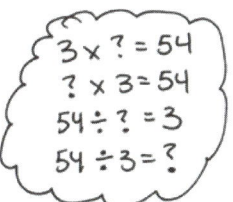

Problem Set (10 minutes)

Students should do their personal best to complete the Problem Set within the allotted 10 minutes. For some classes, it may be appropriate to modify the assignment by specifying which problems they work on first. Some problems do not specify a method for solving. Students should solve these problems using the RDW approach used for Application Problems.

Lesson 36: Represent the multiplication of n times a/b as (n × a)/b using the associative property and visual models.

Student Debrief (10 minutes)

Lesson Objective: Represent the multiplication of n times a/b as $(n \times a)/b$ using the associative property and visual models.

The Student Debrief is intended to invite reflection and active processing of the total lesson experience.

Invite students to review their solutions for the Problem Set. They should check work by comparing answers with a partner before going over answers as a class. Look for misconceptions or misunderstandings that can be addressed in the Debrief. Guide students in a conversation to debrief the Problem Set and process the lesson.

Any combination of the questions below may be used to lead the discussion.

- Problem 4(d) is a good example of how multiplication is more efficient than repeated addition. Explain.
- Explain to your partner the method that you used to solve Problem 4(a–d).
- What was challenging about Problem 4(d)?
- Problem 4(b) results in a fraction greater than 1 with a large numerator. Watch as the fraction is renamed before multiplying. Discuss what you see with your partner. How does this method simplify the work done after the product is found?

$$12 \times \frac{3}{4} = \frac{12 \times 3}{4} = \frac{3 \times 3}{1} = \frac{9}{1} = 9$$

- Try solving Problem 4(c) using a method similar to the one used above. (Note: Simplification is not a requirement in Grade 4 standards.)

Exit Ticket (3 minutes)

After the Student Debrief, instruct students to complete the Exit Ticket. A review of their work will help with assessing students' understanding of the concepts that were presented in today's lesson and planning more effectively for future lessons. The questions may be read aloud to the students.

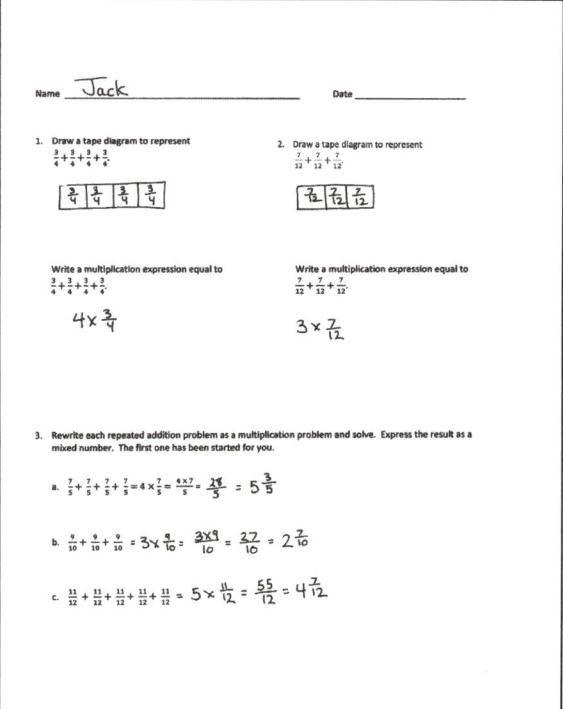

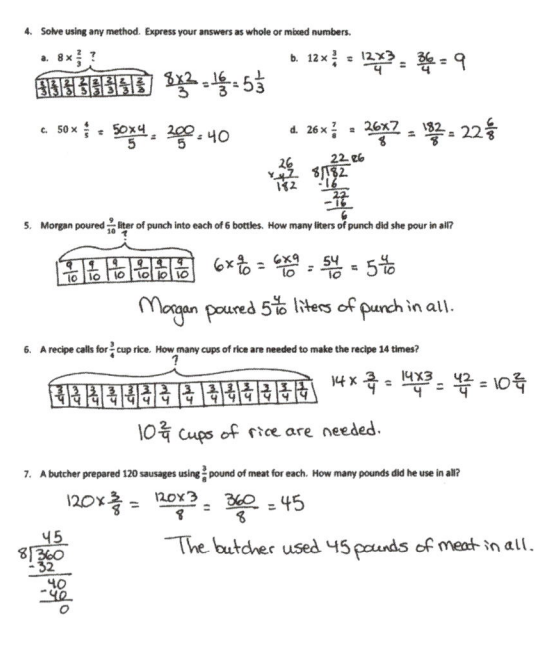

A STORY OF UNITS Lesson 36 Problem Set 4•5

Name _____ Date _____

1. Draw a tape diagram to represent
$\frac{3}{4} + \frac{3}{4} + \frac{3}{4} + \frac{3}{4}$.

2. Draw a tape diagram to represent
$\frac{7}{12} + \frac{7}{12} + \frac{7}{12}$.

Write a multiplication expression equal to $\frac{3}{4} + \frac{3}{4} + \frac{3}{4} + \frac{3}{4}$.

Write a multiplication expression equal to $\frac{7}{12} + \frac{7}{12} + \frac{7}{12}$.

3. Rewrite each repeated addition problem as a multiplication problem and solve. Express the result as a mixed number. The first one has been started for you.

   a. $\frac{7}{5} + \frac{7}{5} + \frac{7}{5} + \frac{7}{5} = 4 \times \frac{7}{5} = \frac{4 \times 7}{5} =$

   b. $\frac{9}{10} + \frac{9}{10} + \frac{9}{10}$

   c. $\frac{11}{12} + \frac{11}{12} + \frac{11}{12} + \frac{11}{12} + \frac{11}{12}$

Lesson 36: Represent the multiplication of n times a/b as $(n \times a)/b$ using the associative property and visual models.

4. Solve using any method. Express your answers as whole or mixed numbers.

   a. $8 \times \frac{2}{3}$

   b. $12 \times \frac{3}{4}$

   c. $50 \times \frac{4}{5}$

   d. $26 \times \frac{7}{8}$

5. Morgan poured $\frac{9}{10}$ liter of punch into each of 6 bottles. How many liters of punch did she pour in all?

6. A recipe calls for $\frac{3}{4}$ cup rice. How many cups of rice are needed to make the recipe 14 times?

7. A butcher prepared 120 sausages using $\frac{3}{8}$ pound of meat for each. How many pounds did he use in all?

A STORY OF UNITS   Lesson 36 Exit Ticket   4•5

Name _____   Date _____

Solve using any method.

1. $7 \times \frac{3}{4}$

2. $9 \times \frac{2}{5}$

3. $60 \times \frac{5}{8}$

A STORY OF UNITS　　　　　　　　　　　　　　　　　　　　　Lesson 36 Homework 4•5

Name _____   Date _____

1. Draw a tape diagram to represent

    $\frac{2}{3}+\frac{2}{3}+\frac{2}{3}+\frac{2}{3}.$

2. Draw a tape diagram to represent

    $\frac{7}{8}+\frac{7}{8}+\frac{7}{8}.$

    Write a multiplication expression equal to $\frac{2}{3}+\frac{2}{3}+\frac{2}{3}+\frac{2}{3}.$

    Write a multiplication expression equal to $\frac{7}{8}+\frac{7}{8}+\frac{7}{8}.$

3. Rewrite each repeated addition problem as a multiplication problem and solve. Express the result as a mixed number. The first one has been completed for you.

    a. $\frac{7}{5}+\frac{7}{5}+\frac{7}{5}+\frac{7}{5} = 4 \times \frac{7}{5} = \frac{4 \times 7}{5} = \frac{28}{5} = 5\frac{3}{5}$

    b. $\frac{7}{10}+\frac{7}{10}+\frac{7}{10}$

    c. $\frac{5}{12}+\frac{5}{12}+\frac{5}{12}+\frac{5}{12}+\frac{5}{12}+\frac{5}{12}$

    d. $\frac{3}{8}+\frac{3}{8}+\frac{3}{8}+\frac{3}{8}+\frac{3}{8}+\frac{3}{8}+\frac{3}{8}+\frac{3}{8}+\frac{3}{8}+\frac{3}{8}+\frac{3}{8}$

4. Solve using any method. Express your answers as whole or mixed numbers.

    a. $7 \times \frac{2}{9}$

    b. $11 \times \frac{2}{3}$

484　　Lesson 36:　Represent the multiplication of n times a/b as (n × a)/b using the associative property and visual models.

c. $40 \times \frac{2}{6}$

d. $24 \times \frac{5}{6}$

e. $23 \times \frac{3}{5}$

f. $34 \times \frac{2}{8}$

5. Coleton is playing with interlocking blocks that are each $\frac{3}{4}$ inch tall. He makes a tower 17 blocks tall. How tall is his tower in inches?

6. There were 11 players on Mr. Maiorani's softball team. They each ate $\frac{3}{8}$ of a pizza. How many pizzas did they eat?

7. A bricklayer places 12 bricks end to end along the entire outside length of a shed's wall. Each brick is $\frac{3}{4}$ foot long. How many feet long is that wall of the shed?

Lesson 37

Objective: Find the product of a whole number and a mixed number using the distributive property.

Suggested Lesson Structure

- Fluency Practice   (10 minutes)
- Application Problem (5 minutes)
- Concept Development (35 minutes)
- Student Debrief   (10 minutes)
- **Total Time**   **(60 minutes)**

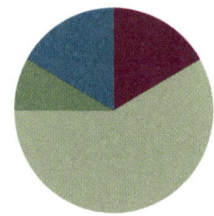

Fluency Practice (10 minutes)

- Add and Subtract **4.NBT.4**   (4 minutes)
- Multiply Fractions **4.NF.4**   (6 minutes)

Add and Subtract (4 minutes)

Materials: (S) Personal white board

Note: This fluency activity reviews adding and subtracting using the standard algorithm.

- T: (Write 547 thousands 869 ones.) On your personal white boards, write this number in standard form.
- S: (Write 547,869.)
- T: (Write 362 thousands 712 ones.) Add this number to 547,869 using the standard algorithm.
- S: (Write 547,869 + 362,712 = 910,581 using the standard algorithm.)

Continue with the following possible sequence: 459,623 + 353,683.

- T: (Write 800 thousands.) On your boards, write this number in standard form.
- S: (Write 800,000.)
- T: (Write 352 thousands 951 ones.) Subtract this number from 800,000 using the standard algorithm.
- S: (Write 800,000 − 352,951 = 447,049 using the standard algorithm.)

Continue with the following possible sequence: 805,813 − 368,265.

A STORY OF UNITS   Lesson 37 4•5

Multiply Fractions (6 minutes)

Materials: (S) Personal white board

Note: This fluency activity reviews Lesson 36.

T: (Write $2 \times \frac{2}{5} = \frac{\times}{5} = \frac{}{5}$.) Write the multiplication sentence, filling in the unknown numbers. You can draw a tape diagram or a number line to help you.

S: (Write $2 \times \frac{2}{5} = \frac{2 \times 2}{5} = \frac{4}{5}$.)

Continue the process for $3 \times \frac{3}{10}$.

T: (Write $3\frac{3}{8} = \frac{\times}{3} = \frac{}{8}$.) Write the multiplication sentence, filling in the unknown number. You can use a tape diagram or a number line to help you.

S: (Write $3 \times \frac{3}{8} = \frac{3 \times 3}{8} = \frac{9}{8}$.)

T: Write $\frac{9}{8}$ as a mixed number.

S: (Write $\frac{9}{8} = 1\frac{1}{8}$.)

Continue with the following possible sequence: $4 \times \frac{2}{3}$, $5 \times \frac{3}{4}$, and $5 \times \frac{5}{12}$.

Application Problem (5 minutes)

The baker needs $\frac{5}{8}$ cup of raisins to make 1 batch of cookies. How many cups of raisins does he need to make 7 batches of cookies?

Solution 1:
$R = 7 \times \frac{5}{8} = 7 \times 5 \times \frac{1}{8}$
$= 35 \times \frac{1}{8}$
$= \frac{35}{8}$
$= 4\frac{3}{8}$

Solution 2:
$R = 7 \times \frac{5}{8} = \frac{7 \times 5}{8}$
$= \frac{35}{8}$
$= 4\frac{3}{8}$

Solution 3:
$R = 7 \times 5$ eighths $= 35$ eighths
$= \frac{35}{8}$
$= 4\frac{3}{8}$

The baker needs $4\frac{3}{8}$ cups raisins.

Note: This Application Problem reviews Lessons 35 and 36 of Topic G, where students learned to represent the product of a whole number and a fraction using the associative property. Notice that, although they can be used, parentheses are not modeled in the solutions. Students have already established that parentheses indicate the changed associations. Since the process has been established, parentheses are not necessary and can make notation cumbersome.

EUREKA MATH   Lesson 37: Find the product of a whole number and a mixed number using the distributive property.   487

Concept Development (35 minutes)

Materials: (S) Personal white board

Problem 1: Draw a tape diagram to show the product of a whole number and a mixed number.

T: With me, draw a tape diagram showing $3\frac{1}{5}$ in two parts, the ones and the fractional part.

S: (Draw.)

T: Point to and say the two parts of your tape diagram.

S: (Point as saying each value.) 3. $\frac{1}{5}$.

T: Draw one more copy of $3\frac{1}{5}$ as two parts on the same tape diagram.

S: (Draw.)

T: There are two copies of $3\frac{1}{5}$. We can record this as $2 \times 3\frac{1}{5}$. (Write $2 \times 3\frac{1}{5}$ on the board.)

T: What are the 4 parts of your tape diagram?

S: 3, $\frac{1}{5}$, 3, and $\frac{1}{5}$. → 2 threes and 2 fifths.

T: Make a new tape diagram of two groups of $3\frac{1}{5}$ the same length as your other tape diagram. This time, draw the threes on the left and the fifths on the right.

T: How many threes do we have?

S: 2 threes.

T: How many fifths do we have?

S: 2 fifths.

T: $2 \times 3\frac{1}{5}$ is equal to 2 threes and 2 fifths. (Write $2 \times 3\frac{1}{5} = (2 \times 3) + (2 \times \frac{1}{5})$.)

T: 2 times 3 is…? (Point to the expression.)

S: 6. (Write their response as shown to the right.)

T: 2 times $\frac{1}{5}$ is…? (Point to the expression.)

S: $\frac{2}{5}$. (Write their response as shown to the right.)

T: The parts are 6 and $\frac{2}{5}$. What is the total?

S: $6 + \frac{2}{5} = 6\frac{2}{5}$.

> **NOTES ON MULTIPLE MEANS OF ACTION AND EXPRESSION:**
>
> A gentle reminder and grid paper may help learners draw appropriately proportioned, though not meticulously precise, tape diagrams. Generally, the bar for 3 should be longer than the bar for $\frac{1}{5}$.

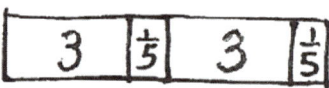

Lesson 37: Find the product of a whole number and a mixed number using the distributive property.

A STORY OF UNITS Lesson 37 4•5

T: Let's try another one. Make a tape diagram to show four units of $5\frac{2}{10}$. Make another tape diagram to show how the whole numbers and fractional parts can be redistributed. Write a multiplication expression to represent your groups of $5\frac{2}{10}$ using the format we used to do two groups of 3 and a fifth.

S: (Write $4 \times 5\frac{2}{10}$.)

Problem 2: Identify the distributive property to multiply a whole number and a mixed number.

T: Express $5\frac{2}{10}$ as an addition expression. (Note that this is a continuation of Problem 1.)

S: $5 + \frac{2}{10}$.

T: (Write $4 \times 5\frac{2}{10} = 4 \times (5 + \frac{2}{10})$.) How many groups of 5 did you draw?

S: Four.

T: How many groups of 2 tenths?

S: Four.

T: There are four groups of 5 and four groups of $\frac{2}{10}$. (Write $(4 \times 5) + (4 \times \frac{2}{10})$.) We distribute our multiplication to both parts of our mixed number.

T: 4×5 is…?

S: 20.

T: $4 \times \frac{2}{10}$ is…?

S: $\frac{8}{10}$.

T: (Write $= 20 + \frac{8}{10}$.) Our total product is…?

S: $20\frac{8}{10}$.

T: (Write $3 \times 7\frac{3}{4}$.) With your partner, write a number sentence to multiply the whole number by each part. (Pause.) What number sentence did you write?

S: $3 \times 7\frac{3}{4} = (3 \times 7) + (3 \times \frac{3}{4})$.

T: Show the products for each part. What are the two products?

S: 21 and $\frac{9}{4}$.

T: Rename $\frac{9}{4}$ as a mixed number. $\frac{9}{4}$ is…?

S: $2\frac{1}{4}$.

> **NOTES ON MULTIPLE MEANS OF REPRESENTATION:**
>
> If students are reversing numerators and denominators, try using a color to distinguish them. For example, write the numerator in red. Have students consistently whisper-read fractions as they solve. Continue to use models for meaning-making. Frequently check for understanding, and guide students to offer personalized solutions.

> **NOTES ON MULTIPLE MEANS OF REPRESENTATION:**
>
> An additional step to solving $3 \times 7\frac{3}{4}$ that may scaffold understanding for students working below grade level is to model the decomposition of $7\frac{3}{4}$ as a number bond, as shown below:
>
>

Lesson 37: Find the product of a whole number and a mixed number using the distributive property.

A STORY OF UNITS   Lesson 37 4•5

T: What is the product of $3 \times 7\frac{3}{4}$?

S: $23\frac{1}{4}$.

T: You used the distributive property when you broke apart $7\frac{3}{4}$ and multiplied each part by 3.

T: Try another. Solve $5 \times 3\frac{2}{3}$. This time, imagine the distributive property in your head. Think out loud if you need to as you solve. Write only as much as you need to.

S: $5 \times 3\frac{2}{3} = 15 + \frac{10}{3} = 18\frac{1}{3}$.

Problem 3: Solve a word problem involving the multiplication of a whole number by a mixed number.

T: In April, Jenny ran in a marathon as part of a relay team. She ran $6\frac{55}{100}$ miles. In September, Jenny ran 4 times as far to complete a marathon on her own. How far did Jenny run in September?

T: Use any strategy we practiced today to solve this problem. Remember to record all of your steps. Be ready to explain your work to your partner.

Jenny ran $26\frac{20}{100}$ miles in September.

Solution 1
$4 \times 6\frac{55}{100} = 4 \times (6 + \frac{55}{100})$
$= (4 \times 6) + (4 \times \frac{55}{100})$
$= 24 + \frac{220}{100}$
$= 26\frac{20}{100}$

Solution 2
$4 \times 6\frac{55}{100} = 24 + \frac{220}{100}$
$= 26\frac{20}{100}$

Problem Set (10 minutes)

Students should do their personal best to complete the Problem Set within the allotted 10 minutes. For some classes, it may be appropriate to modify the assignment by specifying which problems they work on first. Some problems do not specify a method for solving. Students should solve these problems using the RDW approach used for Application Problems.

Student Debrief (10 minutes)

Lesson Objective: Find the product of a whole number and a mixed number using the distributive property.

The Student Debrief is intended to invite reflection and active processing of the total lesson experience.

Invite students to review their solutions for the Problem Set. They should check work by comparing answers with a partner before going over answers as a class. Look for misconceptions or misunderstandings that can be addressed in the Student Debrief. Guide students in a conversation to debrief the Problem Set and process the lesson. Any combination of the questions below may be used to lead the discussion.

- How could your tape diagram from Problem 1 help you solve Problem 2(b)? Explain your thinking.
- We can use the distributive property to show 3×24 as $(3 \times 2 \text{ tens}) + (3 \times 4 \text{ ones})$. Explain how this relates to solving $3 \times 2\frac{4}{10}$.
- Which strategy did you use to solve Problem 3? Why do you prefer this strategy?
- Problem 2(h) shows the expression $5\frac{6}{8} \times 4$ instead of $4 \times 5\frac{6}{8}$. Why are we able to write it either way and still get the same product?
- Look at differences in the solutions for Problem 3 of the Concept Development. In Solution 2, which step was not explicitly written? How did the student move from $4 \times 6\frac{55}{100}$ to $24 + \frac{220}{100}$ in one step? Discuss with a partner.
- Were you able to omit the step expressed in line 2 of Problem 2(a)? Explain.

Exit Ticket (3 minutes)

After the Student Debrief, instruct students to complete the Exit Ticket. A review of their work will help with assessing students' understanding of the concepts that were presented in today's lesson and planning more effectively for future lessons. The questions may be read aloud to the students.

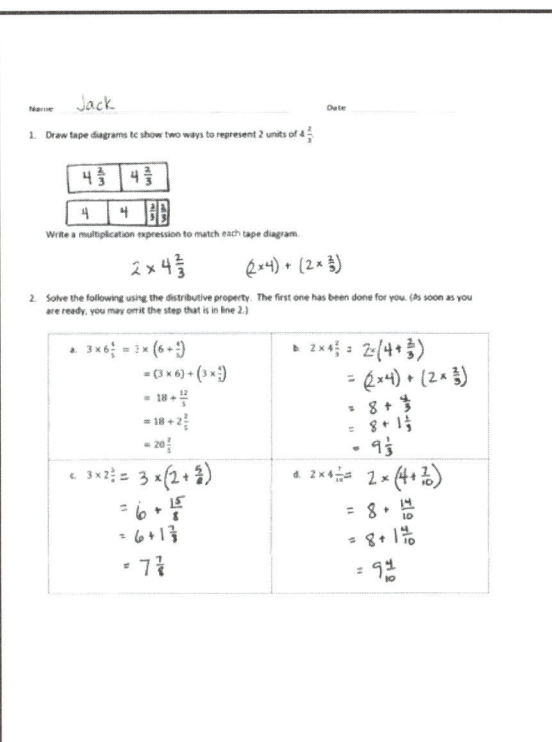

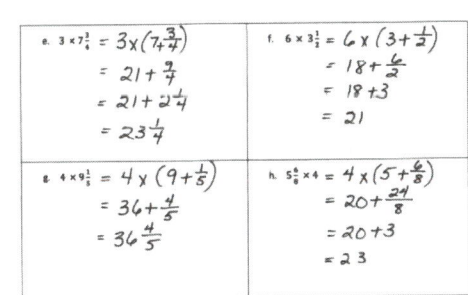

A STORY OF UNITS   Lesson 37 Problem Set   4•5

Name _____   Date _____

1. Draw tape diagrams to show two ways to represent 2 units of $4\frac{2}{3}$.

   Write a multiplication expression to match each tape diagram.

2. Solve the following using the distributive property. The first one has been done for you. (As soon as you are ready, you may omit the step that is in line 2.)

   a. $3 \times 6\frac{4}{5} = 3 \times \left(6 + \frac{4}{5}\right)$

   $\phantom{3 \times 6\frac{4}{5}} = (3 \times 6) + \left(3 \times \frac{4}{5}\right)$

   $\phantom{3 \times 6\frac{4}{5}} = 18 + \frac{12}{5}$

   $\phantom{3 \times 6\frac{4}{5}} = 18 + 2\frac{2}{5}$

   $\phantom{3 \times 6\frac{4}{5}} = 20\frac{2}{5}$

   b. $2 \times 4\frac{2}{3}$

   c. $3 \times 2\frac{5}{8}$

   d. $2 \times 4\frac{7}{10}$

e. $3 \times 7\frac{3}{4}$	f. $6 \times 3\frac{1}{2}$
g. $4 \times 9\frac{1}{5}$	h. $5\frac{6}{8} \times 4$

3. For one dance costume, Saisha needs $4\frac{2}{3}$ feet of ribbon. How much ribbon does she need for 5 identical costumes?

Name _____ Date _____

Multiply. Write each product as a mixed number.

1. $4 \times 5\frac{3}{8}$

2. $4\frac{3}{14} \times 3$

A STORY OF UNITS

Lesson 37 Homework 4•5

Name _____ Date _____

1. Draw tape diagrams to show two ways to represent 3 units of $5\frac{1}{12}$.

   Write a multiplication expression to match each tape diagram.

2. Solve the following using the distributive property. The first one has been done for you. (As soon as you are ready, you may omit the step that is in line 2.)

   a. $3 \times 6\frac{4}{5} = 3 \times \left(6 + \frac{4}{5}\right)$
   $= (3 \times 6) + \left(3 \times \frac{4}{5}\right)$
   $= 18 + \frac{12}{5}$
   $= 18 + 2\frac{2}{5}$
   $= 20\frac{2}{5}$

   b. $5 \times 4\frac{1}{6}$

   c. $6 \times 2\frac{3}{5}$

   d. $2 \times 7\frac{3}{12}$

Lesson 37: Find the product of a whole number and a mixed number using the distributive property.

495

e. $8 \times 7\frac{1}{4}$	f. $3\frac{3}{8} \times 12$

3. Sara's street is $2\frac{3}{10}$ miles long. She ran the length of the street 6 times. How far did she run?

4. Kelly's new puppy weighed $4\frac{7}{10}$ pounds when she brought him home. Now, he weighs six times as much. How much does he weigh now?

Lesson 38

Objective: Find the product of a whole number and a mixed number using the distributive property.

Suggested Lesson Structure

- **Fluency Practice** (10 minutes)
- **Application Problem** (4 minutes)
- **Concept Development** (36 minutes)
- **Student Debrief** (10 minutes)

Total Time (60 minutes)

Fluency Practice (10 minutes)

- Multiply Fractions **4.NF.4**   (5 minutes)
- Multiply Mixed Numbers **4.NF.4**   (5 minutes)

Multiply Fractions (5 minutes)

Materials: (S) Personal white board

Note: This fluency activity reviews Lesson 36.

T: (Write $3 \times \frac{3}{10} = \frac{3 \times 3}{10} = \frac{}{10}$.) Write the multiplication sentence and product. You can draw a tape diagram or number line to help you.

S: (Write $\frac{3 \times 3}{10} = \frac{9}{10}$.)

Continue with the following possible sequence: $7 \times \frac{2}{15}$ and $2 \times \frac{3}{8}$.

T: (Write $4 \times \frac{2}{5} = \frac{4 \times 2}{5} = \frac{}{5}$.) Write the multiplication sentence and product. You can use a tape diagram or number line to help you.

S: (Write $\frac{4 \times 2}{5} = \frac{8}{5}$.)

T: (Write $4 \times \frac{2}{5} = \frac{4 \times 2}{5} = \frac{8}{5}$. Beneath it, write $\frac{8}{5} = $ ___.) Write $\frac{8}{5}$ as a mixed number.

S: (Beneath $\frac{8}{5}$, write $= 1\frac{3}{5}$.)

Continue with the following possible sequence: $4 \times \frac{3}{4}$, $5 \times \frac{3}{8}$, and $7 \times \frac{2}{3}$.

Multiply Mixed Numbers (5 minutes)

Materials: (S) Personal white board

Note: This fluency activity reviews Lesson 37.

T: (Write $2 \times 4\frac{3}{5} = \_\_$.) Break apart $4\frac{3}{5}$ as an addition expression.

S: (Write $4 + \frac{3}{5}$.)

T: (Write $2 \times (4 + \frac{3}{5})$. Beneath it, write $(2 \times \_\_) + (2 \times \_\_)$.) Fill in the unknown numbers.

S: (Write $(2 \times 4) + (2 \times \frac{3}{5})$.)

T: (Write $(2 \times 4) + (2 \times \frac{3}{5})$. Beneath it, write $\_\_ + \_\_$.) Fill in the unknown numbers.

S: (Write $8 + \frac{6}{5}$.)

T: (Write $8 + \frac{6}{5}$. Beneath it, write $8 + \_\_$.) Rename $\frac{6}{5}$ as a mixed number.

S: (Write $8 + 1\frac{1}{5}$.)

T: (Write $8 + 1\frac{1}{5}$.) Write the answer.

S: (Write $9\frac{1}{5}$.)

T: (Point to $2 \times 4\frac{3}{5} = \_\_$.) Say the multiplication sentence.

S: $2 \times 4\frac{3}{5} = 9\frac{1}{5}$.

Continue with the following possible sequence: $3 \times 2\frac{2}{3}$ and $4 \times 2\frac{3}{8}$.

Application Problem (4 minutes)

Eight students are on a relay team. Each runs $1\frac{3}{4}$ kilometers. How many total kilometers does their team run?

$8 \times 1\frac{3}{4} = 8 + \frac{24}{4}$
$= 14$

$8 \times 1\frac{3}{4} = 8 \times \frac{7}{4}$
$= \frac{56}{4}$
$= 14$

The team ran 14 kilometers.

Note: This Application Problem reviews Lesson 37, where students used the distributive property to multiply a whole number and a mixed number.

A STORY OF UNITS Lesson 38 4•5

Concept Development (36 minutes)

Materials: (S) Personal white board

Note: This lesson reviews what students learned from Lesson 37.

Problem 1: Identify the unknown factors.

T: Write $5 \times 8\frac{1}{5} = (\underline{} \times 8) + (\underline{} \times \frac{1}{5})$. Use the distributive property to fill in the unknown numbers. Turn and discuss your answer with your partner. Draw or write as you explain your thinking.

S: Both parts need to be multiplied by 5. → I used a tape diagram to show my partner that there are 5 eights and 5 one-fifths.

Problem 2: Use and share strategies for using the distributive property to find the product of a whole number and a mixed number.

T: (Write $4 \times 9\frac{3}{4} = \underline{}$.) Solve the problem on your personal white boards.

Allow students about one to two minutes to solve.

T: What is $4 \times 9\frac{3}{4}$?

S: 39.

T: Share your work with your partner.

S: I made a tape diagram showing four units of $9\frac{3}{4}$. → I used the distributive property by writing four groups of 9 and four groups of $\frac{3}{4}$. Then, I added those products and got 39. → I took a shortcut and wrote $36 + \frac{12}{4}$.

$4 \times 9\frac{3}{4} = 36 + \frac{12}{4}$
$= 36 + 3$
$= 39$

Have students work with a partner to solve the following problems: $5\frac{6}{8} \times 4$, $12\frac{2}{6} \times 3$, and $9 \times 7\frac{5}{7}$.

> **NOTES ON MULTIPLE MEANS OF REPRESENTATION:**
>
> Scaffold understanding with visual models. Students working below grade level may benefit from connecting, for example, $\frac{2}{6}$ is the same as $2 \times \frac{1}{6}$, to a tape diagram or number line.

> **NOTES ON MULTIPLE MEANS OF ENGAGEMENT:**
>
> Give everyone a fair chance to share their work and solutions by providing appropriate scaffolds. Demonstrating students may use translators, interpreters, or sentence frames to present. If the pace of the lesson is a consideration, prepare presenters beforehand.

MP.3

Lesson 38: Find the product of a whole number and a mixed number using the distributive property.

Problem 3: Solve multiplication of a mixed number and a whole number when embedded in word problems.

T: (Write or project, "Robin rides for $3\frac{1}{2}$ miles round trip to get to and from school. How many miles would Robin ride in 5 days?") Use the RDW process to solve this story problem.

Circulate and note student work that might be beneficial to share with the class.

S: (Solve on a personal white board.)

T: In 5 days, how many miles would Robin ride to and from school?

S: $17\frac{1}{2}$ miles.

Invite each selected student to come to the board and share strategies and solutions. Ask students to share their tape diagrams with the labels and identify all of the referents.

S: I made a tape diagram showing 5 units of $3\frac{1}{2}$. Every time I saw 2 halves, I counted them as 1. I added $(3 \times 5) + 1 + 1 + \frac{1}{2}$. (See image at the right.)
→ I multiplied the whole number of miles by 5 and then multiplied the $\frac{1}{2}$ mile by 5. I added the products together. That's the distributive property.

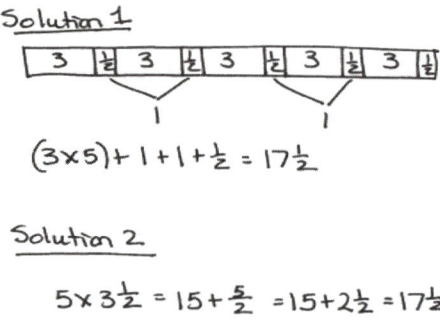

Problem Set (10 minutes)

Students should do their personal best to complete the Problem Set within the allotted 10 minutes. For some classes, it may be appropriate to modify the assignment by specifying which problems they work on first. Some problems do not specify a method for solving. Students should solve these problems using the RDW approach used for Application Problems.

Student Debrief (10 minutes)

Lesson Objective: Find the product of a whole number and a mixed number using the distributive property.

The Student Debrief is intended to invite reflection and active processing of the total lesson experience.

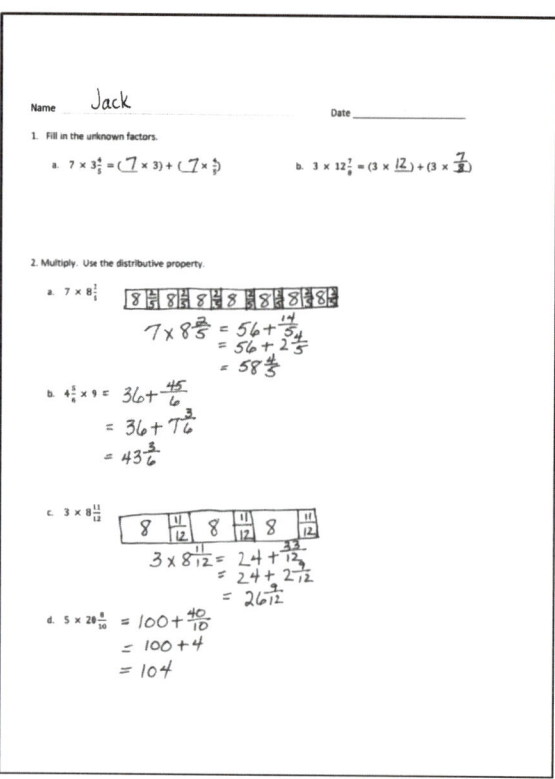

A STORY OF UNITS   Lesson 38   4•5

Invite students to review their solutions for the Problem Set. They should check work by comparing answers with a partner before going over answers as a class. Look for misconceptions or misunderstandings that can be addressed in the Student Debrief. Guide students in a conversation to debrief the Problem Set and process the lesson.

Any combination of the questions below may be used to lead the discussion.

- Explain how you knew what number was unknown from Problem 1.
- What method for solving did you use in Problem 2? Use a specific example from your Problem Set to explain.
- What did you do to solve the problems when the first factor was a mixed number?
- How did you solve Problem 2(e)? Turn and share with your partner.
- Why is it sometimes useful to see both a tape diagram and the numbers?
- How might you improve your work from today's Application Problem?

Exit Ticket (3 minutes)

After the Student Debrief, instruct students to complete the Exit Ticket. A review of their work will help with assessing students' understanding of the concepts that were presented in today's lesson and planning more effectively for future lessons. The questions may be read aloud to the students.

Lesson 38: Find the product of a whole number and a mixed number using the distributive property.

A STORY OF UNITS   Lesson 38 Problem Set   4•5

Name _____   Date _____

1. Fill in the unknown factors.

   a. $7 \times 3\frac{4}{5} = (\underline{} \times 3) + (\underline{} \times \frac{4}{5})$

   b. $3 \times 12\frac{7}{8} = (\overline{A} \times \underline{}) + (3 \times \underline{})$

2. Multiply. Use the distributive property.

   a. $7 \times 8\frac{2}{5}$

   b. $4\frac{5}{6} \times 9$

   c. $3 \times 8\frac{11}{12}$

   d. $5 \times 20\frac{8}{10}$

Lesson 38: Find the product of a whole number and a mixed number using the distributive property.

e. $25\frac{4}{100} \times 4$

3. The distance around the park is $2\frac{5}{10}$ miles. Cecilia ran around the park 3 times. How far did she run?

4. Windsor the dog ate $4\frac{3}{4}$ snack bones each day for a week. How many bones did Windsor eat that week?

Name _____ Date _____

1. Fill in the unknown factors.

   $8 \times 5\frac{2}{3} = (\underline{} \times 5) + (\underline{} \times \frac{2}{3})$

2. Multiply. Use the distributive property.

   $6\frac{5}{8} \times 7$

Name _____ Date _____

1. Fill in the unknown factors.

   a. $8 \times 4\frac{4}{7} = (\underline{} \times 4) + (\underline{} \times \frac{4}{7})$

   b. $9 \times 7\frac{7}{10} = (9 \times \underline{}) + (9 \times \underline{})$

2. Multiply. Use the distributive property.

   a. $6 \times 8\frac{2}{7}$

   b. $7\frac{3}{4} \times 9$

   c. $9 \times 8\frac{7}{9}$

   d. $25\frac{7}{8} \times 3$

Lesson 38: Find the product of a whole number and a mixed number using the distributive property.

e. $4 \times 20\frac{8}{12}$

f. $30\frac{3}{100} \times 12$

3. Brandon is cutting 9 boards for a woodworking project. Each board is $4\frac{5}{8}$ feet long. What is the total length of the boards?

4. Rocky the collie ate $3\frac{1}{4}$ cups of dog food each day for two weeks. How much dog food did Rocky eat in that time?

5. At the class party, each student will be given a container filled with $8\frac{5}{8}$ ounces of juice. There are 25 students in the class. How many ounces of juice does the teacher need to buy?

Lesson 39

Objective: Solve multiplicative comparison word problems involving fractions.

Suggested Lesson Structure

■ Fluency Practice   (12 minutes)
■ Concept Development (38 minutes)
■ Student Debrief   (10 minutes)
   Total Time   **(60 minutes)**

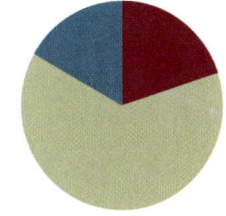

Fluency Practice (12 minutes)

- Sprint: Multiply Whole Numbers Times Fractions **4.NF.4**   (8 minutes)
- Multiply Mixed Numbers **4.NF.4**   (4 minutes)

Sprint: Multiply Whole Numbers Times Fractions (8 minutes)

Materials: (S) Multiply Whole Numbers Times Fractions Sprint

Note: This fluency activity reviews Lesson 35.

Multiply Mixed Numbers (4 minutes)

Materials: (S) Personal white board

Note: This fluency activity reviews Lesson 37.

T: Break apart $3\frac{3}{8}$ using addition.

S: (Write $3\frac{3}{8}$ as $3 + \frac{3}{8}$.)

T: (Write $3 \times 3\frac{3}{8} = $ ___ . Beneath it, write ___ + ─.) Fill in the unknown numbers.

S: (Write $9 + \frac{9}{8}$.)

T: (Write $9 + \frac{9}{8}$. Beneath it, write $9 + $ ___.) Record a mixed number for $\frac{9}{8}$.

S: (Write $9 + 1\frac{1}{8}$.)

Lesson 39:   Solve multiplicative comparison word problems involving fractions.

A STORY OF UNITS Lesson 39 4•5

T: (Write $9 + 1\frac{1}{8}$. Beneath it, write = ___.) Write the answer.

S: (Write = $10\frac{1}{8}$.)

T: (Point at $3 \times 3\frac{3}{8}$ = ___.) Say the multiplication sentence.

S: $3 \times 3\frac{3}{8} = 10\frac{1}{8}$.

Continue with the following possible sequence: $6 \times 3\frac{2}{3}$ and $4 \times 3\frac{7}{8}$.

Concept Development (38 minutes)

Materials: (S) Problem Set

Suggested Delivery of Instruction for Solving Lesson 39 Word Problems

1. Model the problem.

Have two pairs of students who can successfully model the problem work at the board while the others work independently or in pairs at their seats. Review the following questions before beginning the first problem.

- Can you draw something?
- What can you draw?
- What conclusions can you make from your drawing?

Circulate as students work. Reiterate the questions above. After two minutes, have the two pairs of students share only their labeled diagrams. For about one minute, have the demonstrating students receive and respond to feedback and questions from their peers.

2. Calculate to solve and write a statement.

Give students two minutes to finish work on that question, sharing their work and thinking with a peer.

All should then write their equations and statements of the answer.

3. Assess the solution for reasonableness.

Give students one to two minutes to assess and explain the reasonableness of their solution.

Note: Problems 1–4 of the Problem Set are used during the Concept Development portion of the lesson.

508 Lesson 39: Solve multiplicative comparison word problems involving fractions.

Problem 1: Tameka ran $2\frac{5}{8}$ miles. Her sister ran twice as far. How far did Tameka's sister run?

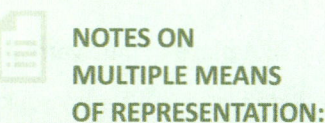

Tameka's sister ran $5\frac{2}{8}$ miles.

Solution 1
$$2 \times 2\frac{5}{8} = 2 \times (2 + \frac{5}{8})$$
$$= (2 \times 2) + (2 \times \frac{5}{8})$$
$$= 4 + \frac{10}{8}$$
$$= 4 + 1\frac{2}{8}$$
$$= 5\frac{2}{8}$$

Solution 2
$$2\frac{5}{8} + 2\frac{5}{8} = 4 + \frac{10}{8}$$
$$= 4 + 1\frac{2}{8}$$
$$= 5\frac{2}{8}$$

Students may choose to multiply or add to solve this problem. Variations in solution strategies may be used to help students see the distributive property at work.

Problem 2: Natasha's sculpture was $5\frac{3}{16}$ inches tall. Maya's was 4 times as tall. How much shorter was Natasha's sculpture than Maya's?

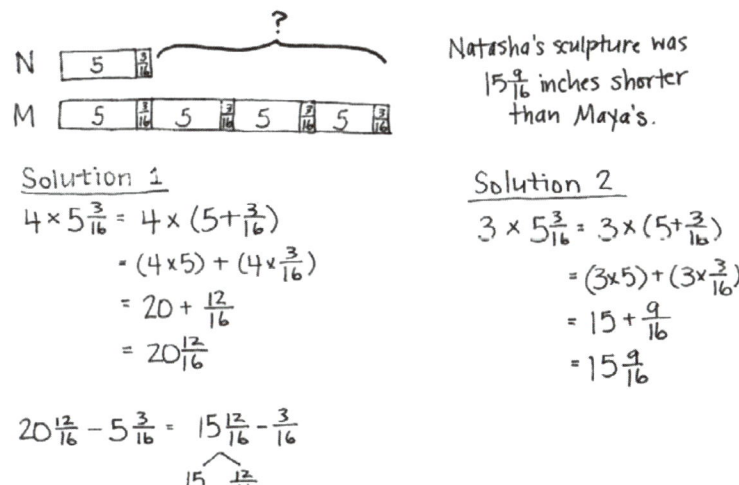

Natasha's sculpture was $15\frac{9}{16}$ inches shorter than Maya's.

Solution 1
$$4 \times 5\frac{3}{16} = 4 \times (5 + \frac{3}{16})$$
$$= (4 \times 5) + (4 \times \frac{3}{16})$$
$$= 20 + \frac{12}{16}$$
$$= 20\frac{12}{16}$$

$$20\frac{12}{16} - 5\frac{3}{16} = 15\frac{12}{16} - \frac{3}{16}$$
$$= 15\frac{9}{16}$$

Solution 2
$$3 \times 5\frac{3}{16} = 3 \times (5 + \frac{3}{16})$$
$$= (3 \times 5) + (3 \times \frac{3}{16})$$
$$= 15 + \frac{9}{16}$$
$$= 15\frac{9}{16}$$

NOTES ON MULTIPLE MEANS OF REPRESENTATION:

Modeling multiplicative comparisons can be tricky. If learners have difficulty representing *y times as much*, take it one step at a time. Ask, "Draw 1 times as much. (That would be the same.) Now, draw 2 times as much." Gradually increase the value until students have modeled, for instance here, 4 times as much.

While some students solve this problem as a two-step process, as shown in Solution 1, others may recognize that it can be solved as a one-step problem. Because the difference between Natasha's tape and Maya's tape is 3 units, students can solve by multiplying 3 and $5\frac{3}{16}$, as shown in Solution 2. Encourage students to reflect on the advantages of Solution 2.

A STORY OF UNITS   Lesson 39   4•5

Problem 3: A seamstress needs $1\frac{5}{8}$ yards of fabric to make a child's dress. She needs 3 times as much fabric to make a woman's dress. How many yards of fabric does she need for both dresses?

C [1 | 5/8]
W [1 | 5/8 | 1 | 5/8 | 1 | 5/8] } ?

She needs $6\frac{4}{8}$ yds of fabric to make both dresses.

Solution 1
$3 \times 1\frac{5}{8} = (3 \times 1) + (3 \times \frac{5}{8})$
$= 3 + \frac{15}{8}$
$= 3 + 1\frac{7}{8}$
$= 4\frac{7}{8}$

$4\frac{7}{8} + 1\frac{5}{8} = 5\frac{7}{8} + \frac{5}{8}$
            $\frac{1}{8}\;\frac{4}{8}$
$= 6\frac{4}{8}$

Solution 2
$4 \times 1\frac{5}{8} = 4 \times (1 + \frac{5}{8})$
$= (4 \times 1) + (4 \times \frac{5}{8})$
$= 4 + \frac{20}{8}$
$= 4 + 2\frac{4}{8}$
$= 6\frac{4}{8}$

While some students solve this problem in two steps, as shown in Solution 1, others may recognize that it can be solved as a one-step problem. In Solution 2, students count the 4 total units in the double tape diagram and multiply 4 by $1\frac{5}{8}$.

Problem 4: A piece of blue yarn is $5\frac{2}{3}$ yards long. A piece of pink yarn is 5 times as long as the blue yarn. Bailey tied them together with a knot that used $\frac{1}{3}$ yard from each piece of yarn. What is the total length of the yarn tied together?

B [5 | 2/3]
P [5 | 2/3 | 5 | 2/3 | 5 | 2/3 | 5 | 2/3 | 5 | 2/3] } y

The length of the yarn tied together was $33\frac{1}{3}$ yards long.

Solution 1
$6 \times 5\frac{2}{3} = (6 \times 5) + (6 \times \frac{2}{3})$
$= 30 + \frac{12}{3}$
$= 34$

$34 - \frac{2}{3} = 33 + \frac{1}{3} = 33\frac{1}{3}$
       $\land$
      $33\;1$

Solution 2
$5 \times 5\frac{2}{3} = (5 \times 5) + (5 \times \frac{2}{3})$
$= 25 + \frac{10}{3}$ $28\frac{1}{3} + 5 = 33\frac{1}{3}$
$= 28\frac{1}{3}$

Solution 1 shows a student's work in modeling the two pieces of yarn using a double tape diagram. The student multiplies to find the total length of the yarn and then subtracts the $\frac{2}{3}$ of a yard that is used in the knot. Solution 2 shows a student's work; the student recognizes that she needs to subtract $\frac{2}{3}$ of a yard at the onset. This student multiplies to identify the length of the pink yarn and then adds on 5, rather than $5\frac{2}{3}$, from the blue yarn. A common error is only subtracting $\frac{1}{3}$ yard instead of $\frac{1}{3}$ yard from each piece of yarn.

Student Debrief (10 minutes)

Lesson Objective: Solve multiplicative comparison word problems involving fractions.

The Student Debrief is intended to invite reflection and active processing of the total lesson experience.

Invite students to review their solutions for the Problem Set. They should check work by comparing answers with a partner before going over answers as a class. Look for misconceptions or misunderstandings that can be addressed in the Student Debrief. Guide students in a conversation to debrief the Problem Set and process the lesson.

Any combination of the questions below may be used to lead the discussion.

- What are some advantages to drawing a double tape diagram as the first step to solve comparison word problems?
- As the number of groups or the whole number in the mixed number gets larger, which strategies seem to be more efficient? Explain your thinking.
- When your peers share their drawings, does it help you understand the problem better? How does seeing your peers' work help you?
- What do you do when you get stuck on a word problem? How do you motivate yourself to persevere?
- When you check for reasonableness, do you look at your number sentences and model? How do you figure out if your answer is reasonable?
- What are some of the words you would use to create a word problem that uses multiplication and comparison?

Exit Ticket (3 minutes)

After the Student Debrief, instruct students to complete the Exit Ticket. A review of their work will help with assessing students' understanding of the concepts that were presented in today's lesson and planning more effectively for future lessons. The questions may be read aloud to the students.

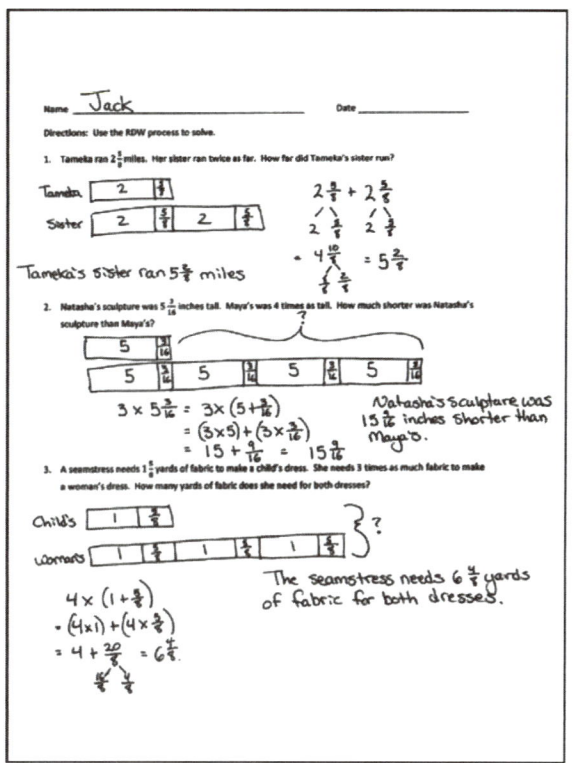

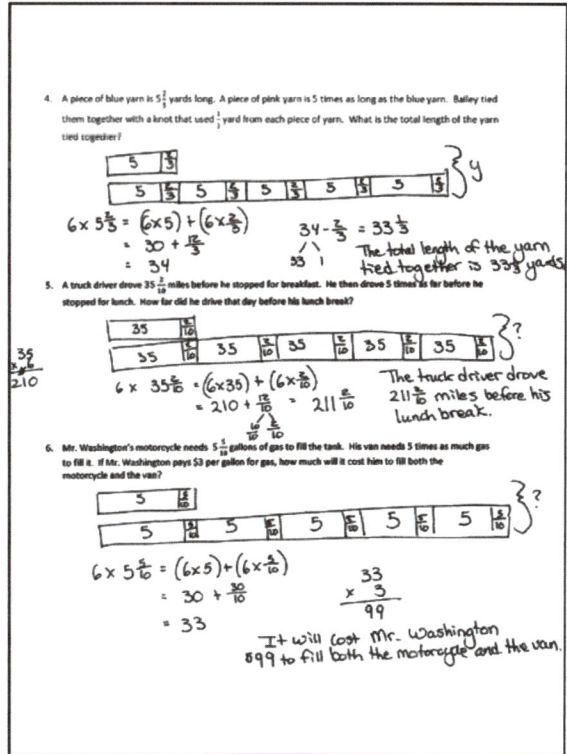

Lesson 39 Sprint 4•5

A

Number Correct: _____

Multiply Whole Numbers Times Fractions

1.	$\frac{1}{3} + \frac{1}{3} =$
2.	$2 \times \frac{1}{3} =$
3.	$\frac{1}{4} + \frac{1}{4} + \frac{1}{4} =$
4.	$3 \times \frac{1}{4} =$
5.	$\frac{1}{5} + \frac{1}{5} =$
6.	$2 \times \frac{1}{5} =$
7.	$\frac{1}{5} + \frac{1}{5} + \frac{1}{5} =$
8.	$3 \times \frac{1}{5} =$
9.	$\frac{1}{5} + \frac{1}{5} + \frac{1}{5} + \frac{1}{5} =$
10.	$4 \times \frac{1}{5} =$
11.	$\frac{1}{10} + \frac{1}{10} + \frac{1}{10} =$
12.	$3 \times \frac{1}{10} =$
13.	$\frac{1}{8} + \frac{1}{8} + \frac{1}{8} =$
14.	$3 \times \frac{1}{8} =$
15.	$\frac{1}{2} + \frac{1}{2} =$
16.	$2 \times \frac{1}{2} =$
17.	$\frac{1}{3} + \frac{1}{3} + \frac{1}{3} =$
18.	$3 \times \frac{1}{3} =$
19.	$\frac{1}{4} + \frac{1}{4} + \frac{1}{4} + \frac{1}{4} =$
20.	$4 \times \frac{1}{4} =$
21.	$\frac{1}{2} + \frac{1}{2} + \frac{1}{2} =$
22.	$3 \times \frac{1}{2} =$
23.	$\frac{1}{3} + \frac{1}{3} + \frac{1}{3} + \frac{1}{3} =$
24.	$4 \times \frac{1}{3} =$
25.	$\frac{5}{6} = \underline{} \times \frac{1}{6}$
26.	$\frac{5}{6} = 5 \times \underline{}$
27.	$\frac{5}{8} = 5 \times \underline{}$
28.	$\frac{5}{8} = \underline{} \times \frac{1}{8}$
29.	$\frac{7}{8} = 7 \times \underline{}$
30.	$\frac{7}{10} = 7 \times \underline{}$
31.	$\frac{7}{8} = \underline{} \times \frac{1}{8}$
32.	$\frac{7}{10} = \underline{} \times \frac{1}{10}$
33.	$\frac{6}{6} = 6 \times \underline{}$
34.	$1 = 6 \times \underline{}$
35.	$\frac{8}{8} = \underline{} \times \frac{1}{8}$
36.	$1 = \underline{} \times \frac{1}{8}$
37.	$9 \times \frac{1}{10} =$
38.	$7 \times \frac{1}{5} =$
39.	$1 = 3 \times \underline{}$
40.	$7 \times \frac{1}{12} =$
41.	$1 = \underline{} \times \frac{1}{5}$
42.	$\frac{3}{5} = \frac{1}{5} + \frac{1}{5} + \underline{}$
43.	$3 \times \frac{1}{4} = \underline{} + \frac{1}{4} + \frac{1}{4}$
44.	$1 = \underline{} + \underline{} + \underline{}$

512 Lesson 39: Solve multiplicative comparison word problems involving fractions.

© 2015 Great Minds. eureka-math.org
G4-M5-TE-B4-1.3.1-01.2016

B

Multiply Whole Numbers Times Fractions

Number Correct: _____
Improvement: _____

1.	$\frac{1}{5} + \frac{1}{5} =$	
2.	$2 \times \frac{1}{5} =$	
3.	$\frac{1}{3} + \frac{1}{3} =$	
4.	$2 \times \frac{1}{3} =$	
5.	$\frac{1}{4} + \frac{1}{4} + \frac{1}{4} =$	
6.	$3 \times \frac{1}{4} =$	
7.	$\frac{1}{5} + \frac{1}{5} + \frac{1}{5} =$	
8.	$3 \times \frac{1}{5} =$	
9.	$\frac{1}{5} + \frac{1}{5} + \frac{1}{5} + \frac{1}{5} =$	
10.	$4 \times \frac{1}{5} =$	
11.	$\frac{1}{8} + \frac{1}{8} + \frac{1}{8} =$	
12.	$3 \times \frac{1}{8} =$	
13.	$\frac{1}{10} + \frac{1}{10} + \frac{1}{10} =$	
14.	$3 \times \frac{1}{10} =$	
15.	$\frac{1}{3} + \frac{1}{3} + \frac{1}{3} =$	
16.	$3 \times \frac{1}{3} =$	
17.	$\frac{1}{4} + \frac{1}{4} + \frac{1}{4} + \frac{1}{4} =$	
18.	$4 \times \frac{1}{4} =$	
19.	$\frac{1}{2} + \frac{1}{2} =$	
20.	$2 \times \frac{1}{2} =$	
21.	$\frac{1}{3} + \frac{1}{3} + \frac{1}{3} + \frac{1}{3} =$	
22.	$4 \times \frac{1}{3} =$	

23.	$\frac{1}{2} + \frac{1}{2} + \frac{1}{2} =$	
24.	$3 \times \frac{1}{2} =$	
25.	$\frac{5}{6} =$	___ $\times \frac{1}{6}$
26.	$\frac{5}{6} =$	$5 \times$ ___
27.	$\frac{5}{8} =$	$5 \times$ ___
28.	$\frac{5}{8} =$	___ $\times \frac{1}{8}$
29.	$\frac{7}{8} =$	$7 \times$ ___
30.	$\frac{7}{10} =$	$7 \times$ ___
31.	$\frac{7}{8} =$	___ $\times \frac{1}{8}$
32.	$\frac{7}{10} =$	___ $\times \frac{1}{10}$
33.	$\frac{8}{8} =$	$8 \times$ ___
34.	$1 =$	$8 \times$ ___
35.	$\frac{6}{6} =$	___ $\times \frac{1}{6}$
36.	$1 =$	___ $\times \frac{1}{6}$
37.	$5 \times \frac{1}{12} =$	
38.	$6 \times \frac{1}{5} =$	
39.	$1 =$	$4 \times$ ___
40.	$9 \times \frac{1}{10} =$	
41.	$1 =$	___ $\times \frac{1}{3}$
42.	$\frac{3}{4} =$	$\frac{1}{4} + \frac{1}{4} +$ ___
43.	$3 \times \frac{1}{5} =$	___ $+ \frac{1}{5} + \frac{1}{5}$
44.	$1 =$	___ $+$ ___ $+$ ___ $+$ ___

Lesson 39: Solve multiplicative comparison word problems involving fractions.

Name _____ Date _____

Use the RDW process to solve.

1. Tameka ran $2\frac{5}{8}$ miles. Her sister ran twice as far. How far did Tameka's sister run?

2. Natasha's sculpture was $5\frac{3}{16}$ inches tall. Maya's was 4 times as tall. How much shorter was Natasha's sculpture than Maya's?

3. A seamstress needs $1\frac{5}{8}$ yards of fabric to make a child's dress. She needs 3 times as much fabric to make a woman's dress. How many yards of fabric does she need for both dresses?

4. A piece of blue yarn is $5\frac{2}{3}$ yards long. A piece of pink yarn is 5 times as long as the blue yarn. Bailey tied them together with a knot that used $\frac{1}{3}$ yard from each piece of yarn. What is the total length of the yarn tied together?

5. A truck driver drove $35\frac{2}{10}$ miles before he stopped for breakfast. He then drove 5 times as far before he stopped for lunch. How far did he drive that day before his lunch break?

6. Mr. Washington's motorcycle needs $5\frac{5}{10}$ gallons of gas to fill the tank. His van needs 5 times as much gas to fill it. If Mr. Washington pays $3 per gallon for gas, how much will it cost him to fill both the motorcycle and the van?

Lesson 39: Solve multiplicative comparison word problems involving fractions.

A STORY OF UNITS

Lesson 39 Exit Ticket 4•5

Name _____   Date _____

Use the RDW process to solve.

Jeff has ten packages that he wants to mail. Nine identical packages weigh $2\frac{7}{8}$ pounds each. A tenth package weighs two times as much as one of the other packages. How many pounds do all ten packages weigh?

A STORY OF UNITS

Lesson 39 Homework 4•5

Name _____   Date _____

Use the RDW process to solve.

1. Ground turkey is sold in packages of $2\frac{1}{2}$ pounds. Dawn bought eight times as much turkey that is sold in 1 package for her son's birthday party. How many pounds of ground turkey did Dawn buy?

2. Trevor's stack of books is $7\frac{7}{8}$ inches tall. Rick's stack is 3 times as tall. What is the difference in the heights of their stacks of books?

3. It takes $8\frac{3}{4}$ yards of fabric to make one quilt. Gail needs three times as much fabric to make three quilts. She already has two yards of fabric. How many more yards of fabric does Gail need to buy in order to make three quilts?

Lesson 39: Solve multiplicative comparison word problems involving fractions.

4. Carol made punch. She used $12\frac{3}{8}$ cups of juice and then added three times as much ginger ale. Then, she added 1 cup of lemonade. How many cups of punch did her recipe make?

5. Brandon drove $72\frac{7}{10}$ miles on Monday. He drove 3 times as far on Tuesday. How far did he drive in the two days?

6. Mrs. Reiser used $9\frac{8}{10}$ gallons of gas this week. Mr. Reiser used five times as much gas as Mrs. Reiser used this week. If Mr. Reiser pays $3 for each gallon of gas, how much did Mr. Reiser pay for gas this week?

A STORY OF UNITS — Lesson 40 4•5

Lesson 40

Objective: Solve word problems involving the multiplication of a whole number and a fraction including those involving line plots.

Suggested Lesson Structure

■ Fluency Practice   (13 minutes)
■ Concept Development   (37 minutes)
■ Student Debrief   (10 minutes)

Total Time   **(60 minutes)**

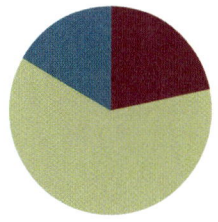

Fluency Practice (13 minutes)

- Make a One **4.NF.3**   (4 minutes)
- Count by Equivalent Fractions **4.NF.1** (5 minutes)
- Multiply Mixed Numbers **4.NF.4**   (4 minutes)

Make a One (4 minutes)

Materials: (S) Personal white board

Note: This fluency activity prepares students for Lesson 41.

T: (Write $\frac{2}{3}$.) Say the fraction.
S: 2 thirds.
T: Say the fraction needed to complete the next one.
S: 1 third.
T: $\frac{3}{4}$.
S: $\frac{1}{4}$.
T: $\frac{3}{5}$.
S: $\frac{2}{5}$.
T: (Write $\frac{5}{8}$ + — = 1.) Complete the number sentence.
S: (Write $\frac{3}{8}$.)

Continue with the following possible sequence: $1\frac{9}{10}, 2\frac{3}{10}, 3\frac{7}{12}, 7\frac{45}{50}, 12\frac{80}{100},$ and $65\frac{290}{400}$.

A STORY OF UNITS   Lesson 40   4•5

Count by Equivalent Fractions (5 minutes)

Note: This fluency activity reviews Lessons 24 and 25. The progression builds in complexity. Work students up to the highest level of complexity in which they can confidently participate.

T: Count by sevens to 70. Start at zero.

S: 0, 7, 14, 21, 28, 35, 42, 49, 56, 63, 70.

T: Count by 7 tenths to 70 tenths, starting at 0 tenths. (Write as students count.)

S: $\frac{0}{10}, \frac{7}{10}, \frac{14}{10}, \frac{21}{10}, \frac{28}{10}, \frac{35}{10}, \frac{42}{10}, \frac{49}{10}, \frac{56}{10}, \frac{63}{10}, \frac{70}{10}$.

$\frac{0}{10}$	$\frac{7}{10}$	$\frac{14}{10}$	$\frac{21}{10}$	$\frac{28}{10}$	$\frac{35}{10}$	$\frac{42}{10}$	$\frac{49}{10}$	$\frac{56}{10}$	$\frac{63}{10}$	$\frac{70}{10}$
0	$\frac{7}{10}$	$\frac{14}{10}$	$\frac{21}{10}$	$\frac{28}{10}$	$\frac{35}{10}$	$\frac{42}{10}$	$\frac{49}{10}$	$\frac{56}{10}$	$\frac{63}{10}$	7
0	$\frac{7}{10}$	$1\frac{4}{10}$	$2\frac{1}{10}$	$2\frac{8}{10}$	$3\frac{5}{10}$	$4\frac{2}{10}$	$4\frac{9}{10}$	$5\frac{6}{10}$	$6\frac{3}{10}$	7

T: Name the fraction that's equal to a whole number.

S: 70 tenths.

T: (Point to $\frac{70}{10}$.) 70 tenths is how many ones?

S: 7 ones.

T: (Beneath $\frac{70}{10}$, write 7.) Count by 7 tenths again. This time, when you come to the whole number, say the whole number. Start at zero. (Write as students count.)

S: $0, \frac{7}{10}, \frac{14}{10}, \frac{21}{10}, \frac{28}{10}, \frac{35}{10}, \frac{42}{10}, \frac{49}{10}, \frac{56}{10}, \frac{63}{10}, 7$.

T: (Point to $\frac{14}{10}$.) Say $\frac{14}{10}$ as a mixed number.

S: $1\frac{4}{10}$.

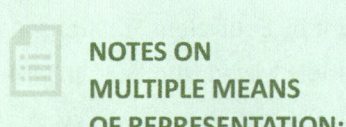

NOTES ON MULTIPLE MEANS OF REPRESENTATION:

One way to scaffold the Count by Equivalent Fractions fluency activity is to repeat the counting until students reach their comfort level. Keep it fun! Try couple counting with music, rhythm, or motion, such as jumping jacks or hops.

Continue the process for $\frac{21}{10}, \frac{28}{10}, \frac{35}{10}, \frac{42}{10}, \frac{49}{10}, \frac{56}{10}$, and $\frac{63}{10}$.

T: Count by 7 tenths again. This time, convert to whole numbers and mixed numbers. Start at zero. (Write as students count.)

S: $0, \frac{7}{10}, 1\frac{4}{10}, 2\frac{1}{10}, 2\frac{8}{10}, 3\frac{5}{10}, 4\frac{2}{10}, 4\frac{9}{10}, 5\frac{6}{10}, 6\frac{3}{10}, 7$.

Lesson 40: Solve word problems involving the multiplication of a whole number and a fraction including those involving line plots.

A STORY OF UNITS   Lesson 40   4•5

Multiply Mixed Numbers (4 minutes)

Materials: (S) Personal white board

Note: This fluency activity reviews Lesson 37.

- T: Break apart $2\frac{4}{5}$ using an addition expression.
- S: (Write $2\frac{4}{5}$ as $2+\frac{4}{5}$.)
- T: (Write $3 \times 2\frac{4}{5}$. Beneath it, write __ + —.) Fill in the unknown numbers.
- S: (Write $6 + \frac{12}{5}$.)
- T: (Write $6 + \frac{12}{5}$. Beneath it, write 6 + __.) Fill in a mixed number for $\frac{12}{5}$.
- S: (Write $6 + 2\frac{2}{5}$.)
- T: (Write $6 + 2\frac{2}{5}$. Beneath it, write = __.) Write the answer.
- S: (Write = $8\frac{2}{5}$.)
- T: (Point to $3 \times 2\frac{4}{5} = $ __.) Say the multiplication sentence.
- S: $3 \times 2\frac{4}{5} = 8\frac{2}{5}$.

Continue the process for $5 \times 3\frac{5}{8}$.

Concept Development (37 minutes)

Materials: (S) Personal white board, Problem Set

Note: Today's Problem Set, in which students construct and interpret a line plot, is used during the Concept Development. As students complete each problem, debrief student solutions. The solutions offered below show variety but are not all-inclusive. Encourage students to discuss their math thinking and accept different strategies and solutions that result in the correct answer.

Suggested Delivery of Instruction for Solving Lesson 40's Word Problems

1. **Model the problem.**

Have two pairs of students who can successfully model the problem work at the board while the others work independently or in pairs at their seats. Review the following questions before beginning the first problem.

- Can you draw something?
- What can you draw?
- What conclusions can you make from your drawing?

Lesson 40:   Solve word problems involving the multiplication of a whole number and a fraction including those involving line plots.   521

A STORY OF UNITS   Lesson 40   4•5

Circulate as students work. Reiterate the questions above. After two minutes, have the two pairs of students share only their labeled diagrams. For about one minute, have the demonstrating students receive and respond to feedback and questions from their peers.

2. Calculate to solve and write a statement.

Give students two minutes to finish work on that question, sharing their work and thinking with a peer. All students should then write their equations and statements of the answer.

3. Assess the solution for reasonableness.

Give students one to two minutes to assess and explain the reasonableness of their solution.

> **NOTES ON MULTIPLE MEANS OF REPRESENTATION:**
>
> Consider breaking word problems into steps for students working below grade level. For example, for Problem 1(a), ask, "What is the height of the tallest player? What is the height of the shortest player? What is the difference in the height of the tallest and shortest players?"

Problem 1

The chart to the right shows the heights, in feet, of some football players. Use the data to create a line plot at the bottom of this page and to answer the questions below.

a. What is the difference in the height of the tallest and shortest players?

Player	Height (in feet)
A	$6\frac{1}{4}$
B	$5\frac{7}{8}$
C	$6\frac{1}{2}$
D	$6\frac{1}{4}$
E	$6\frac{2}{8}$
F	$5\frac{7}{8}$
G	$6\frac{1}{8}$
H	$6\frac{5}{8}$
I	$5\frac{6}{8}$
J	$6\frac{1}{8}$

MP.4

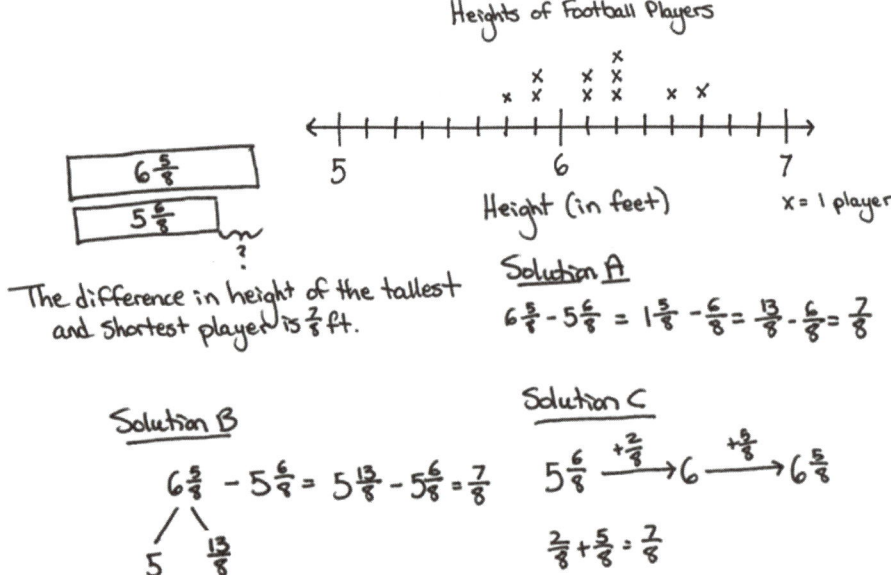

In Solution A, the student subtracts the whole numbers first and then converts to actions greater than 1 to solve. In Solution B, the student decomposes a whole before solving. In Solution C, the student counts up to find the solution.

Lesson 40: Solve word problems involving the multiplication of a whole number and a fraction including those involving line plots.

Player I and Player B have a combined height that is $1\frac{1}{8}$ feet taller than a school bus. What is the height of a school bus?

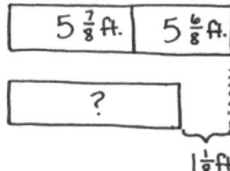

The bus has a height of $10\frac{4}{8}$ feet.

Solution A

$5\frac{7}{8} + 5\frac{6}{8} = 10\frac{13}{8} = 11\frac{5}{8}$

$\phantom{5\frac{7}{8} + 5\frac{6}{8} = 10\frac{13}{8} = 11}\wedge$
$\phantom{5\frac{7}{8} + 5\frac{6}{8} = 10\frac{13}{8} = 11}\frac{8}{8}\ \frac{5}{8}$

$11\frac{5}{8} - 1\frac{1}{8} = 10\frac{4}{8}$

Solution B

$5\frac{7}{8} + 5\frac{6}{8} = 11\frac{5}{8}$

$\phantom{5\frac{7}{8} + 5\frac{6}{8} = 1}\wedge$
$\phantom{5\frac{7}{8} + 5\frac{6}{8} = 1}\frac{1}{8}\ \frac{5}{8}$

$11\frac{5}{8} - 1\frac{1}{8} = 10\frac{4}{8}$

Solution C

$5\frac{6}{8} - 1\frac{1}{8} = 4\frac{5}{8}$

$4\frac{5}{8} + 5\frac{7}{8} = 10\frac{4}{8}$

$\phantom{4\frac{5}{8} + 5\frac{7}{8} = 10}\wedge$
$\phantom{4\frac{5}{8} + 5\frac{7}{8} = 10}\frac{3}{8}\ \frac{4}{8}$

In Solutions A and B, students find the sum and then convert the fraction greater than 1 to find the height of $11\frac{5}{8}$. In the final step, students subtract $1\frac{1}{8}$ to solve for the height of the bus. Solution C subtracts the difference from Player I's height and adds Player B's height as a final step.

Problem 2

One of the players on the team is now 4 times as tall as he was at birth, when he measured $1\frac{5}{8}$ feet. Who is the player?

Birth: $1\frac{5}{8}$
Now: $1\frac{5}{8}\ 1\frac{5}{8}\ 1\frac{5}{8}\ 1\frac{5}{8}$

Player C was $1\frac{5}{8}$ ft at birth.

Solution A

$4 \times 1\frac{5}{8} = 4 \times (1 + \frac{5}{8})$
$= (4 \times 1) + (4 \times \frac{5}{8})$
$= 4 + \frac{4 \times 5}{8}$
$= 4 + \frac{20}{8}$
$\phantom{= 4 + \frac{20}{8}}\wedge$
$\phantom{= 4 + \frac{20}{8}}\frac{16}{8}\ \frac{4}{8}$
$= 6\frac{4}{8}$

Solution B

$4 \times 1\frac{5}{8} = 4 + \frac{20}{8}$
$= 6\frac{4}{8}$

$6\frac{4}{8} = 6\frac{1}{2}$

In this solution, students use the distributive property to compute the current height of the player. Students then look back at the line plot to determine which player's height is equivalent to $6\frac{4}{8}$. When using the distributive property, students may complete some of the computations mentally, as shown in Solution B.

Problem 3

Six of the players on the team weigh over 300 pounds. Doctors recommend that players of this weight drink at least $3\frac{3}{4}$ quarts of water each day. At least how much water should be consumed per day by all 6 players?

Solution A
$6 \times 3\frac{3}{4} = (6 \times 3) + (6 \times \frac{3}{4}) = 18 + \frac{18}{4} = 22\frac{2}{4}$
$\frac{18}{4} \rightarrow \frac{16}{4} \text{ and } \frac{2}{4}$

Solution B
$6 \times 3\frac{3}{4} = 18 + \frac{18}{4} = 22\frac{2}{4}$
$\frac{18}{4} \rightarrow 4 \text{ and } \frac{2}{4}$

[Tape diagram: $3\frac{3}{4}$ | $3\frac{3}{4}$ | $3\frac{3}{4}$ | $3\frac{3}{4}$ | $3\frac{3}{4}$ | $3\frac{3}{4}$]

Solution C
$6 \times \frac{15}{4} = \frac{6 \times 15}{4} = \frac{90}{4}$

All 6 players should consume $22\frac{2}{4}$ quarts.

Students use the distributive property in Solutions A and B. In Solution B, students no longer write out each step. Some of the computations are done mentally. In Solution C, students convert to a fraction greater than 1. Those who do not convert back to a mixed number should be encouraged to use the context of the problem to consider if their answer is in a reasonable form.

Problem 4

Nine of the players on the team weigh about 200 pounds. Doctors recommend that people of this weight each eat about $3\frac{7}{10}$ grams of carbohydrates per pound each day. About how many combined grams of carbohydrates should these 9 players eat per pound each day?

Solution A
$9 \times 3\frac{7}{10} = (9 \times 3) + (9 \times \frac{7}{10}) = 27 + \frac{9 \times 7}{10} = 27 + \frac{63}{10} = 33\frac{3}{10}$
$\frac{63}{10} \rightarrow 6 \text{ and } \frac{3}{10}$

Solution B
$9 \times 3\frac{7}{10} = 27 + \frac{63}{10} = 33\frac{3}{10}$
$\frac{63}{10} \rightarrow 6 \text{ and } \frac{3}{10}$

[Tape diagram: $3\frac{7}{10}$ | $3\frac{7}{10}$ | $3\frac{7}{10}$ | $3\frac{7}{10}$ | $3\frac{7}{10}$ | $3\frac{7}{10}$ | $3\frac{7}{10}$ | $3\frac{7}{10}$ | $3\frac{7}{10}$]

Solution C
$9 \times \frac{37}{10} = \frac{9 \times 37}{10} = \frac{333}{10}$

$\begin{array}{r} 37 \\ \times 9 \\ \hline 333 \end{array}$

The 9 players should eat a combined $33\frac{3}{10}$ grams of carbohydrates per pound each day.

In Solutions A and B, students use the distributive property to solve. Students may choose to solve using Solution C, which is not the most efficient method and does not provide a realistic form of an answer, considering the context. Provide students who are not quick to select the distributive property scaffolds to support their understanding.

A STORY OF UNITS   Lesson 40   4•5

Student Debrief (10 minutes)

Lesson Objective: Solve word problems involving the multiplication of a whole number and a fraction including those involving line plots.

The Student Debrief is intended to invite reflection and active processing of the total lesson experience.

Invite students to review their solutions for the Problem Set. They should check work by comparing answers with a partner before going over answers as a class. Look for misconceptions or misunderstandings that can be addressed in the Student Debrief. Guide students in a conversation to debrief the Problem Set and process the lesson.

Any combination of the questions below may be used to lead the discussion.

- For Problem 1(a), how was the line plot helpful in finding the height of the tallest and shortest players?
- For Problem 1(b), did you refer back to the line plot or chart to find the information necessary to solve? Explain.
- Did you determine the answers to Problems 2, 3, and 4 using the same math strategy? Explain to a partner how you determined your answers.
- How was the *draw* step of the RDW approach helpful in solving Problem 2?
- What information can we gather simply by looking at the line plot? Write one statement about the football players based on the information in the line plot.
- What information about the football players is easier to see when the data is represented using a line plot rather than the chart? A chart rather than the line plot?

Exit Ticket (3 minutes)

After the Student Debrief, instruct students to complete the Exit Ticket. A review of their work will help with assessing students' understanding of the concepts that were presented in today's lesson and planning more effectively for future lessons. The questions may be read aloud to the students.

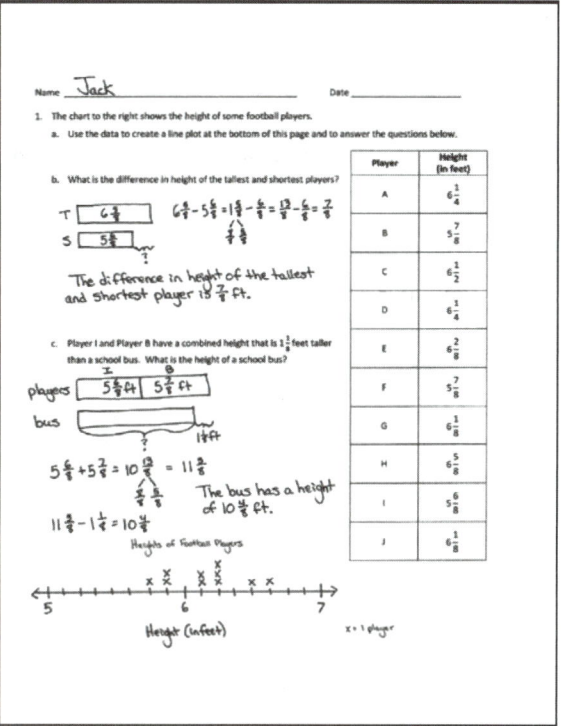

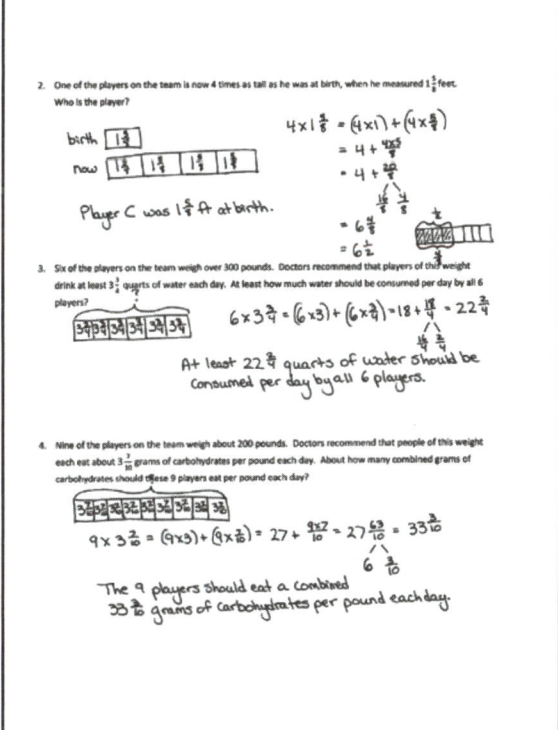

Lesson 40: Solve word problems involving the multiplication of a whole number and a fraction including those involving line plots.

Name _____ Date _____

1. The chart to the right shows the height of some football players.

   a. Use the data to create a line plot at the bottom of this page and to answer the questions below.

   b. What is the difference in height of the tallest and shortest players?

   c. Player I and Player B have a combined height that is $1\frac{1}{8}$ feet taller than a school bus. What is the height of a school bus?

Player	Height (in feet)
A	$6\frac{1}{4}$
B	$5\frac{7}{8}$
C	$6\frac{1}{2}$
D	$6\frac{1}{4}$
E	$6\frac{2}{8}$
F	$5\frac{7}{8}$
G	$6\frac{1}{8}$
H	$6\frac{5}{8}$
I	$5\frac{6}{8}$
J	$6\frac{1}{8}$

2. One of the players on the team is now 4 times as tall as he was at birth, when he measured $1\frac{5}{8}$ feet. Who is the player?

3. Six of the players on the team weigh over 300 pounds. Doctors recommend that players of this weight drink at least $3\frac{3}{4}$ quarts of water each day. At least how much water should be consumed per day by all 6 players?

4. Nine of the players on the team weigh about 200 pounds. Doctors recommend that people of this weight each eat about $3\frac{7}{10}$ grams of carbohydrates per pound each day. About how many combined grams of carbohydrates should these 9 players eat per pound each day?

Name _____ Date _____

Coach Taylor asked his team to record the distance they ran during practice. The distances are listed in the table.

1. Use the table to locate the incorrect data on the line plot.
   Circle any incorrect points.
   Mark any missing points.

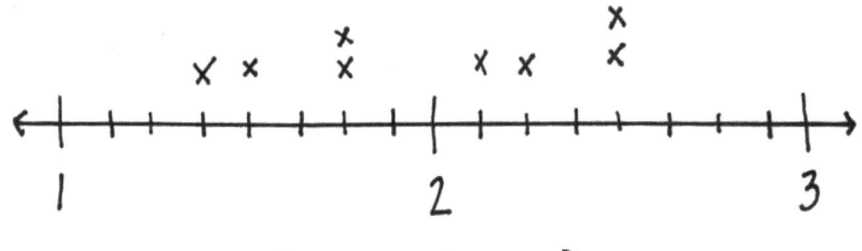

2. Of the team members who ran $1\frac{6}{8}$ miles, how many miles did those team members run combined?

Team Members	Distance (in miles)
Alec	$1\frac{3}{4}$
Henry	$1\frac{1}{2}$
Charles	$2\frac{1}{8}$
Steve	$1\frac{3}{4}$
Pitch	$2\frac{2}{4}$
Raj	$1\frac{6}{8}$
Pam	$2\frac{1}{2}$
Tony	$1\frac{3}{8}$

Name _____   Date _____

The chart to the right shows the total monthly rainfall for a city.

1. Use the data to create a line plot at the bottom of this page and to answer the following questions.

Month	Rainfall (in inches)
January	$2\frac{2}{8}$
February	$1\frac{3}{8}$
March	$2\frac{3}{8}$
April	$2\frac{5}{8}$
May	$4\frac{1}{4}$
June	$2\frac{1}{4}$
July	$3\frac{7}{8}$
August	$3\frac{1}{4}$
September	$1\frac{5}{8}$
October	$3\frac{2}{8}$
November	$1\frac{3}{4}$
December	$1\frac{5}{8}$

Lesson 40: Solve word problems involving the multiplication of a whole number and a fraction including those involving line plots.

2. What is the difference in rainfall from the wettest and driest months?

3. How much more rain fell in May than in April?

4. What is the combined rainfall amount for the summer months of June, July, and August?

5. How much more rain fell in the summer months than the combined rainfall for the last 4 months of the year?

6. In which months did it rain twice as much as it rained in December?

7. Each inch of rain can produce ten times that many inches of snow. If all of the rainfall in January was in the form of snow, how many inches of snow fell in January?

A STORY OF UNITS

GRADE 4

Mathematics Curriculum

GRADE 4 • MODULE 5

Topic H
Exploring a Fraction Pattern

4.OA.5

Focus Standard:	4.OA.5	Generate a number or shape pattern that follows a given rule. Identify apparent features of the pattern that were not explicit in the rule itself. *For example, given the rule "Add 3" and the starting number 1, generate terms in the resulting sequence and observe that the terms appear to alternate between odd and even numbers. Explain informally why the numbers will continue to alternate in this way.*
Instructional Days:	1	
Coherence -Links from:	G3–M5	Fractions as Numbers on the Number Line
-Links to:	G5–M3	Addition and Subtraction of Fractions

Topic H is an exploration lesson in which students find the sum of all like denominators from $\frac{0}{n}$ to $\frac{n}{n}$.

Students first work, in teams, with fourths, sixths, eighths, and tenths. For example, they might find the sum of all sixths from $\frac{0}{6}$ to $\frac{6}{6}$. Students discover that they can make pairs with a sum of 1 to add more efficiently, e.g., $\frac{0}{6} + \frac{6}{6}, \frac{1}{6} + \frac{5}{6}, \frac{2}{6} + \frac{4}{6}$, and there will be one fraction, $\frac{3}{6}$, without a pair. As students make this discovery, they share and compare their strategies within their teams. They then extend this to similarly find sums of thirds, fifths, sevenths, and ninths, observing patterns when finding the sum of odd and even denominators (**4.OA.5**). Through discussion of their strategies, students determine which are most efficient.

Advanced students can be challenged to find the sum of all hundredths from 0 hundredths to 100 hundredths.

A Teaching Sequence Toward Mastery of Exploring a Fraction Pattern
Objective 1: Find and use a pattern to calculate the sum of all fractional parts between 0 and 1. Share and critique peer strategies. (Lesson 41)

Topic H: Exploring a Fraction Pattern

A STORY OF UNITS · Lesson 41 · 4•5

Lesson 41

Objective: Find and use a pattern to calculate the sum of all fractional parts between 0 and 1. Share and critique peer strategies.

Suggested Lesson Structure

- **Fluency Practice** (12 minutes)
- **Application Problem** (4 minutes)
- **Concept Development** (34 minutes)
- **Student Debrief** (10 minutes)

Total Time **(60 minutes)**

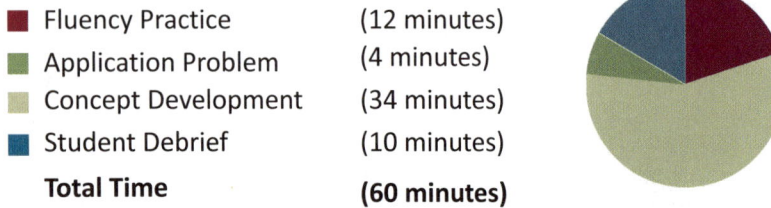

Fluency Practice (12 minutes)

- Add and Subtract **4.NBT.4** (4 minutes)
- Multiply Mixed Numbers **4.NF.4** (4 minutes)
- Make a One **4.NF.3** (4 minutes)

Add and Subtract (4 minutes)

Materials: (S) Personal white board

Note: This fluency activity reviews adding and subtracting using the standard algorithm.

- T: (Write 643 thousands 857 ones.) On your personal white boards, write this number in standard form.
- S: (Write 643,857.)
- T: (Write 247 thousands 728 ones.) Add this number to 643,857 using the standard algorithm.
- S: (Write 643,857 + 247,728 = 891,585 using the standard algorithm.)

Continue the process for 658,437 + 144,487.

- T: (Write 400 thousands.) On your boards, write this number in standard form.
- S: (Write 400,000.)
- T: (Write 346 thousands 286 ones.) Subtract this number from 400,000 using the standard algorithm.
- S: (Write 400,000 – 346,286 = 53,714 using the standard algorithm.)

Continue the process for 609,428 – 297,639.

A STORY OF UNITS   Lesson 41   4•5

Multiply Mixed Numbers (4 minutes)

Materials: (S) Personal white board

Note: This fluency activity reviews Lesson 37.

- T: Write $5\frac{3}{4}$.
- S: (Write $5\frac{3}{4}$.)
- T: Break apart $5\frac{3}{4}$ using a number bond.
- S: (Break apart $5\frac{3}{4}$ into 5 and $\frac{3}{4}$.)
- T: (Write $3 \times 5\frac{3}{4}$. Beneath it, write __ + __.) Fill in the unknown numbers.
- S: (Beneath $3 \times 5\frac{3}{4}$, write $15 + \frac{9}{4}$.)
- T: (Write $15 + \frac{9}{4}$. Beneath it, write $15 +$ __.) Fill in a mixed number for $\frac{9}{4}$.
- S: (Beneath $15 + \frac{9}{4}$, write $15 + 2\frac{1}{4}$.)
- T: (Write $15 + 2\frac{1}{4}$. Beneath it, write = __.) Write the answer.
- S: (Beneath $15 + 2\frac{1}{4}$, write $= 17\frac{1}{4}$).
- T: (Point at $3 \times 5\frac{3}{4} =$ __.) Say the multiplication sentence.
- S: $3 \times 5\frac{3}{4} = 17\frac{1}{4}$.

Continue the process for $5 \times 2\frac{7}{8}$.

Make a One (4 minutes)

Materials: (S) Personal white board

Note: This fluency activity prepares students for Lesson 41.

- T: (Write $\frac{3}{4}$.) Say the fraction.
- S: 3 fourths.
- T: Say the fraction that needs to be added to 3 fourths to make one.
- S: 1 fourth.
- T: $\frac{2}{3}$.
- S: $\frac{1}{3}$.

Lesson 41:   Find and use a pattern to calculate the sum of all fractional parts between 0 and 1. Share and critique peer strategies.

533

A STORY OF UNITS
Lesson 41 4•5

T: $\frac{7}{8}$.

S: $\frac{1}{8}$.

T: (Write $\frac{3}{8} + \underline{} = 1$.) Complete the number sentence.

S: (Write $\frac{3}{8} + \frac{5}{8} = 1$.)

Continue with the following possible sequence: $\frac{7}{10}, \frac{7}{6}$, and $\frac{5}{12}$.

Application Problem (4 minutes)

Jackie's paper chain was 5 times as long as Sammy's, which measured $2\frac{75}{100}$ meters. What was the total length of both their chains?

$6 \times 2\frac{75}{100} = 12 + \frac{450}{100}$

$= 4\frac{50}{100}$

$= 16\frac{50}{100}$

$= 16\frac{1}{2}$

The total length of their chain is $16\frac{1}{2}$ m.

Note: This Application Problem anticipates Module 6's work with decimal numbers.

Concept Development (34 minutes)

Materials: (S) Index cards cut in halves or fourths (20 cards per student)

Problem 1: Explore patterns for sums of fractions.

In groups of four, have students record a set of fractions from 0 to 1 for a given unit on cards.

Part 1: Assign each member of the group to make a different set of fraction cards for the following even denominators: fourths, sixths, eighths, and tenths.

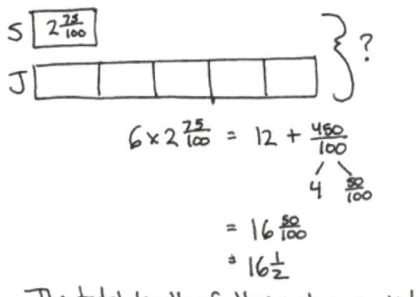

MP.3

1. Lay the cards in order from least to greatest.

2. Solve for the sum of the fractional units $\frac{0}{n}$ to $\frac{n}{n}$. Express the sum as a mixed number.

$\frac{0}{6} + \frac{1}{6} + \frac{2}{6} + \frac{3}{6} + \frac{4}{6} + \frac{5}{6} + \frac{6}{6} = \frac{21}{6} = 3\frac{3}{6}$

$\frac{18}{6} \quad \frac{3}{6}$

534   Lesson 41:   Find and use a pattern to calculate the sum of all fractional parts between 0 and 1. Share and critique peer strategies.

A STORY OF UNITS   Lesson 41   4•5

3. Invite students to share their ways of finding the sum within their teams.
4. Solve for the sum again. This time, group pairs of fractions that equal 1.
5. Each team looks for patterns within their sums.

MP.3

Part 2: Assign each member of the group to make a different set of fraction cards for the following odd denominators: thirds, fifths, sevenths, and ninths. Repeat Steps 1–5 from Part 1.

Part 3: Reconvene as a class, having groups compare and contrast the results when adding pairs of numbers with even denominators to adding numbers with odd denominators. Challenge them to clearly state their thinking using words, pictures, or numbers.

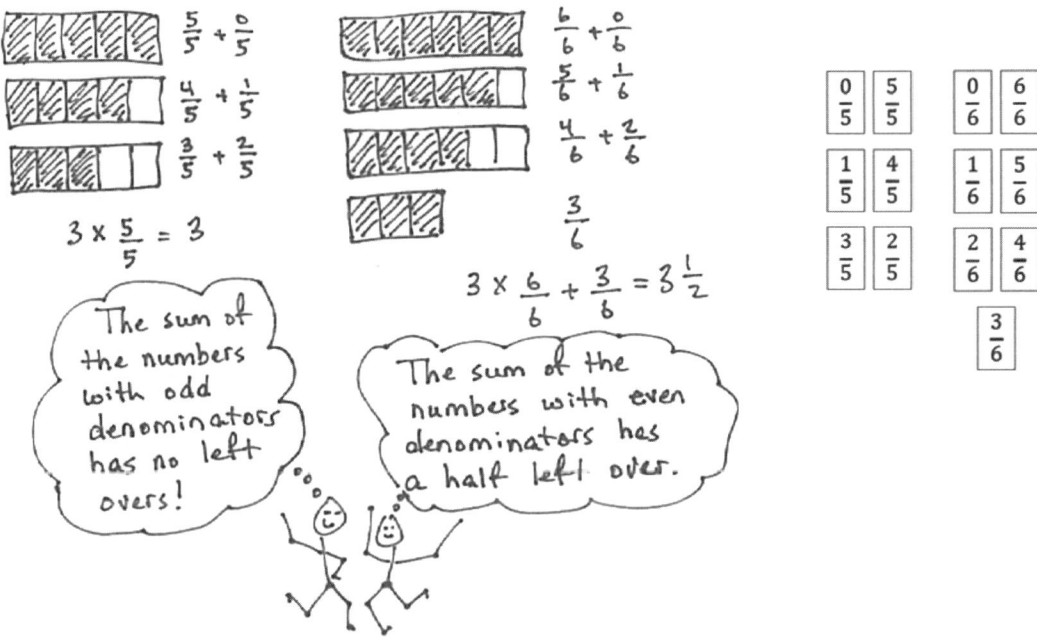

Problem 2: Apply the pattern to find the sum of consecutive fractions with large denominators.

- Each team member chooses at least one large even denominator (above 20) and finds the sum of $\frac{0}{n}$ to $\frac{n}{n}$.

- Each team member chooses at least one large odd denominator (above 20) and finds the sum of $\frac{0}{n}$ to $\frac{n}{n}$.

- Team members share results and look for patterns in their sums. Can they describe a way to find the sum of any set of fractions from $\frac{0}{n}$ to $\frac{n}{n}$?

EUREKA MATH

Lesson 41: Find and use a pattern to calculate the sum of all fractional parts between 0 and 1. Share and critique peer strategies.

535

©2015 Great Minds. eureka-math.org
G4-M5-TE-B4-1.3.1-01.2016

A STORY OF UNITS — Lesson 41 4•5

Problem Set (10 minutes)

Students should do their personal best to complete the Problem Set within the allotted 10 minutes. For some classes, it may be appropriate to modify the assignment by specifying which problems they work on first. Some problems do not specify a method for solving. Students should solve these problems using the RDW approach used for Application Problems.

Student Debrief (10 minutes)

Lesson Objective: Find and use a pattern to calculate the sum of all fractional parts between 0 and 1. Share and critique peer strategies.

The Student Debrief is intended to invite reflection and active processing of the total lesson experience.

Invite students to review their solutions for the Problem Set. They should check work by comparing answers with a partner before going over answers as a class. Look for misconceptions or misunderstandings that can be addressed in the Debrief. Guide students in a conversation to debrief the Problem Set and process the lesson.

Any combination of the questions below may be used to lead the discussion.

- Discuss the difference in the sums between even and odd denominators. Why is this?
- How did the pattern found in Problem 2 work for solving in Problem 4? In what ways did your pattern need revision?
- Is it necessary to test your answer for Problem 6? Why or why not?
- How might you find the sum of all the whole numbers up to 10 using an array?

$$\begin{array}{cccccccccc} 1 & 2 & 3 & 4 & 5 & 6 & 7 & 8 & 9 & 10 \\ 10 & 9 & 8 & 7 & 6 & 5 & 4 & 3 & 2 & 1 \end{array}$$

$$\frac{10 \times 11}{2}$$

- Can you find a shortcut to calculate the sum of all the whole numbers from 0 to 50? To 100? Explain how. (An explanation of one method is found in the Notes in the box above.)

> **A NOTE ON MATH HISTORY:**
>
> The story goes that, in 1885, when 8 years old, Carl Friedrich Gauss aborted his teacher's attempt to keep him busy for an hour. The teacher had assigned the class the tedious task of finding the sum of all the whole numbers up to 100. Quick as a wink, Gauss said 5,050 and also explained his solution strategy. He paired one set of numbers from 1 to 100 with another set of the same numbers:
>
> 1 2 3 4 5 6 7 8...
> 100 99 98 97 96 95 94 93...
>
> Each of 100 pairs had a sum of 101. However, that was double the answer, so the product needed to be divided by 2.
>
> $(100 \times 101) \div 2 = \frac{10,100}{2} = 5,050$

1 10 XOOOOOOOOOO
2   9 XXOOOOOOOOO
3   8 XXXOOOOOOOO
4   7 XXXXOOOOOOO
5   6 XXXXXOOOOOO

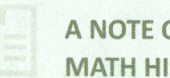

1 2 3 4 5 6 7 8 9 10
50 49 48 47 46 45 44 43 42 41...

Lesson 41   4•5

Exit Ticket (3 minutes)

After the Student Debrief, instruct students to complete the Exit Ticket. A review of their work will help with assessing students' understanding of the concepts that were presented in today's lesson and planning more effectively for future lessons. The questions may be read aloud to the students.

Lesson 41: Find and use a pattern to calculate the sum of all fractional parts between 0 and 1. Share and critique peer strategies.

Name _____ Date _____

1. Find the sums.

   a. $\dfrac{0}{3}+\dfrac{1}{3}+\dfrac{2}{3}+\dfrac{3}{3}$

   b. $\dfrac{0}{4}+\dfrac{1}{4}+\dfrac{2}{4}+\dfrac{3}{4}+\dfrac{4}{4}$

   c. $\dfrac{0}{5}+\dfrac{1}{5}+\dfrac{2}{5}+\dfrac{3}{5}+\dfrac{4}{5}+\dfrac{5}{5}$

   d. $\dfrac{0}{6}+\dfrac{1}{6}+\dfrac{2}{6}+\dfrac{3}{6}+\dfrac{4}{6}+\dfrac{5}{6}+\dfrac{6}{6}$

   e. $\dfrac{0}{7}+\dfrac{1}{7}+\dfrac{2}{7}+\dfrac{3}{7}+\dfrac{4}{7}+\dfrac{5}{7}+\dfrac{6}{7}+\dfrac{7}{7}$

   f. $\dfrac{0}{8}+\dfrac{1}{8}+\dfrac{2}{8}+\dfrac{3}{8}+\dfrac{4}{8}+\dfrac{5}{8}+\dfrac{6}{8}+\dfrac{7}{8}+\dfrac{8}{8}$

2. Describe a pattern you notice when adding the sums of fractions with even denominators as opposed to those with odd denominators.

3. How would the sums change if the addition started with the unit fraction rather than with 0?

4. Find the sums.

   a. $\dfrac{0}{10} + \dfrac{1}{10} + \dfrac{2}{10} + \cdots + \dfrac{10}{10}$

   b. $\dfrac{0}{12} + \dfrac{1}{12} + \dfrac{2}{12} + \cdots + \dfrac{12}{12}$

   c. $\dfrac{0}{15} + \dfrac{1}{15} + \dfrac{2}{15} + \cdots + \dfrac{15}{15}$

   d. $\dfrac{0}{25} + \dfrac{1}{25} + \dfrac{2}{25} + \cdots + \dfrac{25}{25}$

   e. $\dfrac{0}{50} + \dfrac{1}{50} + \dfrac{2}{50} + \cdots + \dfrac{50}{50}$

   f. $\dfrac{0}{100} + \dfrac{1}{100} + \dfrac{2}{100} + \cdots + \dfrac{100}{100}$

5. Compare your strategy for finding the sums in Problems 4(d), 4(e), and 4(f) with a partner.

6. How can you apply this strategy to find the sum of all the whole numbers from 0 to 100?

Name _____ Date _____

Find the sums.

1. $\dfrac{0}{20} + \dfrac{1}{20} + \dfrac{2}{20} + \cdots + \dfrac{20}{20}$

2. $\dfrac{0}{200} + \dfrac{1}{200} + \dfrac{2}{200} + \cdots + \dfrac{200}{200}$

A STORY OF UNITS

Lesson 41 Homework 4•5

Name _____ Date _____

1. Find the sums.

   a. $\dfrac{0}{5}+\dfrac{1}{5}+\dfrac{2}{5}+\dfrac{3}{5}+\dfrac{4}{5}+\dfrac{5}{5}$

   b. $\dfrac{0}{6}+\dfrac{1}{6}+\dfrac{2}{6}+\dfrac{3}{6}+\dfrac{4}{6}+\dfrac{5}{6}+\dfrac{6}{6}$

   c. $\dfrac{0}{7}+\dfrac{1}{7}+\dfrac{2}{7}+\dfrac{3}{7}+\dfrac{4}{7}+\dfrac{5}{7}+\dfrac{6}{7}+\dfrac{7}{7}$

   d. $\dfrac{0}{8}+\dfrac{1}{8}+\dfrac{2}{8}+\dfrac{3}{8}+\dfrac{4}{8}+\dfrac{5}{8}+\dfrac{6}{8}+\dfrac{7}{8}+\dfrac{8}{8}$

   e. $\dfrac{0}{9}+\dfrac{1}{9}+\dfrac{2}{9}+\dfrac{3}{9}+\dfrac{4}{9}+\dfrac{5}{9}+\dfrac{6}{9}+\dfrac{7}{9}+\dfrac{8}{9}+\dfrac{9}{9}$

   f. $\dfrac{0}{10}+\dfrac{1}{10}+\dfrac{2}{10}+\dfrac{3}{10}+\dfrac{4}{10}+\dfrac{5}{10}+\dfrac{6}{10}+\dfrac{7}{10}+\dfrac{8}{10}+\dfrac{9}{10}+\dfrac{10}{10}$

2. Describe a pattern you notice when adding the sums of fractions with even denominators as opposed to those with odd denominators.

3. How would the sums change if the addition started with the unit fraction rather than with 0?

Lesson 41: Find and use a pattern to calculate the sum of all fractional parts between 0 and 1. Share and critique peer strategies.

4. Find the sums.

   a. $\dfrac{0}{20} + \dfrac{1}{20} + \dfrac{2}{20} + \cdots + \dfrac{20}{20}$

   b. $\dfrac{0}{35} + \dfrac{1}{35} + \dfrac{2}{35} + \cdots + \dfrac{35}{35}$

   c. $\dfrac{0}{36} + \dfrac{1}{36} + \dfrac{2}{36} + \cdots + \dfrac{36}{36}$

   d. $\dfrac{0}{75} + \dfrac{1}{75} + \dfrac{2}{75} + \cdots + \dfrac{75}{75}$

   e. $\dfrac{0}{100} + \dfrac{1}{100} + \dfrac{2}{100} + \cdots + \dfrac{100}{100}$

   f. $\dfrac{0}{99} + \dfrac{1}{99} + \dfrac{2}{99} + \cdots + \dfrac{99}{99}$

5. How can you apply this strategy to find the sum of all the whole numbers from 0 to 50? To 99?

Name _____ Date _____

1. a. Partition the tape diagram to show $5 \times \frac{2}{3}$. Partition the number line to show $10 \times \frac{1}{3}$.

   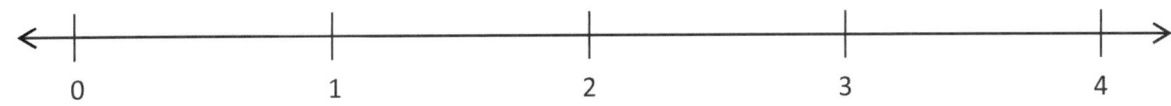

   b. Use the models above to explain why $5 \times \frac{2}{3} = 10 \times \frac{1}{3}$.

2. Fill in the circles below with <, =, or > to make true number sentences. Use decomposition or multiplication to justify your answer.

   a. $7 \bigcirc \frac{43}{6}$

   b. $11\frac{1}{3} \bigcirc \frac{34}{3}$

   c. $\frac{13}{6} \bigcirc \frac{38}{12}$

3. Generate a pattern of at least 13 fractions by adding $\frac{4}{3}$ to $\frac{1}{3}$ and then continuing to add $\frac{4}{3}$ to each fraction. Circle each fraction equal to a whole number. Write what you notice about the pattern of whole numbers. The first two fractions are written for you.

$\frac{1}{3}, \frac{5}{3},$

4. Find each sum or difference.

   a. $6\frac{4}{10} + 7\frac{7}{10}$

   b. $3\frac{3}{8} + 6\frac{5}{8} + 1\frac{7}{8}$

   c. $1\frac{9}{12} - 1\frac{4}{12}$

   d. $5\frac{2}{5} - 1\frac{3}{5}$

5. a. Rewrite $3 \times \frac{6}{8}$ as the product of a unit fraction and a whole number. Solve.

   b. Rewrite $4 \times 6\frac{2}{3}$ as the product of a unit fraction and a whole number. Solve.

6. Determine if the following are true or false. Explain how you know using models or words. Make false problems true by rewriting the right side of the number sentence.

   a. $7\frac{1}{3} = 7 + \frac{1}{3}$

   b. $\frac{5}{3} = \frac{3}{3} + \frac{2}{3}$

   c. $\frac{13}{6} - \frac{5}{6} = \frac{13-5}{6}$

   d. $\frac{11}{3} = 11 + \frac{1}{3}$

   e. $\frac{7}{8} + \frac{7}{8} + \frac{7}{8} + \frac{7}{8} = 4 \times \frac{7}{8}$

   f. $5 \times 3\frac{3}{4} = 15 + \frac{3}{4}$

7. The chart to the right shows data Amashi collected about butterfly wingspans.

   a. At the bottom of this page, create a line plot to display the data in the table.

   b. What is the difference in wingspan between the widest and narrowest butterflies on the chart?

   c. Three butterflies have the same wingspan. Explain how you know the measurements are equal.

Butterfly	Wingspan (inches)
Monarch	$3\frac{7}{8}$
Milbert's Tortoiseshell	$2\frac{5}{8}$
Zebra Swallowtail	$2\frac{1}{2}$
Viceroy	$2\frac{6}{8}$
Postman	$3\frac{3}{8}$
Purple Spotted Swallowtail	$2\frac{2}{8}$
Julia	$3\frac{2}{4}$
Southern Dogface	$2\frac{3}{8}$
Tiger Swallowtail	$3\frac{1}{2}$
Regal Fritillary	$3\frac{4}{8}$

Solve each problem. Draw a model, write an equation, and write a statement for each.

d. Amashi wants to display a Postman and Viceroy side by side in a photo box with a width of 6 inches. Will these two butterflies fit? Explain how you know.

e. Compare the wingspan of the Milbert's Tortoiseshell and the Zebra Swallowtail using >, <, or =.

f. The Queen Alexandra Birdwing can have a wingspan that is 5 times as wide as the Southern Dogface's. How many inches can the Birdwing's wingspan be?

g. Amashi discovered a pattern. She started with $2\frac{2}{8}$ inches and added $\frac{1}{8}$ inch to each measurement. List the next four measurements in her pattern. Name the five butterflies whose wingspans match the measurements in her pattern.

End-of-Module Assessment Task

Standards Addressed — Topics A–H

Generate and analyze patterns.

4.OA.5 Generate a number or shape pattern that follows a given rule. Identify apparent features of the pattern that were not explicit in the rule itself. *For example, given the rule "Add 3" and the starting number 1, generate terms in the resulting sequence and observe that the terms appear to alternate between odd and even numbers. Explain informally why the numbers will continue to alternate in this way.*

Extend understanding of fraction equivalence and ordering.

4.NF.1 Explain why a fraction a/b is equivalent to a fraction $(n \times a)/(n \times b)$ by using visual fraction models, with attention to how the number and size of the parts differ even though the two fractions themselves are the same size. Use this principle to recognize and generate equivalent fractions.

4.NF.2 Compare two fractions with different numerators and different denominators, e.g., by creating common denominators or numerators, or by comparing to a benchmark fraction such as 1/2. Recognize that comparisons are valid only when the two fractions refer to the same whole. Record the results of comparisons with symbols >, =, or <, and justify the conclusions, e.g., by using a visual fraction model.

Build fractions from unit fractions by applying and extending previous understandings of operations on whole numbers.

4.NF.3 Understand a fraction a/b with $a > 1$ as a sum of fractions $1/b$.

   a. Understand addition and subtraction of fractions as joining and separating parts referring to the same whole.

   b. Decompose a fraction into a sum of fractions with the same denominator in more than one way, recording each decomposition by an equation. Justify decompositions, e.g., by using a visual fraction model. *Examples: 3/8 = 1/8 + 1/8 + 1/8; 3/8 = 1/8 + 2/8; 2 1/8 = 1 + 1 + 1/8 = 8/8 + 8/8 + 1/8.*

   c. Add and subtract mixed numbers with like denominators, e.g., by replacing each mixed number with an equivalent fraction, and/or by using properties of operations and the relationship between addition and subtraction.

   d. Solve word problems involving addition and subtraction of fractions referring to the same whole and having like denominators, e.g., by using visual fraction models and equations to represent the problem.

4.NF.4 Apply and extend previous understandings of multiplication to multiply a fraction by a whole number.

   a. Understand a fraction a/b as a multiple of $1/b$. *For example, use a visual fraction model to represent 5/4 as the product 5 × (1/4), recording the conclusion by the equation 5/4 = 5 × (1/4).*

End-of-Module Assessment Task
Standards Addressed — Topics A–H

> b. Understand a multiple of *a/b* as a multiple of *1/b*, and use this understanding to multiply a fraction by a whole number. *For example, use a visual fraction model to express 3 × (2/5) as 6 × (1/5), recognizing this product as 6/5. (In general, n × (a/b) = (n × a)/b.)*
>
> c. Solve word problems involving multiplication of a fraction by a whole number, e.g., by using visual fraction models and equations to represent the problem. *For example, if each person at a party will eat 3/8 of a pound of roast beef, and there will be 5 people at the party, how many pounds of roast beef will be needed? Between what two whole numbers does your answer lie?*

Represent and interpret data.

4.MD.4   Make a line plot to display a data set of measurements in fractions of a unit (1/2, 1/4, 1/8). Solve problems involving addition and subtraction of fractions by using information presented in line plots. *For example, from a line plot find and interpret the difference in length between the longest and shortest specimens in an insect collection.*

Evaluating Student Learning Outcomes

A Progression Toward Mastery is provided to describe steps that illuminate the gradually increasing understandings that students develop *on their way to proficiency.* In this chart, this progress is presented from left (Step 1) to right (Step 4). The learning goal for students is to achieve Step 4 mastery. These steps are meant to help teachers and students identify and celebrate what the students CAN do now and what they need to work on next.

Module 5:   Fraction Equivalence, Ordering, and Operations

End-of-Module Assessment Task

A STORY OF UNITS — 4•5

A Progression Toward Mastery				
Assessment Task Item and Standards Assessed	STEP 1 Little evidence of reasoning without a correct answer. (1 Point)	STEP 2 Evidence of some reasoning without a correct answer. (2 Points)	STEP 3 Evidence of some reasoning with a correct answer or evidence of solid reasoning with an incorrect answer. (3 Points)	STEP 4 Evidence of solid reasoning with a correct answer. (4 Points)
1 4.NF.4ab	The student incorrectly partitions the models and provides little to no reasoning.	The student incorrectly partitions the models but provides some reasoning for equivalence.	The student correctly partitions the models, providing some reasoning.	The student correctly does the following: a. Partitions the tape diagram and number line. b. Explains the equivalence using the models and number sentences.
2 4.NF.1 4.NF.2	The student correctly answers fewer than two of the three parts with little to no reasoning.	The student is able to correctly compare one or two of the three number pairs with some reasoning.	The student is able to correctly compare two of the three number pairs and offers solid reasoning to support correct answers or correctly compares all three numbers and offers some reasoning.	The student correctly compares all three number pairs and offers appropriate modeling or reasoning to justify answers: a. < b. = c. <

550   Module 5:   Fraction Equivalence, Ordering, and Operations

A Story of Units — End-of-Module Assessment Task 4•5

A Progression Toward Mastery

3 **4.OA.5**	The student is unable to complete a majority of the problem.	The student is able to generate most of the pattern and find some whole numbers. The student provides little reasoning about the whole numbers.	The student is able to generate the pattern and find at least three whole numbers but cannot reason about the whole numbers.	The student correctly does the following: • Generates the following pattern: $\frac{1}{3}, \frac{5}{3}, \frac{9}{3}, \frac{13}{3}, \frac{17}{3}, \frac{21}{3},$ $\frac{25}{3}, \frac{29}{3}, \frac{33}{3}, \frac{37}{3}, \frac{41}{3},$ $\frac{45}{3}, \frac{49}{3}, \frac{5\bar{A}}{3}, \frac{5\bar{A}}{3}$ • Circles $\frac{9}{3}, \frac{21}{3}, \frac{33}{3}, \frac{45}{3}, \frac{57}{3}$ • Observes that whole numbers repeat every three fractions, determines that all whole numbers are odd numbers (3, 7, 11, 15, 19), or provides another acceptable response.
4 **4.NF.3c**	The student correctly evaluates one or fewer expressions.	The student correctly evaluates two expressions.	The student correctly evaluates three expressions.	The student correctly evaluates all four expressions: a. $14\frac{1}{10}$ b. $11\frac{7}{8}$ c. $\frac{5}{12}$ d. $3\frac{4}{5}$
5 **4.NF.4ab**	The student is unable to correctly complete either of the two parts.	The student is able to correctly complete one of the two parts.	The student correctly rewrites the expressions in Parts (a) and (b) but does not solve the expressions. OR The student correctly solves for an incorrect expression.	The student correctly rewrites and solves the expressions in both parts of the problem: a. $18 \times \frac{1}{8} = \frac{18}{8}$ or $2\frac{2}{8}$ b. $4 \times \frac{20}{3} = \frac{4 \times \bar{A}0}{3} =$ $80 \times \frac{1}{3} = \frac{80}{3}$ or $26\frac{2}{3}$

Module 5: Fraction Equivalence, Ordering, and Operations

A STORY OF UNITS

End-of-Module Assessment Task 4•5

A Progression Toward Mastery					
6 4.NF.3a 4.NF.4b	The student correctly solves less than four problems with little to no reasoning.	The student correctly solves three or four of the six problems with some reasoning.	The student correctly solves five of the six problems with solid reasoning.	The student correctly analyzes all six problems, revises the incorrect number sentences (answers may vary), and provides solid reasoning using models or words: a. True b. True c. True d. False, $\frac{11}{3} = 11 \times \frac{1}{3}$ e. True f. False, $5 \times 3\frac{3}{4} = 5 \times (3 + \frac{3}{4})$
7 4.NF.1 4.NF.2 4.NF.3 4.NF.4 4.OA.5 4.MD.4	The student correctly completes three or fewer parts with little to no reasoning.	The student correctly completes at least three parts of the question with some reasoning.	The student correctly completes five or six parts of the question, providing solid reasoning. OR The student correctly answers all parts with only some reasoning.	The student correctly completes all seven parts of the question: a. Creates an accurate line plot with 10 data points from the table. b. $1\frac{5}{8}$ inches. c. Shows $3\frac{1}{2} = 3\frac{2}{4} = 3\frac{4}{8}$ using a model or showing equivalences through number sentences. d. No. The combined wingspan is $6\frac{1}{8}$ inches, which is longer than the box. e. $2\frac{5}{8}$ inches > $2\frac{1}{2}$ inches f. $11\frac{7}{8}$ inches. g. $2\frac{3}{8}, 2\frac{1}{2} \left(2\frac{4}{8}\right), 2\frac{5}{8}, 2\frac{6}{8}$. Purple Spotted Swallowtail, Southern Dogface, Zebra Swallowtail, Milbert's Tortoiseshell, and Viceroy.

552   Module 5:   Fraction Equivalence, Ordering, and Operations

A STORY OF UNITS   End-of-Module Assessment Task   4•5

Name __Jack_____   Date _____

1. a. Partition the tape diagram to show $5 \times \frac{2}{3}$. Partition the number line to show $10 \times \frac{1}{3}$.

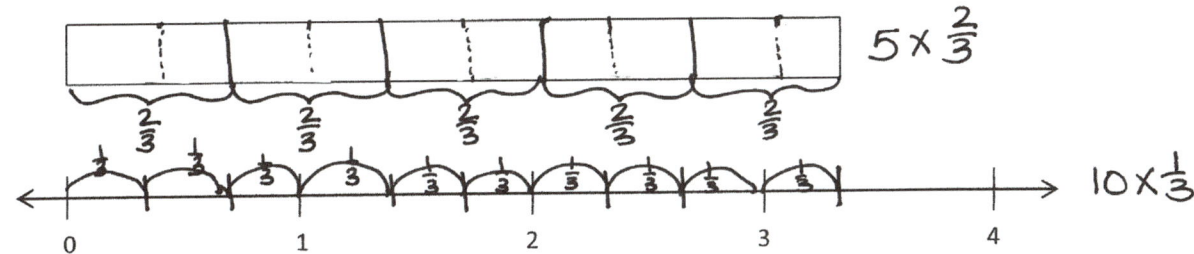

b. Use the models above to explain why $5 \times \frac{2}{3} = 10 \times \frac{1}{3}$.

When you double the size of the piece, you only need half as many to be the same length.

2. Fill in the circles below with <, =, or > to make true number sentences. Use decomposition or multiplication to justify your answer.

a. $7 \; \bigcirc< \; \frac{43}{6}$ $\frac{43}{6} = 7 \times \frac{6}{6} + \frac{1}{6} = 7\frac{1}{6}$

b. $11\frac{1}{3} \; \bigcirc= \; \frac{34}{3}$ $\frac{34}{3} = 11 \times \frac{3}{3} + \frac{1}{3} = 11\frac{1}{3}$

c. $\frac{13}{6} \; \bigcirc< \; \frac{38}{12}$ $\frac{13 \times 2}{6 \times 2} = \frac{26}{12}$ $\frac{26}{12} < \frac{38}{12}$

Module 5:   Fraction Equivalence, Ordering, and Operations

3. Generate a pattern of at least 13 fractions by adding $\frac{4}{3}$ to $\frac{1}{3}$ and then continuing to add $\frac{4}{3}$ to each fraction. Circle each fraction equal to a whole number. Write what you notice about the pattern of whole numbers. The first two fractions are written for you.

$\frac{1}{3}$, $\frac{5}{3}$, ⓐ$\frac{9}{3}$, $\frac{13}{3}$, $\frac{17}{3}$, ⓐ$\frac{21}{3}$, $\frac{25}{3}$, $\frac{29}{3}$, ⓐ$\frac{33}{3}$, $\frac{37}{3}$, $\frac{41}{3}$, ⓐ$\frac{45}{3}$, $\frac{49}{3}$, $\frac{53}{3}$, ⓐ$\frac{57}{3}$

3, 7, 11, 15, 19

I noticed that the pattern of the whole numbers increases by 4 each time and they are all odd numbers.

4. Find each sum or difference.

a. $6\frac{4}{10} + 7\frac{7}{10} = 14\frac{1}{10}$

$6 + 7 = 13$
$\frac{4}{10} + \frac{7}{10} = \frac{11}{10} = 1\frac{1}{10}$
$13 + 1\frac{1}{10} = 14\frac{1}{10}$

b. $3\frac{3}{8} + 6\frac{5}{8} + 1\frac{7}{8} = 11\frac{7}{8}$

$3 + 6 + 1 = 10$
$\frac{3}{8} + \frac{5}{8} + \frac{7}{8} = \frac{15}{8} = 1\frac{7}{8}$
$10 + 1\frac{7}{8} = 11\frac{7}{8}$

c. $1\frac{9}{12} - 1\frac{4}{12} = \frac{5}{12}$

$1 - 1 = 0$
$\frac{9}{12} - \frac{4}{12} = \frac{5}{12}$

d. $5\frac{2}{5} - 1\frac{3}{5} = 3\frac{4}{5}$

$5\frac{2}{5} = 4\frac{7}{5}$
$4 - 1 = 3$
$\frac{7}{5} - \frac{3}{5} = \frac{4}{5}$ } $3\frac{4}{5}$

5. a. Rewrite $3 \times \frac{6}{8}$ as the product of a unit fraction and a whole number. Solve.

$$3 \times \frac{6}{8} = 18 \times \frac{1}{8} = \frac{18}{8} = \frac{8}{8} + \frac{8}{8} + \frac{2}{8} = 1 + 1 + \frac{2}{8} = 2\frac{2}{8}$$

b. Rewrite $4 \times 6\frac{2}{3}$ as the product of a unit fraction and a whole number. Solve

$$4 \times 6\frac{2}{3} = 4 \times \frac{20}{3} = \frac{4 \times 20}{3} = 80 \times \frac{1}{3} = \frac{80}{3} = 26\frac{2}{3}$$

(with $6\frac{2}{3}$ decomposed into $\frac{18}{3}$ and $\frac{2}{3}$)

6. Determine if the following are true or false. Explain how you know using models or words. Make false problems true by rewriting the right side of the number sentence.

a. $7\frac{1}{3} = 7 + \frac{1}{3}$

True $\underbrace{7 + \frac{1}{3}}_{7\frac{1}{3}}$

b. $\frac{5}{3} = \frac{3}{3} + \frac{2}{3}$

True.

c. $\frac{13}{6} - \frac{5}{6} = \frac{13-5}{6}$

True.

$\frac{13-5}{6}$ is the same as $\frac{13}{6} - \frac{5}{6}$

d. $\frac{11}{3} = 11 + \frac{1}{3}$ → $\frac{11}{3} = 11 \times \frac{1}{3}$

False.

$11 + \frac{1}{3} = 11\frac{1}{3}$

$\frac{11}{3} \neq 11\frac{1}{3}$

e. $\frac{7}{8} + \frac{7}{8} + \frac{7}{8} + \frac{7}{8} = 4 \times \frac{7}{8}$

True. $\frac{28}{8}$ $\frac{28}{8}$

f. $5 \times 3\frac{3}{4} = 15 + \frac{3}{4}$ → $5 \times 3\frac{3}{4} = 5 \times (3 + \frac{3}{4})$

$5 \times \frac{15}{4} = 75 \times \frac{1}{4} = \frac{75}{4} = 18\frac{3}{4}$

False. $18\frac{3}{4} \neq 15\frac{3}{4}$

7. The chart to the right shows data Amashi collected about butterfly wingspans.

   a. At the bottom of this page, create a line plot to display the data in the table.

Butterfly	Wingspan (inches)
Monarch	$3\frac{7}{8}$
Milbert's Tortoiseshell	$2\frac{5}{8}$
Zebra Swallowtail	$2\frac{1}{2}$
Viceroy	$2\frac{6}{8}$
Postman	$3\frac{3}{8}$
Purple Spotted Swallowtail	$2\frac{2}{8}$
Julia	$3\frac{2}{4}$
Southern Dogface	$2\frac{3}{8}$
Tiger Swallowtail	$3\frac{1}{2}$
Regal Fritillary	$3\frac{4}{8}$

   b. What is the difference in wingspan between the widest and narrowest butterflies on the chart?

   $3\frac{7}{8}$ inches − $2\frac{2}{8}$ inches = $1\frac{5}{8}$ inches

   3 − 2 = 1
   $\frac{7}{8} - \frac{2}{8} = \frac{5}{8}$

   c. Three butterflies have the same wingspan. Explain how you know the measurements are equal.

   The Julia, Tiger Swallowtail, and Regal Fritillary all have the same wingspan. I know because
   $3\frac{2}{4} = 3\frac{1}{2} = 3\frac{4}{8}$

   $\frac{1 \times 2}{2 \times 2} = \frac{2 \times 2}{4 \times 2} = \frac{4}{8}$

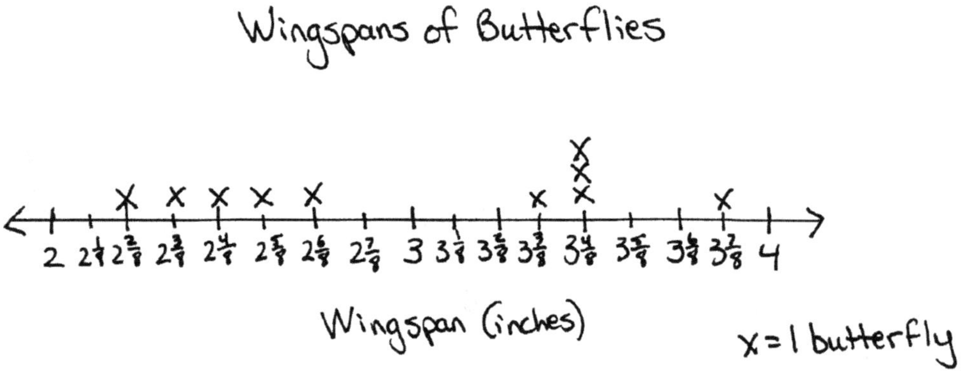

   Wingspans of Butterflies

   Wingspan (inches)

   x = 1 butterfly

Solve each problem. Draw a model, write an equation, and write a statement for each.

d. Amashi wants to display a Postman and Viceroy side by side in a photo box with a width of 6 inches. Will these two butterflies fit? Explain how you know.

Postman $3\frac{3}{8}$ inches

Viceroy $2\frac{6}{8}$ inches

$3\frac{3}{8}$ inches + $2\frac{6}{8}$ inches = $5\frac{9}{8}$ inches = $6\frac{1}{8}$ inches

$\frac{8}{8}$ $\frac{1}{8}$

The two butterflies will <u>not</u> fit because when I added their wingspans together I got $6\frac{1}{8}$ inches which is greater than 6 inches.

e. Compare the wingspan of the Milbert's Tortoiseshell and the Zebra Swallowtail using >, <, or =.

$2\frac{5}{8}$ inches $\boxed{>}$ $2\frac{1}{2}$ inches

Milbert's Tortoiseshell   Zebra Swallowtail

$\frac{1 \times 4}{2 \times 4} = \frac{4}{8}$

$2\frac{1}{2} = 2\frac{4}{8}$

The Milbert's Tortoiseshell has a larger wingspan than the Zebra Swallowtail.

f. The Queen Alexandra Birdwing can have a wingspan that is 5 times as wide as the Southern Dogface's. How many inches can the Birdwing's wingspan be?

$5 \times 2\frac{3}{8} = (5 \times 2) + (5 \times \frac{3}{8}) = 10 + \frac{5 \times 3}{8} = 10 + \frac{15}{8} = 11\frac{7}{8}$

$\frac{8}{8}$ $\frac{7}{8}$

The Queen Alexandra Birdwing's wingspan can be $11\frac{7}{8}$ inches.

g. Amashi discovered a pattern. She started with $2\frac{2}{8}$ inches and added $\frac{1}{8}$ inch to each measurement. List the next four measurements in her pattern. Name the five butterflies whose wingspans match the measurements in her pattern.

$2\frac{2}{8}$ inches, $2\frac{3}{8}$ inches, $2\frac{4}{8}$ inches, $2\frac{5}{8}$ inches, $2\frac{6}{8}$ inches

Purple Spotted Swallowtail | Southern Dogface | Zebra Swallowtail | Milbert's Tortoiseshell | Viceroy

Answer Key

Eureka Math® Grade 4 Module 5

Special thanks go to the Gordon A. Cain Center and to the Department of Mathematics at Louisiana State University for their support in the development of *Eureka Math*.

A STORY OF UNITS

4 GRADE

Mathematics Curriculum

GRADE 4 • MODULE 5

Answer Key
GRADE 4 • MODULE 5

Fraction Equivalence, Ordering, and Operations

A STORY OF UNITS — Lesson 1 Answer Key — 4•5

Lesson 1

Problem Set

1. a. Answer provided

    b. Whole: $\frac{4}{5}$, parts: $\frac{1}{5}, \frac{3}{5}$; $\frac{1}{5} + \frac{3}{5} = \frac{4}{5}$

    c. Whole: $\frac{3}{4}$, parts: $\frac{1}{4}, \frac{1}{4}, \frac{1}{4}$; $\frac{1}{4} + \frac{1}{4} + \frac{1}{4} = \frac{3}{4}$

    d. Whole: $\frac{4}{6}$, parts: $\frac{2}{6}, \frac{2}{6}$; $\frac{2}{6} + \frac{2}{6} = \frac{4}{6}$

    e. Whole: $\frac{6}{8}$, parts: $\frac{2}{8}, \frac{2}{8}, \frac{2}{8}$; $\frac{2}{8} + \frac{2}{8} + \frac{2}{8} = \frac{6}{8}$

    f. Whole: $\frac{5}{4}$, parts: $\frac{3}{4}, \frac{1}{4}, \frac{1}{4}$; $\frac{3}{4} + \frac{1}{4} + \frac{1}{4} = \frac{5}{4}$

    g. Whole: $1\frac{2}{3}$, parts: $\frac{2}{3}, \frac{2}{3}, \frac{1}{3}$; $\frac{2}{3} + \frac{2}{3} + \frac{1}{3} = 1\frac{2}{3}$

    h. Whole: $1\frac{3}{8}$, parts: $\frac{2}{8}, \frac{2}{8}, \frac{2}{8}, \frac{4}{8}, \frac{1}{8}$; $\frac{2}{8} + \frac{2}{8} + \frac{4}{8} + \frac{1}{8} = 1\frac{3}{8}$

2. a. Tape diagram models number sentence.

    b. Tape diagram models number sentence.

    c. Tape diagram models number sentence.

    d. Tape diagram models number sentence.

    e. Tape diagram models number sentence.

    f. Tape diagram models number sentence.

    g. Tape diagram models number sentence.

    h. Tape diagram models number sentence.

Exit Ticket

1. Whole: 1, parts: $\frac{1}{4}, \frac{1}{4}, \frac{1}{4}, \frac{1}{4}$; $\frac{1}{4} + \frac{1}{4} + \frac{1}{4} + \frac{1}{4} = 1$

2. a. Tape diagram models number sentence.

    b. Tape diagram models number sentence.

Homework

1. a. Answer provided
   b. Whole: $\frac{2}{4}$, parts: $\frac{1}{4}, \frac{1}{4}$; $\frac{2}{4} = \frac{1}{4} + \frac{1}{4}$
   c. Whole: $\frac{3}{5}$, parts: $\frac{1}{5}, \frac{2}{5}$; $\frac{3}{5} = \frac{1}{5} + \frac{2}{5}$
   d. Whole: $\frac{5}{6}$, parts: $\frac{3}{6}, \frac{2}{6}$; $\frac{5}{6} = \frac{3}{6} + \frac{2}{6}$
   e. Whole: $\frac{3}{8}$, parts: $\frac{2}{8}, \frac{1}{8}$; $\frac{3}{8} = \frac{2}{8} + \frac{1}{8}$
   f. Whole: $1\frac{1}{5}$, parts: $\frac{5}{5}, \frac{1}{5}$; $1\frac{1}{5} = \frac{5}{5} + \frac{1}{5}$
   g. Whole: $1\frac{2}{4}$, parts: $\frac{3}{4}, \frac{2}{4}, \frac{1}{4}$; $1\frac{2}{4} = \frac{3}{4} + \frac{2}{4} + \frac{1}{4}$
   h. Whole: $1\frac{4}{8}$, parts: $\frac{3}{8}, \frac{2}{8}, \frac{1}{8}, \frac{3}{8}, \frac{3}{8}$; $1\frac{4}{8} = \frac{3}{8} + \frac{2}{8} + \frac{1}{8} + \frac{3}{8} + \frac{3}{8}$

2. a. Tape diagram models number sentence.
   b. Tape diagram models number sentence.
   c. Tape diagram models number sentence.
   d. Tape diagram models number sentence.
   e. Tape diagram models number sentence.

Lesson 2

Problem Set

1. a. Answer provided
    b. Tape diagram models fraction; decompositions will vary.
    c. Tape diagram models fraction; decompositions will vary.
2. a. Tape diagram models fraction; decompositions will vary.
    b. Tape diagram models fraction; decompositions will vary.
    c. Tape diagram models fraction; decompositions will vary.
    d. Tape diagram models fraction; decompositions will vary.

Exit Ticket

Tape diagram models fraction; decompositions will vary.

Homework

1. a. Answer provided
    b. Tape diagram models fraction; decompositions will vary.
    c. Tape diagram models fraction; decompositions will vary.
2. a. Tape diagram models fraction; decompositions will vary.
    b. Tape diagram models fraction; decompositions will vary.
    c. Tape diagram models fraction; decompositions will vary.
    d. Tape diagram models fraction; decompositions will vary.

Lesson 3

Problem Set

1. a. Answer provided
   b. $\frac{2}{5} = \frac{1}{5} + \frac{1}{5}; \frac{2}{5} = 2 \times \frac{1}{5}$
   c. $\frac{5}{6} = \frac{1}{6} + \frac{1}{6} + \frac{1}{6} + \frac{1}{6} + \frac{1}{6}; \frac{5}{6} = 5 \overline{A} \frac{1}{6}$
   d. $\frac{6}{8} = \frac{1}{8} + \frac{1}{8} + \frac{1}{8} + \frac{1}{8} + \frac{1}{8} + \frac{1}{8}; \frac{6}{8} = 6 \overline{A} \frac{1}{8}$
   e. $\frac{4}{3} = \frac{1}{3} + \frac{1}{3} + \frac{1}{3} + \frac{1}{3}; \frac{4}{3} = 4 \times \frac{1}{3}$

2. a. $\frac{5}{3} = \left(3 \times \frac{1}{3}\right) + \left(2 \times \frac{1}{3}\right)$
   b. $\frac{6}{4} = \left(4 \times \frac{1}{4}\right) + \left(2 \times \frac{1}{4}\right)$

3. a. Tape diagram models fraction; $\frac{4}{5} = 4 \times \frac{1}{5}$
   b. Tape diagram models fraction; $\frac{5}{8} = 5 \overline{A} \frac{1}{8}$
   c. Tape diagram models fraction; $\frac{7}{9} = 7 \times \frac{1}{9}$
   d. Tape diagram models fraction; $\frac{7}{4} = 7 \times \frac{1}{4}$
   e. Tape diagram models fraction; $\frac{7}{6} = 7 \times \frac{1}{6}$

Exit Ticket

1. a. $\frac{2}{3} = \frac{1}{3} + \frac{1}{3}; \frac{2}{3} = 2 \times \frac{1}{3}$
   b. $\frac{5}{3} = \frac{1}{3} + \frac{1}{3} + \frac{1}{3} + \frac{1}{3} + \frac{1}{3}; \frac{5}{3} = 5 \overline{A} \frac{1}{3}$

2. Tape diagram models fraction; $\frac{6}{9} = 6 \overline{A} \frac{1}{9}$

Homework

1. a. Answer provided
    b. $\frac{3}{4} = \frac{1}{4} + \frac{1}{4} + \frac{1}{4}; \frac{3}{4} = 3 \times \frac{1}{4}$
    c. $\frac{4}{5} = \frac{1}{5} + \frac{1}{5} + \frac{1}{5} + \frac{1}{5}; \frac{4}{5} = 4 \times \frac{1}{5}$
    d. $\frac{5}{6} = \frac{1}{6} + \frac{1}{6} + \frac{1}{6} + \frac{1}{6} + \frac{1}{6}; \frac{5}{6} = 5 \times \frac{1}{6}$

2. a. $\frac{4}{3} = \left(3 \times \frac{1}{3}\right) + \left(1 \times \frac{1}{3}\right)$
    b. $\frac{8}{6} = \left(6 \times \frac{1}{6}\right) + \left(2 \times \frac{1}{6}\right)$

3. a. Tape diagram models fraction; $\frac{3}{5} = 3 \times \frac{1}{5}$
    b. Tape diagram models fraction; $\frac{3}{8} = 3 \times \frac{1}{8}$
    c. Tape diagram models fraction; $\frac{5}{9} = 5 \times \frac{1}{9}$
    d. Tape diagram models fraction; $\frac{8}{5} = 8 \times \frac{1}{5}$
    e. Tape diagram models fraction; $\frac{12}{4} = 12 \times \frac{1}{4}$

Lesson 4

Problem Set

1. a. Answer provided
   b. $\frac{1}{3} = \frac{1}{6} + \frac{1}{6}$; $\frac{1}{3} = \frac{1}{12} + \frac{1}{12} + \frac{1}{12} + \frac{1}{12}$
   c. Answers will vary.
   d. Answers will vary.

2. a. Answers will vary.
   b. Answers will vary.

3. a. Answer provided
   b. Tape diagram models number sentence.
   c. Tape diagram models number sentence.
   d. Tape diagram models number sentence.

4. Tape diagram models number sentence.
5. Tape diagram models number sentence.
6. Tape diagram models number sentence.

Exit Ticket

1. Answers will vary.
2. Tape diagram models number sentence.

Homework

1. a. Answer provided
   b. $\frac{1}{4} = \frac{1}{8} + \frac{1}{8}$; $\frac{1}{4} = \frac{1}{16} + \frac{1}{16} + \frac{1}{16} + \frac{1}{16}$
2. a. Answers will vary.
   b. Answers will vary.
   c. Answers will vary.
3. a. Answer provided
   b. Tape diagram models number sentence.
   c. Tape diagram models number sentence.
   d. Tape diagram models number sentence.
4. Tape diagram models number sentence.
5. Tape diagram models number sentence.
6. Tape diagram models number sentence.

Lesson 5

Problem Set

1. a. $8, \frac{1}{8}, \frac{2}{8}, \frac{1}{8}, \frac{2}{8}$
   b. 2 rows drawn; $\frac{1}{5} = \frac{2}{10}, \frac{1}{5} = \frac{1}{10} + \frac{1}{10} = \frac{2}{10}, \frac{1}{5} = 2 \times \frac{1}{10} = \frac{2}{10}$
   c. 4 rows drawn; $\frac{1}{3} = \frac{4}{12}, \frac{1}{3} = \frac{1}{12} + \frac{1}{12} + \frac{1}{12} + \frac{1}{12} = \frac{4}{12}, \frac{1}{3} = 4 \times \frac{1}{12} = \frac{4}{12}$

2. a. Area model shows $\frac{1}{2} = \frac{3}{6}; \frac{1}{2} = \frac{1}{6} + \frac{1}{6} + \frac{1}{6} = \frac{3}{6}, \frac{1}{2} = 3 \times \frac{1}{6} = \frac{3}{6}$
   b. Area model shows $\frac{1}{2} = \frac{4}{8}; \frac{1}{2} = \frac{1}{8} + \frac{1}{8} + \frac{1}{8} + \frac{1}{8} = \frac{4}{8}, \frac{1}{2} = 4 \times \frac{1}{8} = \frac{4}{8}$
   c. Area model shows $\frac{1}{2} = \frac{5}{10}; \frac{1}{2} = \frac{1}{10} + \frac{1}{10} + \frac{1}{10} + \frac{1}{10} + \frac{1}{10} = \frac{5}{10}, \frac{1}{2} = 5 \overline{A} \frac{1}{10} = \frac{5}{10}$
   d. Area model shows $\frac{1}{3} = \frac{2}{6}; \frac{1}{3} = \frac{1}{6} + \frac{1}{6} = \frac{2}{6}, \frac{1}{3} = 2 \times \frac{1}{6} = \frac{2}{6}$
   e. Area model shows $\frac{1}{3} = \frac{4}{12}; \frac{1}{3} = \frac{1}{12} + \frac{1}{12} + \frac{1}{12} + \frac{1}{12} = \frac{4}{12}, \frac{1}{3} = 4 \times \frac{1}{12} = \frac{4}{12}$
   f. Area model shows $\frac{1}{4} = \frac{3}{12}; \frac{1}{4} = \frac{1}{12} + \frac{1}{12} + \frac{1}{12} = \frac{3}{12}, \frac{1}{4} = 3 \times \frac{1}{12} = \frac{3}{12}$

3. Explanations will vary.

Exit Ticket

1. a. 2 rows drawn; $\frac{1}{4} = \frac{1}{8} + \frac{1}{8} = \frac{2}{8}, \frac{1}{4} = 2 \times \frac{1}{8} = \frac{2}{8}$
   b. 3 rows drawn; $\frac{1}{4} = \frac{1}{12} + \frac{1}{12} + \frac{1}{12} = \frac{3}{12}, \frac{1}{4} = 3 \times \frac{1}{12} = \frac{3}{12}$

2. Area model shows $\frac{3}{5} = \frac{6}{10}; \frac{3}{5} = \frac{1}{10} + \frac{1}{10} + \frac{1}{10} + \frac{1}{10} + \frac{1}{10} + \frac{1}{10} = \frac{6}{10}, \frac{3}{5} = 6 \, A \frac{1}{10} = \frac{6}{10}$

Homework

1. a. $6, \frac{1}{6}, \frac{1}{6}, \frac{1}{6}$
    b. 2 rows drawn; $\frac{1}{4} = \frac{2}{8}, \frac{1}{4} = \frac{1}{8} + \frac{1}{8} = \frac{2}{8}, \frac{1}{4} = 2 \times \frac{1}{8} = \frac{2}{8}$
    c. 4 rows drawn; $\frac{1}{4} = \frac{4}{16}, \frac{1}{4} = \frac{1}{16} + \frac{1}{16} + \frac{1}{16} + \frac{1}{16} = \frac{4}{16}, \frac{1}{4} = 4 \times \frac{1}{16} = \frac{4}{16}$

2. a. Area model shows $\frac{1}{3} = \frac{2}{6}; \frac{1}{3} = \frac{1}{6} + \frac{1}{6} = \frac{2}{6}, \frac{1}{3} = 2 \times \frac{1}{6} = \frac{2}{6}$
    d. Area model shows $\frac{1}{3} = \frac{3}{9}; \frac{1}{3} = \frac{1}{9} + \frac{1}{9} + \frac{1}{9} = \frac{3}{9}, \frac{1}{3} = 3 \times \frac{1}{9} = \frac{3}{9}$
    e. Area model shows $\frac{1}{3} = \frac{4}{12}; \frac{1}{3} = \frac{1}{12} + \frac{1}{12} + \frac{1}{12} + \frac{1}{12} = \frac{4}{12}, \frac{1}{3} = 4 \times \frac{1}{12} = \frac{4}{12}$
    f. Area model shows $\frac{1}{3} = \frac{5}{15}; \frac{1}{3} = \frac{1}{15} + \frac{1}{15} + \frac{1}{15} + \frac{1}{15} + \frac{1}{15} = \frac{5}{15}, \frac{1}{3} = 5 \times \frac{1}{15} = \frac{5}{15}$
    g. Area model shows $\frac{1}{5} = \frac{2}{10}; \frac{1}{5} = \frac{1}{10} + \frac{1}{10} = \frac{2}{10}, \frac{1}{5} = 2 \times \frac{1}{10} = \frac{2}{10}$
    h. Area model shows $\frac{1}{5} = \frac{3}{15}; \frac{1}{5} = \frac{1}{15} + \frac{1}{15} + \frac{1}{15} = \frac{3}{15}, \frac{1}{5} = 3 \times \frac{1}{15} = \frac{3}{15}$

3. Explanations will vary.

Lesson 6

Sprint

Side A

1. $\frac{2}{3}$
2. $\frac{2}{3}$
3. $\frac{3}{4}$
4. $\frac{3}{4}$
5. $\frac{2}{5}$
6. $\frac{2}{5}$
7. $\frac{3}{5}$
8. $\frac{3}{5}$
9. $\frac{4}{5}$
10. $\frac{4}{5}$
11. $\frac{3}{10}$
12. $\frac{3}{10}$
13. $\frac{3}{8}$
14. $\frac{3}{8}$
15. 1
16. $\frac{2}{2}$
17. $\frac{3}{3}$
18. $\frac{3}{3}$
19. $\frac{4}{4}$
20. $\frac{4}{4}$
21. $\frac{3}{2}$
22. $\frac{3}{2}$
23. $\frac{4}{3}$
24. $\frac{4}{3}$
25. 5
26. $\frac{1}{6}$
27. $\frac{1}{8}$
28. 5
29. $\frac{1}{8}$
30. $\frac{1}{10}$
31. 7
32. 7
33. $\frac{1}{6}$
34. $\frac{1}{6}$
35. 8
36. 8
37. $\frac{9}{10}$
38. $\frac{7}{5}$
39. $\frac{1}{3}$
40. $\frac{7}{12}$
41. 5
42. $\frac{1}{5}$
43. $\frac{1}{4}$
44. $\frac{1}{3},\frac{1}{3},\frac{1}{3}$

Side B

1. $\frac{2}{5}$
2. $\frac{2}{5}$
3. $\frac{2}{3}$
4. $\frac{2}{3}$
5. $\frac{3}{4}$
6. $\frac{3}{4}$
7. $\frac{3}{5}$
8. $\frac{3}{5}$
9. $\frac{4}{5}$
10. $\frac{4}{5}$
11. $\frac{3}{8}$
12. $\frac{3}{8}$
13. $\frac{3}{10}$
14. $\frac{3}{10}$
15. $\frac{3}{3}$
16. 1
17. $\frac{4}{4}$
18. 1
19. $\frac{2}{2}$
20. 1
21. $\frac{4}{3}$
22. $\frac{4}{3}$
23. $\frac{3}{2}$
24. $\frac{3}{2}$
25. 5
26. $\frac{1}{6}$
27. $\frac{1}{8}$
28. 5
29. $\frac{1}{8}$
30. $\frac{1}{10}$
31. 7
32. 7
33. $\frac{1}{8}$
34. $\frac{1}{8}$
35. 6
36. 6
37. $\frac{5}{12}$
38. $\frac{6}{5}$
39. $\frac{1}{4}$
40. $\frac{9}{10}$
41. 3
42. $\frac{1}{4}$
43. $\frac{1}{5}$
44. $\frac{1}{4},\frac{1}{4},\frac{1}{4},\frac{1}{4}$

A STORY OF UNITS
Lesson 6 Answer Key 4•5

Problem Set

1. a. 6; 1; 1; 6; $\frac{1}{6}$; $\frac{1}{6}$; 6; $\frac{1}{6}$; 6

   b. Decomposed horizontally to show tenths; $\frac{1}{5} + \frac{1}{5} = \left(\frac{1}{10} + \frac{1}{10}\right) + \left(\frac{1}{10} + \frac{1}{10}\right) = \frac{4}{10}$; $\left(\frac{1}{10} + \frac{1}{10}\right) + \left(\frac{1}{10} + \frac{1}{10}\right) = \left(2 \times \frac{1}{10}\right) + \left(2 \times \frac{1}{10}\right)$, $\frac{2}{5} = 4 \times \frac{1}{10} = \frac{4}{10}$

   c. Decomposed horizontally to show twelfths; $\frac{1}{4} + \frac{1}{4} + \frac{1}{4} = \left(\frac{1}{12} + \frac{1}{12} + \frac{1}{12}\right) + \left(\frac{1}{12} + \frac{1}{12} + \frac{1}{12}\right) + \left(\frac{1}{12} + \frac{1}{12} + \frac{1}{12}\right) = \frac{9}{12}$; $\left(\frac{1}{12} + \frac{1}{12} + \frac{1}{12}\right) + \left(\frac{1}{12} + \frac{1}{12} + \frac{1}{12}\right) + \left(\frac{1}{12} + \frac{1}{12} + \frac{1}{12}\right) = \left(3 \times \frac{1}{12}\right) + \left(3 \times \frac{1}{12}\right) + \left(3 \times \frac{1}{12}\right) = \frac{9}{12}$, $\frac{3}{4} = 9 \times \frac{1}{12} = \frac{9}{12}$

2. a. Area model shows $\frac{3}{5} = \frac{6}{10}$; $\frac{3}{5} = \frac{1}{5} + \frac{1}{5} + \frac{1}{5} = \left(\frac{1}{10} + \frac{1}{10}\right) + \left(\frac{1}{10} + \frac{1}{10}\right) + \left(\frac{1}{10} + \frac{1}{10}\right) = \frac{6}{10}$, $\frac{3}{5} = \left(\frac{1}{10} + \frac{1}{10}\right) + \left(\frac{1}{10} + \frac{1}{10}\right) + \left(\frac{1}{10} + \frac{1}{10}\right) = \left(2 \times \frac{1}{10}\right) + \left(2 \times \frac{1}{10}\right) + \left(2 \times \frac{1}{10}\right) = \frac{6}{10}$, $\frac{3}{5} = 6 \,\overline{A}\, \frac{1}{10} = \frac{6}{10}$

   b. Area model shows $\frac{3}{4} = \frac{6}{8}$; $\frac{3}{4} = \frac{1}{4} + \frac{1}{4} + \frac{1}{4} = \left(\frac{1}{8} + \frac{1}{8}\right) + \left(\frac{1}{8} + \frac{1}{8}\right) + \left(\frac{1}{8} + \frac{1}{8}\right) = \frac{6}{8}$, $\left(2 \times \frac{1}{8}\right) + \left(2 \times \frac{1}{8}\right) + \left(2 \times \frac{1}{8}\right) = \frac{6}{8}$, $\frac{3}{4} = 6 \,\overline{A}\, \frac{1}{8} = \frac{6}{8}$

3. Answers will vary.

Exit Ticket

1. Decomposed horizontally to show eighths; $\frac{3}{4} = \frac{1}{4} + \frac{1}{4} + \frac{1}{4} = \left(\frac{1}{8} + \frac{1}{8}\right) + \left(\frac{1}{8} + \frac{1}{8}\right) + \left(\frac{1}{8} + \frac{1}{8}\right) = \frac{6}{8}$, $\left(2 \times \frac{1}{8}\right) + \left(2 \times \frac{1}{8}\right) + \left(2 \times \frac{1}{8}\right) = \frac{6}{8}$, $\frac{3}{4} = 6 \,\overline{A}\, \frac{1}{8} = \frac{6}{8}$

2. Area model shows $\frac{4}{5} = \frac{8}{10}$

Lesson 6 Answer Key 4•5

Homework

1. a. 4, 10, 1, 1, 10, $\frac{1}{10}$, $\frac{1}{10}$, 10, $\frac{1}{10}$, 10

   b. Decomposed horizontally to show eighths; $\frac{1}{4}+\frac{1}{4} = \left(\frac{1}{8}+\frac{1}{8}\right) + \left(\frac{1}{8}+\frac{1}{8}\right) = \frac{4}{8}$, $\left(\frac{1}{8}+\frac{1}{8}\right) + \left(\frac{1}{8}+\frac{1}{8}\right) = \left(2 \times \frac{1}{8}\right) + \left(2 \times \frac{1}{8}\right) = \frac{4}{8}$, $\frac{2}{4} = 4 \times \frac{1}{8} = \frac{4}{8}$

   c. Decomposed horizontally to show fifteenths; $\frac{1}{5}+\frac{1}{5}+\frac{1}{5}+\frac{1}{5} = \left(\frac{1}{15}+\frac{1}{15}+\frac{1}{15}\right) + \left(\frac{1}{15}+\frac{1}{15}+\frac{1}{15}\right) + \left(\frac{1}{15}+\frac{1}{15}+\frac{1}{15}\right) + \left(\frac{1}{15}+\frac{1}{15}+\frac{1}{15}\right) = \frac{12}{15}$; $\left(\frac{1}{15}+\frac{1}{15}+\frac{1}{15}\right) + \left(\frac{1}{15}+\frac{1}{15}+\frac{1}{15}\right) + \left(\frac{1}{15}+\frac{1}{15}+\frac{1}{15}\right) + \left(\frac{1}{15}+\frac{1}{15}+\frac{1}{15}\right) = \left(3 \times \frac{1}{15}\right) + \left(3 \times \frac{1}{15}\right) + \left(3 \times \frac{1}{15}\right) + \left(3 \times \frac{1}{15}\right) = \frac{12}{15}$, $\frac{4}{5} = 12 \times \frac{1}{15} = \frac{12}{15}$

2. a. Area model shows $\frac{2}{3} = \frac{4}{6}$; $\frac{1}{3}+\frac{1}{3} = \left(\frac{1}{6}+\frac{1}{6}\right) + \left(\frac{1}{6}+\frac{1}{6}\right) = \frac{4}{6}$, $\left(\frac{1}{6}+\frac{1}{6}\right) + \left(\frac{1}{6}+\frac{1}{6}\right) = \left(2 \times \frac{1}{6}\right) + \left(2 \times \frac{1}{6}\right) = \frac{4}{6}$, $\frac{2}{3} = 4 \times \frac{1}{6} = \frac{4}{6}$

   b. Area model shows $\frac{4}{5} = \frac{8}{10}$; $\frac{1}{5}+\frac{1}{5}+\frac{1}{5}+\frac{1}{5} = \left(\frac{1}{10}+\frac{1}{10}\right) + \left(\frac{1}{10}+\frac{1}{10}\right) + \left(\frac{1}{10}+\frac{1}{10}\right) + \left(\frac{1}{10}+\frac{1}{10}\right) = \frac{8}{10}$, $\left(\frac{1}{10}+\frac{1}{10}\right) + \left(\frac{1}{10}+\frac{1}{10}\right) + \left(\frac{1}{10}+\frac{1}{10}\right) + \left(\frac{1}{10}+\frac{1}{10}\right) = \left(2 \times \frac{1}{10}\right) + \left(2 \times \frac{1}{10}\right) + \left(2 \times \frac{1}{10}\right) + \left(2 \times \frac{1}{10}\right) = \frac{8}{10}$, $\frac{4}{5} = 8 \times \frac{1}{10} = \frac{8}{10}$

3. Answers will vary.

Lesson 7

Problem Set

1. a. Answer provided
   b. $\frac{1}{2} = \frac{1 \times 3}{2 \times 3} = \frac{3}{6}$
   c. $\frac{1}{2} = \frac{1 \times 4}{2 \times 4} = \frac{4}{8}$
   d. $\frac{1}{2} = \frac{1 \times 5}{2 \times 5} = \frac{5}{10}$

2. a. Answers will vary.
   b. Answers will vary.
   c. Answers will vary.
   d. Answers will vary.
   e. The size of the fractional units decreased.
   f. The number of total units increased.

3. a. Area model represents $\frac{1}{3}$ and is decomposed into sixths; $\frac{1}{3} = \frac{1 \times 2}{3 \times 2} = \frac{2}{6}$
   b. Area model represents $\frac{1}{3}$ and is decomposed into ninths; $\frac{1}{3} = \frac{1 \times 3}{3 \times 3} = \frac{3}{9}$
   c. Area model represents $\frac{1}{3}$ and is decomposed into twelfths; $\frac{1}{3} = \frac{1 \times 4}{3 \times 4} = \frac{4}{12}$

Exit Ticket

a. Area model represents $\frac{1}{4}$ and is decomposed into eighths; $\frac{1}{4} = \frac{1 \times 2}{4 \times 2} = \frac{2}{8}$

b. Area model represents $\frac{1}{4}$ and is decomposed into twelfths; $\frac{1}{4} = \frac{1 \times 3}{4 \times 3} = \frac{3}{12}$

Homework

1. a. Answer provided
   b. $\frac{1}{2} = \frac{1 \times 4}{2 \times 4} = \frac{4}{8}$
   c. $\frac{1}{2} = \frac{1 \times 6}{2 \times 6} = \frac{6}{12}$
   d. $\frac{1}{2} = \frac{1 \times 7}{2 \times 7} = \frac{7}{14}$

2. a. Answers will vary.
   b. Answers will vary.
   c. Answers will vary.
   d. Answers will vary.

3. a. Area model shows $\frac{1}{4}$ and is decomposed into eighths; $\frac{1}{4} = \frac{1 \times 2}{4 \times 2} = \frac{2}{8}$
   b. Area model shows $\frac{1}{4}$ and is decomposed into twelfths; $\frac{1}{4} = \frac{1 \times 3}{4 \times 3} = \frac{3}{12}$
   c. Area model shows $\frac{1}{4}$ and is decomposed into sixteenths; $\frac{1}{4} = \frac{1 \times 4}{4 \times 4} = \frac{4}{16}$

Lesson 8

Problem Set

1. a. Answer provided
   b. $\frac{3}{4} = \frac{3 \times 3}{4 \times 3} = \frac{9}{12}$
   c. $\frac{4}{5} = \frac{4 \times 2}{5 \times 2} = \frac{8}{10}$
   d. $\frac{5}{6} = \frac{5 \times 2}{6 \times 2} = \frac{10}{12}$

2. a. Area model shows $\frac{3}{5} = \frac{3 \times 2}{5 \times 2} = \frac{6}{10}$
   b. Area model shows $\frac{3}{5} = \frac{3 \times 3}{5 \times 3} = \frac{9}{15}$

3. a. Area model proves $\frac{2}{5} = \frac{4}{10}$
   b. Area model proves $\frac{2}{3} = \frac{8}{12}$
   c. Area model proves $\frac{3}{6} = \frac{6}{12}$
   d. Area model proves $\frac{4}{6} = \frac{8}{12}$

4. a. Answers will vary.
   b. Answers will vary.
   c. Answers will vary.
   d. Answers will vary.

5. a. False; answers will vary.
   b. True
   c. False; answers will vary.
   d. True

Exit Ticket

1. Answers will vary.

2. False; answers will vary.

Homework

1. a. Answer provided
   b. $\frac{3}{4} = \frac{3 \times 2}{4 \times 2} = \frac{6}{8}$
   c. $\frac{4}{5} = \frac{4 \times 3}{5 \times 3} = \frac{12}{15}$
   d. $\frac{7}{8} = \frac{7 \times 2}{8 \times 2} = \frac{14}{16}$

2. a. Area model shows $\frac{3}{6} = \frac{3 \times 2}{6 \times 2} = \frac{6}{12}$
   b. Area model shows $\frac{2}{4} = \frac{2 \times 3}{4 \times 3} = \frac{6}{12}$

3. a. Area model proves $\frac{1}{3} = \frac{2}{6}$
   b. Area model proves $\frac{2}{5} = \frac{4}{10}$
   c. Area model proves $\frac{5}{7} = \frac{10}{14}$
   d. Area model proves $\frac{3}{6} = \frac{9}{18}$

4. a. Answers will vary.
   b. Answers will vary.
   c. Answers will vary.
   d. Answers will vary.

5. a. False; answers will vary.
   b. True
   c. False; answers will vary.
   d. True

Lesson 9

Problem Set

1. a. Answer provided
    b. Area model shows composed fractions;
       $\frac{3}{6} = \frac{3 \div 3}{6 \div 3} = \frac{1}{2}$
    c. Area model shows composed fractions;
       $\frac{5}{10} = \frac{5 \div 5}{10 \div 5} = \frac{1}{2}$
    d. Area model shows composed fractions;
       $\frac{4}{8} = \frac{4 \div 4}{8 \div 4} = \frac{1}{2}$ or
       $\frac{4}{8} = \frac{4 \div 2}{8 \div 2} = \frac{2}{4}$

2. a. Area model shows composed fractions;
       $\frac{2}{6} = \frac{2 \div 2}{6 \div 2} = \frac{1}{3}$
    b. Area model shows composed fractions;
       $\frac{2}{8} = \frac{2 \div 2}{8 \div 2} = \frac{1}{4}$
    c. Area model shows composed fractions;
       $\frac{2}{10} = \frac{2 \div 2}{10 \div 2} = \frac{1}{5}$
    d. Area model shows composed fractions;
       $\frac{2}{12} = \frac{2 \div 2}{12 \div 2} = \frac{1}{6}$
    e. The size of the fractional units increased.
    f. The number of total units decreased.

3. a. Area models prove $\frac{2}{6} = \frac{1}{3}$ and $\frac{3}{9} = \frac{1}{3}$
    b. $\frac{2}{6} = \frac{2 \div 2}{6 \div 2} = \frac{1}{3}, \frac{3}{9} = \frac{3 \div 3}{9 \div 3} = \frac{1}{3}$

4. a. Area models prove $\frac{2}{8} = \frac{1}{4}$ and $\frac{3}{12} = \frac{1}{4}$
    b. $\frac{2}{8} = \frac{2 \div 2}{8 \div 2} = \frac{1}{4}, \frac{3}{12} = \frac{3 \div 3}{12 \div 3} = \frac{1}{4}$

Exit Ticket

a. Area models prove $\frac{2}{6} = \frac{1}{3}$ and $\frac{4}{12} = \frac{1}{3}$

b. $\frac{2}{6} = \frac{2 \div 2}{6 \div 2} = \frac{1}{3}, \frac{4}{12} = \frac{4 \div 4}{12 \div 4} = \frac{1}{3}$

A STORY OF UNITS

Lesson 9 Answer Key 4•5

Homework

1. a. Answer provided
   b. Area model shows composed fractions; $\frac{4}{8} = \frac{4 \div 4}{8 \div 4} = \frac{1}{2}$ or $\frac{4}{8} = \frac{4 \div 2}{8 \div 2} = \frac{2}{4}$
   c. Area model shows composed fractions; $\frac{6}{12} = \frac{6 \div 6}{12 \div 6} = \frac{1}{2}$ or $\frac{6}{12} = \frac{6 \div 3}{12 \div 3} = \frac{2}{4}$ or $\frac{6}{12} = \frac{6 \div 2}{12 \div 2} = \frac{3}{6}$
   d. Area model shows composed fractions; $\frac{7}{14} = \frac{7 \div 7}{14 \div 7} = \frac{1}{2}$

2. a. Area model shows composed fractions; $\frac{2}{12} = \frac{2 \div 2}{12 \div 2} = \frac{1}{6}$
   b. Area model shows composed fractions; $\frac{2}{10} = \frac{2 \div 2}{10 \div 2} = \frac{1}{5}$
   c. Area model shows composed fractions; $\frac{2}{8} = \frac{2 \div 2}{8 \div 2} = \frac{1}{4}$
   d. Area model shows composed fractions; $\frac{2}{6} = \frac{2 \div 2}{6 \div 2} = \frac{1}{3}$
   e. The size of the fractional units increased.
   f. The number of total units decreased.

3. a. Area models prove $\frac{4}{8} = \frac{1}{2}$ and $\frac{6}{12} = \frac{1}{2}$
   b. $\frac{4}{8} = \frac{4 \div 4}{8 \div 4} = \frac{1}{2}, \frac{6}{12} = \frac{6 \div 6}{12 \div 6} = \frac{1}{2}$

4. a. Area models prove $\frac{4}{8} = \frac{1}{2}$ and $\frac{8}{16} = \frac{1}{2}$
   b. $\frac{4}{8} = \frac{4 \div 4}{8 \div 4} = \frac{1}{2}, \frac{8}{16} = \frac{8 \div 8}{16 \div 8} = \frac{1}{2}$

EUREKA MATH

Module 5: Fraction Equivalence, Ordering, and Operations

Lesson 10

Problem Set

1. a. Answer provided
   b. Area model shows composed fractions; $\frac{9}{12} = \frac{9 \div 3}{12 \div 3} = \frac{3}{4}$
   c. Area model shows composed fractions; $\frac{6}{10} = \frac{6 \div 2}{10 \div 2} = \frac{3}{5}$
   d. Area model shows composed fractions; $\frac{6}{8} = \frac{6 \div 2}{8 \div 2} = \frac{3}{4}$
2. a. Area model shows composed fractions; $\frac{4}{6} = \frac{4 \div 2}{6 \div 2} = \frac{2}{3}$
   b. Area model shows composed fractions; $\frac{8}{12} = \frac{8 \div 4}{12 \div 4} = \frac{2}{3}$ or $\frac{8}{12} = \frac{8 \div 2}{12 \div 2} = \frac{4}{6}$
3. a. Area model shows $\frac{4}{10}$ composed as $\frac{2}{5}$
   b. Area model shows $\frac{6}{9}$ composed as $\frac{2}{3}$
4. a. Answers will vary.
   b. Answers will vary.
   c. Answers will vary.
   d. Answers will vary.

Exit Ticket

Area model proves $\frac{4}{10} = \frac{2}{5}$; $\frac{4}{10} = \frac{4 \div 2}{10 \div 2} = \frac{2}{5}$

Homework

1. a. Answer provided
   b. Area model shows composed fractions; $\frac{4}{10} = \frac{4 \div 2}{10 \div 2} = \frac{2}{5}$
   c. Area model shows composed fractions; $\frac{6}{9} = \frac{6 \div 3}{9 \div 3} = \frac{2}{3}$
   d. Area model shows composed fractions; $\frac{9}{15} = \frac{9 \div 3}{15 \div 3} = \frac{3}{5}$

2. a. Area model shows composed fractions; $\frac{6}{8} = \frac{6 \div 2}{8 \div 2} = \frac{3}{4}$
   b. Area model shows composed fractions; $\frac{12}{16} = \frac{12 \div 4}{16 \div 4} = \frac{3}{4}$ or $\frac{12}{16} = \frac{12 \div 2}{16 \div 2} = \frac{6}{8}$

3. a. Area model shows $\frac{6}{15}$ composed as $\frac{2}{5}$
   b. Area model shows $\frac{6}{18}$ composed as $\frac{2}{6}$

4. a. Answers will vary.
   b. Answers will vary.
   c. Answers will vary.
   d. Answers will vary.

Lesson 11

Problem Set

1. a. $\frac{0}{4}, \frac{1}{4}, \frac{2}{4}, \frac{3}{4}, \frac{4}{4}; \frac{1}{4}$ circled

   b. $\frac{0}{8}, \frac{1}{8}, \frac{2}{8}, \frac{3}{8}, \frac{4}{8}, \frac{5}{8}, \frac{6}{8}, \frac{7}{8}, \frac{8}{8}; \frac{2}{8}$ circled

   c. $\frac{0}{12}, \frac{1}{12}, \frac{2}{12}, \frac{3}{12}, \frac{4}{12}, \frac{5}{12}, \frac{6}{12}, \frac{7}{12}, \frac{8}{12}, \frac{9}{12}, \frac{10}{12}, \frac{11}{12}, \frac{12}{12}; \frac{3}{12}$ circled

2. a. $\frac{1}{4} = \frac{1 \times 2}{4 \times 2} = \frac{2}{8}$

   b. $\frac{1}{4} = \frac{1 \times 3}{4 \times 3} = \frac{3}{12}$

3. a. Number line drawn for $\frac{0}{3}, \frac{1}{3}, \frac{2}{3}, \frac{3}{3}; \frac{2}{3}$ circled

   b. Number line drawn for $\frac{0}{6}, \frac{1}{6}, \frac{2}{6}, \frac{3}{6}, \frac{4}{6}, \frac{5}{6}, \frac{6}{6}; \frac{4}{6}$ circled

   c. Number line drawn for $\frac{0}{12}, \frac{1}{12}, \frac{2}{12}, \frac{3}{12}, \frac{4}{12}, \frac{5}{12}, \frac{6}{12}, \frac{7}{12}, \frac{8}{12}, \frac{9}{12}, \frac{10}{12}, \frac{11}{12}, \frac{12}{12}; \frac{8}{12}$ circled

4. a. $\frac{4}{6} = \frac{4 \div 2}{6 \div 2} = \frac{2}{3}$

   b. $\frac{8}{12} = \frac{8 \div 4}{12 \div 4} = \frac{2}{3}$

5. a. Number line drawn appropriately

   b. $\frac{2}{5} = \frac{2 \times 2}{5 \times 2} = \frac{4}{10}$

   c. $\frac{4}{10} = \frac{4 \div 2}{10 \div 2} = \frac{2}{5}$

Exit Ticket

1. Number line drawn appropriately

2. $\frac{2}{6} = \frac{2 \times 2}{6 \times 2} = \frac{4}{12}$

3. $\frac{4}{12} = \frac{4 \div 2}{12 \div 2} = \frac{2}{6}$

Homework

1. a. $\frac{0}{3}, \frac{1}{3}, \frac{2}{3}, \frac{3}{3}; \frac{1}{3}$ circled

    b. $\frac{0}{6}, \frac{1}{6}, \frac{2}{6}, \frac{3}{6}, \frac{4}{6}, \frac{5}{6}, \frac{6}{6}; \frac{2}{6}$ circled

    c. $\frac{0}{12}, \frac{1}{12}, \frac{2}{12}, \frac{3}{12}, \frac{4}{12}, \frac{5}{12}, \frac{6}{12}, \frac{7}{12}, \frac{8}{12}, \frac{9}{12}, \frac{10}{12}, \frac{11}{12}, \frac{12}{12}; \frac{4}{12}$ circled

2. a. $\frac{1}{3} = \frac{1 \times 2}{3 \times 2} = \frac{2}{6}$

    b. $\frac{1}{3} = \frac{1 \times 4}{3 \times 4} = \frac{4}{12}$

3. a. Number line drawn for $\frac{0}{4}, \frac{1}{4}, \frac{2}{4}, \frac{3}{4}, \frac{4}{4}; \frac{2}{4}$ circled

    b. Number line drawn for $\frac{0}{8}, \frac{1}{8}, \frac{2}{8}, \frac{3}{8}, \frac{4}{8}, \frac{5}{8}, \frac{6}{8}, \frac{7}{8}, \frac{8}{8}; \frac{4}{8}$ circled

    c. Number line drawn for $\frac{0}{10}, \frac{1}{10}, \frac{2}{10}, \frac{3}{10}, \frac{4}{10}, \frac{5}{10}, \frac{6}{10}, \frac{7}{10}, \frac{8}{10}, \frac{9}{10}, \frac{10}{10}; \frac{5}{10}$ circled

4. $\frac{4}{8} = \frac{4 \div 2}{8 \div 2} = \frac{2}{4}$

5. a. Number line drawn appropriately

    b. $\frac{3}{4} = \frac{3 \times 2}{4 \times 2} = \frac{6}{8}$

    c. $\frac{6}{8} = \frac{6 \div 2}{8 \div 2} = \frac{3}{4}$

Lesson 12

Problem Set

1. a. Points plotted appropriately for $\frac{1}{3}, \frac{5}{6}, \frac{7}{12}$
    b. i. >
       ii. <
2. a. Points plotted appropriately for $\frac{11}{12}, \frac{1}{4}, \frac{3}{8}$
    b. Answers will vary.
    c. Explanations will vary.

3. a. <; explanations will vary.
    b. <; explanations will vary.
    c. >; explanations will vary.
    d. >; explanations will vary.
    e. <; explanations will vary.
    f. <; explanations will vary.
    g. >; explanations will vary.
    h. >; explanations will vary.
    i. <; explanations will vary.
    j. <; explanations will vary.

Exit Ticket

1. Points plotted appropriately for $\frac{8}{10}, \frac{3}{5}, \frac{1}{4}$
2. a. <
    b. >
    c. <
    d. <

Homework

1. a. Points plotted appropriately for $\frac{2}{3}, \frac{1}{6}, \frac{4}{10}$
   b. i. >
      ii. >
2. a. Points plotted appropriately for $\frac{5}{12}, \frac{3}{4}, \frac{2}{6}$
   b. Answers will vary.
   c. Explanations will vary.
3. a. >; explanations will vary.
   b. >; explanations will vary.
   c. >; explanations will vary.
   d. <; explanations will vary.
   e. >; explanations will vary.
   f. >; explanations will vary.
   g. <; explanations will vary.
   h. >; explanations will vary.
   i. >; explanations will vary.
   j. >; explanations will vary.

Lesson 13

Problem Set

1. Points plotted appropriately for $\frac{4}{3}, \frac{11}{6}, \frac{17}{12}$
2. a. >
   b. <
3. Points plotted appropriately for $\frac{11}{8}, \frac{7}{4}, \frac{15}{12}$
4. Explanations will vary.

5. a. <; explanations will vary.
   b. <; explanations will vary.
   c. >; explanations will vary.
   d. >; explanations will vary.
   e. >; explanations will vary.
   f. >; explanations will vary.
   g. >; explanations will vary.
   h. <; explanations will vary.
   i. <; explanations will vary.
   j. <; explanations will vary.

Exit Ticket

1. Points plotted appropriately for $\frac{5}{4}, \frac{10}{7}, \frac{16}{9}$
2. a. <
   b. <
   c. >

Homework

1. Points plotted appropriately for $\frac{3}{2}, \frac{9}{5}, \frac{14}{10}$
2. a. <
   b. <
3. Points plotted appropriately for $\frac{12}{9}, \frac{6}{5}, \frac{18}{15}$
4. Explanations will vary.

5. a. <; explanations will vary.
   b. <; explanations will vary.
   c. >; explanations will vary.
   d. <; explanations will vary.
   e. <; explanations will vary.
   f. <; explanations will vary.
   g. <; explanations will vary.
   h. <; explanations will vary.
   i. <; explanations will vary.
   j. >; explanations will vary.

Lesson 14

Problem Set

1. a. >
   b. >
   c. >
   d. >
2. a. <; explanations will vary.
   b. Answer provided
   c. >; explanations will vary.
   d. >; explanations will vary.
3. a. Tape diagrams model fractions; <
   b. Tape diagrams model fractions; <
   c. Tape diagrams model fractions; >
4. a. Number line models fractions; <
   b. Number line models fractions; >
   c. Number line models fractions; <
   d. Number line models fractions; >

5. a. >
   b. >
   c. >
   d. <
   e. <
   f. >
   g. >
   h. >
6. Evan; picture supports answer

Exit Ticket

1. Tape diagrams model fractions; >
2. Number line models fractions; >

Homework

1. a. >
   b. >
   c. >
   d. >
2. a. >; explanations will vary.
   b. Answer provided
   c. >; explanations will vary.
   d. <; explanations will vary.
3. a. Tape diagrams model fractions; >
   b. Tape diagrams model fractions; >
   c. Tape diagrams model fractions; <
4. a. Number line models fractions; >
   b. Number line models fractions; >
   c. Number line models fractions; >
   d. Number line models fractions; >
5. a. <
   b. <
   c. >
   d. >
   e. =
   f. <
   g. >
   h. >
6. Simon; picture supports answer

A STORY OF UNITS

Lesson 15 Answer Key 4•5

Lesson 15

Problem Set

1. a. Area models prove $\frac{1}{2} < \frac{2}{3}$
    b. Area models prove $\frac{4}{5} > \frac{3}{4}$
    c. Area models prove $\frac{3}{5} > \frac{4}{7}$
    d. Area models prove $\frac{3}{7} > \frac{2}{6}$
    e. Area models prove $\frac{5}{8} < \frac{6}{9}$
    f. Area models prove $\frac{2}{3} < \frac{3}{4}$

2. a. <
    b. <
    c. >
    d. >

3. a. <
    b. >
    c. >
    d. >

4. Explanations will vary.

Exit Ticket

1. Area models prove $\frac{3}{4} < \frac{4}{5}$
2. Area models prove $\frac{2}{6} < \frac{3}{5}$

Homework

1. a. Area models prove $\frac{1}{2} < \frac{3}{5}$
    b. Area models prove $\frac{2}{3} < \frac{3}{4}$
    c. Area models prove $\frac{4}{6} > \frac{5}{8}$
    d. Area models prove $\frac{2}{7} < \frac{3}{5}$
    e. Area models prove $\frac{4}{6} = \frac{6}{9}$
    f. Area models prove $\frac{4}{5} < \frac{5}{6}$

2. a. >
    b. >
    c. >
    d. >

3. a. >
    b. <
    c. =
    d. <

4. Explanations will vary.

Lesson 16

Problem Set

1.
   a. 2 fifths
   b. 2 fifths
   c. 1 half
   d. 3 fourths

2.
   a. $\frac{2}{6}$
   b. $\frac{2}{8}$
   c. $\frac{0}{10}$
   d. $\frac{1}{5}$
   e. $\frac{1}{4}$
   f. $\frac{2}{4}$

3.
   a. Answer provided
   b. Number bond shows $\frac{7}{6}$ is $\frac{6}{6}$ and $\frac{1}{6}$; $1\frac{1}{6}$
   c. Number bond shows $\frac{6}{5}$ is $\frac{5}{5}$ and $\frac{1}{5}$; $1\frac{1}{5}$
   d. Number bond shows $\frac{11}{8}$ is $\frac{8}{8}$ and $\frac{3}{8}$; $1\frac{3}{8}$
   e. Number bond shows $\frac{6}{4}$ is $\frac{4}{4}$ and $\frac{2}{4}$; $1\frac{2}{4}$
   f. Number bond shows $\frac{12}{10}$ is $\frac{10}{10}$ and $\frac{2}{10}$; $1\frac{2}{10}$

4.
   a. 3 fourths
   b. 7 fifths

5.
   a. $\frac{7}{8}$
   b. $\frac{9}{12}$

6.
   a. Answer provided
   b. Number bond shows $\frac{7}{4}$ is $\frac{4}{4}$ and $\frac{3}{4}$; $1\frac{3}{4}$
   c. Number bond shows $\frac{12}{9}$ is $\frac{9}{9}$ and $\frac{3}{9}$; $1\frac{3}{9}$
   d. Number bond shows $\frac{13}{10}$ is $\frac{10}{10}$ and $\frac{3}{10}$; $1\frac{3}{10}$
   e. Number bond shows $\frac{12}{6}$ is $\frac{6}{6}$ and $\frac{6}{6}$; 2
   f. Number bond shows $\frac{14}{8}$ is $\frac{8}{8}$ and $\frac{6}{8}$; $1\frac{6}{8}$

7.
   a. Number line models $\frac{7}{4} - \frac{5}{4} = \frac{2}{4}$
   b. Number line models $\frac{5}{4} + \frac{2}{4} = \frac{7}{4}$

Exit Ticket

1. Number bond shows $\frac{11}{9}$ is $\frac{9}{9}$ and $\frac{2}{9}$; $1\frac{2}{9}$
2. Number bond shows $\frac{15}{12}$ is $\frac{12}{12}$ and $\frac{3}{12}$; $1\frac{3}{12}$

A STORY OF UNITS — Lesson 16 Answer Key 4•5

Homework

1. a. 1 sixth
   b. 2 tenths
   c. 1 fourth
   d. 3 thirds

2. a. $\frac{1}{5}$
   b. $\frac{4}{9}$
   c. $\frac{4}{12}$
   d. $\frac{2}{6}$
   e. $\frac{3}{3}$
   f. $\frac{2}{4}$

3. a. Answer provided
   b. Number bond shows $\frac{11}{8}$ is $\frac{8}{8}$ and $\frac{3}{8}$; $1\frac{3}{8}$
   c. Number bond shows $\frac{6}{5}$ is $\frac{5}{5}$ and $\frac{1}{5}$; $1\frac{1}{5}$
   d. Number bond shows $\frac{5}{4}$ is $\frac{4}{4}$ and $\frac{1}{4}$; $1\frac{1}{4}$
   e. Number bond shows $\frac{8}{7}$ is $\frac{7}{7}$ and $\frac{1}{7}$; $1\frac{1}{7}$
   f. Number bond shows $\frac{12}{10}$ is $\frac{10}{10}$ and $\frac{2}{10}$; $1\frac{2}{10}$

4. a. 6 fifths
   b. 7 eighths

5. a. $\frac{9}{11}$
   b. $\frac{9}{10}$

6. a. Number bond shows $\frac{6}{4}$ is $\frac{4}{4}$ and $\frac{2}{4}$; $1\frac{2}{4}$
   b. Number bond shows $\frac{14}{12}$ is $\frac{12}{12}$ and $\frac{2}{12}$; $1\frac{2}{12}$
   c. Number bond shows $\frac{12}{8}$ is $\frac{8}{8}$ and $\frac{4}{8}$; $1\frac{4}{8}$
   d. Number bond shows $\frac{13}{10}$ is $\frac{10}{10}$ and $\frac{3}{10}$; $1\frac{3}{10}$
   e. Number bond shows $\frac{9}{5}$ is $\frac{5}{5}$ and $\frac{4}{5}$; $1\frac{4}{5}$
   f. Number bond shows $\frac{6}{3}$ is $\frac{3}{3}$ and $\frac{3}{3}$; 2

7. a. Number line accurately models $\frac{11}{9} - \frac{5}{9} = \frac{6}{9}$
   b. Number line accurately models $\frac{13}{12} + \frac{4}{12} = \frac{17}{12}$

A STORY OF UNITS — Lesson 17 Answer Key — 4•5

Lesson 17

Problem Set

1. a. $\frac{8}{5}+\frac{2}{5}=\frac{10}{5}, \frac{2}{5}+\frac{8}{5}=\frac{10}{5}, \frac{10}{5}-\frac{2}{5}=\frac{8}{5}, \frac{10}{5}-\frac{8}{5}=\frac{2}{5}$

   b. $\frac{7}{8}+\frac{8}{8}=\frac{15}{8}, \frac{8}{8}+\frac{7}{8}=\frac{15}{8}, \frac{15}{8}-\frac{8}{8}=\frac{7}{8}, \frac{15}{8}-\frac{7}{8}=\frac{8}{8}$

2. a. Answer provided

   b. $\frac{2}{10}$; number line models solution; solved by counting up and subtracting

   c. $\frac{2}{5}$; number line models solution; solved by counting up and subtracting

   d. $\frac{3}{8}$; number line models solution; solved by counting up and subtracting

   e. $\frac{5}{10}$; number line models solution; solved by counting up and subtracting

   f. $\frac{3}{5}$; number line models solution; solved by counting up and subtracting

3. a. Answer provided

   b. $\frac{6}{6}+\frac{3}{6}=\frac{9}{6}, \frac{9}{6}-\frac{4}{6}=\frac{5}{6}; \frac{6}{6}-\frac{4}{6}=\frac{2}{6}, \frac{2}{6}+\frac{3}{6}=\frac{5}{6}$; number bond shows $1\frac{3}{6}$ is $\frac{6}{6}$ and $\frac{3}{6}$

   c. $\frac{8}{8}+\frac{6}{8}=\frac{14}{8}, \frac{14}{8}-\frac{7}{8}=\frac{7}{8}; \frac{8}{8}-\frac{7}{8}=\frac{1}{8}, \frac{1}{8}+\frac{6}{8}=\frac{7}{8}$; number bond shows $1\frac{6}{8}$ is $\frac{8}{8}$ and $\frac{6}{8}$

   d. $\frac{10}{10}+\frac{1}{10}=\frac{11}{10}, \frac{11}{10}-\frac{7}{10}=\frac{4}{10}; \frac{10}{10}-\frac{7}{10}=\frac{3}{10}, \frac{3}{10}+\frac{1}{10}=\frac{4}{10}$; number bond shows $1\frac{1}{10}$ is $\frac{10}{10}$ and $\frac{1}{10}$

   e. $\frac{12}{12}+\frac{3}{12}=\frac{15}{12}, \frac{15}{12}-\frac{6}{12}=\frac{9}{12}; \frac{12}{12}-\frac{6}{12}=\frac{6}{12}, \frac{6}{12}+\frac{3}{12}=\frac{9}{12}$; number bond shows $1\frac{3}{12}$ is $\frac{12}{12}$ and $\frac{3}{12}$

Exit Ticket

1. $\frac{3}{5}$; number line models solution; solved by counting up and subtracting

2. $\frac{7}{7}+\frac{2}{7}=\frac{9}{7}, \frac{9}{7}-\frac{5}{7}=\frac{4}{7}; \frac{7}{7}-\frac{5}{7}=\frac{2}{7}, \frac{2}{7}+\frac{2}{7}=\frac{4}{7}$; number bond shows $1\frac{2}{7}$ is $\frac{7}{7}$ and $\frac{2}{7}$

Homework

1. a. $\frac{5}{6} + \frac{4}{6} = \frac{9}{6}, \frac{4}{6} + \frac{5}{6} = \frac{9}{6}, \frac{9}{6} - \frac{5}{6} = \frac{4}{6}, \frac{9}{6} - \frac{4}{6} = \frac{5}{6}$
   b. $\frac{5}{9} + \frac{8}{9} = \frac{13}{9}, \frac{8}{9} + \frac{5}{9} = \frac{13}{9}, \frac{13}{9} - \frac{5}{9} = \frac{8}{9}, \frac{13}{9} - \frac{8}{9} = \frac{5}{9}$

2. a. $\frac{3}{8}$; number line models solution; solved by counting up and subtracting
   b. $\frac{3}{5}$; number line models solution; solved by counting up and subtracting
   c. $\frac{4}{6}$; number line models solution; solved by counting up and subtracting
   d. $\frac{3}{4}$; number line models solution; solved by counting up and subtracting
   e. $\frac{2}{3}$; number line models solution; solved by counting up and subtracting
   f. $\frac{4}{5}$; number line models solution; solved by counting up and subtracting

3. a. Answer provided
   b. $\frac{8}{8} + \frac{3}{8} = \frac{11}{8}, \frac{11}{8} - \frac{7}{8} = \frac{4}{8}; \frac{8}{8} - \frac{7}{8} = \frac{1}{8}, \frac{1}{8} + \frac{3}{8} = \frac{4}{8}$; number bond shows $1\frac{3}{8}$ is $\frac{8}{8}$ and $\frac{3}{8}$
   c. $\frac{4}{4} + \frac{1}{4} = \frac{5}{4}, \frac{5}{4} - \frac{3}{4} = \frac{2}{4}; \frac{4}{4} - \frac{3}{4} = \frac{1}{4}, \frac{1}{4} + \frac{1}{4} = \frac{2}{4}$; number bond shows $1\frac{1}{4}$ is $\frac{4}{4}$ and $\frac{1}{4}$
   d. $\frac{7}{7} + \frac{2}{7} = \frac{9}{7}, \frac{9}{7} - \frac{5}{7} = \frac{4}{7}; \frac{7}{7} - \frac{5}{7} = \frac{2}{7}, \frac{2}{7} + \frac{2}{7} = \frac{4}{7}$; number bond shows $1\frac{2}{7}$ is $\frac{7}{7}$ and $\frac{2}{7}$
   e. $\frac{10}{10} + \frac{3}{10} = \frac{13}{10}, \frac{13}{10} - \frac{7}{10} = \frac{6}{10}; \frac{10}{10} - \frac{7}{10} = \frac{3}{10}, \frac{3}{10} + \frac{3}{10} = \frac{6}{10}$; number bond shows $1\frac{3}{10}$ is $\frac{10}{10}$ and $\frac{3}{10}$

Lesson 18

Practice Sheet

a. $\frac{8}{8}$

b. $\frac{7}{6}$

c. $\frac{6}{10}$

d. $\frac{4}{12}$

e. $\frac{10}{8}$

f. $\frac{1}{5}$

Problem Set

1. a. $1\frac{1}{5}$
   b. $1\frac{1}{6}$
   c. 2
   d. $\frac{3}{8}$
   e. $1\frac{3}{9}$
   f. 2
   g. $\frac{5}{12}$
   h. 1
   i. 2

2. Answers will vary.
3. Answers will vary.

Exit Ticket

1. $1\frac{2}{9}$
2. $\frac{2}{8}$

A STORY OF UNITS

Lesson 18 Answer Key 4•5

Homework

1. a. $1\frac{1}{3}$
   b. $1\frac{5}{8}$
   c. $1\frac{5}{6}$
   d. $\frac{11}{12}$
   e. $1\frac{3}{7}$
   f. 2
   g. $\frac{6}{10}$
   h. $\frac{3}{5}$
   i. 2

2. Answers will vary.
3. Answers will vary.

Lesson 19

Problem Set

1. $1\frac{6}{10}$ mi
2. $\frac{5}{8}$
3. $\frac{3}{7}$
4. $\frac{5}{8}$
5. $\frac{3}{4}$
6. $1\frac{5}{8}$ gal

Exit Ticket

1. $1\frac{7}{10}$ lb
2. $\frac{3}{4}$

Homework

1. $1\frac{2}{4}$ mi
2. $\frac{2}{3}$ hr
3. $1\frac{7}{8}$ lb
4. $1\frac{7}{8}$ c
5. $\frac{1}{6}$
6. $\frac{2}{4}$ page

Lesson 20

Problem Set

1. a. 2, 1, 3
   b. Tape diagrams model
   $\frac{1}{4} + \frac{1}{12} = \frac{3}{12} + \frac{1}{12} = \frac{4}{12}$
   c. Tape diagrams model $\frac{2}{6} + \frac{1}{3} = \frac{2}{6} + \frac{2}{6} = \frac{4}{6}$
   d. Tape diagrams model $\frac{1}{2} + \frac{3}{8} = \frac{4}{8} + \frac{3}{8} = \frac{7}{8}$
   e. Tape diagrams model
   $\frac{3}{10} + \frac{3}{5} = \frac{3}{10} + \frac{6}{10} = \frac{9}{10}$
   f. Tape diagrams model $\frac{2}{3} + \frac{2}{9} = \frac{6}{9} + \frac{2}{9} = \frac{8}{9}$

2. a. Answer provided
   b. Number line models $\frac{1}{2} + \frac{4}{10}$; $\frac{5}{10} + \frac{4}{10} = \frac{9}{10}$
   c. Number line models $\frac{6}{10} + \frac{1}{2}$; $\frac{6}{10} + \frac{5}{10} = \frac{11}{10}$
   d. Number line models $\frac{2}{3} + \frac{3}{6}$; $\frac{4}{6} + \frac{3}{6} = \frac{7}{6}$
   e. Number line models $\frac{3}{4} + \frac{6}{8}$; $\frac{6}{8} + \frac{6}{8} = \frac{12}{8}$
   f. Number line models $\frac{4}{10} + \frac{6}{5}$; $\frac{4}{10} + \frac{12}{10} = \frac{16}{10}$

3. $\frac{8}{6}$

Exit Ticket

1. Number line models $\frac{5}{8} + \frac{2}{4}$; $\frac{5}{8} + \frac{4}{8} = \frac{9}{8}$
2. $\frac{5}{4}$

Homework

1. a. Tape diagrams model $\frac{1}{3} + \frac{1}{6} = \frac{2}{6} + \frac{1}{6} = \frac{3}{6}$
   b. Tape diagrams model $\frac{1}{2} + \frac{1}{4} = \frac{2}{4} + \frac{1}{4} = \frac{3}{4}$
   c. Tape diagrams model $\frac{3}{4} + \frac{1}{8} = \frac{6}{8} + \frac{1}{8} = \frac{7}{8}$
   d. Tape diagrams model $\frac{1}{4} + \frac{5}{12} = \frac{3}{12} + \frac{5}{12} = \frac{8}{12}$
   e. Tape diagrams model $\frac{3}{8} + \frac{1}{2} = \frac{3}{8} + \frac{4}{8} = \frac{7}{8}$
   f. Tape diagrams model $\frac{3}{5} + \frac{3}{10} = \frac{6}{10} + \frac{3}{10} = \frac{9}{10}$

2. a. Answer provided
   b. Number line models $\frac{3}{5} + \frac{7}{10}$; $\frac{6}{10} + \frac{7}{10} = \frac{13}{10}$
   c. Number line models $\frac{5}{12} + \frac{1}{4}$; $\frac{5}{12} + \frac{3}{12} = \frac{8}{12}$
   d. Number line models $\frac{3}{4} + \frac{5}{8}$; $\frac{6}{8} + \frac{5}{8} = \frac{11}{8}$
   e. Number line models $\frac{7}{8} + \frac{3}{4}$; $\frac{7}{8} + \frac{6}{8} = \frac{13}{8}$
   f. Number line models $\frac{1}{6} + \frac{5}{3}$; $\frac{1}{6} + \frac{10}{6} = \frac{11}{6}$

3. $\frac{7}{6}$

A STORY OF UNITS — Lesson 21 Answer Key — 4•5

Lesson 21

Sprint

Side A

1. 1
2. $\frac{1}{2}$
3. $\frac{1}{2}$
4. 2
5. $\frac{2}{3}$
6. $\frac{2}{3}$
7. 7
8. $\frac{7}{8}$
9. $\frac{7}{8}$
10. 4
11. $\frac{4}{5}$
12. $\frac{4}{5}$
13. $\frac{3}{5}$
14. $\frac{1}{5}$
15. $\frac{2}{5}$
16. $\frac{3}{4}$
17. $\frac{1}{4}$
18. $\frac{9}{10}$
19. $\frac{1}{10}$
20. $\frac{7}{10}$
21. $\frac{3}{10}$
22. 2
23. $\frac{2}{3}$
24. $\frac{2}{3}$
25. $\frac{4}{3}$
26. 3
27. $\frac{3}{5}$
28. $\frac{3}{5}$
29. $\frac{7}{5}$
30. 2
31. $\frac{2}{4}$
32. $\frac{2}{4}$
33. $\frac{6}{4}$
34. $\frac{5}{8}$
35. $\frac{1}{8}$
36. $\frac{12}{8}$
37. $\frac{4}{8}$
38. $\frac{5}{6}$
39. $\frac{1}{6}$
40. $\frac{10}{6}$
41. $\frac{2}{6}$
42. $\frac{7}{12}$
43. $\frac{6}{12}$
44. $\frac{6}{15}$

Side B

1. 2
2. $\frac{2}{3}$
3. $\frac{2}{3}$
4. 1
5. $\frac{1}{2}$
6. $\frac{1}{2}$
7. 5
8. $\frac{5}{6}$
9. $\frac{5}{6}$
10. 9
11. $\frac{9}{10}$
12. $\frac{9}{10}$
13. $\frac{8}{10}$
14. $\frac{6}{10}$
15. $\frac{7}{10}$
16. $\frac{4}{5}$
17. $\frac{1}{5}$
18. $\frac{7}{8}$
19. $\frac{1}{8}$
20. $\frac{5}{8}$
21. $\frac{3}{8}$
22. 2
23. $\frac{2}{4}$
24. $\frac{2}{4}$
25. $\frac{6}{4}$
26. 4
27. $\frac{4}{5}$
28. $\frac{4}{5}$
29. $\frac{6}{5}$
30. 2
31. $\frac{2}{6}$
32. $\frac{2}{6}$
33. $\frac{10}{6}$
34. $\frac{3}{8}$
35. $\frac{1}{8}$
36. $\frac{10}{8}$
37. $\frac{6}{8}$
38. $\frac{3}{4}$
39. $\frac{1}{4}$
40. $\frac{6}{4}$
41. $\frac{2}{4}$
42. $\frac{5}{12}$
43. $\frac{8}{12}$
44. $\frac{11}{15}$

Lesson 21 Answer Key 4•5

Problem Set

1. a. Tape diagrams represent $\frac{3}{4}$ and $\frac{2}{4}$; $\frac{3}{4} + \frac{2}{4} = \frac{5}{4}$; number bond shows $\frac{5}{4}$ as $\frac{4}{4}$ and $\frac{1}{4}$; $1\frac{1}{4}$
    b. Tape diagrams represent $\frac{4}{6}$ and $\frac{3}{6}$; $\frac{4}{6} + \frac{3}{6} = \frac{7}{6}$; number bond shows $\frac{7}{6}$ as $\frac{6}{6}$ and $\frac{1}{6}$; $1\frac{1}{6}$
    c. Tape diagrams represent $\frac{5}{6}$ and $\frac{2}{6}$; $\frac{5}{6} + \frac{2}{6} = \frac{7}{6}$; number bond shows $\frac{7}{6}$ as $\frac{6}{6}$ and $\frac{1}{6}$; $1\frac{1}{6}$
    d. Tape diagrams represent $\frac{8}{10}$ and $\frac{7}{10}$; $\frac{8}{10} + \frac{7}{10} = \frac{15}{10}$; number bond shows $\frac{15}{10}$ as $\frac{10}{10}$ and $\frac{5}{10}$; $1\frac{5}{10}$

2. a. Number line models $\frac{2}{4} + \frac{3}{4}$; $\frac{2}{4} + \frac{3}{4} = \frac{5}{4}$; number bond shows $\frac{5}{4}$ as $\frac{4}{4}$ and $\frac{1}{4}$; $1\frac{1}{4}$
    b. Number line models $\frac{4}{8} + \frac{6}{8}$; $\frac{4}{8} + \frac{6}{8} = \frac{10}{8}$; number bond shows $\frac{10}{8}$ as $\frac{8}{8}$ and $\frac{2}{8}$; $1\frac{2}{8}$
    c. Number line models $\frac{7}{10} + \frac{6}{10}$; $\frac{7}{10} + \frac{6}{10} = \frac{13}{10}$; number bond shows $\frac{13}{10}$ as $\frac{10}{10}$ and $\frac{3}{10}$; $1\frac{3}{10}$
    d. Number line models $\frac{4}{6} + \frac{5}{6}$; $\frac{4}{6} + \frac{5}{6} = \frac{9}{6}$; number bond shows $\frac{9}{6}$ as $\frac{6}{6}$ and $\frac{3}{6}$; $1\frac{3}{6}$

3. a. $\frac{6}{8} + \frac{2}{8} = \frac{8}{8} = 1$
    b. $\frac{4}{6} + \frac{3}{6} = \frac{7}{6} = 1\frac{1}{6}$
    c. $\frac{4}{6} + \frac{4}{6} = \frac{8}{6} = 1\frac{2}{6}$
    d. $\frac{8}{10} + \frac{6}{10} = \frac{14}{10} = 1\frac{4}{10}$
    e. $\frac{5}{8} + \frac{6}{8} = \frac{11}{8} = 1\frac{3}{8}$
    f. $\frac{5}{8} + \frac{4}{8} = \frac{9}{8} = 1\frac{1}{8}$
    g. $\frac{4}{8} + \frac{5}{8} = \frac{9}{8} = 1\frac{1}{8}$
    h. $\frac{3}{10} + \frac{8}{10} = \frac{11}{10} = 1\frac{1}{10}$

Exit Ticket

1. $\frac{1}{4} + \frac{7}{8} = \frac{2}{8} + \frac{7}{8} = \frac{9}{8}$; number bond shows $\frac{9}{8}$ as $\frac{8}{8}$ and $\frac{1}{8}$; $1\frac{1}{8}$
2. $\frac{2}{3} + \frac{7}{12} = \frac{8}{12} + \frac{7}{12} = \frac{15}{12}$; number bond shows $\frac{15}{12}$ as $\frac{12}{12}$ and $\frac{3}{12}$; $1\frac{3}{12}$

Module 5: Fraction Equivalence, Ordering, and Operations

Homework

1. a. Tape diagrams represent $\frac{7}{8}$ and $\frac{2}{8}$; $\frac{7}{8} + \frac{2}{8} = \frac{9}{8}$; number bond shows $\frac{9}{8}$ as $\frac{8}{8}$ and $\frac{1}{8}$; $1\frac{1}{8}$
   b. Tape diagrams represent $\frac{4}{8}$ and $\frac{4}{8}$; $\frac{4}{8} + \frac{4}{8} = \frac{8}{8}$; 1
   c. Tape diagrams represent $\frac{4}{6}$ and $\frac{3}{6}$; $\frac{4}{6} + \frac{3}{6} = \frac{7}{6}$ number bond shows $\frac{7}{6}$ as $\frac{6}{6}$ and $\frac{1}{6}$; $1\frac{1}{6}$
   d. Tape diagrams represent $\frac{6}{10}$ and $\frac{8}{10}$; $\frac{6}{10} + \frac{8}{10} = \frac{14}{10}$; number bond shows $\frac{14}{10}$ as $\frac{10}{10}$ and $\frac{4}{10}$; $1\frac{4}{10}$

2. a. Number line models $\frac{4}{8} + \frac{5}{8}$; $\frac{4}{8} + \frac{5}{8} = \frac{9}{8}$; number bond shows $\frac{9}{8}$ as $\frac{8}{8}$ and $\frac{1}{8}$; $1\frac{1}{8}$
   b. Number line models $\frac{6}{8} + \frac{3}{8}$; $\frac{6}{8} + \frac{3}{8} = \frac{9}{8}$; number bond shows $\frac{9}{8}$ as $\frac{8}{8}$ and $\frac{1}{8}$; $1\frac{1}{8}$
   c. Number line models $\frac{4}{10} + \frac{8}{10}$; $\frac{4}{10} + \frac{8}{10} = \frac{12}{10}$; number bond shows $\frac{12}{10}$ as $\frac{10}{10}$ and $\frac{2}{10}$; $1\frac{2}{10}$
   d. Number line models $\frac{2}{6} + \frac{5}{6}$; $\frac{2}{6} + \frac{5}{6} = \frac{7}{6}$; number bond shows $\frac{7}{6}$ as $\frac{6}{6}$ and $\frac{1}{6}$; $1\frac{1}{6}$

3. a. $\frac{4}{8} + \frac{6}{8} = \frac{10}{8} = 1\frac{2}{8}$
   b. $\frac{7}{8} + \frac{6}{8} = \frac{13}{8} = 1\frac{5}{8}$
   c. $\frac{5}{6} + \frac{2}{6} = \frac{7}{6} = 1\frac{1}{6}$
   d. $\frac{9}{10} + \frac{4}{10} = \frac{13}{10} = 1\frac{3}{10}$
   e. $\frac{4}{12} + \frac{9}{12} = \frac{13}{12} = 1\frac{1}{12}$
   f. $\frac{3}{6} + \frac{5}{6} = \frac{8}{6} = 1\frac{2}{6}$
   g. $\frac{3}{12} + \frac{10}{12} = \frac{13}{12} = 1\frac{1}{12}$
   h. $\frac{7}{10} + \frac{8}{10} = \frac{15}{10} = 1\frac{5}{10}$

Lesson 22

Sprint

Side A

1. 2
2. $\frac{2}{5}$
3. 3
4. $\frac{3}{5}$
5. 4
6. $\frac{4}{5}$
7. 5
8. 5 fifths
9. 1
10. $\frac{5}{5}$
11. 5
12. $\frac{5}{8}$
13. 7
14. $\frac{7}{8}$
15. 8 eighths
16. 1
17. $\frac{8}{8}$
18. 4
19. 4 thirds
20. $1\frac{1}{3}$
21. 6
22. 6 fifths
23. $1\frac{1}{5}$
24. 9
25. 9 eighths
26. $1\frac{1}{8}$
27. $1\frac{7}{8}$
28. 3
29. 3 halves
30. $1\frac{1}{2}$
31. 12
32. 12 tenths
33. $1\frac{2}{10}$
34. $1\frac{8}{10}$
35. 6
36. 6 sixths
37. $\frac{6}{6}$
38. $1\frac{3}{6}$
39. $\frac{11}{12}$
40. $\frac{12}{12}$
41. $1\frac{5}{12}$
42. $1\frac{11}{12}$
43. $1\frac{7}{15}$
44. $1\frac{14}{15}$

Side B

1. 2
2. $\frac{2}{6}$
3. 4
4. $\frac{4}{6}$
5. 5
6. $\frac{5}{6}$
7. 6
8. 6 sixths
9. 1
10. $\frac{6}{6}$
11. 7
12. $\frac{7}{8}$
13. 7
14. $\frac{7}{8}$
15. 8 eighths
16. 1
17. $\frac{8}{8}$
18. 4
19. 4 thirds
20. $1\frac{1}{3}$
21. 9
22. 9 eighths
23. $1\frac{1}{8}$
24. 3
25. 3 halves
26. $1\frac{1}{2}$
27. 6
28. 6 fifths
29. $1\frac{1}{5}$
30. $1\frac{4}{5}$
31. 18
32. 18 tenths
33. $1\frac{8}{10}$
34. $1\frac{5}{10}$
35. 6
36. 6 sixths
37. $\frac{6}{6}$
38. $1\frac{3}{6}$
39. $\frac{11}{12}$
40. $\frac{12}{12}$
41. $1\frac{5}{12}$
42. $1\frac{11}{12}$
43. $1\frac{7}{15}$
44. $1\frac{14}{15}$

Lesson 22 Answer Key 4•5

Problem Set

1. a. Tape diagram drawn; $3\frac{1}{3}$
   b. Tape diagram drawn; $4\frac{3}{4}$
   c. Tape diagram drawn; $2\frac{3}{4}$
   d. Tape diagram drawn; $4\frac{3}{5}$

2. a. $6\frac{3}{8} - \frac{3}{8} = 6, 6\frac{3}{8} - 6 = \frac{3}{8}, 6\overline{A}\frac{3}{8} = 6\frac{3}{8}, \frac{3}{8} + 6 = 6\frac{3}{8}$
   b. $9 - \frac{4}{7} = 8\frac{3}{7}, 9 - 8\frac{3}{7} = \frac{4}{7}, 8\frac{3}{7} + \frac{4}{7} = 9, \frac{4}{7} + 8\frac{3}{7} = 9$

3. a. Answer provided
   b. $4\frac{1}{3}$; number bond shows 5 as 4 and $\frac{3}{3}$; number line drawn
   c. $6\frac{5}{8}$; number bond shows 7 as 6 and $\frac{8}{8}$; number line drawn
   d. $9\frac{6}{10}$; number bond shows 10 as 9 and $\frac{10}{10}$; number line drawn

4. a. $2\frac{9}{10}$; number bond shows 3 as 2 and $\frac{10}{10}$
   b. $4\frac{1}{4}$; number bond shows 5 as 4 and $\frac{4}{4}$
   c. $5\frac{3}{8}$; number bond shows 6 as 5 and $\frac{8}{8}$
   d. $6\frac{6}{9}$; number bond shows 7 as 6 and $\frac{9}{9}$
   e. $7\frac{4}{10}$; number bond shows 8 as 7 and $\frac{10}{10}$
   f. $28\frac{3}{12}$; number bond shows 29 as 28 and $\frac{12}{12}$

Exit Ticket

1. $5\frac{4}{5}$; number bond shows 6 as 5 and $\frac{5}{5}$.
2. $7\frac{1}{6}$; number bond shows 8 as 7 and $\frac{6}{6}$.
3. $6\frac{3}{8}$; number bond shows 7 as 6 and $\frac{8}{8}$.

Homework

1. a. Tape diagram drawn; $2\frac{1}{4}$

   b. Tape diagram drawn; $3\frac{2}{3}$

   c. Tape diagram drawn; $1\frac{4}{5}$

   d. Tape diagram drawn; $2\frac{1}{4}$

2. a. $4\frac{5}{8} - \frac{5}{8} = 4, 4\frac{5}{8} - 4 = \frac{5}{8}, 4 + \frac{5}{8} = 4\frac{5}{8}, \frac{5}{8} + 4 = 4\frac{5}{8}$

   b. $6 - \frac{2}{7} = 5\frac{5}{7}, 6 - 5\frac{5}{7} = \frac{2}{7}, 5\frac{5}{7} + \frac{2}{7} = 6, \frac{2}{7} + 5\frac{5}{7} = 6$

3. a. Answer provided

   b. $7\frac{1}{6}$; number bond shows 8 as 7 and $\frac{6}{6}$; number line drawn

   c. $6\frac{1}{5}$; number bond shows 7 as 6 and $\frac{5}{5}$; number line drawn

   d. $2\frac{7}{10}$; number bond shows 3 as 2 and $\frac{10}{10}$; number line drawn

4. a. $5\frac{3}{4}$; number bond shows 6 as 5 and $\frac{4}{4}$

   b. $6\frac{8}{10}$; number bond shows 7 as 6 and $\frac{10}{10}$

   c. $4\frac{1}{6}$; number bond shows 5 as 4 and $\frac{6}{6}$

   d. $5\frac{2}{8}$; number bond shows 6 as 5 and $\frac{8}{8}$

   e. $2\frac{1}{8}$; number bond shows 3 as 2 and $\frac{8}{8}$

   f. $25\frac{3}{10}$; number bond shows 26 as 25 and $\frac{10}{10}$

Lesson 23

Problem Set

1. a. $\frac{0}{3}, \frac{1}{3}, \frac{2}{3}, \frac{3}{3}, \frac{4}{3}, \frac{5}{3}, \frac{6}{3}; \frac{0}{3}, \frac{3}{3}, \frac{6}{3}$ circled; 0, 1, 2 recorded

   b. $\frac{0}{2}, \frac{1}{2}, \frac{2}{2}, \frac{3}{2}, \frac{4}{2}, \frac{5}{2}, \frac{6}{2}, \frac{7}{2}, \frac{8}{2}; \frac{0}{2}, \frac{2}{2}, \frac{4}{2}, \frac{6}{2}, \frac{8}{2}$ circled; 0, 1, 2, 3, 4 recorded

2. $\left(\frac{1}{4} + \frac{1}{4} + \frac{1}{4} + \frac{1}{4}\right) + \left(\frac{1}{4} + \frac{1}{4} + \frac{1}{4} + \frac{1}{4}\right) + \left(\frac{1}{4} + \frac{1}{4} + \frac{1}{4} + \frac{1}{4}\right) = 3$

3. a. Answer provided

   b. $6 \times \frac{1}{2} = 3 \times \frac{2}{2} = 3$; number line supports answer.

   c. $12 \times \frac{1}{4} = 3 \times \frac{4}{4} = 3$; number line supports answer.

4. a. Answer provided

   b. $7 \times \frac{1}{2} = \left(3 \times \frac{2}{2}\right) + \frac{1}{2} = 3 + \frac{1}{2} = 3\frac{1}{2}$; number line supports answer.

   c. $10 \times \frac{1}{4} = \left(2 \times \frac{4}{4}\right) + \frac{2}{4} = 2 + \frac{2}{4} = 2\frac{2}{4}$; number line supports answer.

   d. $14 \times \frac{1}{3} = \left(4 \times \frac{3}{3}\right) + \frac{2}{3} = 4 + \frac{2}{3} = 4\frac{2}{3}$; number line supports answer.

Exit Ticket

1. $8 \times \frac{1}{2} = 4 \times \frac{2}{2} = 4$; number line supports answer.

2. $7 \times \frac{1}{4} = \left(1 \times \frac{4}{4}\right) + \frac{3}{4} = 1\frac{3}{4}$; number line supports answer.

3. $13 \times \frac{1}{3} = \left(4 \times \frac{3}{3}\right) + \frac{1}{3} = 4\frac{1}{3}$; number line supports answer.

Homework

1. a. $\frac{0}{4}, \frac{1}{4}, \frac{2}{4}, \frac{3}{4}, \frac{4}{4}, \frac{5}{4}, \frac{6}{4}; \frac{0}{4}, \frac{4}{4}$ circled; 0, 1 recorded

   b. $\frac{0}{6}, \frac{1}{6}, \frac{2}{6}, \frac{3}{6}, \frac{4}{6}, \frac{5}{6}, \frac{6}{6}, \frac{7}{6}, \frac{8}{6}, \frac{9}{6}, \frac{10}{6}, \frac{11}{6}, \frac{12}{6}, \frac{13}{6}, \frac{14}{6}; \frac{0}{6}, \frac{6}{6}, \frac{12}{6}$ circled; 0, 1, 2 recorded

2. $\left(\frac{1}{3}+\frac{1}{3}+\frac{1}{3}\right) + \left(\frac{1}{3}+\frac{1}{3}+\frac{1}{3}\right) + \left(\frac{1}{3}+\frac{1}{3}+\frac{1}{3}\right) + \left(\frac{1}{3}+\frac{1}{3}+\frac{1}{3}\right) = 4$

3. a. Answer provided

   b. $10 \times \frac{1}{2} = 5 \overline{A} \frac{2}{2} = 5$; number line supports answer.

   c. $8 \times \frac{1}{4} = 2 \times \frac{4}{4} = 2$; number line supports answer.

4. a. Answer provided

   b. $7 \times \frac{1}{4} = \left(1 \times \frac{4}{4}\right) + \frac{3}{4} = 1 + \frac{3}{4} = 1\frac{3}{4}$; number line supports answer.

   c. $11 \times \frac{1}{5} = \left(2 \times \frac{5}{5}\right) + \frac{1}{5} = 2 + \frac{1}{5} = 2\frac{1}{5}$; number line supports answer.

   d. $7 \times \frac{1}{2} = \left(3 \times \frac{2}{2}\right) + \frac{1}{2} = 3 + \frac{1}{2} = 3\frac{1}{2}$; number line supports answer.

   e. $9 \times \frac{1}{5} = \left(1 \times \frac{5}{5}\right) + \frac{4}{5} = 1 + \frac{4}{5} = 1\frac{4}{5}$; number line supports answer.

Lesson 24

Problem Set

1. a. Answer provided
   b. $2\frac{2}{5}$; number bond shows $\frac{12}{5}$ as $\frac{10}{5}$ and $\frac{2}{5}$; number line drawn
   c. $6\frac{1}{2}$; number bond shows $\frac{13}{2}$ as $\frac{12}{2}$ and $\frac{1}{2}$; number line drawn
   d. $3\frac{3}{4}$; number bond shows $\frac{15}{4}$ as $\frac{12}{4}$ and $\frac{3}{4}$; number line drawn

2. a. Answer provided
   b. $\frac{9}{2} = \frac{2 \times 4}{2} + \frac{1}{2} = 4 + \frac{1}{2} = 4\frac{1}{2}$; number line drawn
   c. $\frac{17}{4} = \frac{4 \times 4}{4} + \frac{1}{4} = 4 + \frac{1}{4} = 4\frac{1}{4}$; number line drawn

3. a. $2\frac{1}{4}$
   b. $3\frac{2}{5}$
   c. $4\frac{1}{6}$
   d. $4\frac{2}{7}$
   e. $4\frac{6}{8}$
   f. $5\frac{3}{9}$
   g. $6\frac{3}{10}$
   h. $8\frac{4}{10}$
   i. $3\frac{1}{12}$

Exit Ticket

1. $3\frac{2}{5}$; number bond shows $\frac{17}{5}$ as $\frac{15}{5}$ and $\frac{2}{5}$; number line drawn
2. $6\frac{1}{3}$; number line drawn
3. $2\frac{3}{4}$

Homework

1.
   a. Answer provided
   b. $3\frac{1}{4}$; number bond shows $\frac{13}{4}$ as 3 and $\frac{1}{4}$; number line drawn
   c. $3\frac{1}{5}$; number bond shows $\frac{16}{5}$ as 3 and $\frac{1}{5}$; number line drawn
   d. $7\frac{1}{2}$; number bond shows $\frac{15}{2}$ as 7 and $\frac{1}{2}$; number line drawn
   e. $5\frac{2}{3}$; number bond shows $\frac{17}{3}$ as 5 and $\frac{2}{3}$; number line drawn

2.
   a. Answer provided
   b. $\frac{13}{2} = \frac{2 \times 6}{2} + \frac{1}{2} = 6 + \frac{1}{2} = 6\frac{1}{2}$; number line drawn
   c. $\frac{18}{4} = \frac{4 \times 4}{4} + \frac{2}{4} = 4 + \frac{2}{4} = 4\frac{2}{4}$; number line drawn

3.
   a. $4\frac{2}{3}$
   b. $4\frac{1}{4}$
   c. $5\frac{2}{5}$
   d. $4\frac{4}{6}$
   e. $3\frac{2}{7}$
   f. $4\frac{5}{8}$
   g. $5\frac{6}{9}$
   h. $7\frac{4}{10}$
   i. $3\frac{9}{12}$

Lesson 25

Problem Set

1. a. Answer provided
    b. $\frac{14}{5}$; number line models work.
    c. $\frac{29}{8}$; number line models work.
    d. $\frac{44}{10}$; number line models work.
    e. $\frac{43}{9}$; number line models work.

2. a. Answer provided
    b. $4\frac{1}{3} = 4 + \frac{1}{3} = \left(4 \times \frac{3}{3}\right) + \frac{1}{3} = \frac{12}{3} + \frac{1}{3} = \frac{13}{3}$
    c. $4\frac{3}{5} = 4 + \frac{3}{5} = \left(4 \times \frac{5}{5}\right) + \frac{3}{5} = \frac{20}{5} + \frac{3}{5} = \frac{23}{5}$
    d. $4\frac{6}{8} = 4 + \frac{6}{8} = \left(4 \times \frac{8}{8}\right) + \frac{6}{8} = \frac{32}{8} + \frac{6}{8} = \frac{38}{8}$

3. a. $\frac{11}{4}$
    b. $\frac{12}{5}$
    c. $\frac{21}{6}$
    d. $\frac{27}{8}$
    e. $\frac{31}{10}$
    f. $\frac{35}{8}$
    g. $\frac{17}{3}$
    h. $\frac{13}{2}$
    i. $\frac{73}{10}$

Exit Ticket

1. $\frac{16}{5}$
2. $\frac{13}{5}$
3. $\frac{38}{9}$

Lesson 25 Answer Key 4•5

Homework

1. a. Answer provided
   b. $\frac{22}{5}$; number line models work.
   c. $\frac{43}{8}$; number line models work.
   d. $\frac{37}{10}$; number line models work.
   e. $\frac{56}{9}$; number line models work.

2. a. Answer provided
   b. $5\frac{2}{3} = 5\text{ A }\frac{2}{3} = \left(5\text{ A }\frac{3}{3}\right) + \frac{2}{3} = \frac{15}{3} + \frac{2}{3} = \frac{17}{3}$
   c. $4\frac{1}{5} = 4 + \frac{1}{5} = \left(4 \times \frac{5}{5}\right) + \frac{1}{5} = \frac{20}{5} + \frac{1}{5} = \frac{21}{5}$
   d. $3\frac{7}{8} = 3 + \frac{7}{8} = \left(3 \times \frac{8}{8}\right) + \frac{7}{8} = \frac{24}{8} + \frac{7}{8} = \frac{31}{8}$

3. a. $\frac{7}{3}$
   b. $\frac{11}{4}$
   c. $\frac{17}{5}$
   d. $\frac{19}{6}$
   e. $\frac{53}{12}$
   f. $\frac{22}{5}$
   g. $\frac{41}{10}$
   h. $\frac{26}{5}$
   i. $\frac{35}{6}$
   j. $\frac{25}{4}$
   k. $\frac{15}{2}$
   l. $\frac{\text{A}5}{12}$

Lesson 26

Problem Set

1. a. $2\frac{7}{8}, \overline{A}\frac{1}{6}, \frac{29}{12}$ plotted
   b. i. <
      ii. <
2. a. $\frac{70}{9}, \overline{A}\frac{2}{4}, \frac{25}{3}$ plotted
   b. i. >
      ii. <
   c. Explanations will vary.

3. a. >; explanations will vary.
   b. <; explanations will vary.
   c. <; explanations will vary.
   d. <; explanations will vary.
   e. >; explanations will vary.
   f. >; explanations will vary.
   g. <; explanations will vary.
   h. >; explanations will vary.
   i. <; explanations will vary.
   j. <; explanations will vary.

Exit Ticket

1. =; explanations will vary.
2. >; explanations will vary.
3. >; explanations will vary.
4. >; explanations will vary.

Homework

1. a. $2\frac{1}{6}, 3\frac{3}{4}, \frac{33}{9}$ plotted
   b. i. >
      ii. <
2. a. $\frac{65}{8}, 8\frac{5}{6}, \frac{29}{4}$ plotted
   b. i. >
      ii. <
   c. Explanations will vary.

3. a. <; explanations will vary.
   b. <; explanations will vary.
   c. <; explanations will vary.
   d. >; explanations will vary.
   e. >; explanations will vary.
   f. >; explanations will vary.
   g. <; explanations will vary.
   h. <; explanations will vary.
   i. <; explanations will vary.
   j. >; explanations will vary.

Lesson 27

Problem Set

1. a. Tape diagram models comparison; <
   b. Tape diagram models comparison; <
   c. Tape diagram models comparison; >
   d. Tape diagram models comparison; <
2. a. Area model shows like units; >
   b. Area model shows like units; >

3. a. >
   b. <
   c. >
   d. >
   e. >
   f. >
   g. >
   h. <
   i. <
   j. >

Exit Ticket

1. >
2. <
3. <
4. <

Homework

1. a. Tape diagram models comparison; <
   b. Tape diagram models comparison; =
   c. Tape diagram models comparison; >
   d. Tape diagram models comparison; <
2. a. Area model shows like units; >
   b. Area model shows like units; <

3. a. >
   b. <
   c. >
   d. <
   e. >
   f. <
   g. >
   h. >
   i. >
   j. >

A STORY OF UNITS — Lesson 28 Answer Key 4•5

Lesson 28

Problem Set

1. Line plot created accurately
2. a. Morgan
   b. Morgan
   c. Saisha and Anson; 9 quarter miles
   d. 2 miles
   e. $1\frac{3}{4}$ miles > $1\frac{5}{8}$ miles
   f. $1\frac{2}{8}$ miles
   g. $1\frac{3}{10}$ miles > $1\frac{2}{8}$ miles; Mr. Reynolds
3. Answers will vary.

Exit Ticket

1. Line plot created accurately
2. $2\frac{1}{4}$ hours; Gail

Homework

1. Line plot created accurately
2. a. Mary
   b. Ben
   c. 31 quarter inches
   d. $\frac{1}{4}$ inch
   e. $7\frac{1}{2}$ inches < $7\frac{3}{4}$ inches
   f. 4
   g. 8
   h. $\frac{25}{2}$ inches > $8\frac{3}{4}$ inches; Mr. Jones
3. Answers will vary.

Lesson 29

Problem Set

1. a. 4; explanations will vary.
    b. $7\frac{1}{2}$ or 8; explanations will vary.
    c. 7; explanations will vary.
    d. 4; explanations will vary.
    e. $8\frac{1}{2}$; explanations will vary.
2. a. 6; explanations will vary.
    b. $3\frac{1}{2}$ or 4; explanations will vary.
    c. $8\frac{1}{2}$ or 9; explanations will vary.
3. Julio's; explanations will vary.
4. a. $44\frac{1}{2}$ or $44\frac{3}{4}$ or 45
    b. 58
    c. 9 or $9\frac{1}{2}$
    d. 2

Exit Ticket

1. 5 or $5\frac{1}{2}$; explanations will vary.
2. $8\frac{1}{2}$; explanations will vary.

Homework

1. a. $4\frac{1}{2}$ or 5; explanations will vary.
    b. 8; explanations will vary.
    c. 5; explanations will vary.
    d. 3; explanations will vary.
    e. 11 or $11\frac{1}{2}$; explanations will vary.
2. a. 7 or $7\frac{1}{2}$; explanations will vary.
    b. $1\frac{1}{2}$ or 2 or $2\frac{1}{2}$; explanations will vary.
    c. 10 or $10\frac{1}{2}$; explanations will vary.
3. Gina's; explanations will vary.
4. a. $23\frac{1}{2}$ or $23\frac{3}{4}$ or 24
    b. 26
    c. $6\frac{1}{2}$ or $6\frac{3}{4}$ or 7
    d. 4

Lesson 30

Sprint

Side A

1. 1	12. 1	23. 2	34. 2
2. 1	13. 5	24. 8	35. 2
3. 1	14. 5	25. 2	36. 4
4. 1	15. 2	26. 2	37. 2
5. 1	16. 2	27. 2	38. 4
6. 1	17. 3	28. 9	39. 3
7. 1	18. 4	29. 9	40. 1
8. 1	19. 1	30. 2	41. 7
9. 3	20. 2	31. 2	42. 12
10. 3	21. 11	32. 6	43. 3
11. 1	22. 11	33. 2	44. 5

Side B

1. 1	12. 1	23. 2	34. 2
2. 1	13. 5	24. 5	35. 2
3. 1	14. 5	25. 1	36. 2
4. 1	15. 3	26. 1	37. 2
5. 1	16. 3	27. 1	38. 2
6. 1	17. 4	28. 11	39. 4
7. 1	18. 1	29. 11	40. 1
8. 1	19. 2	30. 2	41. 3
9. 4	20. 2	31. 2	42. 12
10. 4	21. 7	32. 4	43. 7
11. 1	22. 7	33. 1	44. 7

Problem Set

1. a. $3\frac{2}{4}$
   b. 8
   c. $5\frac{5}{8}$
   d. 7

2. a. $\frac{1}{8}$
   b. $\frac{3}{5}$
   c. $\frac{5}{6}$
   d. $\frac{11}{12}$

3. a. Number bond and arrow way used to make one; $3\frac{1}{4}$
   b. Number bond and arrow way used to make one; $4\frac{1}{5}$

4. a. $5\frac{1}{3}$
   b. $4\frac{2}{5}$
   c. $6\frac{3}{6}$
   d. $7\frac{3}{8}$
   e. $8\frac{6}{10}$
   f. $10\frac{6}{12}$
   g. $3\frac{57}{100}$
   h. $17\frac{28}{100}$

5. Explanations will vary.

Exit Ticket

1. $\frac{3}{5}$
2. $3\frac{2}{8}$

Homework

1. a. $4\frac{2}{3}$
    b. $5\frac{3}{4}$
    c. 4
    d. 8

2. a. $\frac{1}{6}$
    b. $\frac{4}{7}$
    c. $\frac{7}{8}$
    d. $\frac{8}{12}$

3. a. Answer provided
    b. Number bond and arrow way used to make one; $4\frac{1}{3}$
    c. Number bond and arrow way used to make one; $5\frac{3}{6}$

4. a. $3\frac{1}{5}$
    b. $4\frac{2}{8}$
    c. $6\frac{1}{6}$
    d. $7\frac{3}{10}$
    e. $9\frac{4}{10}$
    f. $8\frac{7}{12}$
    g. $4\frac{48}{100}$
    h. $15\frac{39}{100}$

5. Explanations will vary.

Lesson 31

Sprint

Side A

1. 4	12. 3	23. 11	34. 11
2. 4	13. 5	24. 1	35. 14
3. 4	14. 5	25. 7	36. 14
4. 4	15. 5	26. 7	37. 14
5. 6	16. 7	27. 7	38. 18
6. 6	17. 6	28. 1	39. 7
7. 6	18. 6	29. 11	40. 19
8. 6	19. 6	30. 11	41. 13
9. 3	20. 8	31. 11	42. 21
10. 3	21. 11	32. 11	43. 14
11. 3	22. 11	33. 11	44. 31

Side B

1. 5	12. 6	23. 13	34. 8
2. 5	13. 4	24. 1	35. 15
3. 5	14. 4	25. 5	36. 15
4. 5	15. 4	26. 5	37. 15
5. 3	16. 5	27. 5	38. 19
6. 3	17. 11	28. 1	39. 9
7. 3	18. 11	29. 9	40. 14
8. 3	19. 11	30. 9	41. 19
9. 6	20. 17	31. 9	42. 23
10. 6	21. 13	32. 8	43. 13
11. 6	22. 13	33. 8	44. 23

Lesson 31 Answer Key

Problem Set

1.
   a. 6
   b. $7\frac{3}{4}$
   c. 9

2.
   a. $4\frac{1}{5}$; number line used
   b. $5\frac{2}{4}$; number line used
   c. $6\frac{1}{8}$; number line used

3.
   a. $4\frac{3}{6}$; arrow way used to make one
   b. $5\frac{2}{4}$; arrow way used to make one
   c. $6\frac{1}{8}$; arrow way used to make one

4.
   a. $5\frac{2}{5}$
   b. $6\frac{5}{8}$
   c. $6\frac{3}{12}$

Exit Ticket

1. 4
2. $6\frac{2}{5}$

Homework

1.
   a. 4
   b. $4\frac{4}{5}$
   c. 5

2.
   a. $4\frac{1}{4}$; number line used
   b. $6\frac{3}{6}$; number line used
   c. $3\frac{4}{12}$; number line used

3.
   a. $4\frac{2}{4}$; arrow way used to make one
   b. $6\frac{3}{8}$; arrow way used to make one
   c. $6\frac{3}{9}$; arrow way used to make one

4.
   a. $3\frac{2}{5}$
   b. $5\frac{3}{10}$
   c. $6\frac{4}{7}$

A STORY OF UNITS

Lesson 32 Answer Key 4•5

Lesson 32

Problem Set

1. a. $3\frac{2}{4}$; number line or arrow way drawn
   b. $4\frac{4}{10}$; number line or arrow way drawn
   c. $4\frac{2}{3}$; number line or arrow way drawn
   d. $8\frac{4}{5}$; number line or arrow way drawn

2. a. $4\frac{4}{5}$; number line or arrow way drawn
   b. $3\frac{3}{4}$; number line or arrow way drawn
   c. $4\frac{2}{3}$; number line or arrow way drawn
   d. $1\frac{6}{8}$; number line or arrow way drawn

3. a. Answer provided
   b. $4\frac{2}{8}$; total decomposed as $4\frac{1}{8}$ and 1
   c. $4\frac{4}{5}$; total decomposed as $4\frac{3}{5}$ and 1
   d. $4\frac{5}{6}$; total decomposed as $4\frac{4}{6}$ and 1
   e. $5\frac{9}{12}$; total decomposed as $5\frac{4}{12}$ and 1
   f. $8\frac{4}{8}$; total decomposed as $8\frac{1}{8}$ and 1
   g. $6\frac{2}{6}$; total decomposed as $6\frac{1}{6}$ and 1
   h. $7\frac{9}{10}$; total decomposed as $7\frac{3}{10}$ and 1
   i. $11\frac{4}{5}$; total decomposed as $11\frac{3}{5}$ and 1
   j. $10\frac{3}{6}$; total decomposed as $10\frac{2}{6}$ and 1

Exit Ticket

1. $10\frac{1}{6}$
2. $7\frac{5}{8}$

Module 5: Fraction Equivalence, Ordering, and Operations

A STORY OF UNITS Lesson 32 Answer Key 4•5

Homework

1. a. $6\frac{2}{5}$; number line or arrow way drawn
    b. $4\frac{2}{12}$; number line or arrow way drawn
    c. $6\frac{2}{4}$; number line or arrow way drawn
    d. $7\frac{6}{8}$; number line or arrow way drawn

2. a. $1\frac{3}{5}$; number line or arrow way drawn
    b. $1\frac{2}{3}$; number line or arrow way drawn
    c. $3\frac{3}{6}$; number line or arrow way drawn
    d. $2\frac{4}{6}$; number line or arrow way drawn
    e. $8\frac{4}{8}$; number line or arrow way drawn
    f. $6\frac{5}{10}$; number line or arrow way drawn
    g. $9\frac{4}{8}$; number line or arrow way drawn
    h. $8\frac{9}{12}$; number line or arrow way drawn
    i. $10\frac{4}{5}$; number line or arrow way drawn
    j. $16\frac{5}{9}$; number line or arrow way drawn

3. a. Answer provided
    b. $4\frac{4}{5}$; total decomposed as $4\frac{2}{5}$ and 1
    c. $6\frac{6}{8}$; total decomposed as $6\frac{1}{8}$ and 1
    d. $2\frac{8}{9}$; total decomposed as $2\frac{3}{9}$ and 1
    e. $5\frac{6}{10}$; total decomposed as $5\frac{3}{10}$ and 1
    f. $1\frac{6}{9}$; total decomposed as $1\frac{5}{9}$ and 1

Lesson 33

Sprint

Side A

1. 3
2. 3
3. 3
4. 3
5. 5
6. 5
7. 5
8. 5
9. 4
10. 4
11. 4
12. 4
13. 6
14. 6
15. 6
16. 7
17. 9
18. 8
19. 7
20. 7
21. 11
22. 11
23. 11
24. 1
25. 5
26. 5
27. 5
28. 1
29. 9
30. 9
31. 9
32. 11
33. 11
34. 11
35. 19
36. 19
37. 19
38. 17
39. 9
40. 15
41. 19
42. 29
43. 19
44. 39

Side B

1. 6
2. 6
3. 6
4. 6
5. 4
6. 4
7. 4
8. 4
9. 5
10. 5
11. 5
12. 5
13. 11
14. 11
15. 11
16. 12
17. 14
18. 13
19. 5
20. 5
21. 15
22. 15
23. 15
24. 1
25. 5
26. 5
27. 5
28. 1
29. 7
30. 7
31. 7
32. 15
33. 15
34. 15
35. 14
36. 14
37. 14
38. 18
39. 11
40. 11
41. 25
42. 19
43. 17
44. 29

Problem Set

1. a. $1\frac{2}{3}, 1\frac{2}{3}$
   b. $2\frac{2}{4}, 2\frac{3}{4} + 2\frac{2}{4} = 5\frac{1}{4}$

2. a. Answer provided
   b. $1\frac{2}{5}$; $\frac{4}{5}$ decomposed as $\frac{1}{5}$ and $\frac{3}{5}$
   c. $1\frac{4}{7}$; $\frac{6}{7}$ decomposed as $\frac{3}{7}$ and $\frac{3}{7}$

3. a. $2\frac{4}{5}$
   b. $\frac{4}{6}$; $1\frac{3}{6}$ decomposed as $\frac{3}{6}$ and 1
   c. $5\frac{6}{10}$; $6\frac{3}{10}$ decomposed as $5\frac{3}{10}$ and 1

4. a. $2\frac{2}{4}$
   b. $2\frac{2}{8}$
   c. $4\frac{7}{12}$
   d. $2\frac{4}{100}$

Exit Ticket

1. $2\frac{1}{3}$
2. $3\frac{6}{8}$

Homework

1. a. $1\frac{3}{5}$, $1\frac{3}{5}$

   b. $2\frac{6}{8}$, $\overline{A}\frac{5}{8} + 2\frac{6}{8} = 5\frac{3}{8}$

2. a. Answer provided

   b. $1\frac{4}{7}$; $\frac{4}{7}$ decomposed as $\frac{1}{7}$ and $\frac{3}{7}$

   c. $1\frac{9}{12}$; $\frac{8}{12}$ decomposed as $\frac{5}{12}$ and $\frac{3}{12}$

3. a. $2\frac{6}{8}$

   b. $\frac{7}{12}$; $1\frac{3}{12}$ decomposed as $\frac{3}{12}$ and 1

   c. $2\frac{2}{10}$; $3\frac{1}{10}$ decomposed as $2\frac{1}{10}$ and 1

4. a. $1\frac{7}{9}$

   b. $1\frac{7}{10}$

   c. $2\frac{10}{12}$

   d. $4\frac{12}{100}$

Lesson 34

Sprint

Side A

1. 1	12. 1	23. 2	34. 3
2. 1	13. 5	24. 11	35. 3
3. 1	14. 5	25. 3	36. 1
4. 1	15. 3	26. 3	37. 2
5. 1	16. 3	27. 3	38. 1
6. 1	17. 2	28. 7	39. 5
7. 1	18. 1	29. 7	40. 1
8. 1	19. 4	30. 2	41. 5
9. 4	20. 2	31. 2	42. 9
10. 4	21. 14	32. 15	43. 12
11. 1	22. 14	33. 2	44. 5

Side B

1. 1	12. 1	23. 2	34. 2
2. 1	13. 5	24. 8	35. 2
3. 1	14. 5	25. 2	36. 1
4. 1	15. 4	26. 2	37. 16
5. 1	16. 4	27. 2	38. 1
6. 1	17. 3	28. 21	39. 5
7. 1	18. 2	29. 21	40. 1
8. 1	19. 1	30. 2	41. 5
9. 8	20. 2	31. 2	42. 3
10. 8	21. 9	32. 12	43. 12
11. 1	22. 9	33. 1	44. 5

Lesson 34 Answer Key 4•5

Problem Set

1. a. $3\frac{2}{3}$
   b. $4\frac{3}{4}$
   c. $7\frac{4}{5}$
2. a. Answer provided
   b. $2\frac{4}{5}$
   c. $1\frac{3}{6}$
   d. $6\frac{4}{5}$

3. a. $4\frac{6}{8}$
   b. $2\frac{6}{10}$
   c. $4\frac{7}{12}$
   d. $7\frac{9}{50}$

Exit Ticket

1. $4\frac{3}{6}$
2. $8\frac{6}{8}$

Homework

1. a. $4\frac{2}{4}$
   b. $5\frac{5}{8}$
   c. $6\frac{5}{6}$
2. a. Answer provided
   b. $1\frac{4}{6}$
   c. $5\frac{6}{8}$
   d. $4\frac{6}{10}$

3. a. $2\frac{6}{12}$
   b. $3\frac{8}{10}$
   c. $7\frac{11}{16}$
   d. $3\frac{11}{100}$

A STORY OF UNITS — Lesson 35 Answer Key 4•5

Lesson 35

Problem Set

1. a. Tape diagram drawn and labeled
   b. Tape diagram drawn and labeled
2. a. 7 × 2 thirds = 14 thirds
   b. 4 × 2 fourths = 8 fourths
   c. 16 × 3 eighths = 48 eighths
   d. 6 × 5 eighths = 30 eighths
3. a. $\frac{28}{9}$
   b. $\frac{18}{5}$
   c. $\frac{24}{4}$
   d. $\frac{48}{8}$
   e. $\frac{84}{10}$
   f. $\frac{162}{100}$
4. $\frac{18}{5}$ yd

Exit Ticket

1. 5 × 2 thirds = 10 thirds
2. $\frac{55}{6}$

Homework

1. a. Tape diagram drawn and labeled
   b. Tape diagram drawn and labeled
2. a. 10 × 2 fifths = 20 fifths
   b. 3 × 5 sixths = 15 sixths
   c. 9 × 4 ninths = 36 ninths
   d. 7 × 3 fourths = 21 fourths
3. a. $\frac{18}{4}$
   b. $\frac{35}{8}$
   c. $\frac{26}{3}$
   d. $\frac{36}{3}$
   e. $\frac{98}{10}$
   f. $\frac{98}{100}$
4. $\frac{10}{3}$ c

Lesson 36

Problem Set

1. Tape diagram drawn; $4 \times \frac{3}{4}$
2. Tape diagram drawn; $3 \times \frac{7}{12}$
3. a. $5\frac{3}{5}$
   b. $3 \times \frac{9}{10} = 2\frac{7}{10}$
   c. $5\overline{A}\frac{11}{12} = 4\frac{7}{12}$

4. a. $5\frac{1}{3}$
   b. 9
   c. 40
   d. $22\frac{6}{8}$
5. $5\frac{4}{10}$ L
6. $10\frac{2}{4}$ c
7. 45 lb

Exit Ticket

1. $5\frac{1}{4}$
2. $3\frac{3}{5}$
3. $37\frac{4}{8}$

Homework

1. Tape diagram drawn; $4 \times \frac{2}{3}$
2. Tape diagram drawn; $3 \times \frac{7}{8}$
3. a. Answer provided
   b. $3 \times \frac{7}{10} = 2\frac{1}{10}$
   c. $6\overline{A}\frac{5}{12} = 2\frac{6}{12}$
   d. $12 \times \frac{3}{8} = 4\frac{4}{8}$

4. a. $1\frac{5}{9}$
   b. $7\frac{1}{3}$
   c. $13\frac{2}{6}$
   d. 20
   e. $13\frac{4}{5}$
   f. $8\frac{4}{8}$
5. $12\frac{3}{4}$ in
6. $4\frac{1}{8}$
7. 9 ft

Lesson 37

Problem Set

1. Two tape diagrams drawn; $2 \times 4\frac{2}{3}$, $(2 \times 4) + \left(2 \times \frac{2}{3}\right)$
2. a. Answer provided
   b. $9\frac{1}{3}$
   c. $7\frac{7}{8}$
   d. $9\frac{4}{10}$
   e. $23\frac{1}{4}$
   f. 21
   g. $3\frac{4}{5}$
   h. 23
3. $23\frac{1}{3}$ feet

Exit Ticket

1. $21\frac{4}{8}$
2. $12\frac{9}{10}$

Homework

1. Two tape diagrams drawn; $3 \times 5\frac{1}{12}$, $(3 \times 5) + \left(3 \times \frac{1}{12}\right)$
2. a. Answer provided
   b. $20\frac{5}{6}$
   c. $15\frac{3}{5}$
   d. $14\frac{6}{10}$
   e. 58
   f. $40\frac{4}{8}$
3. $13\frac{8}{10}$ mi
4. $28\frac{2}{10}$ lb

Lesson 38

Problem Set

1. a. 7, 7
   b. 12, $\frac{7}{8}$
2. a. $58\frac{4}{5}$
   b. $43\frac{3}{6}$
   c. $26\frac{9}{12}$
   d. 104
   e. $100\frac{16}{100}$

3. $7\frac{5}{10}$ mi
4. $33\frac{1}{4}$

Exit Ticket

1. 8, 8
2. $46\frac{3}{8}$

Homework

1. a. 8, 8
   b. 7, $\frac{7}{10}$
2. a. $49\frac{5}{7}$
   b. $69\frac{3}{4}$
   c. 79
   d. $77\frac{5}{8}$
   e. $82\frac{8}{12}$
   f. $360\frac{36}{100}$

3. $41\frac{5}{8}$ ft
4. $45\frac{2}{4}$ c
5. $215\frac{5}{8}$ oz

Lesson 39

Sprint

Side A

1. $\frac{2}{3}$
2. $\frac{2}{3}$
3. $\frac{3}{4}$
4. $\frac{3}{4}$
5. $\frac{2}{5}$
6. $\frac{2}{5}$
7. $\frac{3}{5}$
8. $\frac{3}{5}$
9. $\frac{4}{5}$
10. $\frac{4}{5}$
11. $\frac{3}{10}$
12. $\frac{3}{10}$
13. $\frac{3}{8}$
14. $\frac{3}{8}$
15. 1
16. 1
17. 1
18. 1
19. 1
20. 1
21. $\frac{3}{2}$
22. $\frac{3}{2}$
23. $\frac{4}{3}$
24. $\frac{4}{3}$
25. 5
26. $\frac{1}{6}$
27. $\frac{1}{8}$
28. 5
29. $\frac{1}{8}$
30. $\frac{1}{10}$
31. 7
32. 7
33. $\frac{1}{6}$
34. $\frac{1}{6}$
35. 8
36. 8
37. $\frac{9}{10}$
38. $\frac{7}{5}$
39. $\frac{1}{3}$
40. $\frac{7}{12}$
41. 5
42. $\frac{1}{5}$
43. $\frac{1}{4}$
44. $\frac{1}{3}, \frac{1}{3}, \frac{1}{3}$

Side B

1. $\frac{2}{5}$
2. $\frac{2}{5}$
3. $\frac{2}{3}$
4. $\frac{2}{3}$
5. $\frac{3}{4}$
6. $\frac{3}{4}$
7. $\frac{3}{5}$
8. $\frac{3}{5}$
9. $\frac{4}{5}$
10. $\frac{4}{5}$
11. $\frac{3}{8}$
12. $\frac{3}{8}$
13. $\frac{3}{10}$
14. $\frac{3}{10}$
15. 1
16. 1
17. 1
18. 1
19. 1
20. 1
21. $\frac{4}{3}$
22. $\frac{4}{3}$
23. $\frac{3}{2}$
24. $\frac{3}{2}$
25. 5
26. $\frac{1}{6}$
27. $\frac{1}{8}$
28. 5
29. $\frac{1}{8}$
30. $\frac{1}{10}$
31. 7
32. 7
33. $\frac{1}{8}$
34. $\frac{1}{8}$
35. 6
36. 6
37. $\frac{5}{12}$
38. $\frac{6}{5}$
39. $\frac{1}{4}$
40. $\frac{9}{10}$
41. 3
42. $\frac{1}{4}$
43. $\frac{1}{5}$
44. $\frac{1}{4}, \frac{1}{4}, \frac{1}{4}, \frac{1}{4}$

Problem Set

1. $5\frac{2}{8}$ mi
2. $15\frac{9}{16}$ in
3. $6\frac{4}{8}$ yd
4. $33\frac{1}{3}$ yd
5. $211\frac{2}{10}$ mi
6. $99

Exit Ticket

$31\frac{5}{8}$ lb

Homework

1. 20 lb
2. $15\frac{6}{8}$ in
3. $24\frac{1}{4}$ yd
4. $50\frac{4}{8}$ c
5. $290\frac{8}{10}$ mi
6. $147

Lesson 40

Problem Set

1. a. Line plot drawn accurately
   b. $\frac{7}{8}$ ft
   c. $10\frac{4}{8}$ ft
2. Player C
3. $22\frac{2}{4}$ qt
4. $33\frac{3}{10}$ g

Exit Ticket

1. One mark added on $1\frac{6}{8}$; $2\frac{2}{8}$ circled
2. $5\frac{2}{8}$ miles

Homework

1. Line plot drawn accurately
2. $2\frac{7}{8}$ in
3. $1\frac{5}{8}$ in
4. $9\frac{3}{8}$ in
5. $1\frac{1}{8}$ in
6. August and October
7. $22\frac{4}{8}$ in

Lesson 41

Problem Set

1. a. 2
   b. $2\frac{1}{2}$
   c. 3
   d. $3\frac{1}{2}$
   e. 4
   f. $4\frac{1}{2}$

2. Answers will vary.

3. The sum would remain the same.

4. a. $5\frac{5}{10}$
   b. $6\frac{6}{12}$
   c. 8
   d. 13
   e. $25\frac{25}{50}$
   f. $50\frac{50}{100}$

5. Answers will vary.

6. Answers will vary.

Exit Ticket

1. $10\frac{10}{20}$
2. $100\frac{100}{200}$

Homework

1. a. 3
   b. $3\frac{1}{2}$
   c. 4
   d. $4\frac{1}{2}$
   e. 5
   f. $5\frac{1}{2}$

2. Answers will vary.

3. The sums would remain the same.

4. a. $10\frac{10}{20}$
   b. 18
   c. $18\frac{18}{36}$
   d. 38
   e. $50\frac{50}{100}$
   f. 50

5. Answers will vary.